BEHAVIOR-MODIFYING CHEMICALS FOR INSECT MANAGEMENT

Applications of Pheromones and Other Attractants

edited by

RICHARD L. RIDGWAY
Agricultural Research Service
U.S. Department of Agriculture
Beltsville, Maryland

ROBERT M. SILVERSTEIN
College of Environmental Science and Forestry
State University of New York
Syracuse, New York

MAY N. INSCOE
Agricultural Research Service
U.S. Department of Agriculture
Beltsville, Maryland

MARCEL DEKKER, INC. New York and Basel

Library of Congress Cataloging-in-Publication Data

Behavior-modifying chemicals for insect management : applications of pheromones and other attractants / edited by Richard L. Ridgway, Robert M. Silverstein, May N. Inscoe.
p. cm.
ISBN 0-8247-8156-2 (alk. paper)
1. Semiochemicals. 2. Insect sex attractants. 3. Insects--Behavior. I. Ridgway, R. L. II. Silverstein, Robert M. (Robert Milton) III. Inscoe, May N.
SB933.5.B44 1990
632'.7--dc20 89-71510
CIP

This book is printed on acid-free paper.

MARCEL DEKKER, INC.
270 Madison Avenue, New York, New York 10016

Current printing (last digit):
10 9 8 7 6 5 4 3 2 1

PRINTED IN THE UNITED STATES OF AMERICA

Preface

The discovery, isolation, and chemical identification of the sex attractant pheromone of the silkworm, *Bombyx mori*, in 1959 marked the beginning of a new era for insect biology and chemistry. In the following decade, pheromones were identified for about 20 additional insects. During the 1970s, the number rose to over 200, and by the end of the 1980s, pheromones and parapheromones (pheromone mimics) for more than 1000 insects were known. Knowledge of the chemical identity of pheromones has made possible a wide range of fundamental and applied studies that have greatly expanded our knowledge of insect behavior and population dynamics.

During the 1970s and early 1980s, great expectations were voiced as to the potential of pheromones and other semiochemicals for use as effective and environmentally benign methods of insect control to replace many of the supposedly less desirable conventional insecticides. These expectations were not realized in the short term, and significant credibility problems concerning the practical application of pheromones began to develop. Despite considerable evidence that pheromones were or could be of substantial practical use, there seemed to be some feeling that pheromones would ultimately be regarded as interesting subjects for scientific investigations, with little real value.

The need to address this issue of "Practical Applications of Insect Pheromones and Other Behavior-Modifying Chemicals" surfaced during information discussions at various international conferences from 1984 to 1986 and led to a fairly broad consensus that there was a real need for an international review to obtain a realistic evaluation of the status of behavioral chemicals in pest control.

A survey of the 20 or so published or planned books dealing with insect pheromones showed that although a number of excellent books were, or would soon be, available, few of them focused specifically on potential applications, and none of these provided a comprehensive international viewpoint. Therefore it was decided that a book such as this, based on an extensive international symposium, was needed to document the present state of the art. Topics were chosen to provide good overall coverage of the field, including principles involved, types of applications that have showed promise in various commodities, and development and regulatory issues. Authors from Europe, Australia, Asia, Canada, and the United States have provided a truly international perspective.

This book begins with a section containing reviews on the principles involved in using behavior-modifying chemicals, their chemistry, and delivery systems for their efficient application. The case studies covered in the second section describe current and potential practical uses of these chemicals against insects in horticultural crops, field crops, forestry, stored products, and insects affecting animals. Discussions on regulatory matters, including a highly authoritative discussion of registration procedures in the United States, follow. The book concludes with descriptions of the commercial development and availability of these chemicals and their future potential. A list of commercial suppliers is provided in the appendix.

Considerable emphasis has been placed on scientific and technical documentation in this volume. However, the subject matter covered in the various chapters is such that the documentation varies considerably from chapter to chapter. For instance, some case studies were included because the magnitude of the insect problems involved and the potential role of pheromones in management of these pests required consideration, even though the scientific and technical bases were somewhat limited. Also the very nature of developmental and regulatory activities is such that scientific and technical literature is limited. Nevertheless, inclusion of these topics appeared to be essential in order to provide an adequate overview of practical applications of pheromones. Thus, this volume should be a useful reference for workers in such areas as insect ecology, insect behavior, chemical communication, chemistry of natural products, commercial development of specialty chemicals, and insect pest management. Further, it is our hope that this book can be used as a basis for researchers, extension personnel, and practitioners to improve existing methodology and to develop new uses for behavior-modifying chemicals.

There has been major expansion in the use of pheromones in monitoring over the past decade. The use of pheromones in traps for detection of specific insects is well established and pheromone traps are important tools in many pest management programs. Lures for over 250 insects are now available (see Chapter 38). Concurrently, there has been modest but real growth in applications of behavioral chemicals for pest suppression. Pheromone formulations for 19 insect species are registered with the U.S. Environmental Protection Agency (see Chapter 35); it must be remembered, however, that registration does not ensure efficacy or commercial viability. At present, pheromones of insects such as grape berry moths, boll weevils, codling moths, pink bollworms, tomato pinworms, and peachtree borers are being used or show substantial evidence that they will find a viable niche in pest management programs.

From the case studies in this book it is clear that development of practical applications of behavior-modifying chemicals is not simply or rapidly accomplished and a number of constraints must be overcome. Extensive knowledge of the behavior, physiology, and population dynamics of the target insect is essential. Reductions in the use of conventional pesticides that will result from applications of semiochemicals introduce a built-in conflict of interest for pesticide manufacturers; the in-depth interest and involvement of several international pesticide producers in planning this book indicate that they are beginning to deal with this question. The specificity of pheromones automatically limits the available market and the resultant lack of major profit incentives will require significant input from the public sector for development of these products. Regulatory procedures have also been perceived as a barrier. Discussions following up issues raised in the symposium on which the book is based have led to specific efforts to obtain some regulatory relief, particularly for experimental uses.

From the material presented in this book, we can conclude that in the area of behavioral chemicals there is reason for optimism that these chemicals can eventually lead to reductions in the use of conventional pesticides and significant expansion in the use of biologically based methods of pest control. We have reached "a watershed and . . . the glass is half-full," not half-empty (see Chapter 1).

This volume, and the symposium from which it derived, would not have been possible without the participation of a great many people. The content was molded by the symposium organizing committee: Henrich Arn, Dennis S. Banasiak, Michael G. Banfield, Reidar Lie, Albert K. Minks, Richard L. Ridgway, Wendell L. Roelofs, Robert M. Silverstein, Jeffrey J. Slocum, and Clive Wall. An enthusiastic group of sponsors provided support that made possible the presence of international speakers and their contributions to this book; these included: Entomological Society of America; International Organization for Biological Control, Charles Valentine Riley Memorial Foundation; Agron, Inc.; BASF Corporation; Biocontrol, Ltd.; Biological Control Systems Limited (now AgriSense BCS Limited); Borregaard Industries Limited; Shell Agrar GmbH & Co. (formerly Celamerck GmbH & Co.); CIBA-GEIGY Corporation, Agricultural Division; Consep Membranes, Inc.; Ecogen, Inc.; Hercon Laboratories; Insects Limited Inc.; KenoGard;

Orsynex Inc.; Mitsubishi Corporation; Phero Tech Inc.; Provesta Corporation; Trece, Incorporated; and Scentry, Inc. (Agron, Inc. and Provesta Corp. interests are now represented by AgriSense). Appreciation is also extended to the members of the 1987 Program Committee of the Entomological Society of America—Kenneth V. Yeargan, Herbert Oberlander, Ring T. Cardé, James E. Roberts, and Charles J. Eckenrode—who made possible the unprecedented integration of two days of symposium programming into the programs of four sections of the society. Particular recognition goes to the authors, whose contributions made possible this general overview of the current status of applications of insect behavioral chemicals.

Finally, a very special acknowledgment is extended to Robert M. Silverstein, a pioneer in the insect pheromone field, who through his years of cooperation and dedication has provided inspiration to many of us and who provided a unique perspective and encouragement, particularly during the initial phases of the organizing activities. Dr. Silverstein's joining us to edit this volume also contributed substantially to the total effort.

Richard L. Ridgway
May N. Inscoe

Contents

Contributors

Heinrich Arn Swiss Federal Agricultural Research Station, Wädenswil, Switzerland

Thomas C. Baker Department of Entomology, University of California, Riverside, California

*Dennis S. Banasiak** Provesta Corporation, Bartlesville, Oklahoma

John H. Borden Department of Biological Sciences, Simon Fraser University, Burnaby, British Columbia, Canada

Lloyd E. Browne Ecogen, Inc., Langhorne, Pennsylvania

Wendell E. Burkholder Stored-Product Insects Research Unit, Agricultural Research Service, U.S. Department of Agriculture, and Department of Entomology, University of Wisconsin, Madison, Wisconsin

James D. Byers† Provesta Corporation, Bartlesville, Oklahoma

Derek G. Campion Overseas Development Natural Resources Institute, Chatham Maritime, Chatham, Kent, England

Ring T. Cardé Department of Entomology, University of Massachusetts, Amherst, Massachusetts

Pierre-Joseph Charmillot Federal Agricultural Research Station of Changins, Nyon, Switzerland

Brian R. Critchley Overseas Development Natural Resources Institute, Chatham Maritime, Chatham, Kent, England

**Current affiliation*: AgriSense, Fresno, California
†*Current affiliation*: Phillips Petroleum Corporation, Bartlesville, Oklahoma

Roy T. Cunningham Agricultural Research Service, U.S. Department of Agriculture, Tropical Fruit and Vegetable Research Laboratory, Hilo, Hawaii

Gary E. Daterman Forest Service, U.S. Department of Agriculture, Corvallis, Oregon

Timothy J. Dennehy Department of Entomology, New York State Agricultural Experiment Station, Cornell University, Geneva, New York

Charles Descoins Laboratoire des Médiateurs Chimiques, Saint-Remy-lès-Chevreuse, France

*Willard A. Dickerson** Boll Weevil Eradication Research Laboratory, Agricultural Research Service, U.S. Department of Agriculture, Raleigh, North Carolina

Charles C. Doane Scentry, Inc., Buckeye, Arizona

Hollis M. Flint Agricultural Research Service, U.S. Department of Agriculture, Phoenix, Arizona

David R. Hall Overseas Development National Resources Institute, Chatham Maritime, Chatham, Kent, England

May N. Inscoe Insect Chemical Ecology Laboratory, Agricultural Research Service, U.S. Department of Agriculture, Beltsville, Maryland

Jack W. Jenkins Scentry, Inc., Buckeye, Arizona

Manuel J. Jimenez University of California, Cooperative Extension Service, County Civic Center, Visalia, California

Philipp Kirsch Biocontrol, Ltd., Davis, California

Richard M. Kobayashi Animal and Plant Health Inspection Service, U.S. Department of Agriculture, Honolulu, Hawaii

Douglas M. Kolodny-Hirsch† Forest Pest Management, Maryland Department of Agriculture, Annapolis, Maryland

Gerald N. Lanier Department of Environmental and Forest Entomology, State University of New York, Syracuse, New York

Barbara A. Leonhardt Agricultural Research Service, U.S. Department of Agriculture, Beltsville, Maryland

B. Staffan Lindgren Phero Tech, Inc., Vancouver, British Columbia, Canada

---

**Current affiliation*: North Carolina Department of Agriculture, Raleigh, North Carolina.

†*Current affiliation*: Espro, Inc., Columbia, Maryland.

Juan D. López, Jr. Pest Management Research Unit, U.S. Department of Agriculture, College Station, Texas

Victor C. Mastro Animal and Plant Health Inspection Service, U.S. Department of Agriculture, Otis Air National Guard Base, Massachusetts

John R. McLaughlin Agricultural Research Service, U.S. Department of Agriculture, Gainesville, Florida

Lawrence J. McVeigh Overseas Development Natural Resources Institute, Chatham Maritime, Chatham, Kent, England

Albert K. Minks Department of Biological and Integrated Control, Research Institute for Plant Protection, Wageningen, The Netherlands

Doris H. Miyashita Agricultural Research Service, U.S. Department of Agriculture, Hilo, Hawaii

Kurt Nabholz Sandoz Ltd., Agro Division, Basel, Switzerland

Ulrich Neumann BASF Aktiengesellschaft, Limburgerhof, Federal Republic of Germany

Charles A. O'Connor III McKenna, Conner & Cuneo, Washington, D.C.

Kinya Ogawa Shin-Etsu Chemical Company Ltd., Tokyo, Japan

Richard E. Rice Department of Entomology, University of California, Davis, California

Richard L. Ridgway Insect Chemical Ecology Laboratory, Agricultural Research Service, U.S. Department of Agriculture, Beltsville, Maryland

Wendell L. Roelofs Department of Entomology, New York State Agricultural Research Station, Cornell University, Geneva, New York

Chris J. Sanders Forestry Canada, Ontario Region, Sault Ste. Marie, Ontario, Canada

David J. Schuster Gulf Coast Research and Education Center, University of Florida, Bradenton, Florida

Charles P. Schwalbe Animal and Plant Health Inspection Service, U.S. Department of Agriculture, Otis Air National Guard Base, Massachusetts

Ted N. Shaver Pest Management Research Unit, U.S. Department of Agriculture, College Station, Texas

Robert M. Silverstein College of Environmental Science and Forestry, State University of New York, Syracuse, New York

*Jeffrey J. Slocum** Orsynex, Inc., Clifton, New Jersey

**Current affiliation*: Marlborough Chemicals, Inc., Charlotte, North Carolina.

J. Wendell Snow Agricultural Research Service, U.S. Department of Agriculture, Byron, Georgia

Robert T. Staten Animal and Plant Health Inspection Service, U.S. Department of Agriculture, Phoenix, Arizona

Theodore N. Taft Department of Entomology, New York State Agricultural Experiment Station, Cornell University, Geneva, New York

Emil F. Taschenberg Department of Entomology, New York State Agricultural Experiment Station, Cornell University, Geneva, New York

*Sadahiro Tatsuki** Institute of Agriculture and Forestry, University of Tsukuba, Tsukuba, Ibaraki, Japan

Edwin F. Tinsworth U.S. Environmental Protection Agency, Washington, D.C.

James H. Tumlinson Agricultural Research Service, U.S. Department of Agriculture, Gainesville, Florida

Richard A. Vickers Division of Entomology, Commonwealth Scientific and Industrial Research Organization, Canberra City, Australia

Clive Wall† Department of Animal Ecology, University of Lund, Lund, Sweden

Iain Weatherston‡ Sault Ste. Marie, Ontario, Canada

**Current affiliation*: Laboratory of Applied Entomology, The University of Tokyo, Tokyo, Japan.

†*Current affiliation*: Bunting Biological Control Ltd., Great Horkesley, Colchester, Essex, England

‡*Current affiliation*: Department of Entomology, University of California, Riverside, California.

Part I
Principles of Research and Development

1

Practical Use of Pheromones and Other Behavior-Modifying Compounds: Overview

ROBERT M. SILVERSTEIN / State University of New York, Syracuse, New York

I. INTRODUCTION

Ever since Butenandt's pioneering study (1), collaborative studies of insect pheromones and other attractants have been propelled mainly by three considerations:

1. For chemists, the availability of instruments and techniques for isolation and structure determination of organic compounds
2. For biologists, the fascination of studying insect behavior on a molecular basis with the molecules provided by chemists
3. For both, the immediately obvious possibilities of controlling insect pests by perverting normal behavior to self-destructive responses, with the same molecules that regulate their normal behavior, and to do so without environmental damage

In discussing the practical application of pheromones and other attractants with investigators directly involved, I received responses ranging from "the glass is half full" to "the glass is half empty." Why the range from contained optimism to contained pessimism? To assess the state of the art and to elicit projections into the near future, the symposium that formed the basis of these proceedings was organized.

One issue seems clear: Survey and monitoring with pheromones and other attractants are practiced worldwide against a broad array of insect pests, and these techniques are integral parts of a growing number of control programs. It is the high promise of large-scale, direct control with pheromones and other attractants—mass trapping, bait-and-kill, or mating disruption—that remains unfulfilled. Why and whither? I shall not attempt a comprehensive exploration of the gap between expectations and reality, but perhaps I can briefly review some of the hopes and doubts as I interpret them.

II. HISTORICAL BACKGROUND

In 1968, David Wood, Minoru Nakajima, and I convened a group of biologists and chemists from the United States and Japan to discuss *Control of Insects by Natural Products*. The proceedings were published under the same title (2). The theme was collaboration between biologists and chemists. The topics that were covered included chemosensory behavior; electrophysiology; isolation, identification and synthesis of bioactive compounds; and applications to pest suppression. The proceedings remain, of course, only as a historical document—an early comprehensive assessment of the art of semiochemistry—but despite the title, one looks in vain for solid evidence of applications to pest suppression, although the promises are clear.

In 1975, H. H. Shorey and J. J. McKelvey convened a conference in Bellagio, Italy under the auspices of the Rockefeller Foundation. The proceedings, published in 1977 (3) contained eight chapters (out of a total of 24 chapters) under the heading, *Status and Prospects for Behavior-Modifying Chemicals in Pest Management*. A cogent, prescient statement in Shorey's opening chapter has been paraphrased many times since. It is worth repeating:

> It cannot be stressed too strongly that the key to devising efficient systems for the management of insect pests by chemically modifying their behavior is the acquisition of an intimate knowledge of the insects' own normal use of chemicals. This important factor is too often overlooked. Once a pheromone or other behaviorally active chemical is identified, there is a tendency to feel that the research is all over, and that the chemical can be used as a bait in traps or perhaps distributed through fields, causing insect control. Rather, the identification of the chemical should open the door to more, necessary research to determine whether the normal behavior of insects can be interfered with and manipulated to our advantage.

The multiauthored summary chapter, *Advancing Toward Operational Behavior-Modifying Chemicals*, made the following plea for perspective:

> Some observers have evidenced impatience with the various programs that have been directed toward the development of behavior-modifying chemicals (especially the pheromones) for insect pest management. However, it must be recalled that almost all of the practical research directed toward operational control programs has been initiated only during the past five years and by a small number of investigators. Additionally, only limited resources have been made available for this research. Thus, on any reasonable basis of expectation in science, progress should be considered quite remarkable.

Thus, we have an eloquent statement from "the-glass-is-half-full" school. These authors also urged that pheromones, in themselves, are not a panacea and must be considered in the context of integrated programs; they reviewed the limited successes and the failures; they projected needed areas of research; they suggested criteria for the selection of target pests; and they acknowledged the problems of involving industry and of persuading regulatory agencies that pheromones are not synonymous with hard insecticides.

Not all of the large-scale field tests reported at Bellagio were completely successful. We can identify some of the major causes of failure.

1. Inadequate knowledge of insect behavior
2. Inadequate definition of chemical communication systems
3. High population density
4. Too small an effort; inadequate resources
5. Inadequate pheromone formulations
6. Improper distribution of traps or release sources
7. Invasion from outside the test area
8. Poor timing

It is easy, with ineffable hindsight, to fault the early investigators who went into the field with inadequate information and resources and a lack of appreciation of the complexities. Columbus had similar problems, and certainly some of the peer reviewers consulted by Queen Isabella carped at premature exploration of a world whose very shape was in question and recommended cuts in funding. Subsequent explorations and investigations have benefited from earlier deficiencies.

In the proceedings (4) of the conference convened by F. Ritter in The Netherlands, in 1978, under NATO auspices, 12 of the 32 chapters addressed issues of practical applications. Added to the growing list of successful applications of survey and monitoring traps, were several reports of successes in large-scale mass-trapping and mating-disruption experiments. At the same time, the complexities and difficulties of direct population suppression were thoroughly aired. As in previous discussions, the need for a thorough understanding of insect behavior, a complete description of the chemical composition of the attractant, improved release devices, and sophisticated field testing was acknowledged. Economic realities and the requirement for demonstration of efficacy were stressed in several chapters. J. B. Siddall discussed the problems faced by industry in developing pheromones for large scale use: registration hurdles, costs of chemical synthesis, lack of adequate patent protection, and user resistance to novel products with the attendant fear of failure to protect a vital crop.

Primary among the legacies of the meeting in The Netherlands was the concerted attempt to confront directly the problems of registration; a list of recommendations emerged and was forwarded to the regulatory agencies of the NATO countries. These recommendations and the follow-up by the ad hoc committee, chaired by F. Ritter, were effective in moving the Environmental Protection Agency (EPA) to categorize behavior-mediating compounds as "biorational pesticides"—surely an infelicitous example of poor language spawned by good motives; behavior-mediating compounds do not kill pests. However, this change recognized that, "...operationally, there is a vast difference between spraying kilograms per hectare of a liquid or solid, toxic, persistent, broad-spectrum pesticide to kill insects, and releasing minute amounts of a biodegradable, species-specific, natural product in the vapor phase to lure an insect to a trap or to disorient the mate-finding process" (5). At least, however, these compounds were relieved of the more rigorous requirements for registration of hard pesticides, although seemingly irrational requirements and lengthy delays remain. It is not unreasonable to expect a regulatory agency to facilitate transfer of a beneficent technology.

In the symposium held in Florida in 1980 on *Management of Insect Pests with Semiochemicals; Concepts and Practice* (proceedings edited by Mitchell, 1981; 6), the 36 speakers addressed their topics under these headings: biomonitoring; mass trapping; mating distruption; formulation, toxicology, and registration; and oviposition disruptants and antiaggregants. Mitchell's preface acknowledged the development and widespread use of biomonitoring and listed the following obstacles to direct control [paraphrased]:

1. High cost of active ingredients
2. Deficiencies in formulation technology
3. Lack of clearly defined marketing strategies and deficiencies in technology transfer

To this list should be added lack of solid patent protection (natural products are not patentable), user reluctance to adopt a novel, unproved technology, and a generally negative attitude on the part of insecticide manufacturers and distributors.

A two-volume treatise edited by Kydonieus and Beroza, in 1982, provided an exhaustive account of *Insect Suppression with Controlled Release Pheromone Systems* (7). In particular, the excellent chapter *Chemical Attractants in Integrated Pest Management Programs* by Klassen, Ridgway, and Inscoe is recommended for study in the context of the present symposium. A brief discussion of marketing and production strategies by Kydonieus and Beroza in Chapter 14 presented some of the problems faced by an industrial company that has already entered the pheromone market.

Campion (8) provided a brief survey of pheromone applications and ranked pest insects in order of suitability for control by mating disruption.

Mitchell (9) again addressed the topic under the provocative title, *Pheromones: As the Glamour and Glitter Fade—the Real Work Begins.* Here, he elaborated on the problems of technology transfer in the underdeveloped countries of the tropics and called for ingenuity, creativity, and *simplicity* in the integration of pheromones into any pest control system.

Even more provocative was Goodell's (10) devastating critique of "international pest management research and extension in the third world." "Integrated pest management," she wrote, "requires the farmer to grasp a far more complex set of data, data which are often anything but self-evident, unitary, and standardized, or amenable to trial-and-error learning." Added to the misguided attempts to transfer high technology to primitive areas without adequate supervision are the disinterest, ineptitude, and corruption endemic in govenment agencies and in local industries [my paraphrase].

III. THE INSECTICIDE INDUSTRY AND INTEGRATED PEST MANAGEMENT

Let me close by confronting the intractable question of the obvious conflict of interest faced by the insecticide industry. This industry has produced extraordinarily useful products which, through overuse caused by resistance of the target insect and destruction of natural controls, now are not only largely ineffective and counterproductive, but also cause unacceptable environmental damage. This industry is now expected voluntarily to reduce its income from insecticides by promoting integrated pest management programs, in particular to produce, formulate, and market insect semiochemicals. This industry is asked to forego materials that are manufactured in large volume and are made available in convenient form through defined channels directly to the farmer who understands their use. Profits are made because, like it or not, the demand is still there. What the semiochemical component of integrated pest management has to offer is beneficial but complex and, at this point, hardly perfected. Only small amounts of chemicals are involved, and the high technology must somehow be transferred through undefined channels to the ultimate user, who is by no means convinced of its efficacy. And out of this, the participants must turn a short-term profit. I made the following statement in 1981 (5):

> The concept of integrated pest management is beginning to take hold, but full implementation is conditional upon the values and thrusts of society and on user acceptance. Can we forego the desire for a quick fix and the need for short-term returns on investments? Can marketing strategies be devised [for pheromones]? Will we support the basic studies needed? What values do we assign to environmental quality? It is exciting to consider the part that pheromones (or semiochemicals in general) may play an intetrated pest management, but it is sobering to consider the obstacles.
>
> Can we decrease the use of a profitable, convenient, though partially discredited tool [insecticides] for an approach that requires a sophisticated understanding of the target insect, that demands precise timing and uses small dosages, that may promise only partial success . . ., and that cannot guarantee large concentrated profits? . . . The difficult transition will make severe demands on administrators in government . . . agencies and in industry.

At this point, it may be useful to list the major components of integrated pest management (IPM):

1. Sound agricultural practices
2. Use of resistant plants
3. Selective use of pesticides
4. Biological control through introduction or manipulation of predators, parasites, and pathogenic organisms
5. Reproductive suppression by radiation or chemical sterilization
6. Introduction of reproductively incompatible strains of pest insects
7. Release of sterile insects [into the pest population in overwhelming numbers]
8. Use of hormones or hormone analogues
9. Use of pheromones and other behavioral chemicals
10. Combinations of any of these

The dilemma faced by industry cannot be resolved by scolding and moralizing; both the stick and the carrot are needed. Further restrictive legislation on pesticides provides the stick. The carrot consists of continued support by government agencies for basic studies and for development of effective, simple, integrated technologies. Transfer of these technologies from the research laboratories, through the extension services, to the ultimate user will result in an enlightened market to whose demands industry will respond. Users are becoming aware of the enormity of the problems of chemical pollution. Note, however, that a market for insecticides will exist for the foreseeable future, albeit in reduced amounts.

A partial list of suppliers of insect pheromones, synthetic attractants, dispensers, formulations, and traps has been compiled by Inscoe and Ridgway and is presented in the Appendix to this volume. The total number of more than 50 companies, worldwide, is impressive, but most are small and presumably struggling, and the turnover through the years has been large. Larger companies, represented characteristically by those with agrochemical interests, have assigned small groups to "watch developments" and select promising areas for exploratory studies. Several of the small entrepreneurial groups sell a package of traps, formulations, advice, service, and follow-up.

IV. CONCLUSIONS

For the past 25 years or so, the goal of using behavioral chemicals as a major tool for insect control has remained elusive. Not only must a new technology be introduced, but an established technology—the use, or rather the overuse, of insecticides—must be displaced. But as I trace the developments over the past quarter century and discuss them with my colleagues, the impression grows that the symposium in Boston, the basis for these proceedings, represents a watershed and that the authors will persuade most of us that the glass is half full.

REFERENCES

1. Butenandt, A., Beckman, R., Stamm, D., and Hecker, E. Über den Sexuallockstoff des Seidenspinners *Bombyx mori*; Reindarstellung und Konstitution. *Z. Naturforsch.* *14b*:283–284, 1959.
2. Wood, D. L., Silverstein, R. M., and Nakajima, M. (eds.). *Control of Insect Behavior by Natural Products*. Academic Press, New York, 1970.
3. Shorey, H. H. and McKelvey, J. J. (eds.). *Chemical Control of Insect Behavior*. John Wiley & Sons, New York, 1977.
4. Ritter, F. J. (ed.). *Chemical Ecology: Odour Communication in Animals*. Elsevier/North-Holland, Amsterdam, 1979.
5. Silverstein, R. M. Pheromones: Background and potential for use in insect pest control. *Science* *213*:1326–1332, 1981.

6. Mitchell, E. R. (ed.). *Management of Insect Pests with Semiochemicals*. Plenum, New York, 1981.
7. Kydonieus, A. F. and Beroza, M. (eds.). *Insect Suppression with Controlled Release Pheromone Systems*. Vols. 1 and 2; CRC Press, Boca Raton, 1982.
8. Campion, D. G. Survey of pheromone use in pest control. In *Techniques in Pheromone Research*. (Hummel, H. E. and Miller, T. A., eds.). Springer-Verlag, New York, 1984, pp. 405–449.
9. Mitchell, E. R. Pheromones: As the glamour and glitter fade—the real work begins. *Fl. Entomol.* *69*:132–139 (1986).
10. Goodell, G. Challenges to international pest management research and extension in the Third World: Do we really want IPM to work? *Bull. Entomol. Soc. Am.* *30*:18–25, 1984.

2

Principles of Monitoring

CLIVE WALL* / University of Lund, Lund, Sweden

I. INTRODUCTION

The large number of insect pheromones currently available (see Chap. 1) make it possible to selectively catch many of the important insect pests in the world. However, the development of monitoring schemes for such pests requires the integration and application of information from a number of sources. This process is outlined in Figure 1. The clear definition of the objective(s) of the scheme will determine what form the monitoring system will take. After development of the system, it should be evaluated, and either the system or the objective(s) may require modification if the latter have not been achieved. Thorough evaluation should result in an effective operational system.

II. DEFINING OBJECTIVE(S)

A large number of factors need to be taken into account, both scientific and practical (Fig. 2).

**Current affiliation:* Bunting Biological Control Ltd., Great Horkesley, Colchester, Essex, England

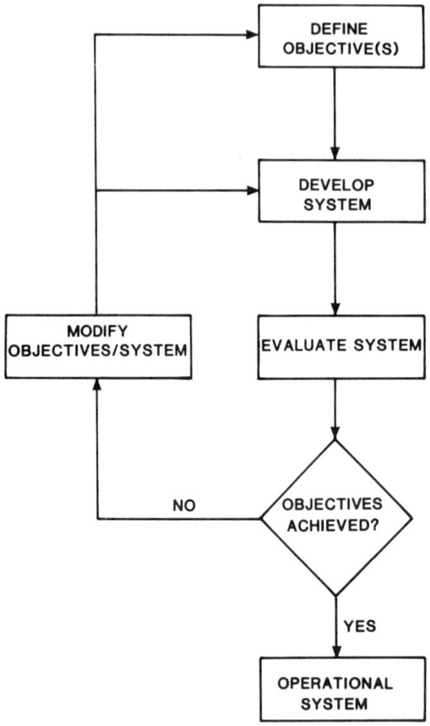

FIGURE 1 Research and development plan for a pheromone-monitoring system.

A. Scientific Factors

1. Identifying the Problem

There will be a need to provide an easy-to-use trapping system that catches a certain species selectively; but for what purpose? In general, there are three main types of problem—detection, timing of control measures, and risk assessment of the need to apply control measures. Detection requires only a sensitive trapping method that provides qualitative information about presence or absence. Timing requires some quantitative information from the trap-catches and, often, additional biological and meteorological data with which to predict the occurrence of the susceptible stage in the life cycle. For adequate risk assessment, there needs to be a quantified relationship between trap-catch and population density or the extent of subsequent damage.

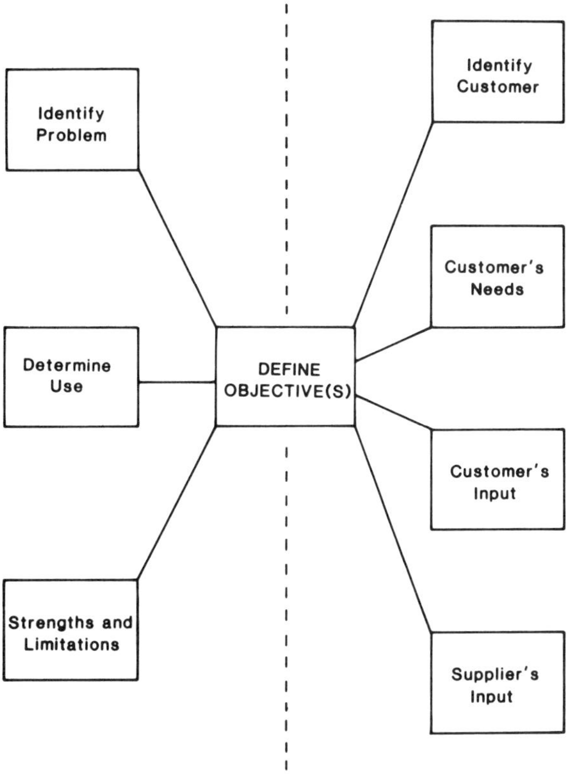

FIGURE 2 Factors that should be taken into account when defining the objective(s) of a pheromone-monitoring system.

2. *Determining the Use*

Table 1 summarizes the main uses to which pheromone traps have been put, together with the information that is required from the trap-catch. For early warning (of emergence, immigration, or migration), survey, and quarantine work, simple detection is all that is required. The timing of control measures often relies on the determination of a threshold catch (1–6). This threshold usually indicates the start of significant emergence or influx of the pest and may be used to optimize the timing of either insecticide sprays (2) or further observations (e.g., egg or larval sampling) to determine the need to spray (7). A threshold catch may be used also to establish whether or not an economic threshold is likely to be exceeded (3,4,7–9). However,

TABLE 1 Uses of Monitoring Traps

Information from trap-catch	Application
Detection	Early-warning Survey Quarantine
"Threshold"	Timing of treatments Timing of other sampling methods Risk assessment
Density estimation	Population trends Dispersion Risk assessment Effects of control measures

such assessment of risk is likely to be far more effective if a quantifiable relationship between trap-catch and population density can be established so that reasonable estimates of the latter can be produced (10–13). Such estimates can be used, not only for risk assessment, but also to follow population trends (12), study the dispersion of a species within the habitat, and establish the effectiveness of control measures.

3. *Strengths and Limitations*

Pheromone traps have five main strengths as tools for monitoring: they are the most sensitive sampling technique known, they are usually species-specific, they require no power and little maintenance, they are not labor-intensive, and they can be operated by nonentomologists.

Their limitations really center around the problems of interpreting the catches. Little is known about their range of action, in particular their sampling range and how this changes with time (14). Climatic effects on trap-catch and changes in trapping efficiency as catches accumulate (15) are further complications, although the former may not be a disadvantage if the result is that the traps indicate the true level of activity (16,17). These, and other factors, such as possible competition with wild females (18–21) and the fact that the traps catch adult insects, which are separated in time from the damaging stage (22), may contribute to the frequent difficulty in relating trap-catch to population density.

B. Practical Factors

1. *Identifying the Customer*

The identity of the people or organizations who will actually use the operational system should have a major influence on the form of the system. If the user is to be an advisory and extension service, the personnel involved will be able to operate a quite sophisticated monitoring system and integrate it with other information and procedures. However, if the grower is to be involved, this may impose certain limitations on what can be achieved. These considerations must be taken into account when deciding what the objectives of the system are, and it may be advantageous to apportion tasks for maximum effectiveness, for example, growers operate the traps, thus ensuring a widespread coverage of local monitoring points, but the interpretation is done centrally by the advisory/extension service (9).

2. *Identifying the Customer's Needs*

The customer (advisory entomologist, grower, processing company) may well perceive the problem and requirements differently from the research entomologist. It is imperative that these differences, if they exist, are discussed at an early stage and a common understanding of the problem reached. Only in this way is an operational system likely to evolve, particularly if it is to be supplied commercially.

3. *The Customer's Input*

What is the customer prepared to do to make the system work? If growers are to run the traps, with what frequency can they be expected to examine them? It may be far better to opt for a slightly lower frequency, which the customer will be prepared to adopt, than risk an inefficient system caused by failure to examine traps at the correct times.

Similarly, the positioning of the traps may need to be a compromise between the scientific and practical ideals; that is, it may not be possible to run traps in a forest or orchard at their optimum height or to have traps positioned in the center of arable crop fields, simply because the operator (grower) will not have the time or be sufficiently motivated in those circumstances to run the traps properly.

If the customer is an advisory organization, the proposed monitoring system should complement the other activities of the personnel involved, and any information transfer should either attempt to make use of the existing infrastructure or to introduce new procedures that are acceptable.

4. *The Supplier's Input*

For an effective operational monitoring system, there must be a supplier of at least the trapping materials (traps, attractant) and preferably detailed instructions on all aspects of trap operation and interpretation. If the supplier is to be a company, the system has to be commercially attractive in its own right or a valuable addition to an existing product (such as an insecticide) that will provide an edge in the market place. Systems can be designed in such a way that they are made more commercially attractive (e.g., traps that need replacing each season, systems requiring large numbers of traps), provided these considerations complement rather than conflict with the scientific requirements. As always, the major commercial problems arise with pests of minor crops, which represent rather small markets for monitoring systems that have to be tailor-made for each pest species.

III. DEVELOPING THE SYSTEM

The basic components of a pheromone-monitoring system are the attractant, the trap, and sufficient knowledge of the pest biology to interpret the trap catches.

A. The Attractant

Perhaps the most important point to stress about the chemical identity of attractants for use in monitoring traps is that it does not necessarily need to reflect the full identity of the natural pheromone system. The stringency of the requirements will vary enormously depending on the species, but divergence from the identity of the natural pheromone may not be a disadvantage and may have certain advantages. Indeed, as reexaminations reveal more and more minor components in the pheromone systems of many pest species, it is evident that most monitoring systems use traps containing attractants that differ from the natural pheromone to varying degrees. In the pea moth, *Cydia nigricana* (Fabricius), it is possible to use either the single-component synthetic pheromone (*E*,*E*)-8,10-dodecadien-1-yl acetate or its analogue (*E*)-10-dodecen-1-yl acetate (23). Even at low doses, the pheromone can attract impractically large numbers of moths, saturating the traps and making interpretation of the catches impossible; it is also rather unstable, causing problems in field use. The analogue, on the other hand, attracts far fewer insects, but a number sufficient to provide a useful range of catches within the extremes of high and low infestations usually encountered in commercial crops.

Adoxophyes orana (Fischer von Röslerstamm) was one of the first species known to use a sex pheromone of more than one compound (24). It was shown by field trapping that maximum attraction was obtained with a 9:1 ratio of the two positional isomers, (*Z*)-9-tetradecenyl acetate and (*Z*)-11-tetradecenyl acetate (25). This species has been monitored successfully using this mixture, but more recent work has shown the blend to be more complex (26–28). In contrast, the main problem with monitoring *Cydia pomonella* (Linnaeus) seems to have been that insufficient moths are caught in the traps. Most of the thresholds used range from two to five moths per trap per week, with the single-component pheromone (*E*,*E*)-8,10-dodecadien-1-ol (29). Recent findings (30–32) that there are other components in the natural female pheromone raise hopes that more attractive lures can be produced for this species, thus leading to more sensitive discrimination between low- and high-risk infestations.

Whatever attractant mixture is used, it should sample a representative proportion of the population (this may be impossible to determine except empirically) and be sufficiently specific to be practical. Here the needs and expertise of the user are crucial. Total specificity may be essential for a system operated by nonentomologists in situations in which similar insect species occur together. However, if species that are attracted to the same traps tend to be separated either in space or time, such specificity may not be important (33). With *C. nigricana* the monoene analogue of the natural pheromone, (*E*)-10-dodecen-1-yl acetate, is also an attractant for *Phyllonorycter blancardella* Fabricus (34), which is sometimes a contaminant in monitoring traps in combining peas; however, the low frequency of such contamination and the ease of distinguishing the two species result in no real problem. (*E*,*E*)-8,10-dodecadien-1-yl acetate, on the other hand, which is used in the more recently developed monitoring system for *C. nigricana* in vining peas, is an attractant for several other species within the genus *Cydia* (2) and, therefore, may pose more of a problem; for instance, during 1987, traps placed in crops close to willow stands caught *C. servillana* (Duponchel), which could confuse a nonentomologist grower (2).

The attractant needs to be only sufficiently attractive to accomplish the main objective of the monitoring system; dispensers that are too attractive can cause problems (35). Most importantly, the sampling range of the traps should be appropriate to the situation (14,36). Adjustment to the sampling range can be achieved by varying both the release rate and the composition of the attractant. Formulation technology has advanced to such a degree that release rates can be tailor-made to requirements (see Chap. 6), but consistency is vital if interpretation of trap-catch is to be successful (35), and reports of variations in release rates from individual commercial lures are disturbing (see Chap. 7). The use of suboptimal attractants may reduce the sampling range, but such use also can result in a deleterious effect on the close-range behavior of attracted insects, leading to lower trap efficiency. It cannot be overemphasized how important it is to have some information on the sampling range of monitoring traps (15,37–41); without it, there is virtually no chance of useful interpretation of trap-catch. Even in simple detection systems, in which it might seem appropriate to utilize the most sensitive trap available, knowledge of the sampling range is important. If this is substantially larger than the area of crop or commodity in question, the traps might attract insects from outside and indicate false-positives. The lack of such information about the range of action of traps has almost certainly been one of the main stumbling blocks to progress in using attractant traps for risk assessment. There are many examples of large trap-catches associated with very low populations or damage levels within the crop being monitored (7,42).

The mean sampling range of the traps must remain constant throughout the trapping period, although there are bound to be fluctuations about the mean caused by variations in wind speed and turbulence. Formulations that give constant release rates over given periods are difficult to produce (see Chap. 1) but would provide this consistency (12); however, many trapping systems still rely on impregnated rubber or polyethylene formulations with which constant attractiveness over long periods can be achieved only if the insects are equally responsive to a wide range of release rates (43).

B. The Trap

If the sampling range of a monitoring trap is to be tailored to the requirements of the crop, it may be necessary to have an efficient trap design that catches a high proportion of the insects attracted. Alternatively, at high-population densities, it may be advantageous to have a relatively inefficient design.

Traps should be easy to use, of standard construction (including the type of retentive material), and inexpensive. Variations in design may well affect both the sampling range and close-range behavior and introduce unnecessary variation in trap data. Other important factors that must be standardized are the height of deployment (usually relative to the vegetation; 44) and trap density. The latter needs to be sufficient to sample the area of crop or commodity adequately, but not so great that traps interact with one another causing distortion of individual trap-catches. Very little work has been done on this, perhaps because the experimental designs involved may be difficult and time-consuming to implement (45) and the results difficult to interpret. All the published studies have been on tortricid moths in either orchards (21,46–48) or arable crops (38–40).

The sampling range of a trap may be considerable; for instance, Riedl (15) established that monitoring traps for *C. pomonella* interact at densities of more than one per hectare (i.e., sampling range of approximately 350 m). If local variations in pest density are important and need to be monitored (49), there may be a conflict between the density of traps required and the possibility of their interacting, with consequent effects on the individual trap-catches (usually reduction). Therefore, at the very least, the density of traps, where possible, ought to be standardized. However, there may be a further complication; the sampling range of a trap includes both the distance over which insects are attracted and the distance over which they may have traveled before attraction (through migration or appetitive behavior) (14). If traps are placed within each other's range of attraction, insects may fly predominantly to the upwind sources, as shown in *C. nigricana* by Wall and Perry (38). In this type of situation, even standardizing trap spacing would not overcome the problem of interactions because individual trap catches would be either increased or reduced depending on the wind direction.

C. Pest Biology

The biology of the species being monitored is going to affect the efficiency of the monitoring system. Because pheromone traps are usually, by far, the most sensitive sampling technique available, considerable confidence can be placed in negative results. Conversely, the problems of relating trap-catch to population density mean that the effective use of pheromone traps to monitor population fluctuations (say, in multivoltine species with overlapping generations) is very limited. Thus, the potentially most successful application of pheromone traps is to monitor the commencement of flight of populations that are clearly separated in time from previous flights/generations in that locality. This has been shown to be possible for both *C. nigricana* (9) and *P. blancardella* (50). However, in the latter, peaks in the numbers of moths trapped did not occur at time of peak emergence, and changes in the size of trap-catches were not a reliable indicator of changes in the rate of moth emergence. Hennebery and Clayton (42) encountered similar problems in monitoring *Pectinophora gossypiella* (Saunders).

The monitoring of population trends can be achieved, but considerable work is required to establish the relationship between trap-catch and population density (51), and only large changes (such as occur in outbreak situations) can be monitored reliably. The variability associated with regressions of damage or population density against trap-catch (51) has often led to the conclusion that such relationships are virtually useless predictively. One of the major contributing factors to that variation is the high frequency of large catches associated with little or no damage. However, if instead of attempting to predict damage or population levels one uses the data to predict the maximum likely damage or population level, progress can be made. This prediction can be expressed in terms of probability and is nearer to the information that the grower really needs. The use of such predictions will undoubtedly lead to some cases of control measures being applied unnecessarily but, on the whole, it will be more effective than any alternative approach (if indeed one exists). This technique has been used to produce predictive tables of damage caused by *C. nigricana* based on cumulative trap-catch (52); a regression of proportion of pea seeds damaged against trap-catch was used to determine confidence limit contours for any future observations within the range of trap-catches obtained during experimental trials. The predictions were then used to produce a predictive table for use by the grower. Sanders (Chap. 21) reports a similar approach for *Choristoneura fumiferana* (Clements).

IV. EVALUATING THE SYSTEM

For evaluating the system, the criteria used should be the same as those involved when the objective(s) of the system are being defined. Thorough critical assessment of the system is essential before it can be declared operational, which usually involves some form of commercial production.

V. THE OPERATIONAL SYSTEM

The success of operational monitoring systems hinges not only on a balanced research and development program, but also on certain key features. They should be produced with adequate, well-researched instructions that deal with both the operational and interpretative aspects *in detail.* Glossy pamphlets that reveal little more than a "typical" flight curve are useless when it comes to decision-making time. The detailed information required must be provided with every trap (or set of traps) to make the task as easy as possible for the growers or the advisor. Worked examples of how to deal with the data from the traps are also particularly valuable, helping to instill confidence in the system.

However detailed instructions are, there will always be borderline cases for which decisions are difficult. In these circumstances, expert guidance is invaluable, and some mechanisms for obtaining it should be set up. This may involve the research worker acting as a guiding influence on the advisor, or one person (associated with the marketing company?) acting as a trouble-shooter. When customers get into trouble, someone should be prepared to help sort out the problem; only in this way can teething problems be overcome and confidence in the system gained.

The system needs to be sold at an attractive price that the customer is prepared to pay and, in this context, some information on the savings in time, money, and pesticides possible with the correct use of the system will help enormously. This type of information is likely to be obtainable only after the system has been launched (9), but effort spent acquiring the necessary data is likely to be well rewarded.

VI. CONCLUSIONS

Traps containing sex pheromones and related chemicals are the basis of many highly sensitive, specific monitoring systems. Their full potential is realized only when the objectives of the monitoring are defined clearly, and the systems are developed with due consideration for the type of customer, his needs and input. The inherent strengths (sensitivity, specificity, ease of use, cost) and limitations (interpretation) of pheromone traps should be recognized and used in the planning of operational systems.

REFERENCES

1. Macaulay, E. D. M., Etheridge, P., Garthwaite, D. G., Greenway, A. R., Wall, C., and Goodchild, R. E. Prediction of optimum spraying dates against pea moth, [*Cydia nigricana* (F.)], using pheromone traps and temperature measurements. *Crop protection 4*:85–98 (1985).
2. Wall, C. The application of sex-attractants for monitoring the pea moth, *Cydia nigricana* (F.) (Lepidoptera: Tortricidae). *J. Chem. Ecol. 14*:1857–1866 (1988).
3. Riedl, H., Croft, B. A., and Howitt, A. J. Forecasting codling moth phenology based on pheromone trap catches and physiological-time models. *Can. Entomol. 108*:449–460 (1976).
4. Glen, D. M. and Brain, P. Pheromone-trap catch in relation to the phenology of codling moth (*Cydia pomonella*). *Ann. Appl. Biol. 101*:429–440 (1982).
5. Minks, A. K. and Jong De, D. J. Determination of spraying dates for *Adoxophyes orana* by sex pheromone traps and temperature recordings. *J. Econ. Entomol. 68*:729–732 (1975).
6. Rice, R. E., Weakley, C. V., and Jones, R. A. Using degree-days to determine optimum spray timing for the oriental fruit moth (Lepidoptera: Tortricidae). *J. Econ. Entomol.* 77:698–700 (1984).

7. Charmillot, P. J. Developpement d'un systeme de prévision et de lutte contre le carpocapse (*Laspeyresia pomonella* L.) en Suisse romande: Role du service regional d'avertissement et de l'arboriculture. *Bull. OEPP 10*:231–239 (1980).
8. Alford, D. V., Carden, P. W., Dennis, E. B., Gould, H. J., and Vernon, J. D. R. Monitoring codling and tortrix moths in United Kingdom apple orchards using pheromone traps. *Ann. Appl. Biol. 91*:165–178 (1979).
9. Wall, C., Garthwaite, D. G., Blood Smyth, J. A., and Sherwood, A. The efficacy of sex-attractant monitoring for the pea moth, *Cydia nigricana*, in England, 1980–85. *Ann. Appl. Biol. 110*:223–229 (1987).
10. Rummel, D. R., White, J. R., Carroll, S. C., and Pruitt, C. R. Pheromone trap index system for predicting need for overwintered boll weevil control. *J. Econ. Entomol. 73*:806–810 (1980).
11. Benedict, J. H., Urban, T. C., George, D. M., Severs, J. C., Anderson, D. J., McWhorter, G. M., and Zummo, G. R. Pheromone trap thresholds for management of overwintering boll weevils (Coleoptera: Curculionidae). *J. Econ. Entomol. 78*:169–171 (1985).
12. Sanders, C. J. Sex pheromone traps and lures for monitoring spruce budworm populations—the Ontario experience (*Choristoneura fumiferana*). USDA Forest Service General Tech. Rep. NE–United States. Northeastern Forest Experiment Station, *88*:17–22 (1984).
13. Daterman, G. E. and Sower, L. L. Douglas-fir tussock moth pheromone research using controlled-release systems. *Proc. Int. Controlled Release Pesticide Symp. 4*:68–77 (1977).
14. Wall. C. and Perry, J. N. Range of action of moth sex-attractant sources. *Entomol. Exp. Appl. 44*:5–14 (1987).
15. Riedl, H. The importance of pheromone trap density and trap maintenance for the development of standardized monitoring procedures for the codling moth (Lepidoptera: Tortricidae). *Can. Entomol. 112*:655–663 (1980).
16. Carroll, S. C. and Rummel, D. R. Relationship between time of boll weevil (Coleoptera, Curculionidae) emergence from winter habitat and response to grandlure-baited pheromone traps. *Environ. Entomol. 14*:447–451 (1985).
17. Graham, J. C. Emergence, dispersal and reproductive biology of *Cydia nigricana* (F.) (Lepidoptera: Tortricidae). Unpublished Ph.D. Thesis, University of London, 1984, 376 pp.
18. Cardé, R. T. Behavioural responses of moths to female-produced pheromone and the utilization of attractant-baited traps for population monitoring. In *Movement of Highly Mobile Insects: Concepts and Methodology in Research*. (Rabb, R. L., and

Kennedy, G. G., eds.). North Carolina State University, 1979, pp. 286–315.
19. Hartstack, A. W. and Witz, J. A. Estimating field populations of tobacco budworm moths from pheromone trap catches. *Environ. Entomol.* *10*:908–914 (1981).
20. Miller, C. A. and McDougall, G. A. Spruce budworm moth trapping using virgin females. *Can. J. Zool.* *51*:853–858 (1973).
21. Riedl, M. and Croft, B. A. A study of pheromone trap catches in relation to codling moth (Lepidoptera: Olethreutidae) damage. *Can. Entomol.* *106*:525–537 (1974).
22. Sower, L. L., Daterman, G. E., and Sartwell, C. Surveying populations of western pine shoot borers (Lepidoptera: Olethreutidae). *J. Econ. Entomol.* 77:715–719 (1984).
23. Greenway, A. R. and Wall, C. Attractant lures for males of the pea moth, *Cydia nigricana* (F.), containing (*E*)-10-dodecen-1-yl acetate and (*E*,*E*)-8,10-dodecadien-1-yl acetate. *J. Chem. Ecol.* *7*:563–573 (1981).
24. Meijer, G. M., Ritter, F. J., Persoons, C. H., Minks, A. K., and Voerman, S. Sex pheromones of summer fruit tortrix moth *Adoxophyes orana*: Two synergistic isomers. *Science* *175*:1469–1470 (1972).
25. Minks, A. K. and Voerman, S. Sex pheromones of the summerfruit tortrix moth, *Adoxophyes orana*: Trapping performance in the field. *Entomol. Exp. Appl.* *16*:541–549 (1973).
26. Den Otter, C. J. and Klijnistra, J. W. Behaviour of male summer fruit tortrix moths, *Adoxophyes orana* (Lepidoptera: Tortricidae), to synthetic and natural female sex pheromone. *Entomol. Exp. Appl.* *28*:15–21 (1980).
27. Charmillot, P. J. Technique de confusion contre la tordeuse de la pelure *Adoxophyes orana* F.v.R (Lep., Tortricidae): II. Deux ans d'essais de lutte en vergers. *Mitt. Schweiz. Entomol. Ges.* *54*:191–204 (1981).
28. Guerin, P. M., Arn, H., Buser, H. R., and Charmillot, P. J. Sex pheromone of *Adoxophyes orana*: Additional components and variability of (*Z*)-9-and (*Z*)-11-tetradecenyl acetate. *J. Chem. Ecol.* *12*:763–772 (1986).
29. Roelofs, W. L., Comeau, A., Hill, A., and Milicevic, G. Sex attractant of the codling moth: Characterisation with electroantennogram technique. *Science* *174*:297–299 (1971).
30. Arn, H., Guerin, P. M., Buser, P. R., Rauscher, S., and Mani, E. Sex pheromone blend of the codling moth, *Cydia pomonella*: Evidence for a behavioural role of dodecan-1-ol. *Experientia* *41*:1482–1484 (1985).
31. Einhorn, J., Beauvais, F., Gallois, M., Descoins, C., and Causse, R. Constituants secondaires de la phéromone sexuelle

du Carpocapse des Pommes, *Cydia pomonella* L. (Lepidoptera, Tortricidae). *C.R. Acad. Sci. Paris Ser. III 19*:773–778 (1984).
32. Einhorn, J., Witzgall, P., Audemard, H., Boniface, B., and Causse, R. Constituants secondaires de la phéromone sexuelle du Carpocapse des Pommes, *Cydia pomonella* L. (Lepidoptera, Tortricidae). II. Première approche des effets comportementaux. *C.R. Acad. Sci. Paris Ser. III 7*:263–266 (1986).
33. Hand, S. C., Ellis, N. W., and Stoakley, J. T. Development of a pheromone monitoring system for the winter moth, *Operophtera brumata* (L.), in apples and in Sitka spruce, *Crop Prot. 6*:191–196 (1987).
34. Roelofs, W. L., Reissig, W. H., and Weires, R. W. Sex attractant for the spotted tentiform leaf miner moth, *Lithocolletis blancardella*. *Environ. Entomol. 63*:373–374 (1977).
35. Sanders, C. J. and Meighen, E. A. Controlled-release sex pheromone lures for monitoring spruce budworm populations. *Can. Entomol. 119*:305–313 (1987).
36. Baker, T. C. and Roelofs, W. L. Initiation and termination of oriental fruit moth male response to pheromone concentrations in the field. *Environ. Entomol. 10*:211–218 (1981).
37. Mitchell, E. B. and Hardee, D. D. Boll Weevils: Attractancy to pheromone in relation to distance and wind direction. *J. Ga. Entomol. Soc. 11*:114–117 (1976).
38. Wall, C. and Perry, J. N. Interactions between pheromone traps for the pea moth, *Cydia nigricana* (F.). *Entomol. Exp. Appl. 24*:155–162 (1978).
39. Wall, C. and Perry, J. N. Effects of spacing and trap number on interactions between pea moth pheromone traps. *Entomol. Exp. Appl. 28*:313–321 (1980).
40. Wall, C. and Perry, J. N. Effects of dose and attractant on interactions between pheromone traps for the pea moth, *Cydia nigricana* (F.). *Entomol. Exp. Appl. 30*:26–30 (1981).
41. Elkinton, J. S. and Cardé, R. T. Distribution, dispersal and apparent survival of male gypsy moths as determined by capture in pheromone-baited traps. *Environ. Entomol. 9*:729–737 (1980).
42. Henneberry, T. J. and Clayton, T. E. Pink bollworm of cotton [*Pectinophora gossypiella* (Saunders)]: Male moth catches in gossyplure-baited traps and relationships to oviposition, boll infestation and moth emergence. *Crop Prot. 1*:497–504 (1982).
43. Wall, C. and Greenway, A. R. An effective lure for use in pheromone monitoring traps for the pea moth, *Cydia nigricana* (F.). *Plant Pathol. 30*:73–76 (1981).
44. Lewis, T. and Macaulay, E. D. M. Design and elevation of sex-attractant traps for the pea moth, *Cydia nigricana* (Steph.), and the effect of plume shape on catches. *Ecol. Entomol. 1*:175–187 (1976).

45. Perry, J. N., Wall, C., and Greenway, A. R. Latin-square designs in field experiments involving insect sex-attractants. *Ecol. Entomol.* *5*:385–396 (1980).
46. Charmillot, P. J. and Schmid, A. Influence de la densité des piègés sexuels sur les captures de capua, la tordeuse de la pelure (*Adoxophyes orana* F.v.R.). *Rev. Suisse Vitic. Arboric. Hortic.* *13*:93–97 (1981).
47. McNally, P. S. and Barnes, M. M. Effects of codling moth pheromone trap placement, orientation and density on trap catches. *Environ. Entomol.* *10*:22–26 (1981).
48. van der Kraan, C. and Van Deventer, P. Range of action and interaction of pheromone traps for the summer fruit tortrix moth, *Adoxophyes orana* (F.v.R.). *J. Chem. Ecol.* *8*:1251–1262 (1982).
49. Perry, J. N. and Wall, C. Local variation between catches of pea moth, *Cydia nigricana* (F.) (Lepidoptera: Tortricidae), in sex-attractant traps, with reference to the monitoring of field populations. *Prot. Ecol.* *6*:43--49 (1984).
50. Trimble, R. M. Assessment of a sex-attractant trap for monitoring the spotted tentiform leafminer, *Phyllonorycter blancardella* (Fabr.) (Lepidoptera: Gracillariidae): Relationship between male and female emergence and between trap catches and emergence. *Can. Entomol.* *118*:1241–1253 (1986).
51. Allen, D. C., Abrahamson, L. P., Eggen, D. A., Lanier, G. N., Swier, S. R., Kelley, R. S., and Auger, M. Monitoring spruce budworm (Lepidoptera: Tortricidae) populations with pheromone-baited traps. *J. Econ. Entomol.* *15*:152–165 (1986).
52. Wall, C., Garthwaite, D. G., Greenway, A. R., and Biddle, A. J. Prospects for pheromone monitoring of the pea moth, *Cydia nigricana* (F.), in vining peas. *Aspects Appl. Biol.* *12*:117–125 (1986).

3

Principles of Attraction–Annihilation: Mass Trapping and Other Means

GERALD N. LANIER / State University of New York, Syracuse, New York

I. HISTORY

The use of odorants to lure insects to an early death probably predates any written record of the practice. In their review of trapping for pest control, Snetsinger and Shelar (1) describe insect traps dating from Roman antiquity to the early twentieth century. Most prevalent were various mechanical devices and sticky boards to be baited with odorous substances for flies. The sticky traps probably originated in Europe (or perhaps Africa; [1]), but they were reinvented several times in North America; an advertisement from the late nineteenth century (Fig. 1) indicates that catching insects on them was an avocation as well as a means of pest control.

The first notable success in direct control of an insect pest of crop or forest may have inadvertently utilized pheromones in a mass trapping system. The spruce bark beetle, *Ips typographus* L., was causing great damage to forests of the Harz mountains of Germany when, in 1772, the Hanover Electoral Chamber commissioned the most respected minds to devise a solution to *Wurmtrokniss*—the desiccation of trees as the beetle grubs mined their inner bark (2). The commission prescribed a remedy that was to be successfully applied in Europe for more than 200 years. Green timber was felled in early spring to invite infestation by bark beetles as they emerged from overwintering on the forest floor; infested material was then processed to destroy the beetle broods. After World War II, DDT and other insecticides were sometimes applied to the bark of the

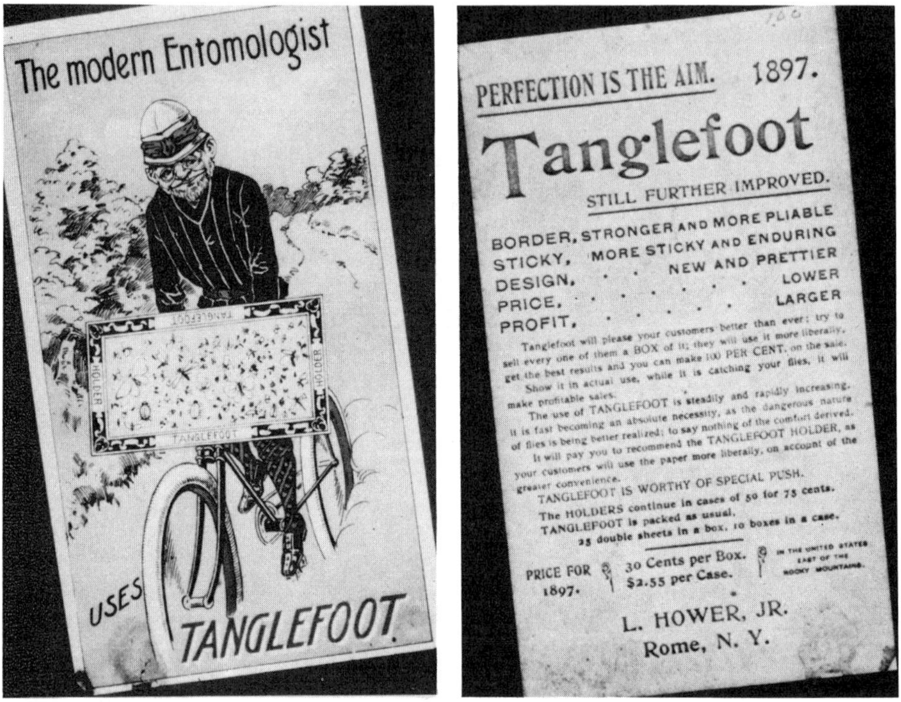

FIGURE 1 A loose-leaf advertisement for sticky insect-trapping material dated 1897.

trap trees in lieu of more laborious processes. In certain cases the insecticide-treated trap trees were ignored by the beetles while they attacked standing timber; the foresters were unaware the mass attraction depended upon aggregation pheromone produced by pioneer males once they fed on the bark of the tree.

Although the role of pheromones in the aggregation of bark beetles was not to be accurately described until the middle of the twentieth century (3), sex attractant pheromones of Lepidoptera were deliberately used in an attempt to exterminate the gypsy moth, *Lymantria dispar* L., from Massachusetts in 1893 (4). The results did not justify the labor of collection or production of the virgin females that were used to attract the males; further attempts to mass trap this species awaited identification and synthesis of its sex attractant pheromone (5) three-quarters of a century later (6).

The early success of utilizing pheromones to attract and control bark beetles and the failure with gypsy moth are indicative of differences in mass trapping of insect pests that have disparate semiochemical-mediated behaviors. These differences and the effects of population density, lure potency, trap distribution, and trap types (or other killing mechanisms) are discussed in this chapter.

II. DEFINITIONS

Attraction—annihilation is pest management by luring and eliminating the target species. A *lure* may employ odorants or visual cues or combinations of these. Odorants may be semiochemicals, such as pheromones (produced by target species), kairomones (produced by a host or prey species), or apneumones (produced by a nonliving substance and used as a cue for aggregation); or they can be empirically derived attractants that have no known role in nature. The insects attracted by a lure are acted upon by an *affector*, which could be a container or sticky trap, or a surface treated with a toxicant, sterilant, or pathogenic organism. The *insect—affector interface* is the area of active surface available to attracted insects; in container traps, it is the cross-sectional area of the trap entrances. *Pheromone mass trapping* is a means of *attractant—annihilation*; the use of the latter term allows inclusion of lures other than pheromones and means of extermination other than trapping. Various types of lures and devices used to eliminate attracted pests are outlined in Table 1.

III. STATUS OF ATTRACTION—ANNIHILATION AND OPERATIONAL PROGRAMS

A. Status

Since the mid-1960s, the identification and synthesis of numerous insect-attractant pheromones has spawned many experiments to test and develop attraction—annihilation for control of insect damage. Results of many of these attempts were discouraging or, when effective control was achieved, the approach was seldom developed for operational use. Commercial development of systems has been inhibited by the degree of expertise required for effective use and the cost of traps and trap placement; mass trapping is often deemed to be uneconomic, or at least noncompetitive with other means of control. Mitchell's (7) contention that mass trapping is useful in a very limited number of

TABLE 1 Lures and Affectors Employed in Attraction–Annihilation

Lures

Semiochemicals
- pheromones: intraspecific attractants
- kairomones: odors of a host or prey
- apneumones: odors from nonliving substances; e.g., carrion-attracting flies
- empirical attractants: e.g., methyl eugenol attracts fruit flies, but its role in nature is uncertain

Light source: e.g., UV attracts moths

Colors: e.g., yellow attracts aphids and whiteflies

Objects: e.g., large dark objects attract Tabanidae

Affectors

Killing mechanisms
- traps
  - sticky surface
  - container with restricted exit
  - container with insecticidal vapor
  - flight barrier (e.g., glass) over water or oil tray
- electric grid
- insecticide-treated surface

Indirect mechanisms
- dislocation of insects from crop or material to be protected; pest succumbs to exhaustion and natural mortality factors
- sterilization of insects
- contamination with pathogens
- harvesting or sanitation of trap crop

specialized cases was probably influenced by the preponderance of work being devoted to sex attractants released by female Lepidoptera. Beroza and Knipling (8) argue that because of the effect of competing natural pheromone sources, chances for control of insects by attraction–annihilation drop dramatically as the population density increases, and that attempts were too often directed against populations that had already reached damaging densities. For species that use female-produced sex attractants, mating disruption (see Chap. 4) is generally regarded to be more practical than mass trapping for control of damage with sex attractants. However, the power and the economics of trapping are quite different when the lure attracts females or both sexes, or when, as with the tephritid fruit flies, an artificial lure is more potent than any known natural source.

B. Operational Programs

1. *Fruit Flies (Tephritidae), Oriental Fruit Fly,* Dacus dorsalis *Hendel, and Others*

Plant derivatives have been used to monitor tephritid fruit flies since Howlett's 1915 (9) report of remarkable attraction of males of three *Dacus* species to methyl eugenol and *iso*-eugenol isolated from oil of citronella. Attraction–annihilation began with a successful attempt to reduce the oriental fruit fly population and damage to guava in Hawaii in 1952 (10). Steiner and colleagues (11), within a single year, eradicated a "heavy population" of the oriental fruit fly from the 65-km^2 Island of Rota by distributing palm fiber pads treated with methyl eugenol plus 3% naled. The entire operation required only 3.4 g of toxicant per acre.

Beginning in 1957, substantial cost savings have been achieved by employing attractant baits laced with insecticide in the eradication of the oriental fruit fly, the Medfly, *Ceratitis capitata* (Wiedemann), and other Tephritidae following their periodic introductions in Florida and California (12). Methyl eugenol and other attractants discovered by empirical screening are now widely used in fruit fly monitoring and attraction–annihilation programs. However, future use of methyl eugenol is clouded by a report that it apparently caused liver cancer in injected mice (13). Screening has identified several possible alternative attractants that are unlikely to be carcinogenic (14). Experience with tephritid fruit flies demonstrates that mass trapping can be economical and effective if the lure is exceptionally potent relative to competitive natural attractants, even when only males are attracted.

2. *Cotton Boll Weevil,* Anthonomus grandis *Boheman*

Volatiles released by male boll weevils act as an aggregation pheromone when adults emerge in spring and in the fall when they seek overwintering sites; once the insects are established on plants, the voltatiles act as a female attractant. Spring attraction–annihilation of adult boll weevils as they emerged from overwintering sites removed 76% to 90% of sparse populations (up to 25 weevils per acre) (15,16). The employment of this strategy can "significantly delay" buildup of the summer population to damaging levels (17). Attraction–annihilation has been an important aspect of integrated management of the boll weevil population and of an experimental program attempting to eradicate it in part of the cotton-growing region of the southeastern United States (18).

3. *Japanese Beetle,* Popillia japonica *Newman*

Japanese beetles of both sexes are attracted in impressive numbers to food lures. The synthetic sex pheromone, japonilure, captures only males, but catches of both sexes are maximized by combining pheromone and food lures [phenethyl propionate, eugenol, and geraniol (PEG)] (19). Traps are widely employed by state and federal authorities to monitor changes in the distribution of Japanese beetles in North America. Traps with food lures or with food lures plus japonilure, are marketed commercially and are frequently seen on urban properties in eastern states. Three years of mass trapping on Nantucket Island, Massachusetts, was reported to have reduced the Japanese beetle population by 50% (20). However, I found a surprising absence of reports about research on the effect of trapping on foliage damage by adults or turf damage by larvae, even though reviews (21, 22) had stressed the need for research relating annihilation of adults with levels of damage.

4. *Spruce Bark Beetle,* Ips typographus *(L.)*

A massive trapping program from 1979 to 1981 combatted a major outbreak of spruce bark beetles in Norway and Sweden. Bakke (23) calculated that 7.4 billion beetles taken in Norway during 1979 and 1980 could have killed an aggregate of 6.1 million spruce tress. Tree mortality declined or stopped in most localities, but the impact of attraction–annihilation was difficult to assess scientifically because the size of the beetle population was not determined and no experimental check areas were excluded from trapping. In a well-controlled study of mass trapping of the western pine beetle, *Dendroctonus brevicomis* Leconte, in central California during 1970, tree mortality dropped from 283 ± 46 trees per year before treatment to < 30 trees per year after treatment (24). In both these cases, collapse of the outbreaks may have resulted from depleting the populations to levels at which the beetles could no longer generate densities of attack sufficient to overcome the resistance of most of the trees, as would be predicted by Berryman's (25) threshold density model (Fig. 2).

5. *Ambrosia Beetles,* Gnathotrichus sulcatus *(LeConte),* G. retusus *(LeConte), and* Trypodendron lineatum *(Olivier) in British Columbia*

Attraction–annihilation of ambrosia beetles has a direct effect on reduction of degrade due to "shot holes" these beetles make in logs and green lumber. In 3 years of trapping Lindgren and Borden (26) increased the portion of the estimated total population of *T. lineatum* killed by pheromone-baited traps from 1% to 3% in 1979 to 44% to 77% in 1981. Direct assessment of economic benefits of removing 2.8

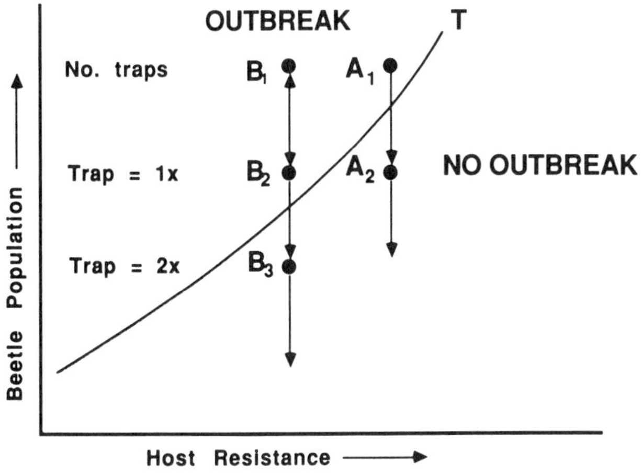

FIGURE 2 Attraction–annihilation of bark beetle at varying population densities and levels of host resistance following Berryman's (25) outbreak threshold concept. Populations in stands A and B are initially above the threshold at which beetles are able to colonize healthy trees and increase in the next generation. In the relatively resistant stand A, depleting 33% of the beetles to A_2 by attraction–annihilation will push the population below the outbreak threshold, after which it will continue to decline because most of the beetles will be killed by host resistance mechanisms. In less resistant stand B, a 33% reduction (to B_2) will not bring the population below the threshold, and the population will rebound. A doubling of effort to reduce B_1 by 50% to B_3 is necessary to drive the population below the outbreak threshold.

million beetles was not possible under conditions of the study area, but later economic analyses (27,28) showed that considerable reduction in the degrade loss in the Vancouver log market area (63.7 million dollars, Canadian, in 1980/1981) could be attained by mass trapping of ambrosia beetles. During 1987 attraction--annihilation of ambrosia beetles was commercially operated at 25 timber-processing locations (29).

6. *European Elm Bark Beetle,* Scolytus multistriatus *(Marshall)*

Sticky traps (45 × 66 cm) baited with the aggregation pheromone (multilure) are affixed like posters on utility poles and trees (not elms) to intercept *S. multistriatus* before they feed on twigs of healthy elms and transmit the Dutch elm disease-causing fungus.

Rabaglia and Lanier (30) found a correlation ($r = 0.89$, $F < 0.10 > 0.05$) between an index of twig feeding and numbers of European elm bark beetles trapped when the population was at low or moderate levels. When breeding material (diseased elm) was very abundant because of an epidemic of elm yellows disease, the actual number of beetles trapped did not increase in proportion to the twig-feeding index. However, the index of twig feeding and Dutch elm disease (DED) infection rate was always lower within the trapping area than outside of it. From 1978 through 1980, the number of elm bark beetles taken was 1.6 to 3.8 times higher than the projected total number of twig-feeding injuries on juvenile elms within the trapping area. Because the DED-causing fungus is transported by adult elm bark beetles, it would seem that eliminating part of the emergent population would have a proportional impact on the infection rate. This relationship was confirmed, indirectly, by positive correlations of DED rate with trap-catches and with twig feeding (both $r = > 90$, $F < 0.01$; 30) and, directly, by consistent decline in infection rates within groves of healthy elms while they were surrounded with traps, contrasted with DED increases after traps were removed (31,32). Concurrently, DED rates continued to fall in another test group in which traps were maintained. Nonetheless, no discernible reduction in DED infection rates was found in mass trapping of elm bark beetles in two cities, even though millions of beetles were killed and populations were demonstrably reduced during the trapping period (33).

During 1987, commercial traps were employed as part of integrated elm management programs by about 25 municipalities or agencies in New England, the Washington, D.C., area, and the Great Lake region. Multilure is also used to bait cacodylic acid-killed trap trees.

IV. OPERATING PRINCIPLES

A. Tabulation of Principles

Most of the key aspects of attraction--annihilation can be grouped under the following five operating principles:

1. Optimal synthetic semiochemical lures are usually identical with the natural structures and bouquets.
2. Effectiveness of attraction–annihilation will increase as the lure and the insect–affector interface are increased, relative to the size of the pest population.
3. When the noxious life stage is removed by an attraction–annihilation system, damage will be reduced in proportion to the percentage of the population affected; when the life stage removed is

not that which causes injury, the proportion of damage reduced will be less than that of the population affected.

4. Among polygamous species, lures that attract females will have greater impact on reproduction, relative to the percentage affected, than attractants that affect only males.
5. The proportion of a population activated by a lure, and the area over which the activation threshold is reached, will increase with release rate, but the proportion of the activated individuals that contact the affector will decrease with increasing release rates.

B. Discussion of Principles

1. Optimization of Lures

Synthetic attractants that are less active than the natural sources are missing component chemicals, have unnatural ratios of components, or contain inhibitory substances (i.e., wrong isomer or enantiomers). Although they are unlikely to be more attractive, the effectiveness of certain analogues of semiochemicals that are less subject to degradation within the controlled-release mechanism could exceed that of the natural lures. An optimal attractant may not be required to achieve acceptable control, but the amount of attractant released or the insect–affector interface required, or both, will be greater than that (those) needed to achieve the same degree of population suppression with the optimal bouquet.

Identification of the structure of the sex attractant pheromone released by the gypsy moth (5), spawned eight experimental attempts to mass trap it between 1971 and 1975 (see Ref. 6 for review), which failed to demonstrate that attraction–annihilation was a viable strategy for the control of gypsy moth. However, all of these trials employed racemic disparlure, the (-)-enantiomer of which inhibits response to the natural (+)-enantiomer. There was *no* dosage at which the racemic material could compete with the attraction of a virgin female (34). The Japanese beetle also requires enantiomeric purity for optimal response of males to the sex attractant pheromone japonilure, but up to 5% of the *S* enantiomer could be used in operational trapping if the empirically derived feeding attractants phenethyl propionate (PEP) and eugenol are used with pheromone (22).

The European elm bark beetle tolerates the presence of opposite (to the natural) isomers and enantiomers in its response to female-released components of its aggregation pheromone, (-)-α-multistriatin (35) and (-)-*threo*-4-methyl-3-heptanol (36,37). A relatively inexpensive synthetic blend of these chemicals plus impure host synergist, α-cubebene, from an alternate natural source is used in multilure.

Specificity of response may create problems in attempts to control a complex of related pests. Operational traps for ambrosia beetles combine racemic sulcatol for *Gnathotrichus sulcatus* and lineatin plus ethanol and α-pinene for *Trypodendron lineatus*. Control of *G. retusus* would require another set of traps baited with the (+)-sucatol. Unfortunately, the expense of the pure (+)-enantiomer limits its operational use (29).

The redbanded leafroller, *Argyrotaenia velutinana* (Walker), and the grape berry moth, *Paralobesia viteana* (Clemens), were controlled at the same time in New York vineyards by sticky traps baited with (*Z*)-11-tetradecenyl acetate (38). (Fortuitously, the baits contained an impurity of about 8% *E* isomer, which was later found to be essential for response by the redbanded leafroller; 39.) A similar trapping scheme was effective on apples but was judged to be impractical for control of the redbanded leafroller because there remained a complex of lepidopterous pests of New York apples that would require treatment with chemical insecticides (40).

The pine engraver beetle, *Ips pini* (Say), and the European corn borer, *Ostrinia nubilalis* (Hübner), have remarkable differences in pheromone systems among populations of different geographic origin. Intraspecific variation in response to enantiomers (41), isomers (39), and host synergists underscore the admonition of Borden et al. (42) that ". . . pest management decisions should be made on the basis of data generated from the target population."

2. *Insect–Affector Interface*

A model describing efficiency of sex pheromone-baited traps, relative to the density and distribution of both the traps and the pest, was described by Knipling and McGuire (43). If the attractant power of each lure is equivalent to one virgin female and the killing efficiency of the affector is 1.0 (assumes that all insects reaching the bait are killed), 9% of the males will be killed when the trap/female ratio is 1:10, whereas 91% will be destroyed when this ratio is 10:1 (44). This model has been affirmed for the boll weevil (18) and the redbanded leafroller (45). Webb (6) and Knipling (44) emphasize that systems employing sex attractants in attraction–annihilation strategies will be most useful against low-level populations. If the lure is a pheromone, competition from natural sources would require many traps for attraction–annihilation to have a significant impact on high-level populations. Thus, use of attraction–annihilation might be feasible against pests, such as the peach tree borer, *Synanthedon exitiosa* (Say), that cause economic damage at low-population densities, or against eruptive outbreak pests, such as the gypsy moth, either before the population has reached damaging levels or after the population has been reduced by insecticide or other means.

Success of attraction–annihilation programs will be enhanced by assessment of the target population, the competitiveness of the lure, and the efficiency of the affector (44). Effectiveness can be improved by increasing the potency of the lure and the size or efficiency of the affector. Increases in attractiveness will be nonlinear with increasing rates of attractant release from individual lures; therefore, the greatest gains in numbers of the target species affected can be realized by increasing the number of points from which the attractant is released. Increases in the number of lures should continue at least until the average kill per affector decreases (this signifies that lures are in competition), but not beyond the point that total kill begins to diminish because of disruptive effects.

Whenever fewer than 100% of the insects lured are killed by the affector, improvements in efficiency or increases in size of the affector should be considered. Increasing the size of openings to container traps will result in admission of a greater portion of the insects attracted, but it will also permit more to escape. The optimal orifice diameter for capture and retention of five Lepidoptera species was about twice the average thorax width (6). Sticky traps are generally more efficient at catching insects attracted than are container traps (6,26,46), and traps with exposed sticky surfaces are more efficient than traps with sticky surfaces enclosed (6,47, 48). Unfortunately, the useful life of sticky traps may be short because they are prone to become saturated with captured insects and debris.

In general, a large increase in insect-affector interface can be achieved most economically by treating tree boles or other objects with insecticides or, in certain cases, by attracting insects to a trap crop that is harvested or sanitized before a new generation is produced.

The efficiency of traps can be strongly affected by their position. Traps for elm bark beetles (47) and spruce beetles (49) placed in open areas were consistently more effective than those near trees. Such traps provide a strong silhouette for visual orientation of insects following the odor plume. Open placement has the added benefits of luring the pest away from the material to be protected and, probably, of causing attrition by exhaustion and heat prostration of insects that may strike the trap and fall to a sun-heated soil surface. A strong tendency to trap more gypsy moths in traps on tree trunks, compared with those hanging away from trunks, was attributed to orientation of males to the tree trunk on which females would normally be found (46).

Other aspects of affectors demonstrated to be factors include color (19,47), height above ground (1–4 m usually optimum; 6,22,50,51), and qualities of the trap surface (52,53).

3. *Life Stage Affected*

The power of attraction–annihilation against damaging adults such as bark beetles, is much greater than against adults of species such as the gypsy moth, whose contribution to damage is delayed and indirect.

4. *Trapping of Females*

In most insect species, the degree of polygamy exceeds the degree of polyandry; thus, the effect on reproductive potential will increase as the relative kill of females increases. Even if a large percentage of males is destroyed, the impact of attraction–annihilation on damage caused by the subsequent generation of larvae will be less than the kill ratio as long as the surviving males inseminate an average of more than one female. To predict the effect of eliminating males, one must ascertain the *male mating potential* (MMP; ability of males to find potential mates; 54) at the relevant population density. At an MMP of 1.0 or higher, all females in the population should be inseminated. If the average male has the potential to mate five females (MMP = 5.0), up to 80% of the males could be removed without affecting the production of fertilized eggs. In very sparse populations, the MMP for the same species will probably be less than 1.0; elimination of any males will have some impact on the next generation (Fig. 3).

Attraction of virgin and mated females to synthetic boll weevil pheromone endows attraction–annihilation strategies with considerably more power than would be possible if males were the attracted sex. The attraction of both sexes to lures as the weevils emerge from overwintering, when the population is generally at its lowest level, can severely erode the base for population buildup in subsequent generations (17).

Food or oviposition attractants can be incorporated with female-produced sex attractant pheromones so that both sexes are affected by the combined lure. Traps baited with food lures, phenethyl propionate and eugenol, plus sex attractant caught more Japanese beetles of both sexes than either attractant alone (19,22). Hamilton et al. (20) reported a 50% reduction in an isolated population after three years of mass trapping.

Removal of potential parent adults will conceptually have an effect on the size of the next generation when the reproductive substrate is not limiting. For ambrosia beetles, which utilize a three-dimensional substrate (sapwood of logs) and rarely, if ever, completely colonize all of the suitable logs in a processing area, it is unlikely that pheromone-based trapping could be offset by a favorable effect on the reproductive success of the beetles that escape capture

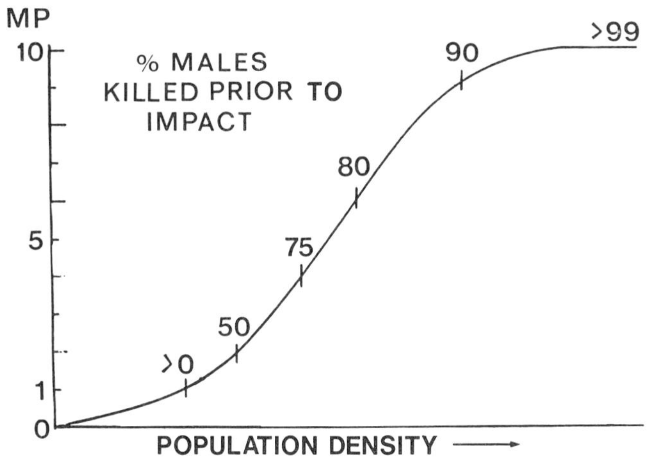

FIGURE 3 The impact of capture of males that have the potential to mate (MP) 10 females each when the population density is high. At high densities 90% or more of the males must be destroyed before any effect on the numbers of fertilized eggs is expected. At very low population densities (MP = 1.0), elimination of males will cause, at least, a proportional decrease in the next generation (concept of Granett; 54).

(26). In contrast, European elm bark beetles breed in the essentially two-dimensional phloem–cambium layer and usually appear to colonize virtually all of the elm in suitable condition at the time of beetle flight. Here, construction of longer egg galleries, increased oviposition, and reduction of competition among larvae can compensate for removal of a large portion of potential parent beetles (55). Dutch elm disease control programs should emphasize the elimination of breeding material (sanitation) to reduce the population and rely on attraction–annihilation to distract elm bark beetles from feeding in twigs of healthy elm trees (32,33,50).

5. Dosage of Attractant

The volume of the plume of odorant molecules within which the activation threshold is exceeded will increase with the release rate under any given set of meteorological conditions. Therefore, the number of individuals likely to become activated by the lure will also increase. Average concentration of odorant molecules within the plume will also increase with release rate; this will ensure that, within a population of individuals with varying thresholds for response, more will be activated at higher release rates. At some very high release rate, many insects will break off their movement toward the lure because their olfactory systems will signal that they have arrived at the source when they actually have not, or they will become habituated to the point of insensitivity. Another factor that will affect the rate of successful arrival is that individuals coming from a longer distance and searching a larger plume will have more opportunity to lose their way than individuals that encounter a smaller plume and are closer to the lure. Individual insects may typically fly directly to lures at distances of 400 m or more (56), and they may be captured several kilometers from their points of origin (51).

The relationship between varying attractant release rates and trap-catches is typically sigmoidal until, at very high release rates (25), the number trapped declines precipitously (Fig. 4). The most

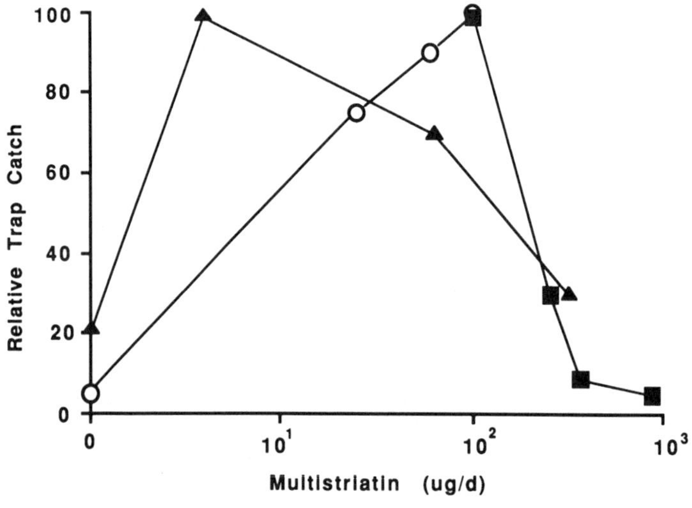

FIGURE 4 The effect on trap-catch of *Scolytus multistriatus* of varying rates of release of α-multistriatin as observed in England by Blight et al. (37; closed triangle) and North America by Lanier (unpublished observations; open circles) and Cuthbert and Peacock (57, closed squares). The numbers of beetles captured drop rapidly when excessive α-multistriatin is released; optimal release rates appear different in England and North America.

efficient release rate for a lure-affector system is at the upper inflection point of the curve. If the release rate is too high, a decreasing percentage of the target individuals activated by the lure will reach the affector. Thus, either the release rate should be reduced or the size of the insect–affector interface should be increased.

V. CONCLUSIONS

Attraction–annihilation can reduce damage by insect pests and even exterminate isolated populations. However, elimination of large numbers of the target species is not sufficient to justify operational use of attraction–annihilation; studies of efficacy and cost-effectiveness are necessary.

The power of attraction–annihilation will be inversely related to the population density and the size of the area over which the pest is present. Therefore, attraction–annihilation is most likely to be an appropriate strategy for control of populations that are immigrant, geographically limited (e.g., island), or at low density. Such populations could be at low levels naturally, or because they have been reduced by integrated pest management (IPM). The most economically opportune situation for use of an attraction–annihilation strategy would be that of a high-value crop on which relatively low pest population densities cause economic damage.

One of the most serious problems for attraction–annihilation is competition from natural sources of attractants. Optimization of the lure is most critical when the competition is a sex lure or pheromone that attracts potentially polygamous males. Synthetic lures that are more attractive than any natural source of attractant (e.g., methyl eugenol or the oriental fruit fly) will probably draw a rather constant proportion of the pest regardless of the population density. Attracting females by incorporation of ovipositional or feeding attractants (kairomones) with sex attractants could considerably increase the power of attraction–annihilation against Lepidoptera.

Attraction–annihilation against beetles and other orders in which the adult is the damaging stage should be cost-effective when the cost of the operation is less than the multiple of proportion killed and the value of the expected damage. For pests in which there is positive feedback between population density and damage, as in tree-killing bark beetles, the benefit (percentage of damage prevented) of attraction–annihilation may exceed the proportion of the population destroyed by the affector.

The density of lures should be between that at which competition between lures can be measured by a decreasing average kill per affector unit and that at which total kill stops increasing. Increasing the insect–affector interface will almost always increase the effect of a given amount of attractant released. This could be accomplished by increasing the number or size of traps, using a sticky surface in the open, rather than container traps or, especially, by using a trap crop or insecticide-treated surface. Use of toxicants usually will greatly improve cost-effectiveness. Entomologists have probably been predisposed to use traps because they allow easy estimation of the number of insects killed and there is gut-level pleasure in presiding over a pile of dead pest insects. The application of an insecticide to a baited surface represents a very limited use compared with broadcast sprays. Nevertheless, the public may perceive unrealistic hazards when pesticides are involved. Traps will probably find continued application in urban areas.

REFERENCES

1. Snetsinger, R. and Sheltar, D. J. Traps for animal pest control. *Melshimer Entomol. Ser.* *32*:12–19 (1982).
2. Schwerdtfager, F. Forest entomology. In *History of Entomology*. (Smith, R. F., Mittler, T. E., and Smith, C. N., eds.). Annual Reviews, Palo Alto, 1973, pp. 361–386.
3. Anderson, R. F. Host selection by the pine engraver. *J. Econ. Entomol.* *41*:596–602 (1948).
4. Forbush, E. H. and Fernald, C. H. *The Gypsy Moth*. Wright & Potter, Boston, 1896, 495 pp.
5. Bierl, B. A., Beroza, M., and Collier, C. W. Potent sex attractant of the gypsy moth: Isolation, identification and synthesis. *Science* *170*:87–89 (1970).
6. Webb, R. E. Mass trapping of the gypsy moth. In *Insect Suppression with Controlled Release Pheromone Systems*, Vol. II. (Kydonieus, A. F. and Beroza, M., eds.). CRC Press, Boca Raton, 1982, pp. 27–56.
7. Mitchell, E. R. Pheromones: As the glamour and glitter fade, the real work begins. *Fla. Entomol.* *69*:132–139 (1986).
8. Beroza, M. and Knipling, E. F. Gypsy moth control with sex attractant pheromone. *Science* *177*:19–27 (1972).
9. Howlett, F. M. Chemical reactions of fruit flies. *Bull. Entomol. Res.* *6*:297–305 (1915).
10. Steiner, L. F. Methyl eugenol as an attractant for oriental fruit fly. *J. Econ. Entomol.* *45*:241–248 (1952).
11. Steiner, L. F., Mitchell, W. C., Harris, E. J., Kozama, T. T., and Fujimoto, M. S. Oriental fruit fly eradication by male annihilation. *J. Econ. Entomol.* *58*:961–964 (1965).

12. Steiner, L. F., Rohwer, G. G., Ayers, E. L., and Christenson, L. D. The role of attractants in the recent Mediterranean fruit fly eradication program in Florida. *J. Econ. Entomol.* *54*:20--35 (1961).
13. Miller, E. C., Swanson, A. B., Phillips, D. H., Fletcher, T. L., Leim, A., and Miller, J. A. Structure—activity studies of the carcinogenicity in the mouse and rat of some naturally occurring synthetic alkenylbenzene derivatives related to safrole and estragol. *Cancer Res.* *43*:1124—1134 (1983).
14. Mitchell, W. C., Metcalf., R. L., Metcalf, E. R., and Mitchell, S. Candidate substitutes for methyl eugenol as attractants for the area-wide monitoring and control of the oriental fruit fly, *Dacus dorsalis* Hendel (Diptera: Tephritidae). *Environ. Entomol.* *14*:176—181 (1985).
15. Mitchell, E. B., Lloyd, E. P., Hardee, D. D., Cross, W. H., and Davich, T. B. In-field traps and insecticides for suppression and elimination of populations of boll weevils. *J. Econ. Entomol.* *69*:83--88 (1976).
16. Lloyd, E. P., McCoy, J. R., and Haynes, J. W. Release of sterile-male boll weevils in the Pilot Boll Weevil Eradication Experiment in 1972--73. In *Proceedings of a Conference on Boll Weevil Suppression, Management, and Elimination Technology*, Feb. 13--15, 1974, Memphis, U.S. Dept. Agr., Agr. Res. Serv. S-71, Washington, D.C., 1976, pp. 95--102.
17. Hardee, D. D. Mass trapping and trap cropping of the boll weevil, *Anthonomus grandis* Boheman. In *Insect Suppression with Controlled Release Pheromone Systems*, Vol. II. (Kydonieus, A. F. and Beroza, M., eds.). CRC Press, Boca Raton, 1982, pp. 65—71.
18. Hardee, D. D. and Boyd, F. J. Trapping during the Pilot Boll Weevil Eradication Experiment, 1971—73. In *Proceedings of a Conference on Boll Weevil Suppression, Management, and Elimination Technology*. Feb. 13—15, 1974, Memphis, U.S. Dept. Agr., Agr. Res. Serv. S-71, Washington, D.C., 1976, pp. 82—89.
19. Ladd. T. L. and Klein, M. G. Japanese beetle (Coleoptera: Scarabididae) response to color traps with phenethyl propionate + eugenol + geraniol (3:7:3) and japonilure. *J. Econ. Entomol.* *79*:84--86 (1986).
20. Hamilton, D. W., Schwartz, P. H., Townshed, B. G., and Jester, C. W. Traps reduce an isolated infestation of Japanese beetle. *J. Econ. Entomol.* *64*:150—153 (1971).
21. Klein, M. G. Mass trapping for suppression of Japanese beetles. In *Management of Insect Pest with Semiochemicals*. (Mitchell, E. R., ed.). Plenum Press, New York, 1981, pp. 181—190.

22. Ladd, T. L. Trapping Japanese beetles with synthetic female sex pheromone and food-type lures. In *Insect Suppression with Controlled Release Pheromone Systems*, Vol. II. (Kydonieus, A. F. and Beroza, M., eds.). CRC Press, Boca Raton, 1982, pp. 57–64.
23. Bakke, A. Utilization of aggregation pheromones for control of the spruce bark beetle. In *Insect Pheromone Technology: Chemistry and Applications*. (Leonhardt, B. A. and Beroza, M., eds.). American Chemical Society, Washington, D.C., 1982, pp. 219–227.
24. Bedard, W. D. and Wood, D. L. Programs utilizing pheromones in survey and control. Bark beetles—the western pine beetle. In *Pheromones*. (Birch, M. C., ed.). North-Holland, Amsterdam, 1974, pp. 441–449.
25. Berryman, A. A. Population dynamics of bark beetles. In *Bark Beetles in North American Conifers*. (Mitton, J. B. and Sturgeon, K. B., eds.). Univ. Texas Press, Austin, 1982, pp. 264–314.
26. Lindgren, B. S. and Borden, J. H. Survey and mass trapping of ambrosia beetles (Coleoptera: Scolytidae) in timber processing areas on Vancouver Island. *Can. J. For. Res. 13*:481–493 (1983).
27. McLean, J. A. Ambrosia beetles: A multimillion dollar degrade problem of sawlogs in coastal British Columbia. *For. Chron. 61*:295–298 (1985).
28. Grey, D. R. and Borden, J. H. Ambrosia beetle attack on logs before and after processing through dry land sorting areas. *For. Chron. 61*:299–302 (1985).
29. McLean, Personal communication, 1987.
30. Rabaglia, R. J. and Lanier, G. N. Twig feeding by *Scolytus multistriatus* (Coleoptera: Scolytidae): Within-tree distribution and use for assessment of mass trapping. *Can. Entomol. 116*:1025–1032 (1984).
31. Peacock, J. W., Cuthbert, R. A., and Lanier, G. N. Deployment of traps in a barrier strategy to reduce populations of the European elm bark beetle, and the incidence of Dutch elm disease. In *Management of Insect Pests with Semiochemicals*. (Mitchell, E. R., ed.). Plenum Press, New York, 1981, pp. 155–174.
32. Lanier, G. N. Behavior-modifying chemicals in Dutch elm disease control. In *Proceedings of the Dutch Elm Disease Symposium and Workshop*. (Kondo, E. S., Hiratsuka, Y., and Denyer, W. B. G., eds.). Manitoba Dept. Natural Resources, Winnipeg, 1982, pp. 371–394.

33. Peacock, J. W. City wide mass trapping of *Scolytus multistriatus* with multilure. In *Proceedings of the Dutch Elm Disease Symposium and Workshop*. Kondo, E. S., Hiratsuka, Y., and Denyer, W. B. G., eds.). Manitoba Dept. Natural Resources, Winnipeg, 1982, pp. 406–426.
34. Miller, J. R. and Roelofs, W. L. Gypsy moth response to pheromone enantiomers as evaluated in a sustained-flight tunnel. *Environ. Entomol.* 7:42–44 (1978).
35. Elliott, W. J., Homnak, G., Fried, J., and Lanier, G. N. Synthesis of multistriatin enantiomers and their action on *Scolytus multistriatus* (Coleoptera: Scolytidae). *J. Chem. Ecol.* 5:279–287 (1979).
36. Lanier, G. N., Gore, W. E., Pearce, G. T., Peacock, J. W., and Silverstein, R. M. Response of the European elm bark beetle, *Scolytus multistriatus* (Coleoptera: Scolytidae), to isomers and components of its pheromone. *J. Chem. Ecol.* 3:1–8 (1977).
37. Blight, M. M. Chemically mediated behavior of *Scolytus scolytus* and *S. multistriatus* in the United Kingdom: Studies on the role of multistriatin and host compounds. In *Proceedings of the Dutch Elm Disease Symposium and Workshop*. (Kondo, E. S., Hiratsuka, Y., and Denyer, W. B. G., eds.). Manitoba Dept. Natural Resources, Winnipeg, 1982, pp. 427–450.
38. Taschenberg, E. F., Cardé, R. T., and Roelofs, W. L. Sex pheromone mass trapping and mating disruption for control of redbanded leafroller and grape berry moth in vineyards. *Environ. Entomol.* 3:239–242 (1974).
39. Klun, J. A. Insect sex pheromones: Minor amount of opposite geometrical isomer critical to attraction. *Science 181*:661–663 (1973).
40. Trammel, K., Roelofs, W. L., and Glass, E. H. Sex pheromone trapping of males for control of redbanded leafroller in apple orchards. *J. Econ. Entomol.* 67:159–164 (1974).
41. Lanier, G. N., Claesson, A., Stewart, T., Piston, J. J., and Silverstein, R. M. *Ips pini*: The basis for interpopulational differences in pheromone biology. *J. Chem. Ecol.* 6:677–687 (1980).
42. Borden, J. H., King, C. J., Lindgren, S., Chong, L., Gray, D. R., Ochlschlager, A. C.,Slessor, K. N., and Pierce, H. D. Jr. Variation in response of *Trypodendron lineatum* from two continents to smiochemicals and trap form. *Environ. Entomol.* *11*:403–408 (1982).
43. Knipling, E. F., and McGuire, J. U., Jr. Population models to test theoretical effects of sex attractants used for insect control. *U.S. Dept. Agric. Inform. Bull.* 308, 1966, 20 pp.

44. Knipling, E. F. *The Basic Principles of Insect Population Suppression and Management*. Agric. Handbook 512, U.S. Dept. Agric., Washington, 1979, 659 pp.
45. Roelofs, W. L., Glass, E. H., Tette, J., and Comeau, A. Sex pheromone trapping for red-banded leaf roller control: Theoretical and actual. *J. Econ. Entomol.* *63*:1162–1167 (1970).
46. Elkinton, J. S. and Childs, R. D. Efficiency of two gypsy moth (Lepidoptera: Lymantridae) pheromone baited traps. *Environ. Entomol.* *12*:1519–1525 (1983).
47. Lanier, G. N., Silverstein, R. M., and Peacock, J. W. Attractant pheromone of the European elm bark beetle (*Scolytus multistriatus*): Isolation, identification, synthesis and utilization studies. In *Perspectives in Forest Entomology*. (Anderson, J. E. and Kaya, H. K., eds.). Academic Press, New York, 1976, pp. 149–175.
48. Holt, S. C., Westigard, P. H., and Rice, R. E. Development of pheromone trapping techniques for male San Jose scale (Homoptera: Diaspidae). *Environ. Entomol.* *12*:371–375 (1983).
49. Bakke, A. and Riege, L. The pheromone of the spruce beetle *Ips typographus* in Norway and its potential use in the suppression of beetle populations. In *Insect Suppression with Controlled Release Pheromone Systems*, Vol. II. (Kydonieus, A. F and Beroza, M., eds.). CRC Press, Boca Raton, 1982, pp. 3–15.
50. Cuthbert, R. A. and Peacock, J. W. Attraction of *Scolytus multistriatus* to pheromone-baited traps at different heights. *Environ. Entomol.* *4*:889–890 (1975).
51. Birch, M. C., Paine, T. D., and Miller, J. C. Effectiveness of pheromone mass-trapping of the smaller European elm bark beetle. *Calif. Agric.* *35*:6–7 (1981).
52. Lie, R. and Bakke, A. Practical results from the mass trapping of *Ips typographus* in Scandinavia. In *Management of Insect Pests with Smiochemicals*. (Mitchell, E. E., ed.). Plenum Press, New York, 1981, pp. 175–181.
53. von Keyserling, H. Control of Dutch elm disease by behavioral manipulation of its vectors. *Med. Fac. Landboww Rijdsuniv. Gent.* *45*:475–488 (1980).
54. Granett, J. Estimation of male mating potential of gypsy moths with disparlure baited traps. *Environ. Entomol.* *3*:383–385 (1974).
55. Svihra, P. The behavior of *Scolytus multistriatus* in California. In *Proceedings of the Dutch Elm Disease Symposium and Workshop*. (Kondo, E. S., Hiratsuka, Y., and Denyer, W. B. G., eds.). Manitoba Dept. Natural Resources, Winnipeg, 1982, pp. 395–405.

56. Wall, C. and Perry, J. N. Range of action of moth sex-attractant sources. *Entomol. Exp. Appl.* *44*:4–14 (1987).
57. Cuthbert, R. A. and Peacock, J. W. Response of *Scolytus multistriatus* (Coleoptera: Scolytidae) to component mixtures and doses of the pheromone multilure. *J. Chem. Ecol.* *4*:363–373 (1978).

4

Principles of Mating Disruption

RING T. CARDÉ / University of Massachusetts, Amherst, Massachusetts

I. INTRODUCTION

The very chemicals utilized by pest insects to communicate location and sexual availability to a prospective mate have long been deemed to be of potential value in their management (1–3), even in advance of the first identification of the pheromone of any pest species. With the advent of synthetic copies of these messages and formulations enabling them to be broadcast across crops and forests, a flurry of field trials, beginning with Gaston et al. (4), established that sexual recruitment could be interrupted. These early demonstrations provided an impetus for the creation of commercial formulations effective in direct population control. To date, the chemical modifiers of behavior that have been successful in disruption of mating have been targeted toward interfering with the in-flight maneuvers that bring the sexes together over distances of several to perhaps hundreds of meters. The particular vulnerability of moths to the manipulation of their olfactory guidance system has dictated that most of the practical efforts aimed toward developing commercially viable products should be directed toward moths, which conveniently also are among our most damaging insects. Formulations effective in achieving direct population control are available for a number of moth pests.

The most typically applied term for the behavioral effect of preventing mate location is *disruption of communication*, a reasoned attempt to avoid implying either a behavioral or physiological mechanism.

Such a calculated avoidance of any mechanistic connotation underscores that, as yet, we do not fully understand how this process works, although theories abound.

Our ignorance stems from three facts. The mechanisms of navigation to a natural chemical emitter and mate recognition at close range remain imperfectly understood for most insects (5), although for several moth species experiments in the wind tunnel and in the field have provided a provisional outline of the probable mechanisms (6). Walking orientation to chemicals, on the other hand, is more fully understood (6,7), but interference with this process seems of minor importance to disruption of communication as it is currently employed in crop protection. A second deficiency is simply that the in-flight maneuvers probably differ among the various groups of insects that fly to pheromone sources. All of the moth species investigated to date seem to employ upwind anemotaxis, but this is a portmanteau concept that, along with "attraction," embraces many possible mechanisms of orientation (6,8). The maneuvers that the 120,000 or so species of moths employ in mate location presumably have a common phylogenetic origin and considerable similarity. However, sawflies and bark beetles, although they too employ long-distance pheromonal communication, may use quite divergent navigational mechanisms. And third, precious few experiments have been undertaken to describe how disruption is achieved.

Instead, pheromones, once identified, are all too frequently formulated and launched forthwith into large-scale and poorly replicated field evaluations, which on fortuitous occasions have yielded a viable management tool. In retrospect, the deficiencies of such approaches have become evident: In many cases only a portion of the pheromone blend was known, an admission that its pheromone-mediated behavior was incompletely described; the rate of application was chosen on such criteria as the size of the field plots available and the amount of synthetic in hand to formulate, rather than on empirically derived dose-response measurements of efficacy; the formulation had largely unknown emission and longevity characteristics; and there was inadequate isolation of the test plots to prevent the invasion of mated females. Many of these costly failures are not available in the literature for scrutiny.

Notwithstanding, there is a profusion of literature describing attempts, successful and otherwise, at disrupting communication and managing pest populations. Yet the theory underlying communication disruption has received little attention, excepting the comprehensive review by Bartell (9) and appraisals by Shorey (10), Sanders (12,13), Rothschild (13), and Cardé (14,15). A fresh consideration is warranted because new information is available both on mechanisms of navigation used by flying insects to locate a pheromone emitter and, as well, on the complexity of the natural communication systems of several moth species that have been most extensively studied in communication disruption.

The approach to understanding communication disruption detailed here will derive from describing the natural process of mate location and how the application of a synthetic pheromone is presumed to modify sensory inputs, sensory perception, and ultimately the behavior manifested. The intended emphasis, then, will be on the way in which natural patterns of behavior and sensory input are interrupted by the omnipresence of synthetic pheromone. Moth examples will be used because these represent the cases for which we have the most detailed knowledge, but many of the generalities proposed should be applicable to other insect groups that employ a pheromone for long-distance mate finding. Considerable speculation will be requisite, but the resulting principles will, it is hoped, point the way to improvements in how these systems are designed and evaluated.

A confounding factor in analyzing the efficacy of communication disruption is the level of disruption needed to achieve management objectives. This value is quite idiosyncratic, being dependent on the reproductive characteristics of the species and the dispersal ecology of the particular population to be manipulated. Knipling (16), among others, has emphasized that population density, the intrinsic rate of population increase, and the rate of influx of mated insects into the application area, all greatly alter the level of mating disruption required to achieve control of a population. As a case in point, early season disruption of the pink bollworm, *Pectinophora gossypiella*, is practical because of initial low-population levels and presumed lowered dispersiveness, whereas in late season the higher population levels and immigration can render this tactic ineffective. Understanding such factors is crucial to knowing when (or if) pheromone should be applied for direct control, but documenting the role of these ingredients in the success of management schemes largely lies beyond the intended boundaries of this review.

However, an increase in population density can diminish the typical distance of communication needed for mate finding by pheromone and elevate the likelihood of mate location by nonpheromonal cues. If disruption is dependent on the competition between numerous point sources of synthetic pheromone and pheromone-emitting insects (as elaborated on later), then the spatial distribution of artificial and natural sources, as well as their ratio, will alter efficacy.

As we proceed through a discussion of the mechanisms proposed to promote disruption of communication and the supporting evidence, it should be borne in mind that most of the experiments cited were not designed with the intent to elucidate these mechanisms. Bartell (9) has provided an incisive summary of the methods used to study natural patterns of response and the limitations of these approaches to the questions at hand. Two principal points of this caveat seem most pertinent.

First, laboratory studies, because they simulate the natural milieu only imperfectly, rarely allow the full expression of a behavioral repertory. Even the now widely used wind tunnel cannot duplicate the range of distributions of pheromone and wind fetches found in the field (17). Male gypsy moths, *Lymantria dispar*, can be induced in a wind tunnel to fly to pheromone for more than 30 min on average by moving a pattern on the floor of the wind tunnel in the direction of the wind flow (18), thereby creating an optical illusion of upwind progress. But the behavior of moths following termination of upwind plume-following cannot be observed for more than a few seconds before the male flies against the side of the tunnel. Field observations can be revealing but challenging to quantify and replicate; the majority of moth pests are nocturnal and nighttime viewing of their behavior, even with image intensification equipment or other devices, exacerbates these difficulties.

Second, we have little idea how coincident an insect's path is with the pheromone plume. That in part derives from our imperfect understanding of the fine-scale structure of natural pheromone plumes, particularly at distances on the order of tens of meters from the emitter (19, 20). Electrophysiological measurements of response suffer for the same reason: The distribution of chemical stimulus cannot be set to match the natural fluxes because these are imperfectly known. These limitations will be amplified as specific experiments are addressed. An overview of the proposed mechanisms for effecting communication disruption will now be outlined.

II. MECHANISMS FOR INTERRUPTING PHEROMONAL COMMUNICATION

Disruption of long-distance communication by pheromone was viewed by Shorey in 1973 (10) as involving three mechanisms: peripheral—sensory adaptation; diminished or lost behavioral response; and "confusion." Drawing upon an improved understanding of the complex nature of the chemical blends used by most moths (and other insects), Bartell's 1982 review (9) listed five factors, a classification scheme largely followed here.

A. Peripheral Reception and Central Nervous System Effects

Laboratory assays demonstrating that exposure to pheromone (followed by a response) causes a loss of subsequent responsiveness are legion, but the precise neuronal factors causing such diminutions in responsiveness are difficult to verify. These behavioral effects typically last for several hours after the initial exposure to the stimulus. Two explanations can be invoked: sensory adaptation at the peripheral (antennal) receptor level and central nervous system habituation.

1. *Sensory Adaptation*

Continuous presentation of a pheromone odor readily reduces the firing rate of an insect's receptors, but a return to clean air usually restores much of the receptor's sensitivity within less than a second (e.g., 21). Given the wide dynamic range of pheromone concentration within which most insects respond behaviorally to pheromone, peripheral adaptation alone seems an unlikely explanation for the loss of responsiveness after pheromone stimulation. Furthermore, as noted earlier, some gypsy moth males will fly upwind to pheromone in a wind tunnel (maintained "in place" by artificial movement of the floor pattern) for tens of minutes (18). Such observations confirm that gypsy males can zigzag along a narrow pheromone plume for extended intervals without receiving sufficient pheromone to become peripherally adapted (or centrally habituated, as discussed later) and thus halt upwind anemotaxis. Males so orienting in a wind tunnel, or especially a forest, would encounter considerable fluctuations in signal strength ("flickering") because of turbulent dispersion of the pheromone signal (19,20), and these discontinuities would be compounded by their zigzags to and fro across the plume. Signal flickering should promote disadaptation. Thus, during flight in a naturally occurring and therefore relatively dilute plume, little peripheral adaptation seems likely and behavioral response need not be interrupted.

Moths flying upwind to a comparatively intense pheromone source, as can emanate from a synthetic point source of pheromone, often cease upwind progress close to the emitter (22–24). In the oriental fruit moth, *Grapholita molesta*, such arrestment can be interpreted as interference with peripheral disadaptation (25). The signal encountered by the male should flicker because of discontinuities in its strength caused by turbulent diffusion and also because a male's zigzag path takes it back and forth across the plume. As the moth closes in on a pheromone source, the mean peak intensities of the signal increases, whereas the mean burst length and the mean return periods do not change significantly (20). These factcrs presumably contribute to a perception at a central neuronal level that the signal has "fused." Peak-to-trough amplitude fluctuations in pheromone concentration (26) as well as how well the pheromone blend mimics the natural rates and ratios could determine when fusion of the neuronal signals occur (25).

The necessity of a phasic signal to the maintenance of upwind flight to pheromone has been shown in two tortricids, *Adoxophyes orana* (27) and the oriental fruit moth (28). In wind tunnel trials, males of both species did not continue to fly upwind after being engulfed by a homogeneous cloud of pheromone. Although this mechanism needs to be verified with other moth species, these observations suggest that an even miasma of pheromone interferes with mate location.

Wind tunnel manipulations with two tortricid species, however, argue against a major contribution of either signal fusion or arrestment of upwind progress to the efficacy of disruption of communication. Males were able to "lock-on" to a narrow pheromone plume superimposed in a flow of somewhat more dilute pheromone-permeated air and follow the plume to its source (12,28). This suggests that, to achieve disruption, the airborne concentration of pheromone obtained from the formulation ought to be intense enough to be near or above the level found immediately downwind of the emitting insect. Such levels can readily be achieved near artificial point-source emitters, but they would be difficult to maintain throughout the area to be manipulated because of the difficulties of even application of pheromone in the plant canopy and vagaries of wind flow; moreover, the dosage requisite for achieving this concentration might be prohibitively expensive.

Thus, sensory adaptation alone seems inadequate to explain disruption, unless extraordinarily massive concentrations of disruptant are employed. But it seems a reasonable speculation that artificially applied pheromone could cause enough peripheral adaptation to raise the threshold of sensitivity and, thereby, reduce the ability of males to follow a natural pheromone plume against a background of disruptant.

2. *Habituation*

When the decline in responsiveness is attributed to a temporary modification at the central neuronal level, it is labeled *habituation*. This phenomenon has been examined mainly in confined bioassay setups that limit the expression of the full behavioral repertory (especially upwind flight) and the analysis has often focused on one or two "key" reactions (e.g., wing fanning) that were presumed to be representative of all behaviors. Traynier (29) concluded in *Anagastia kuehniella* that the proportion of males that were habituated was not markedly influenced by the concentration of pheromone employed in the first exposure, provided sufficient pheromone was given to evoke a behavioral response. However, because of the long periods that the moths were continually exposed to pheromone (in most instances 10 min), a contribution of sensory adaptation to the observed reduction in wing fanning cannot be ruled out.

Conversely, Bartell and Lawrence (30) demonstrated that within a diel cycle the extent of habituation in *Epiphyas postvittana*, the light-brown apple moth, was related to the concentration of pheromone extract and duration of the first exposure. Furthermore, preexposure to discrete pulses of pheromone decreased subsequent responsiveness more than exposure to the same total dosage provided continuously (31). A succinct interpretation of these pulsed-stimulus experiments is not feasible here. Bartell and Rumbo (32) concluded that adaptation, subsequent disadaptation, and habituation contributed to the reduction in responsiveness. These interactions may be complex: A disadapted cell should provide more central input, enhancing habituation.

Although adaptation and habituation could severely reduce the propensity of males to engage in plume-oriented flight, their relative contribution to disruption should be heavily dependent on the airborne concentration of disruptant achieved, the evenness of its distribution in the habitat and, as emphasized by Weatherston (Chap. 6), whether it emanates from either point sources or from microdispersible formulations. We may speculate that males that have oriented to point sources of synthetic pheromone are more apt to be subject to alterations in threshold and subsequent propensity to respond to a calling female than males exposed to an equivalent dose of pheromone applied in a microdispersible formulation. Males in the former situation have encountered relatively intense concentrations of pheromone (levels most likely to cause some adaptation), and undergoing a behavioral response is likely to facilitate habituation (30).

B. Competition Between Natural and Synthetic Sources

Given that responsiveness remains unabated, another possible mechanism invokes competition between native calling (pheromone-emitting) insects and the synthetic disruptant. This mechanism is only applicable to formulations that produce point sources of pheromone that mimic the release rate and chemical features of the pest insect well enough to elicit normal aerial "trail" following. As Sanders has emphasized (11), the efficacy of the disruptant formulation is thus dependent on (a) the emission rate of each point source (perhaps ideally at a level between a natural emitter and the upper rate effective in luring a male); (b) the number and distribution of point sources relative to calling females; and (c) the population level. The management objective may be to achieve a finite population per unit area which, in turn, necessitates fewer females being mated as population density is raised. Yet, at a given rate of distribution of point sources of artificial pheromone, the likelihood of a male flying to a female instead of an artificial pheromone source increases as the females become more numerous. Furthermore, the average distance that a male must navigate to locate a female is reduced. Notwithstanding, such a false trail-following scheme remains appealing, and its contribution to disruption of mating may be synergized by some of the mechanisms enumerated in the previous and next section.

C. Camouflage of the Natural Plumes

Given the application of a sufficiently high level of disruptant that is identical, or nearly so, with the naturally emitted pheromone, an aerial trail from a calling female should be rendered indiscernible from the background of synthetic at some distance from the natural plume's origin (11,12,15). An insect would simply be incapable of detecting in a consistent fashion the added pheromone from the female among the ubiquitous miasma generated by the disruptant formulation. As yet, models describing the flux of pheromone in a plume remain tentative and unvalidated (19,20), so that, even if the aerial concentration of disruptant and emission rate from a female were known, we could not estimate the distance from the female at which masking occurred. But the mechanism predicts that the distance from the female at which the camouflage becomes effective should decrease with either increased levels of emission from the female or decreased release from the formulation. Bartell (9) noted that the complexities of most natural environments and the attendant wind patterns would seem to rule out even permeation with disruptant. Furthermore, uniform application of formulation throughout the habitat seems operationally impossible. At some distance from the natural emitter, nonetheless, the natural aerial trail should be indistinguishable from the background of artificially applied odor, regardless of the discontinuous intensity of the plume's strength and uneven permeation of disruptant.

D. Imbalance in Sensory Input

Although there remain a few moth species that evidently have long-distance mate finding that is mediated by a single chemical, most species appear to utilize blends of two to six, and perhaps more, components. Among these more typical communication systems, individual insects of a given species may show a very tight regulation of the production of pheromone blend ratio, as in the pink bollworm (33,34), or at the other extreme, a wide and seemingly indiscriminate variability, as in the potato tuberworm, *Phthorimaea operculella*, with a range of its diene/triene component ratios of 1:9 to 7:3 (35). The pattern of male response can involve a rather precise discrimination of component ratio, as in the (*Z*)/(*E*) ratio of the acetates in the oriental fruit moth (22–24), or a promiscuous response, as in the male potato tuberworm moth, which shows equal attraction to ratios spanning 1:9 to 9:1 (36). Narrow tuning of the signal emitted, as by the female pink bollworm, however, need not be matched by a comparably restricted male response. The male pink bollworm shows a nearly equivalent attraction to proportions of *Z*,*E* isomer that exceed the natural blend by 15% (37).

The processing of these complex signals and the means for their integration in the central nervous system (CNS) are far from elucidated, and their effect on the organization of the behavioral sequence is under some debate. In one view, the entire pheromone blend mediates all phases of the behavioral sequence, from initial turning upwind some distance from the source (locking-on) to courtship. Its corollary is that the lowest threshold for all behaviors in the sequence are elicited by the complete blend in the natural ratio (38, 39). Other possibilities are that the earliest behaviors either are evoked by a portion of the blend or that ratio discrimination is less critical than at later stages in the response sequence, when, as the insect approaches the emitter, airborne concentrations of pheromone are high enough to enable ratio and minor component discrimination (38). This organization envisages two (or more) successive stages of behavior having different "active spaces," each mediated either by increasingly complex mixtures of components or by discrimination of proportions. Resolution of these propositions awaits detailed wind tunnel and field tests, such as those of Linn et al. (39) on the oriental fruit moth, with a number of species. To date there is no moth species for which a "nested" system of active spaces has been unequivocally established.

Because it is the final stages of orientation behavior that must be disrupted if mating is to be prevented, it might be argued that these questions, although theoretically intriguing, are of no relevance to the possible mechanisms of disruption. If the initial reactions in mate finding, however, are dependent upon only a portion of the blend, then a degree of interruption could be achieved by liberation of that part of the blend. What might be most easily accomplished would be a camouflage of the dilute aerial trial at some distance from the emitter. In such cases, disruption might be far less effective than would be effected by the complete bouquet because, when close to the natural emitter, a male should be able to sense the natural emitter (the complete blend) and then successfully locate its source.

Alternatively, release of either only a portion of the blend or the incorrect ratio of components might create an imbalance to the sensory input. Males encountering such an abnormal input from the permeated disruptant might experience a "real-time" modification of the ratios in the natural plume at the peripheral input level, such that they would not detect the ratio in the female's plume as normal, and their response to females would be lessened or even abolished. A related possibility has been elaborated on by Roelofs (40) as the *threshold hypothesis*, a major tenet of which is that the optimum ratio of components for attraction can shift from the natural ratio at higher than normal concentrations of odorant. Evidence for such a modification of specificity has emerged from some trapping studies using lures of differing strengths and ratios, rather than from disruption trials, hence, some of the predictions of the hypothesis on disruption have yet to be explicitly tested.

Another possibility is that an imbalance in the input ratio could cause a relatively permanent alteration in the specificity of the behavioral response to ratio, a phenomenon demonstrated in the oriental fruit moth by preexposure of males to the *E* acetate component of its pheromone. This modification seemed attributable to short-term changes at the central neuronal level (41).

The question of whether the entire natural blend of a pheromone, an unnatural ratio of components, or a portion of the blend is the most efficacious disruptant has not been given adequate experimental attention. The current evidence, summarized by Minks and Cardé (42) for a number of moth species, suggests that the complete blend in its natural ratio has the most potential for communication disruption.

E. Antagonists to Attraction and Pheromone Mimics

Before the first pheromone of any insect pest was characterized, Wright (43) proposed that compounds, probably quite similar in structure to the pheromone, might be efficacious disruptants. Such compounds have been more generally known as either inhibitors, masking agents, or antipheromones (9). Usually, the discovery of their effect on attraction stemmed from explorations of structure–activity relationships. Could very similar compounds replace pheromones? (Analogues were not as effective as the natural attractants.) Admixed with the pheromone did they either synergize? (no) or reduce attraction? (sometimes). The reduction in attraction was expressed when the antagonist and the attractant were emitted from the same locus, as from a single bait source in a trap. In some examples the antagonist was a compound that was emitted by another species sharing a portion of the same chemical communication channel; thus, the antagonistic effect relied on the natural detection of these compounds to maintain reproductive isolation and to create exclusive communication channels (e.g., 44,45). These reactions could be mediated by receptor cells specifically coded to detect these compounds (44,45), or the antipheromones might compete for the same acceptor sites on the antennal sensilla as the pheromones.

Field tests of antagonists defined by the criterion of diminution of attraction when emitted with the pheromone and released, not from a point source with the attractant but, instead, from a widely dispersed formulation, showed at best a weak disruptive effect upon attraction to point source lures (e.g., 46–49) and, in one (50), an enhancement of attraction! The natural pheromone, emitted at similar levels, was a more efficacious disruptant. This approach may yet identify a candidate chemical for disruption that is not the natural pheromone, but based on the studies available, this approach as yet has remained unfruitful.

A very different behavioral reaction is involved in the so-called male–male inhibitory compounds that seem to be used by courting males to compete for a female (e.g., 51). How widespread this sort of signal is among the moths (or other insects) is not at all clear, and the prospect of successful exploitation of this behavior is dubious.

A more profitable approach may be to employ pheromone mimics, analogues designed to substitute for the configuration and moiety of the pheromone, but which need not have the antagonistic effect just described. An advantage of this approach is that many of these mimics may be produced at lower cost than the natural pheromone, and they may afford proprietary protection as well. Moderately encouraging results have been obtained with several species (e.g., 52–54), but there is not yet convincing evidence that mimics effect direct population control of any moth pest.

F. Combinations of Pheromone and Insecticide

A recent approach has been to combine pheromone and conventional insecticides. One formulation in use against the pink bollworm utilizes point sources of pheromone applied to cotton foliage with an insecticide-laced sticker. Males attracted to these release sites contact the formulation and receive either a lethal dose of insecticide or a sublethal dose that diminishes the ability of males to reorient to pheromone over a single diel cycle (55). The pheromone formulation used, however, is efficacious in controlling the pink bollworm without the added insecticide, and its probable mode of action in disrupting mating is discussed in the following section. The combined insecticide--pheromone combination ought to augment population control over the levels obtained by pheromone alone. In such approaches care should be taken to avoid insecticides that have a repellent effect (55).

A second use pattern mixes pheromone formulated in a microdispersible matrix with contact insecticide (see Chap. 6). Its projected mode of action is to induce male movement and, therefore, enhance direct male mortality attributable to insecticide. The contribution of such concoctions to disruption of communication is speculative, and their efficacy in management remains undocumented.

With these provisional mechanisms in mind, it is now profitable to consider several specific examples. It would not be feasible within the confines of the present review to consider every case for which information is available. Instead three moth species, for which much is known about their pheromone communication systems and for which there is more than ample empirical evidence demonstrating interruption of their mating by atmospheric permeation with disruptant, will be examined in some detail. Support for each of the first five mechanisms will be evaluated.

III. CASE HISTORIES

A. Pink Bollworm

The most celebrated example, both from historical and practical perspectives, of direct population control by disruption of mating remains the pink bollworm. Its history and current status are reviewed in depth by Baker et al. (Chap. 25) and Henneberry et al. (56). The natural blend of the two components in the pheromone-producing gland is a 44:56 ratio of the *Z*,*E* and *Z*,*Z* isomers of 7,11-hexadecadienyl acetate (33), a ratio closely matched in the airborne effluvium (34). Although the signal emitted has a low variance, males are rather more indiscriminate in the range of ratios to which they are attracted. The blend of these components commonly used for commercial disruption is roughly a 1:1 isomer mix, a choice stemming from the earliest characterizations of the supposed natural ratio. Most of the disruption tests available in the literature report on the efficacy of either hollow-fiber or plastic-laminate flake formulations, each of which in field use provide point sources that create an overall background of disruptant, and that serve as potential sources of competition with a female. At some distance from a calling female, the natural plume should be rendered indistinguishable, whereas relatively close to the emitter a male should be able to discern the boundaries of the natural plume and, thereby, begin upwind anemotaxis. Successful orientation along a comparatively intense plume superimposed on a less intense background of pheromone has been demonstrated in two tortricid species (27,28), and the occurrence of this reaction in the pink bollworm seems likely.

Indirect evidence of such a behavioral phenomenon is provided by Doane and Brooks (57). Traps baited with increasing numbers of fiber pheromone dispensers caught increased numbers of males when arrayed in a field permeated with pheromone, whereas the same series of traps in an untreated field produced no increase in catch over a 16-fold range of increases in the number of fiber lures in traps. The number of males trapped in the two fields, of course, cannot be contrasted, but the evident trends suggest that males in the site permeated with disruptant are attracted to more concentrated lures because the more intense plumes generated are not entirely camouflaged.

A factor not specifically addressed in these trials, or in other observations, is whether males orienting to such disruptant lures (58) receive a sufficiently intense stimulation by pheromone to reduce their subsequent responsiveness to plumes of lower concentration, or possibly, to render them refractory to such responses for a long enough interval to reduce their mating success. Gypsy moth males in a wind tunnel setting do not fly toward a lure emitting little pheromone after an encounter with an intense source, whereas the reverse presentation of stimuli elicits continued plume following (59). Such short-term habituation (*sensu lato*) is a mechanism of potential importance in the pink bollworm, particularly when the disruptant formulation comprises numerous point sources that mimic a female. The efficacy of each of these mechanisms is likely dependent on population density. The emission rate of each point source and the number deployed in a given plot determines the level of competition with calling females and possibly the extent to which habituation can occur after attraction to these sources.

A hollow "rope" formulation of pheromone has been demonstrated to show excellent control of the pink bollworm (60). It emits very high rates of pheromone, levels well above those to which a male will orient at close range. Point-source competition would thus seem an improbable mode of action with the rope formulation. The likely mechanism of action would involve a camouflaging effect and some adaptation/habituation of response.

In contrast, microdispersible formulations, some of which also have been verified as effective in management (61,62), yield innumerable release sites. Emitted at the same rate per unit area as the point-source formulation, a microdispersible matrix offers the possibilities of a camouflage, and perhaps some degree of habituation, but probably no opportunity of following a false trail, because no discernible plumes are produced. A microdispersible formulation does present the advantage of a more ubiquitous atmospheric permeation with disruptant.

The use of disruptant formulations creating a sensory imbalance in the signal has been explored by Flint and Merkle (63–66). The nearly natural 1:1 mix of the two components was contrasted with essentially pure *Z*,*Z* isomer in plastic laminate (flake) and microencapsulated formulations. The *Z*,*Z* isomer is of course unattractive, but as a disruptant of attraction measured by suppression of male catch in traps baited with a 1:1 lure and by mating of females held in mating stations, this isomer alone appears to be roughly comparable in activity with the 1:1 disruptant mix. Evidence on the mode of action of the *Z*,*Z* isomer can be inferred from the trap catch with a spectrum of lure ratios of 1:0, 9:1, 3:1, 1:1, 1:3, 1:9, and 0:1 of the *Z*,*Z*/*Z*,*E* isomers. In the plot treated with the 1:1 disruptant, the catch was suppressed some 10-fold over the untreated plot, but in both fields the maximum catch was recorded in the traps baited with the 1:1 mix. In the *Z*,*Z* disruptant plot, total catch was diminished by twofold and the maximal catch shifted toward blends rich in this isomer. The 9:1 ratio was the most attractive, whereas catch with the nearly natural ratio was comparable with levels achieved with this lure in the 1:1 disruptant plot.

The mechanism responsible for this effect would seem to be a sensory imbalance in which either peripheral adaptation of the receptors attuned to the *Z*,*Z* isomer or a CNS habituation of ratio discrimination (or both) would cause males to perceive the 9:1 ratio as the natural blend (see also Ref. 41). Males in the *Z*,*Z* disruptant plots thus may remain quite capable of attraction to pheromone (perhaps more so than males in the 1:1 disruptant plots), but the ratio that appears most attractive is a distorted blend, not the one emitted by females. Whether the natural blend, the 1:1 mix, or the *Z*,*Z* isomer is most efficacious at the lowest rate of application should establish which approach is the most useful commercially. At the least, the tests of Flint and Merkle have verified that the sensory imbalance approach to disruption is feasible.

The pink bollworm has thus provided not only the first case of commercial application of pheromone to protect an agricultural crop but also considerable insight into the mechanisms of disruption and several distinctive approaches to development of effective formulation strategies.

B. Oriental Fruit Moth

This tortricid moth is an exceptionally well-studied species. Its female-emitted attractant consists of three important components: (*Z*)-8- and (*E*)-8-dodecenyl acetates and (*Z*)-8-dodecen-1-ol (23,39,67).

To evoke upwind flight and all of the behaviors leading to courtship the two acetates must be present in a ratio very close to the natural blend of 5% *E* isomer. The alcohol also affects all stages of behavior, but its presence at 1% to 10% of the total mix induces a large increase in the proportion of males that give a courtship display. These three components thus act as a unit, mediating the initial response (locking-on to the plume) through courtship (23). When presented to males held in open-ended cages positioned along the midline of a plume in an open field, the three-component blend has the lowest threshold for inducing upwind flight (39).

Unlike many moth species, the oriental fruit moth is particularly sensitive to pheromone concentration. Males will proceed upwind, land, and court only when the emission rate falls within a narrow range (22,23). Arrestment of upwind flight at pheromone concentrations no more than 10-fold elevated over that released by a calling female means that for point source competition to be effective in drawing the moth directly to the dispenser it must mimic both the natural ratio of the three components and a calling female's emission rate.

Plastic-laminate, hollow-fiber, microencapsulated, and plastic rope formulations have been tested, and the latter formulation is clearly efficacious in protection of peaches (68; see Chap. 13). It will be most useful to review the hollow-fiber tests (69) because the release rates employed spanned a 100-fold range, and both the acetates alone and the acetates plus alcohol combinations were evaluated. The number of sites in the orchard from which disruptant was released was held constant, and the release rates were varied by the number of hollow-fiber dispensers at each site. All combinations of release rate and two versus three components proved effective in disrupting male attraction to females and synthetic pheromone sources, but not surprisingly the higher rates seemed most efficacious. The three-component blend, however, seemed most disruptive to long-distance attraction when the criterion was the proportion of pheromone-baited traps luring at least one male per sample interval. This measure is a stringent test that is comparable with a female attracting (and mating with) a single male.

An individual fiber emitted substantially more acetate than a calling female (69). Thus, males would not approach individual fibers and attempt copulation, even with the admixture of the correct proportion of the alcohol. Source competition would not seem an obvious mechanism, except that males could orient to such sources well downwind and terminate plume-following some meters from the source, as shown in direct observations of male behavior in other experiments (24). At some distance downwind from the female, both blends should provide a camouflage, with the alcohol-containing disruptant presumably providing this effect at the longest distance downwind of the female because the complete blend has the lowest threshold (39). Caged males and females held in these plots readily mate (69), so that adaptation/habituation of mating is not achieved. The possibility of raising the threshold is not excluded by these tests. Indeed, even brief encounters with sources emanating pheromone well above levels achieved by females may render males less apt to orient to females.

The recent use of plastic rope, high-emission dispensers of the acetates plus alcohol dispensed one per peach tree (68 and Chap. 13) would not seem to rely on point-source competition because the high-release rates would produce unattractive dispensers (23). A downwind camouflage is likely, and some adaptation/habituation also is presumed to contribute to interruption.

The use of the acetates alone as a disruptant also could cause sensory imbalance, possibly causing a shift in the perception of the ratio of acetate/alcohol, although this ratio need not be emitted precisely (23). Exposure of oriental fruit moth males to the *E* acetate shifted the males' subsequent perception of the optimal *Z/E* ratio (41). The potential of shifting the male's preference away from the natural ratio by using an off ratio of acetates in a disruptant formulation remains to be tested but, given the success of the natural blend in the rope formulation, imminent tests of these alternative strategies seem improbable.

C. Gypsy Moth

The gypsy moth differs from most moths in using a single-component attractant, *cis*-7,8-epoxy-2-methloctadecane, named disparlure (70). This chemical was originally synthesized as a racemate, a 1:1 mix of the (+) and (-) enantiomers, and was found to be an effective attractant. Eventual syntheses and behavioral evaluation of the pure enantiomers of disparlure, however, showed (+)-disparlure to be some 10-fold more attractive than the racemate. The (-) enantiomer has thus proved to be an antagonist to attraction when both forms were released from the same point source. This response seems attributable to the release of both enantiomers of disparlure by *Lymantria monarcha*, the nun moth, which is widely sympatric and synchronic with the gypsy moth (45). The male gypsy moth's aversion to (-)-disparlure aids in maintaining reproductive isolation between these congeners.

Field tests of the disruption have employed microencapsulated, plastic-laminate flakes, and hollow-fiber formulations, almost always using the racemate as the disruptant because of its very low cost and ready availability compared with the (+) enantiomer. These tests are most recently summarized by Kolodny-Hirsch and Schwalbe (Chap. 22) and Plimmer (71). The racemate is most effective in preventing mating when dispensed at relatively high levels and at the lowest population densities (71,72, and Chap. 22).

Following the logic advanced in the previous two case histories, a camouflage at some distance from the female is probable, but competition of point sources with females is not an appealing mechanism because racemic disparlure sources are but weakly attractive.

Adaptation/habituation may be important. In the wind tunnel, males flown to 1000-ng pheromone sources will not immediately follow plumes from 10-ng sources, but the same lures presented in the reverse order elicit continued plume following (59). However, it is not (+)-disparlure that is routinely used as a disruptant. Rapid dispersal of males from the field test area after encounters with racemic disparlure dispensers also have been noted (49).

In the small plot tests conducted to date, the (+) enantiomer is no more effective than the racemate in achieving mating disruption (see Chap. 22). Whether the antagonistic action of the (-) enantiomer, so evident in inhibiting attraction to a single locus of (+)-disparlure, contributes to disruption is unknown.

The behavioral events allowing some male gypsy moths to locate a female in disruptant plots can be inferred from field and laboratory investigations of natural mate-finding behaviors. Because females typically call and mate on tree boles, the plumes issuing from the sources are diffused by vortices shed as the wind flows on the tree trunk's lee side, typically creating a plume about as wide as the tree's diameter. Males have difficulty following such diffuse plumes to their source. Instead they land on the tree within about 1 m of the female and walk a somewhat circuitous course with most movement in the vertical plane. Males contact females following a "preprogrammed" walk, and only rarely by walking upwind under guidance of the pheromone, because they seldom end up directly downwind of a female (73). The male's recognition of the female and the induction of a mating attempt is due to tactile cues perceived tarsally. Visual cues, so evident to the human eye, play no discernible role in the mate-recognition process (74). Thus, the final stages of mate location and recognition after landing often are not mediated by disparlure.

In dense populations, a small proportion of males is to be found walking on tree trunks where they may contact virgin females and mate. In population flushes, females may be mated before they commence calling or even before wing hardening (75). Disruption of long-distance pheromone communication thus may be insufficient to eliminate all mate-finding behavior, and the difficulty of achieving population management should be directly related to moth density, which sets the probability of males walking on a tree encountering a female.

IV. CONCLUSIONS

As amply demonstrated in a number of moths, mating disruption can be effected by interruption of the long-distance chemical communication system by atmospheric permeation with pheromone. In several major pests this principle has been developed into crop protection formulations that are in commercial use. Yet their precise mode of action remains speculative. It is not surprising that these chemicals were deployed before it was fully understood how they disrupted mating. The need to justify applied research has dictated that workable formulations be field tested as rapidly as possible, and the perceived necessity of evaluating efficacy in large field plots has limited the variety of formulations, rates of application, and blends that could be screened. A parallel certainly has existed with conventional insecticides, most of which have been registered before their mode of action was fully documented.

Much of what has been learned about how disruptants work has been gleaned from experiments not specifically aimed toward this goal. Undoubtedly, for most disruption systems, several mechanisms are operative, and they may act synergistically. Different formulations, such as the microdispersible versus the multiple point source, should generate differing mechanisms. Each species, because of potentially distinctive characteristics of its pheromone and its response pattern, may pose new opportunities for disruption and, as well, necessitate a fresh investigation. As shown in the three moth species used here to consider the behavioral mechanisms responsible for disruption, each possesses unique features in its mate-finding strategies. Perhaps the most outstanding problem is knowing the behaviors that are to be modified and the sensory inputs that modulate these reactions. Disruption will never be fully understood without this knowledge.

The design of disruptant systems has, thus far, proceeded largely on an empirical, trial-and-error basis. The not inconsiderable costs of field tests have typically prevented the systematic study needed to establish optimal formulation and deployment strategies. Ignorance of how interruption works has camouflaged clear vistas to optimal design of disruptant systems.

REFERENCES

1. Wright, R. H. Insect control by nontoxic means. *Science 144*: 487 (1964).
2. Wright, R. H. After pesticides—what? *Nature 204*:123–124 (1964).
3. Beroza, M. Insect attractants are taking hold. *Agric. Chem. 15*(7):37–40 (1960).

4. Gaston, L. K., Shorey, H. H., and Saario, C. A. Insect population control by the use of sex pheromones to inhibit orientation between the sexes. *Nature 213*:1155 (1967).
5. Cardé, R. T. Epilogue: Behavioural mechanisms. In *Mechanisms in Insect Olfaction*. (Payne, T. L., Birch, M. C., and Kennedy, C. J. E., eds.). Clarendon Press, Oxford, 1986, pp. 175--186.
6. Kennedy, J. S. Some current issues in orientation to odour-sources. In *Mechanisms in Insect Olfaction*. (Payne, T. L., Birch, M. C., and Kennedy, C. J. E., eds.). Clarendon Press, Oxford, 1986, pp. 11--25.
7. Bell, W. J. Chemo-orientation in walking insects. In *Chemical Ecology of Insects*. (Bell, W. J. and Cardé, R. T., eds.). Chapman and Hall, London, 1984, pp. 93--109.
8. Kennedy, J. S. The concepts of olfactory "arrestment" and "attraction." *Physiol. Entomol. 3*:91–98 (1978).
9. Bartell, R. J. Mechanisms of communication disruption by pheromone in the control of Lepidoptera: A review. *Physiol. Entomol. 7*:353--364 (1982).
10. Shorey, H. H. Manipulation of insect pests of agricultural crops. In *Chemical Control of Insect Behavior: Theory and Application*. (Shorey, H. H. and McKelvey, J. J., eds.). John Wiley & Sons, New York, 1973, pp. 353–367.
11. Sanders, C. J. Disruption of spruce budworm mating–the state of the art. In *Management of Insect Pests with Semiochemicals*. (Mitchell, E. R., ed.). Plenum Press, New York, 1981, pp. 339--349.
12. Sanders, C. J. Disruption of spruce budworm, *Choristoneura fumiferana* (Lepidoptera: Tortricidae), mating in a wind tunnel by synthetic pheromone: Role of habituation. *Can. Entomol. 117*:391--393 (1985).
13. Rothschild, G. H. L. Mating disruption of lepidopterous pests: Current status and future prospects. In *Management of Insect Pests with Semiochemicals*. (Mitchell, E. R., ed.). Plenum Press, New York, 1981, pp. 207--228.
14. Cardé, R. T. Utilization of pheromones in the population management of moth pests. *Environ. Health Persp. 14*:133--144 (1976).
15. Cardé, R. T. Disruption of long distance pheromone communication in the oriental fruit moth: Camouflaging the natural aerial trails from females? In *Management of Insect Pests with Semiochemicals*. (Mitchell, E. R., ed.), Plenum Press, New York, 1981, pp. 385--398.
16. Knipling, E. F. The basic principles of insect population suppression and management. *U.S.D.A. Agric. Handbook 512*, 1979, 659 pp.

17. Elkinton, J. S., Schal, C., Ono, T., and Cardé, R. T. Pheromone puff trajectory and upwind flight of male gypsy moths in a forest. *Physiol. Entomol. 12*:399--406 (1987).
18. Miller, J. R. and Roelofs, W. L. Gypsy moth responses to pheromone enantiomers as evaluated in a sustained-flight wind tunnel. *Environ. Entomol. 7*:742–744 (1978).
19. Murlis, J. and Jones, C. D. Fine-scale structure of odour plumes in relation to insect orientation to distant pheromone and other attractant sources. *Physiol. Entomol. 6*:71–86 (1981).
20. Murlis, J. The structure of odour plumes. In *Mechanisms in Insect Olfaction*. (Payne, T. L., Birch, M. C., and Kennedy, C. E. J., eds.). Clarendon Press, Oxford, 1986, pp. 27–38.
21. Rumbo, E. R. Study of single sensillum response to pheromone in the light-brown apple moth, *Epiphyas postvittana*, using an averaging technique. *Physiol. Entomol. 6*:87–98 (1981).
22. Cardé, R. T., Baker, T. C., and Roelofs, W. L. Ethological function of components of a sex attractant system for oriental fruit moth males, *Grapholitha molesta* (Lepidoptera: Tortricidae). *J. Chem. Ecol. 1*:475–491 (1975).
23. Baker, T. C. and Cardé, R. T. Analysis of pheromone-mediated behavior in male *Grapholitha molesta*, the oriental fruit moth (Lepidoptera: Tortricidae). *Environ. Entomol. 8*:956--958 (1979).
24. Baker, T. C. and Roelofs, W. L. Initiation and termination of oriental fruit moth male response to phermone concentrations in the field. *Environ. Entomol. 10*:211–218 (1981).
25. Baker, T. C. Pheromone-modulated movements of flying moths. In *Mechanisms in Insect Olfaction*. (Payne, T. L., Birch, M. C., and Kennedy, C. J. E., eds.). Clarendon Press, Oxford, 1986, pp. 39–48.
26. Kennedy, J. S. Zigzagging and casting as preprogrammed responses to wind-borne odour: A review. *Physiol. Entomol. 8*:109–120 (1983).
27. Kennedy, J. S., Ludlow, A. R., and Sanders, C. J. Guidance of flying male moths by wind-borne sex pheromone. *Physiol. Entomol. 6*:395–412 (1981).
28. Willis, M. A. and Baker, T. C. Effects of intermittent and continuous stimulation on the flight behaviour of the oriental fruit moth, *Grapholita molesta*. *Physiol. Entomol. 9*:343–358 (1984).
29. Traynier, R. M. M. Habituation of the response to sex pheromone in two species of Lepidoptera, with reference to a method of control. *Entomol. Exp. Appl. 13*:179--187 (1970).
30. Bartell, R. J. and Lawrence, L. A. Reduction in sexual responsiveness of male light-brown apple moth following previous

brief pheromonal exposure is concentration dependent. *J. Aust. Entomol. Soc. 15*:236 (1976).
31. Bartell, R. J. and Lawrence, L. A. Reduction in responsiveness of light-brown apple moths, *Epiphyas postvittana*, to sex pheromone following pulsed pre-exposure to pheromone components. *Physiol. Entomol. 2*:89--95 (1977).
32. Bartell, R. J. and Rumbo, E. R. Correlations between electrophysiological and behavioural responses elicited by pheromone. In *Mechanisms in Insect Olfaction.* (Payne, T. L., Birch, M. C., and Kennedy, C. J. E., eds.). Clarendon Press, Oxford, 1986, pp. 169–174.
33. Collins, R. D. and Cardé, R. T. Variation in and heritability of aspects of pheromone production in the pink bollworm moth, *Pectinophora gossypiella* (Lepidoptera: Gelechiidae). *Ann. Entomol. Soc. Am. 78*:229–234 (1985).
34. Haynes, K. F., Gaston, L. K., Pope, M. M., and Baker, T. C. Potential for evolution of resistance to pheromone: Interindividual and interpopulation variation in chemical communication system of the pink bollworm moth. *J. Chem. Ecol. 10*:1551–1565 (1984).
35. Ono, T., Charlton, R. E., and Cardé, R. T. Variability in pheromone composition and periodicity of pheromone titer in the potato tuberworm moth, *Phthorimaea operculella* (Lepidoptera: Gelechiidae). *J. Chem. Ecol.* (in press).
36. Voerman, S. and Rothschild, G. H. L. Synthesis of two components of the sex pheromone system of the potato tuberworm moth, *Phthorimaea opercullela* (Zeller) (Lepidoptera: Gelechiidae) and field experiments with them. *J. Chem. Ecol. 4*:531--542 (1979).
37. Flint, H. M., Balasubramanian, M., Campero, J., Strickland, G. R., Ahmad, Z., Barral, J., Barbosa, J., and Khail, A. F. Pink Bollworm: Response of native males to ratios of *Z,Z*- and *Z,E*-isomers of gossyplure in several cotton growing areas of the world. *J. Econ. Entomol. 72*:758--761 (1979).
38. Cardé, R. T. and Charlton, R. E. Olfactory sexual communication in Lepidoptera: Strategy, sensitivity, and selectivity. In *Insect Communication.* (Lewis, T., ed.). Academic Press, London, 1984, pp. 241–265.
39. Linn, C. E., Campbell, M. G., and Roelofs, W. L. Pheromone components and active spaces: What do moths smell and where do they smell it? *Science 237*:650–652 (1987).
40. Roelofs, W. L. Threshold hypothesis for pheromone perception. *J. Chem. Ecol. 4*:658–699 (1978).
41. Linn, C. E. and Roelofs, W. L. Modification of sex pheromone blend discrimination in male oriental fruit moths by pre-exposure to (*E*)-dodecenyl acetate. *Physiol. Entomol. 6*:421–429 (1981).

42. Minks, A. K. and Cardé, R. T. Disruption of pheromone communication in moths: is the natural blend really most efficacious? *Entomol. Exp. Appl. 49*:25–36 (1988).
43. Wright, R. H. Finding metarchons for pest control. *Nature 207*:103–104 (1965).
44. Grant, A. J., O'Connell, R. J., and Hammond, A. M. A comparative study of pheromone perception in two species of noctuid moths. *J. Insect Behav. 1*:75–96 (1988).
45. Hansen, K. Discrimination and production of disparlure enantiomers by the gypsy moth and the nun moth. *Physiol. Entomol. 9*:9–18 (1984).
46. McLaughlin, J. R., Gaston, L. K., Shorey, H. H., Hummel, H. H., and Stewart, F. D. Sex pheromones of Lepidoptera. XXXIII. Evaluation of the disruptive effect of tetradecyl acetate on sex pheromone communication in *Pectinophora gossypiella. J. Econ. Entomol. 65*:1592–1593 (1972).
47. Kaae, R. S., Shorey, H. H., Gaston, L. K., and Hummel, H. H. Sex pheromones of Lepidoptera: Disruption of pheromone communication in *Trichoplusia ni* and *Pectinophora gossypiella* by permeation of the air with nonpheromone chemicals. *Environ. Entomol. 3*:87–89 (1974).
48. Hathaway, D. O., Moffitt, H. R., and George, D. A. Codling moth (Lepidoptera: Tortricidae) disruption of sexual communication with an antipheromone, *E*,*E*-8,10-dodecadien-1-ol acetate. *J. Entomol. Soc. B. C. 82*:18–22 (1985).
49. Cardé, R. T., Doane, C. C., Granett, J., and Roelofs, W. L. Disruption of pheromone communication in the gypsy moth: Some behavioral effects of disparlure and an attractant modifier. *Environ. Entomol. 4*:448–450 (1975).
50. Rothschild, G. H. L. Problems in defining synergists and inhibitors of the oriental fruit moth by field experimentation. *Entomol. Exp. Appl. 17*:294–302 (1974).
51. Hirari, K., Shorey, H. H., and Gaston, L. K. Competition among courting male moths: Male-to-male inhibitory pheromone. *Science 202*:644–645 (1978).
52. Beevor, P. S. and Campion, D. The field use of "inhibitory" components of lepidopterous sex pheromones and pheromone mimics. In *Chemical Ecology: Odour Communication in Animals*. (Ritter, F. J., ed.). Elsevier/North-Holland, Amsterdam, 1979, pp. 313–325.
53. Curtis, C. E., Clark, J. D., Carlson, D. A., and Coffelt, J. A. A pheromone mimic: Disruption of mating communication in the navel orangeworm, *Amyelois transitella*, with *Z*,*Z*-1,12,14-heptadecatriene. *Entomol. Exp. Appl. 44*:249–255 (1987).

54. Kanno, H., Hattori, M., Sato, A., Tatsuki, S., Uchiumi, K., Kurihara, M., Ohguchi, Y., and Fukami, J. I. Release rate and distance effects of evaporators containing (*Z*)-11-hexadecenal and (*Z*)-5-hexadecene on disruption of orientation in the rice stem moth, *Chilo suppressalis* Walker (Lepidoptera: Pyralidae). *Appl. Entomol. Zool.* *17*:432–438 (1982).
55. Haynes, K. F., Li, W.-G., and Baker, T. C. Control of pink bollworm moth (Lepidoptera: Gelechiidae) with insecticides and pheromones (attracticide): Lethal and sublethal effects. *J. Econ. Entomol.* *79*:1466–1471 (1986).
56. Henneberry, T. J., Gillespie, J. M., Bariola, L. A., Flint, H. M., Butler, G. D., Lingren, P. D., and Kydonieus, A. F. Mating disruption as a method of suppressing pink bollworm (Lepidoptera: Gelechiidae) and tobacco budworm (Lepidoptera: Noctuidae) populations on cotton. In *Insect Suppression with Controlled Release Pheromone Systems*, Vol. 2. (Kydonieus, A. F. and Beroza, M., eds.). CRC Press, Boca Raton, 1982, pp. 75–98.
57. Doane, C. C. and Brooks, T. W. Research and development of pheromones for insect control with emphasis on the pink bollworm. In *Management of Insect Pests with Semiochemicals.* (Mitchell, E. R., ed.). Plenum, New York, 1981, pp. 285–303.
58. Miller, E., Jones, E., and Staten, R. The use of moth scales to determine male pink bollworm (*Pectinophora gossypiella*) visitation to individual pheromone dispensers in a mating disruption system. *Southwest. Entomol.* *11*:42–44 (1986).
59. Willis, M. A. and Cardé, R. T., (Unpublished).
60. Staten, R. T., Flint, H. M., Weddle, R. C., Quintero, E., Zarate, R. E., Finnell, C. M., Hernandes, H., and Yamamoto, A. Pink bollworm (Lepidoptera: Gelechiidae): Large-scale field trials with a high-rate gossyplure formulation. *J. Econ. Entomol.* *80*:1267–1271 (1987).
61. Critchley, B. R., Campion, D. G., McVeigh, L. J., Hunter-Jones, P., Hall, D. R., Cork, A., Nesbitt, B. F., Marrs, G. J., Jutsum, A. R., Hosny, M. M., and Nasr, E.-S. A. Control of pink bollworm, *Pectinophora gossypiella* (Saunders) (Lepidoptera: Gelechiidae), in Egypt by mating disruption using an aerially applied microencapsulated pheromone formulation. *Bull. Entomol. Res.* *73*:289–299 (1983).
62. Critchley, B. R., Campion, D. G., McVeigh, L. J., Cavanagh, G. C., Hosny, M. M., Nasr, E.-S. A., Khidr, A. A., and Naguib, M. Control of pink bollworm, *Pectinophora gossypiella* (Saunders) (Lepidoptera: Gelechiidae), in Egypt by mating

disruption using hollow-fibre, laminate-flake and microencapsulated formulations of synthetic pheromone. *Bull. Entomol. Res.* *75*:329–345 (1985).
63. Flint, H. M. and Merkle, J. R. Pink bollworm (Lepidoptera: Gelechiidae): Communication disruption by pheromone composition imbalance. *J. Econ. Entomol.* *76*:40–46 (1983).
64. Flint, H. M. and Merkle, J. R. The pink bollworm (Lepidoptera: Gelechiidae): Alteration of male response to gossyplure by release of its component *Z,Z*-isomer. *J. Econ. Entomol.* *77*:1099–1104 (1984).
65. Flint, H. M. and Merkle, J. R. Pink bollworm: Disruption of sexual communication by release of the *Z,Z*-isomer of gossyplure. *Southwest. Entomol.* *9*:58–61 (1984).
66. Flint, H. M. and Merkle, J. R. Studies on disruption of sexual communication in the pink bollworm, *Pectinophora gossypiella* (Saunders) (Lepidoptera: Gelechiidae), with microencapsulated gossyplure or its component *Z,Z*-isomer. *Bull. Entomol. Res.* *74*:25–32 (1984).
67. Cardé, A. M., Baker, T. C., and Cardé, R. T. Identification of a four-component sex pheromone of the female oriental fruit moth, *Grapholitha molesta* (Lepidoptera: Tortricidae). *J. Chem. Ecol.* 5:423–427 (1980).
68. Vickers, R. A., Rothschild, G. H. L., and Jones, E. L. Control of the oriental fruit moth, *Cydia molesta* (Busck) (Lepidoptera: Tortricidae), at a district level by mating disruption with synthetic sex pheromone. *Bull. Entomol. Res.* 75:625–634 (1985).
69. Charlton, R. E. and Cardé, R. T., Comparing the effectiveness of sexual communication disruption in the oriental fruit moth (*Grapholitha molesta*) using different combinations and dosages of its pheromone blend. *J. Chem. Ecol.* *7*:501–508 (1982).
70. Bierl, B. A., Beroza, M., and Collier, C. W. Potent sex attractant of the gypsy moth: Its isolation, identification and synthesis. *Science* *170*:87–89 (1970).
71. Plimmer, J. R. Disruption of mating in the gypsy moth. In *Insect Suppression with Controlled Release Pheromone Systems*, Vol. 2. (Kydonieus, A. F. and Beroza, M., eds.). CRC Press, Boca Raton, 1982, pp. 135–154.
72. Schwalbe, C. P., Paszek, E. C., Bierl-Leonhardt, B. A., and Plimmer, J. R. Disruption of gypsy moth *Lymantria dispar* (Lepidoptera: Lymantriidae) mating with disparlure. *J. Econ. Entomol.* *76*:841–844 (1983).
73. Charlton, R. E. and Cardé, R. T. Orientation of male gypsy moths, *Lymantria dispar* (L.) to pheromone sources: The role of olfactory and visual cues. *J. Insect Behav.* (in press).

74. Charlton, R. E. and Cardé, R. T. Factors mediating mate recognition and copulatory behaviour in the male gypsy moth, *Lymantria dispar* (L.). *Canad. J. Zool.* (in press).
75. Cardé, R. T. and Hagaman, T. E. Mate location strategies of gypsy moth in dense populations. *J. Chem. Ecol. 10*:25–31 (1984).

5

Chemical Analysis and Identification of Pheromones

JAMES H. TUMLINSON / Agricultural Research Service, U.S. Department of Agriculture, Gainesville, Florida

I. INTRODUCTION

If we wish to use pheromones more effectively in insect pest management systems, it is imperative that we learn as much as possible about the chemistry, behavior, physiology, and biochemistry of the insects' pheromonal communication systems. Although in the last 20 years our knowledge of these systems, especially within the Lepidoptera and Coleoptera, has increased tremendously, we still understand relatively little about chemical communication in most insects. We have progressed from what, by present technical standards, were simple identifications of single compounds that elicited wing-fanning responses by male moths, to sophisticated analyses of the chemistry, biochemistry, and associated behavioral and electrophysiological responses of multichemical communication systems. However, it is still difficult to determine whether or not all components of a pheromone blend have been identified for a particular insect. Although the "complete" pheromone may not be necessary for trapping or monitoring a particular species, or even, in some cases, for management by communication desruption, it is highly likely that in most cases more effective control or management systems could be developed with more complete knowledge of the pheromone system.

In any investigation for which the goal is the elucidation of the complete pheromone blend there are four essential components: (a) A bioassay must be developed that accurately measures the behaviors elicited by the targeted pheromone. (b) Analytical chemical methods must be available, or must be developed, to separate the pheromone components from all the other compounds present in the insect, its effluvia, or its by-products. Each component must be isolated, purified, identified, and the amounts and ratios in which the components are produced must be ascertained. (c) Methods of organic chemical synthesis must be available or be developed to synthesize the candidate pheromonal compounds in a high degree of stereoisomeric purity. (d) The pheromone must be formulated to be released at the rate and in the ratio of components that duplicates that released by the insect so that the identification and synthesis can be verified.

Although adherence to these four precepts may not guarantee the complete and correct identification of the pheromone of an insect, the likelihood of success is much greater than if one of them is ignored. The following is a more detailed discussion of each of these areas. The discussions are illustrated by examples taken primarily from the sex pheromones of Lepidoptera and Coleoptera, but the principles should apply to most semiochemical investigations.

II. THE BIOASSAY

The bioassay is the pheromone detector. It tells the chemist which extracts, fractions, or compounds are biologically active and which are not. Without an accurate bioassay, it is impossible for the analytical chemist to select the active pheromonal compounds from the complex mixture of compounds produced by the insect.

The type of assay employed will dictate the type of pheromone identified. For example, the male excitation or wing-fanning tests employed in the early days of pheromone investigations were perfectly adequate if one desired to isolate and identify a compound that would elicit excitation or wing-fanning in males. However, this assay often did not lead to compounds that were effective lures for trapping males in the field. Therefore, other types of bioassays were developed in attempts to measure all the behaviors elicited by pheromones.

In addition to behavioral bioassays it is also possible to use electrophysiological assays to monitor pheromone isolations. The electroantennogram (EAG) can be used very effectively to screen fractions or pure compounds rapidly for biological activity (1). Additionally, the EAG can be coupled with the gas chromatograph (GC) so that EAG responses and flame ionization detector responses can be recorded simultaneously as compounds elute from a GC column (2). It is even possible to couple single-cell recording with GC (3). However, it should be noted that these techniques measure detection of the compounds by the antennae and not behavioral responses of the insects to the compounds. Additionally, it is not always possible to detect minor components of a pheromone blend with electrophysiological tests, even though these components may have important roles in eliciting behavioral responses by the insect. Thus, the EAG and single-cell electrophysiological assays are important and useful methods, but they cannot be solely relied on for pheromone identification.

A variety of olfactometers have been developed to bioassay pheromones and monitor isolations and identifications (4). However, it should be stressed that a bioassay developed for one particular insect or type of pheromone may not be suitable for other situations. Thus, each bioassay must be developed to measure the particular behavioral responses that are being studied and to meet the objectives of the research. One illustration of this is the bioassay developed to monitor the isolation and identification of the boll weevil sex pheromone (5; Fig. 1). In this bioassay, the weevils were required to traverse a distance in response to the pheromone and, thus, attraction was measured. This was important because one of the goals of this research was to develop an attractant that could be used to trap weevils in the field. However, the weevils were not required to fly to reach the source and, thus, the full repertoire of behaviors elicited by the pheromone was not measured. Perhaps because the weevil is not a strong flyer like the moths, this assay allowed the successful isolation and identification of a pheromone (6) that was an effective lure in traps in the field. Therefore, it is possible to use olfactometers or other methods to monitor the isolation and identification of a pheromone. However, if the bioassay does not permit or require the full expression of the insect's "normal" response to the pheromone, the complete pheromone blend may not be elucidated.

Where sufficient material can be obtained to bait traps in the field, this is a very effective method of monitoring an isolation and identification. This is particularly true if the objective of the investigation is to develop a lure for traps because it ensures that the compounds identified will be effective for this purpose. However, trap design is critically important in this type of bioassay because a pheromone that is effective as a lure for one type of trap may not be for others (7). Additionally, the complete pheromone blend may not be required to lure the insects into a trap, and components of the pheromone important in eliciting the full range of behaviors may be missed. In particular, close-range courtship behaviors may not occur in many traps because the insect is caught in a sticky material or electrocuted at a distance from the pheromone source. It is also common for males of more than one species to be captured with a single blend in a trap, although females of these species may only attract conspecific males.

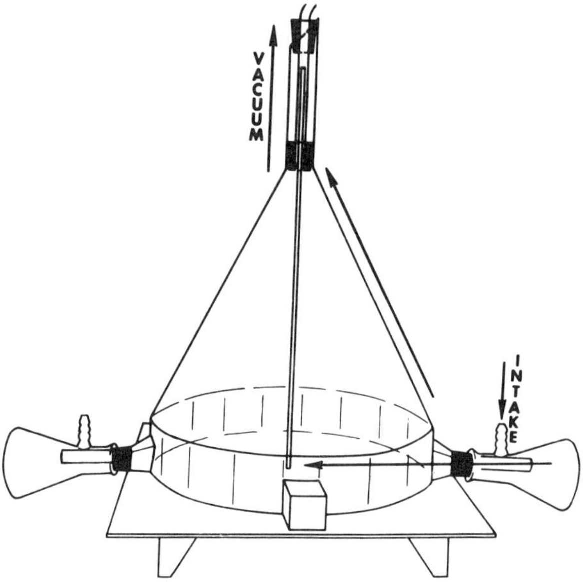

FIGURE 1 Apparatus for laboratory bioassay of the boll weevil sex pheromone. An inverted glass funnel (30-cm diam, 12.7-cm stem) was glued to a circular wall composed of 5-cm upright glass squares glued together and to a 35.5 × 40.6-cm glass base. Two 2.5-cm openings at the base at 180° angles from each other were covered with No. 5 rubber stoppers containing a glass tube (1.6-cm diam × 7.6-cm long). One of the 125-ml suction flasks connected to the rubber stoppers at opposite sides of the chamber contained the sample (live insects, extracts, or fractions thereof) being assayed for attraction. The other flask remained empty and served as a control. Air, pulled into the side arms of the suction flasks, entrained the pheromone and carried it into the center chamber into which were released 20 female boll weevils. Responding weevils migrated into the flask of their choice (from Ref. 5).

Thus, pheromones identified with this type of bioassay may lack components that are important to the reproductive behavior of the insect and that may be important when developing methods such as communication disruption for control of a particular species.

The wind tunnel (8) is generally accepted as the most effective apparatus for evaluating the responses of insects to semiochemicals of various types. It has been used successfully to determine the role of various components of pheromone blends in eliciting behavioral responses and thus to elucidate the complete blend (9,10). Wind tunnels of various designs have been developed to accommodate insects with different behavioral responses to pheromones. Yet even the behaviors exhibited in a wind tunnel may differ to a degree from responses to the pheromone in nature. For example, pheromone plumes are never as uniform in nature as in a wind tunnel. Also, in nature other odors from plants or other insects may impinge on the pheromone plume and affect the insect's response. Additionally, some insects, particularly some beetles, are not strong flyers like the moths, and they may not fly well enough in a wind tunnel to allow a useful bioassay to be conducted. Nonetheless, the wind tunnel is probably the most useful bioassay method presently available.

The most important consideration in designing any bioassay is that it must allow evaluation and measurement of the natural behavior of the insect in response to the pheromone being investigated. It also must be designed to allow the objectives of the research to be attained and it must give consistent results that will allow the chemist to differentiate between active and inactive compounds.

III. ISOLATION AND ANALYSIS

Pheromones are produced in minute amounts in various glands or in other structures in insects. Even when only the pheromone gland is excised and extracted, the pheromone components usually comprise only a small fraction of the total chemical constituents. Occasionally, a preliminary analysis of a solvent rinse of a gland surface may facilitate the investigation by indicating the types of compounds produced and excreted from the pheromone gland. However, all the compounds excreted by the pheromone gland may not be pheromones. The task of the chemist is to separate the active components from all the other compounds and to isolate, purify, and identify them. This usually requires several sequential fractionations and purifications monitored by an accurate and suitable bioassay. This task can be simplified somewhat by collecting the volatiles released into the air during the period when the insect is sending its chemical signal, or "calling." Collection of volatiles from calling insects has several advantages. First, as indicated earlier, most of the other compounds in the insect are eliminated and, therefore, do not have to be separated from the pheromone. More importantly, however, many pheromonal compounds are released as soon as their biosynthesis is complete and are not stored. Also, the ratio of components released in the volatiles may differ considerably from that found in the gland. Table 1 shows the compounds

TABLE 1 Composition of Pentane Extracts of Excised Pheromone Glands and of Volatiles Collected from Calling *Heliothis virescens* Females

Compound	Percentage[a] in gland extract	Percentage[a] in female volatiles
Tetradecanal	2.5 (± 0.13)	13.0 (± 0.65)
(Z)-9-Tetradecenal	3.3 (± 0.38)	18.1 (± 1.09)
Hexadecanal	8.6 (± 0.71)	7.3 (± 0.59)
(Z)-7-Hexadecenal	0.6 (± 0.31)	0.6 (± 0.1)
(Z)-9-Hexadecenal	0.8 (± 0.15)	1.0 (± 0.1)
(Z)-11-Hexadecenal	79.5 (± 0.84)	60.0 (± 1.32)
Tetradecanol	0.3 (± 0.12)	
(Z)-9-Tetradecenol	0.3 (± 0.22)	
Hexadecanol	0.4 (± 0.25)	
(Z)-11-Hexadecenol	4.8 (± 0.71)	

Source: From Ref. 9.
[a]Data in parentheses ± SE of the mean.

and ratios found in gland extracts and in volatiles from *Heliothis virescens* (Fab.) females (9). Knowledge of the composition of, and the rate and ratio of components in volatiles from a calling insect is of critical importance in subsequently formulating synthesized pheromones for behavioral evaluation. Hence, collection and analysis of volatiles released during calling should provide more accurate knowledge of the true composition of the pheromone. Techniques for collecting volatiles from insects and other sources have been reviewed (11) and, more recently, an improved method for collecting volatiles during different intervals of the photoperiod, without disturbing the insects, has been developed (12).

Although obtaining pheromone by volatile collection offers many advantages, as noted, there are disadvantages with this method. A major problem is that usually insufficient amounts of pheromone are collected to allow bioassay of the collected material or fractions thereof. This is because nanogram quantities of relatively pure volatile compounds cannot easily be formulated to be released in the proper ratios and over a sufficient period of time to conduct a bioassay. Also, all of the volatile compounds released by the insect may not be pheromones or pheromone components. Therefore, this method requires that all the components of the volatiles be identified and blends of the synthesized components be formulated to duplicate the release ratio and rate from the insect. Then blends containing various combinations of the components are bioassayed to determine which are active. Figure 2 shows a chromatogram of the volatiles collected from fall armyworm, *Spodoptera frugiperda* (J. E. Smith), females. Of the five compounds identified in these volatiles, only two, (Z)-7-dodecen-1-ol acetate (Z7-12:Ac) and (Z)-9-tetradecen-1-ol acetate (Z9-14:Ac) were

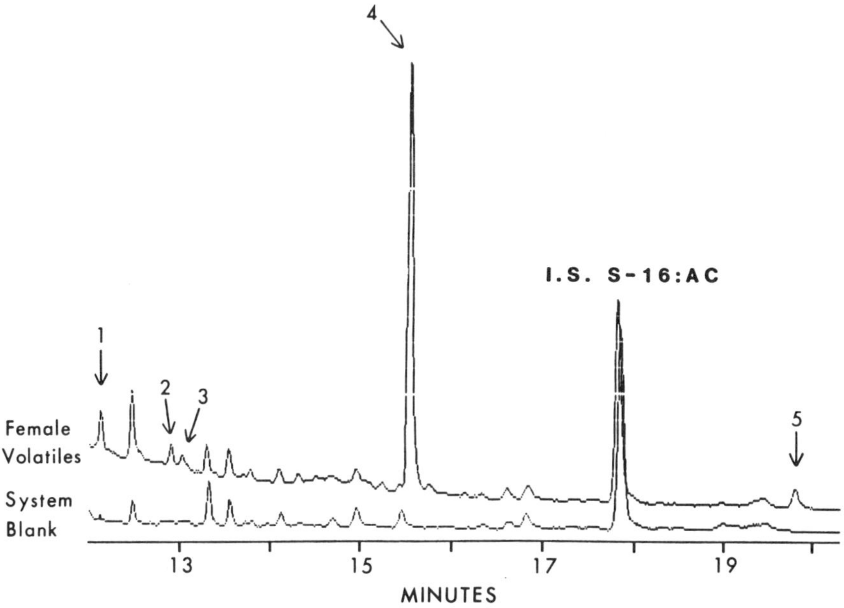

FIGURE 2 Volatiles collected from calling, laboratory-reared fall armyworm female moths were analyzed on a 60-m × 0.25-mm (i.d.) fused silica SP-2330 capillary GC column. Peaks not present in the blank correspond in retention time to: (1) dodecan-1-ol acetate, (2) (Z)-7-dodecen-1-ol acetate, (3) 11-dodecen-1-ol acetate, (4) (Z)-9-tetradecen-1-ol acetate, and (5) (Z)-11-hexadecen-1-ol acetate; I.S., internal standard (from Ref. 13).

demonstrated to have activity in field trapping tests (13). Of course, the remaining other components may influence behavior in more subtle ways not detected by trapping experiments.

Regardless of the method of extraction or collection of the pheromone, its identification requires that all the components be separated from each other and from any other compounds present. Purification of each component to 99.9% or higher purity is required to be certain that a subsequent identification is correct. There are many examples of impurities of 1% or less interfering with the activity of a pheromone (14,15). Similarly, it is possible for the "impurity" to be the active compound, and the major component, when identified and synthesized, may be found to be inactive.

Techniques for purifying, analyzing, and identifying pheromones and factors that affect the separation of compounds by chromatography have been reviewed (16). There are many different types of high-performance liquid chromatographic (HPLC) and GC columns and sets of conditions that can be used to separate these compounds. The type of compounds and the matrix in which they are contained usually dictate the methods employed. For example, when volatiles are collected, they may be injected directly onto a high-resolution capillary GC column. Usually, analysis of a volatile blend on two or three capillary columns coated with stationary phases that separate by different mechanisms is sufficient to separate all the components. Here, correlation of retention times on the three columns and mass spectra with those of authentic standards may be sufficient for identification (see later). Typically, stationary phases like OV-101, a methyl silicone of low polarity; SP-2340, a highly polar cyanosilicone; and cholesteryl *p*-chlorocinnamate, a liquid-crystal phase, may be used for these analyses. Figure 3 illustrates the complementary nature of cyanosilicone and liquid-crystal phases in separating functional and geometrical isomers of typical lepidopteran pheromonal compounds.

When the pheromone is obtained in a crude extract of a gland or whole insect, a more lengthy and complex separation scheme is required. This may involve several fractionations by HPLC and GC on different types of columns (16). Ultimately, the components of the pheromone must be isolated in pure (>99%) form, and the pure compounds, either singly or in combination with others, should be demonstrated to be biologically active before proceeding with identification.

Elucidation of the pheromone structures can be as simple as correlating the mass spectrum of a component and its retention time on two or more capillary GC columns with that of an authentic standard. For this, the GC columns should be capable of resolving all positional and geometrical isomers of that particular type of compound. This is usually possible with lepidopteran pheromones, which are nearly all

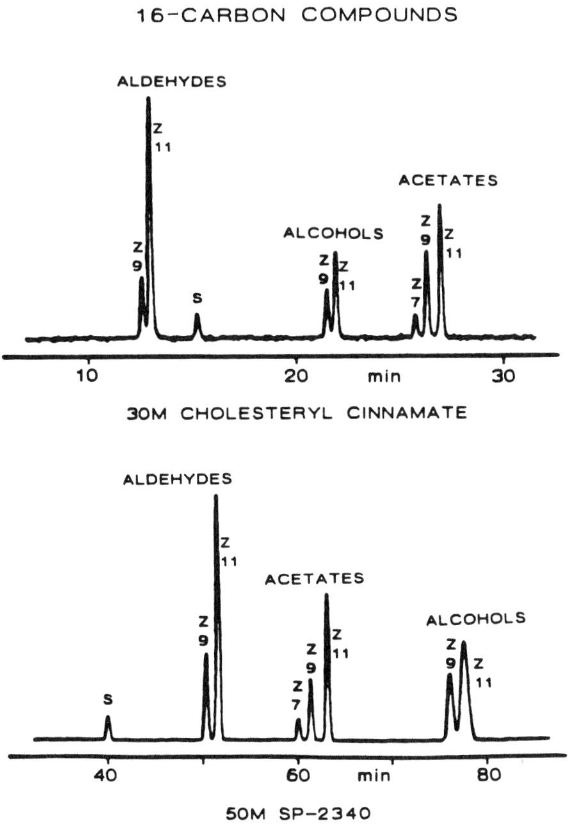

FIGURE 3 Separation of blends of 16-carbon compounds on a liquid crystal and a cyanosilicon capillary column (from Ref. 16).

fatty acid-like compounds with zero to three olefinic bonds and either aldehyde, alcohol, or ester functional groups. With more complex or less predictable structures, like those of many of the beetle pheromones, a more rigorous analysis is usually required. This may include infrared (IR) and nuclear magnetic resonance (NMR) spectral analyses and possibly microdegradative reactions such as hydrogenolysis or ozonolysis, followed by spectral analysis of the products. Fortunately, we now have instrumentation and techniques with sufficient resolution and sensitivity to identify minute quantities of most pheromones (16).

The importance of determining the structure of a compound precisely and accurately is emphasized by the many instances in which a very subtle change in the structure greatly decreases the activity of the molecule or even makes it inhibitory. This is common for geometrical isomers of many lepidopteran pheromones. Also, sometimes an analogue of the pheromonal compound will be active when tested at high concentrations and thus may be incorrectly identified as the pheromone. For example (Z)-9-dodecen-1-ol acetate (Z9-12:Ac) was reported to be a sex pheromone for the fall armyworm (17) and, in fact, it is an effective trap bait when used at high doses in the field (18). However, a subsequent investigation revealed that Z9-12:Ac was not produced by fall armyworm females, but that two components, Z7-12:Ac and Z9-14:Ac, of the volatiles collected from females (see Fig. 2) were attractive to males in the field at a much lower concentration than Z9-12:Ac (13). Similarly, changes in the chirality of a pheromone can also affect its activity. For example, the racemic form of the gypsy moth pheromone, *cis*-7,8-epoxy-2-methyl-octadecane (I), is much less active than the pure (+) enantiomer (19,20), and the *S*-(+) enantiomer of the Japanese beetle pheromone, (*Z*)-5-(1-decenyl)-dihydro-2(3H)-furanone (II), inhibits the response of the male beetles to the active *R*-(-) enantiomer (15). Mori (21) has recently reviewed methods for determining the absolute configuration and optical purity of pheromones.

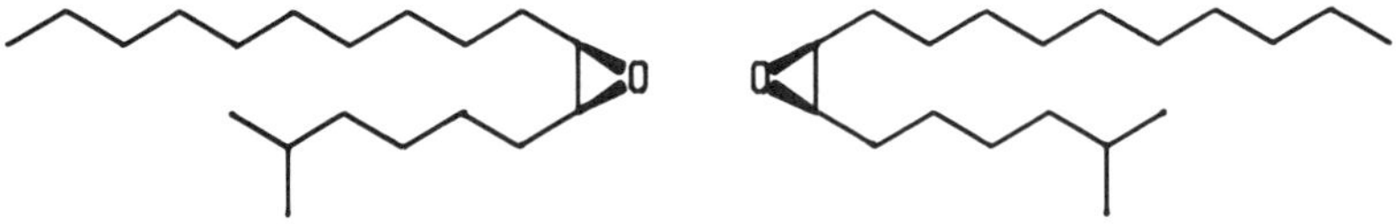

7S,8R (-) 7R,8S (+)

I

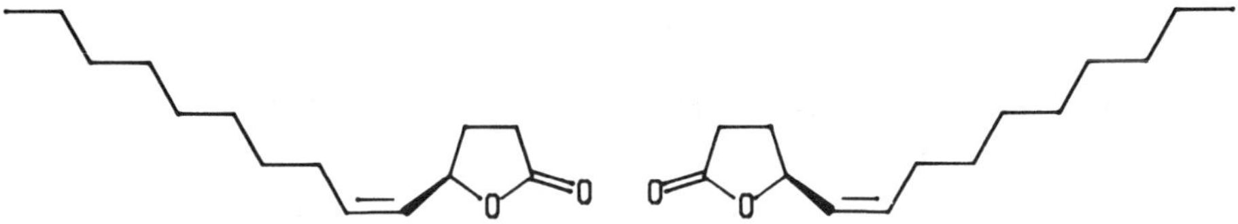

R(-) S(+)

II

IV. SYNTHESIS

The synthesis is usually the final proof of the structure of a pheromone. The synthesized compound(s) must be identical both in chromatographic, spectroscopic, and other chemical properties and in biological activity with the naturally produced pheromone, when tested at the same release rate and in the same context (see formulations later). Therefore, synthetic procedures must be selected or developed to construct the specific molecule in precisely the same configuration as the natural material. When insufficient natural material can be obtained to determine the chirality of a pheromone molecule, synthesis must be relied upon to determine which enantiomer(s) or stereoisomer(s) is active and whether or not any of the other isomers are inhibitory or synergistic. This is illustrated by the recent discovery of the pheromones of several species of beetles in the *Diabrotica virgifera* species group (Chrysomelidae). The pheromone of the western corn rootworm, *D. virgifera virgifera* LeConte, was identified as 8-methyl-2-decanol propanoate (22), and the racemic compound was synthesized and found to be active in attracting males of this species in the field. However, when all four stereoisomers were synthesized in a high degree of isomeric purity and field tested, a very interesting pattern of activity emerged (Fig. 4). The *D. virgifera virgifera* and *D. virgifera zeae* Krysan and Smith males are attracted to both the *RR*- and *SR* isomers (23), whereas northern corn rootworm, *D. barberi* Smith and Lawrence, males are attracted to the *RR* and inhibited by the *SR* isomer (24). Conversely, *D. lemniscata* LeConte is attracted strongly to the *SR* and weakly to the *RR* isomer, whereas *D. longicornis* (Say) is attracted by the *SR* and inhibited by the *RR* isomer (25). Thus, reproductive isolation in this closely related group of beetles appears to be due to the stereochemistry of a single pheromonal compound. These discoveries were greatly facilitated by the availability of the pure stereoisomers for field tests.

V. FORMULATION

Probably the most critical, but often the most neglected, phase of any pheromone investigation is the formulation of the synthesized compound(s) for bioassay and field testing. Many incorrect or incomplete identifications have resulted from poor formulations. One of the major faults of many past investigations has been the failure to test the synthetic compounds at a release rate comparable with that at which the insect releases the pheromone. Similarly, it is common to load onto a substrate the ratio of compounds found in the extract of the insect, the gland, or other source without considering either

SPECIES	STEREOCHEMISTRY	STRUCTURE
D. virgifera virgifera (WCR)	2R,8R(++), 2S,8R(+)	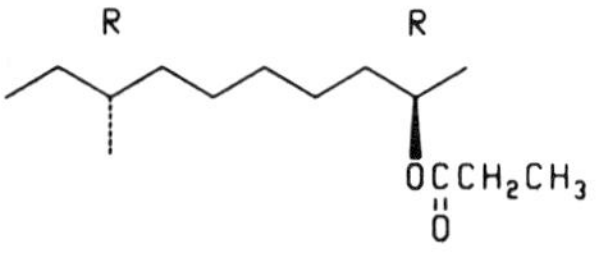
D. virgifera zeae (MCR)	2R,8R(++), 2S,8R(+)	
D. barberi (NCR)	2R,8R(++), 2S,8R(--), 2S,8S(-)	
D. longicornis	2S,8R(++), 2R,8R(--)	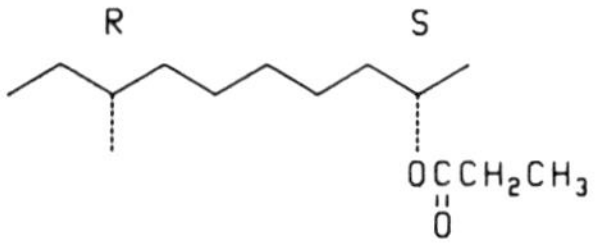
D. lemniscata	2S,8R(++), 2R,8R(+)	
D. porracea	2S,8R(++)	

FIGURE 4 Structures and stereochemistry of *Diabrotica* sex pheromones or attractants. Structures shown are the main attractants of the corresponding species. (++) = strong attraction; (+) = weak attraction; (--) = strong inhibition; (-) = weak inhibition (from Ref. 31).

the ratio of components actually released by the insect or the ratio that will be released from the substrate loaded with a given ratio. If the synthetic pheromone is to be adequately tested, it must be released in the same ratio and at the same rate as the natural pheromone is released by the insect. It has been demonstrated many times that incomplete blends, incorrect ratios, and even incorrect compounds will attract insects if tested at high enough concentrations but, in nearly every case, the correct pheromone blend is more effective at lower concentrations. Developing formulations for pheromones, particularly for blends with many components differing greatly in volatility, can be difficult. It is very important in developing pheromone formulations to be able to measure the actual release ratios and rates. Particular care must be taken when estimating release rates and ratios on the basis of residual pheromone in a formulation because of many factors that can introduce errors. For example, increases in moisture content can interfere substantially with measurements of pheromone depletion by weighing. We have recently developed a method for predicting the release ratios of components of most lepidopteran pheromones, and other compounds of similar volatility, from rubber septa (26,27). This method allows calculation of the percentage load of each component required to achieve a desired release ratio. The release rate from these formulations is determined by the amount loaded, the volatility of the blend, the temperature, and the wind speed over the septa and, thus, absolute control of this factor is difficult to achieve in the field. Measured release ratios from these septa correlated very highly with those predicted by calculations (27). Highly volatile compounds, such as 2-methyl-2-vinylpyrazine, the pheromone of the papaya fruit fly (28), cannot be formulated on rubber septa and require other methods of formulation.

Recent work with the pheromone of the velvetbean caterpillar moth (VBC), *Anticarsia gemmatalis* Hübner, illustrates the effects of both dose and ratio of components on captures of male moths in the field. As shown in Figure 5, changes in the ratio of (*Z*,*Z*,*Z*)-3,6,9-eicosatriene (C_{20}) and (*Z*,*Z*,*Z*)-3,6,9-heneicosatriene (C_{21})

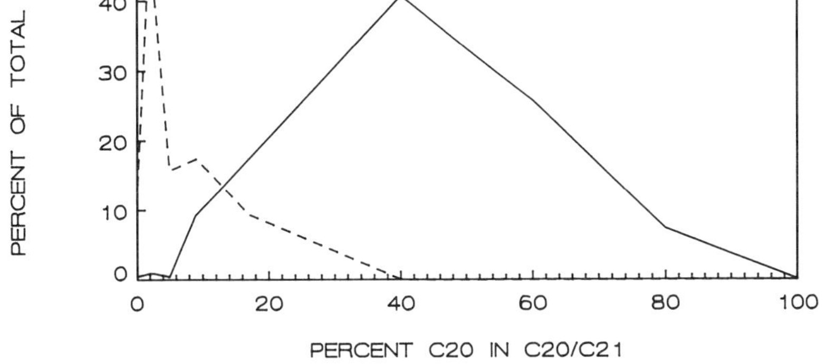

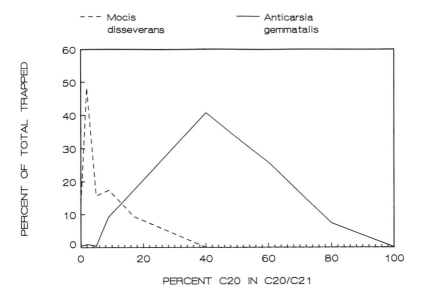

FIGURE 5 Percentage of total captures of males of *M. disseverans* and *A. gemmatalis* (VBC) in traps baited with different ratios of (*Z*,*Z*,*Z*)-3,6,9-eicosatriene (C_{20}) and (*Z*,*Z*,*Z*)-3,6,9-heneicosatriene (C_{21}), loaded on a rubber septum at a combined dose of 1 mg (from Ref. 29).

greatly affected capture of VBC males, with optimum captures at a 40:60 ratio of C_{20}/C_{21} (29). At very low percentages of C_{20}, an entirely different species, *Mocis disseverans* (Walker), was captured. Even when the 40:60 ratio of C_{20}/C_{21} was maintained, changes in dose had a great effect on captures (Fig. 6; 30). Thus VBC males respond to this pheromone blend over a very narrow range of component ratios and release rates. Had these compounds been tested at 5 mg or more per septum, as is common for many lepidopteran pheromone tests, or at a less than optimum ratio, the activity of this blend might have been missed.

Precise formulations are also required to accurately assess the role of all components, especially minor components, of a blend in eliciting the complete repertoire of an insect's behavioral responses

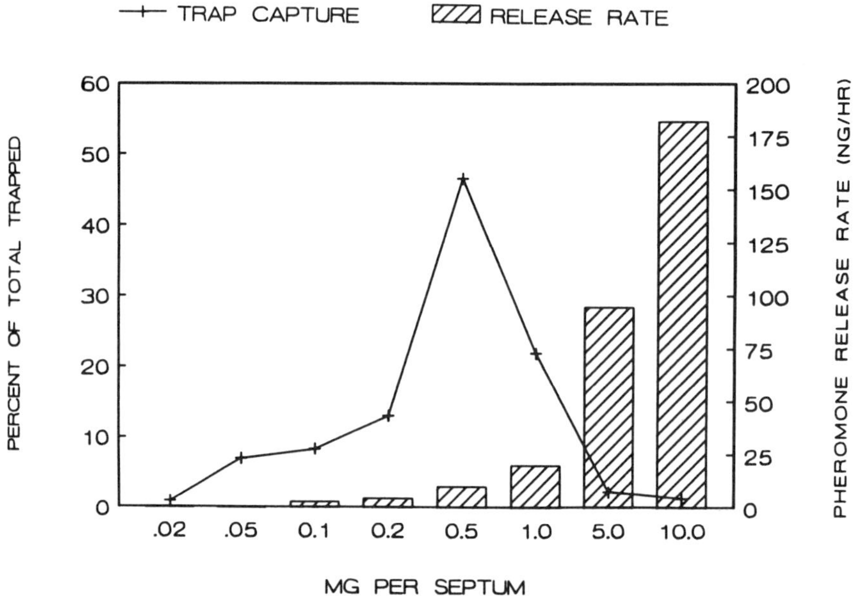

FIGURE 6 Response of male velvetbean caterpillar moths to bucket traps baited with rubber septa impregnated with different doses of sex pheromone, a 40:60 blend of (*Z*,*Z*,*Z*)-3,6,9-eicosatriene and (*Z*,*Z*,*Z*)-3,6,9-heneicosatriene. Each dose was replicated 10 times; a total of 18,710 moths were captured in the test. Pheromone release rate data were determined at a constant wind speed of 0.225 m/sec. All release rate data were converted to nanograms of total pheromone per hour (from Ref. 30).

in a wind tunnel. Figure 7 illustrates the effects on male response, in a wind tunnel, of deletion of one or more of the six aldehydes from the *H. virescens* pheromone blend (9). This six-component aldehyde blend was formulated on rubber septa to release at the precise rate and in the same ratio as the pheromone is released by calling *H. virescens* females. When males were allowed to choose between two blends, or a blend and calling females, it was found that all six components are required to equal the pheromone released by the females (see Fig. 7).

VI. CONCLUSION

A thorough knowledge of insect pheromonal communication systems is essential if we hope to develop more widespread practical uses of pheromones for control of insect pests. In addition to the identification of bioactive chemicals, there are many aspects of these systems that need more investigation to develop a sufficient base of knowledge from which to develop practical, effective applications. These include mechanisms of pheromone biosynthesis and perception and behavioral mechanisms that mediate the insects' responses. Increased knowledge from these areas will aid in more accurately defining the pheromone systems of insects. However, the accurate and precise identification of the chemical components of these systems is essential for meaningful studies in many of the other areas and for eventual development of practical applications. Successful identification depends on the integration of high-resolution and sensitive chemical analyses, highly selective syntheses, and appropriate formulations with definitive behavioral analyses.

REFERENCES

1. Roelofs, W. L. Electroantennogram assays: Rapid and convenient screening procedures for pheromones. In *Techniques in Pheromone Research.* (Hummel, H. E. and Miller, T. A., eds.). Springer-Verlag, New York, 1984, pp. 131–159.
2. Struble, D. L. and Arn, H. Combined gas chromatography and electroantennogram recording of insect olfactory responses. In *Techniques in Pheromone Research.* (Hummel, H. E. and Miller, T. A., eds.). Springer-Verlag, New York 1984, pp. 161--177.
3. Wadhams, L. J. The coupled gas chromatography-single cell recording technique. In *Techniques in Pheromone Research.* (Hummel, H. E. and Miller, T. A., eds.). Springer-Verlag, New York, 1984, pp. 179–189.

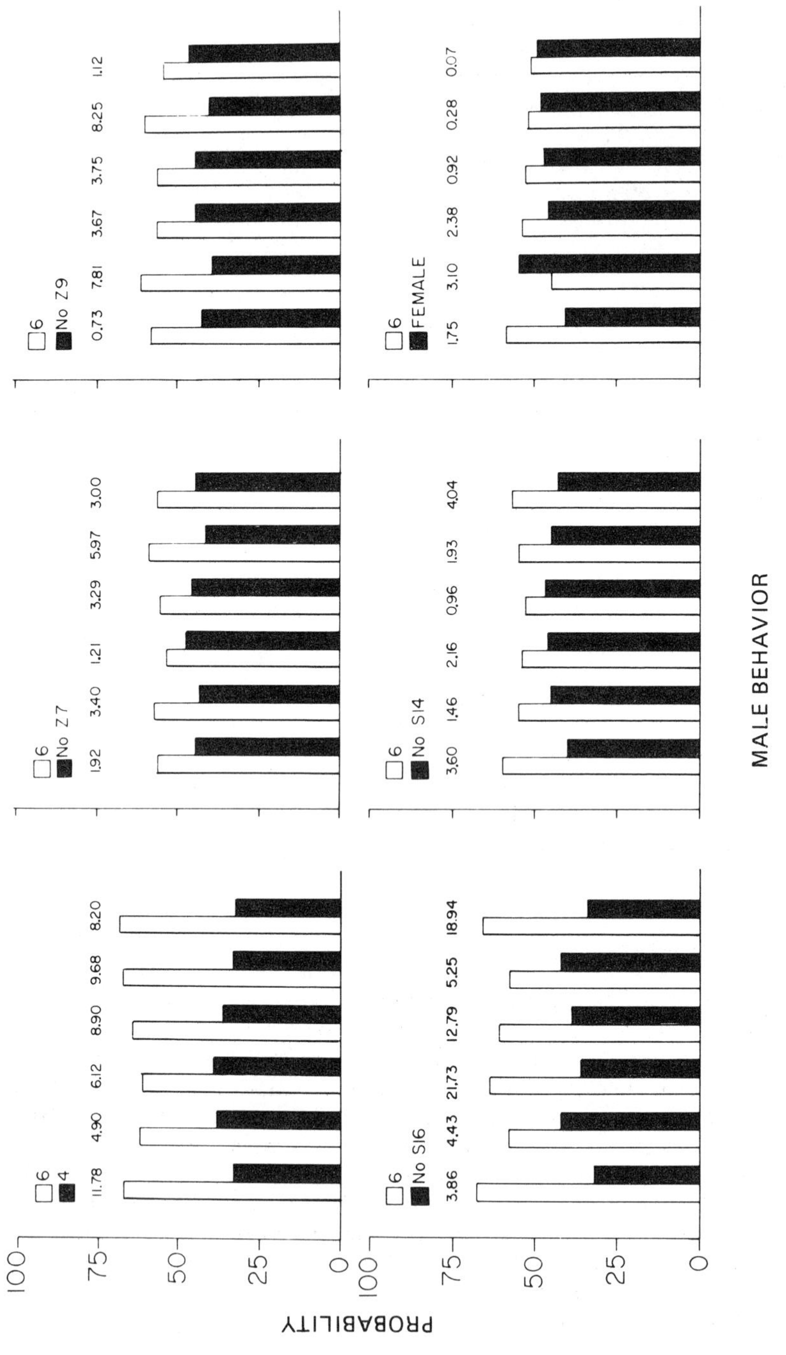
PROBABILITY
100
75
50
25
0
6
4
11.78 4.90 6.12 8.90 9.68 8.20
6
No S16
3.86 4.43 21.73 12.79 5.25 18.94
6
No Z7
1.92 3.40 1.21 3.29 5.97 3.00
6
No S14
3.60 1.46 2.16 0.96 1.93 4.04
6
No Z9
0.73 7.81 3.67 3.75 8.25 1.12
6
FEMALE
1.75 3.10 2.38 0.92 0.28 0.07
MALE BEHAVIOR

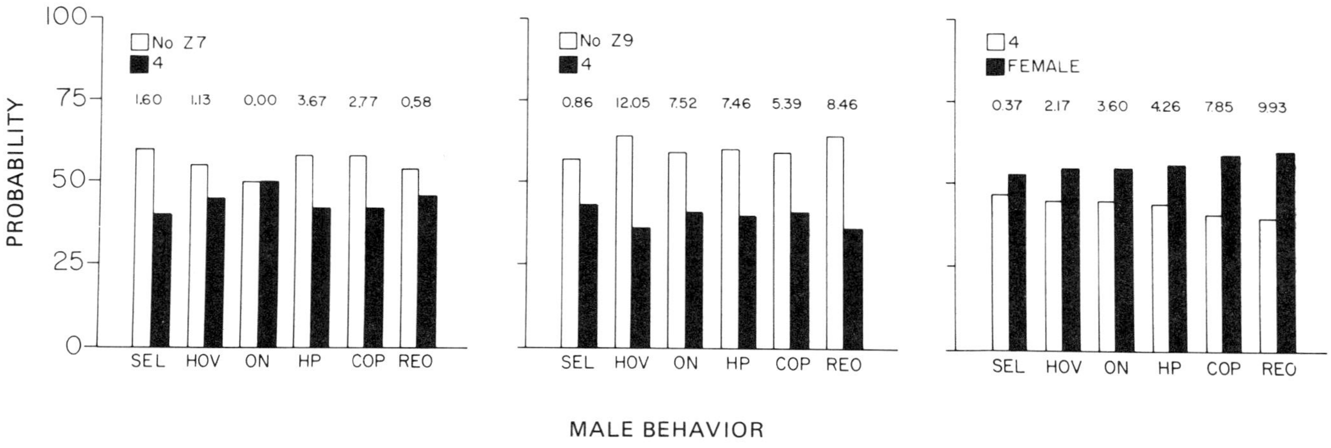

FIGURE 7 Results of dual selection studies in a wind tunnel. Numbers above figures are χ^2 values calculated using Yates' correction. SEL, selection; HOV, hover at source; ON, lands on source; HP, full hair-pencil exposure after landing; COP, copulation attempts at the source; REO, close-range reorientation. The six-component blend is identical with that found in female volatiles (see Table 1). The four-component blend consists of tetradecanal, (*Z*)-9-tetradecenal, hexadecanal, and (*Z*)-11-hexadecenal released in the same ratio that they occur in female volatiles (see Table 1). Blends marked "No Z7," "No Z9," "No S16," and "No S14" are five-component blends that do not contain (*Z*)-7-hexadecenal, (*Z*)-9-hexadecenal, hexadecanal, or tetradecanal, respectively. Probabilities are based on the number of times individual behaviors were performed in response to each blend and depend upon the performance of the previous behavior (from Ref. 9).

4. Baker, T. C. and Cardé, R. T. Techniques for behavioral bioassays. In *Techniques in Pheromone Research.* (Hummel, H. E. and Miller, T. A., eds.). Springer-Verlag, New York, 1984, pp. 45–73.
5. Hardee, D. D., Mitchell, E. B., and Huddleston, P. M. Procedure for bioassaying the sex attractant of the boll weevil. *J. Econ. Entomol.* *60*:169–171 (1967).
6. Tumlinson, J. H., Hardee, D. D., Gueldner, R. C., Thompson, A. C., Hedin, P. A., and Minyard, J. P. Sex pheromones produced by the male boll weevil: Isolation, identification and synthesis. *Science* *166*:1010–1012 (1969).
7. Cardé, R. T. and Elkinton, J. S. Field trapping with attractants: Methods and interpretation. In *Techniques in Pheromone Research.* (Hummel, H. E. and Miller, T. A., eds.). Springer-Verlag, New York, 1984, pp. 111–139.
8. Baker, T. C. and Linn, C. E. Wind tunnels in pheromone research. In *Techniques in Pheromone Research.* (Hummel, H. E. and Miller, T. A., eds.). Springer-Verlag, New York, 1984, pp. 75–109.
9. Teal, P. E. A., Tumlinson, J. H., and Heath, R. R. Chemical and behavioral analyses of volatile sex pheromone components released by calling *Heliothis virescens* (F.) females (Lepidoptera: Noctuidae). *J. Chem. Ecol.* *12*:107–126 (1986).
10. Bjostad, L. B., Linn, C. E., Du, J.-W., and Roelofs, W. L. Identification of new sex pheromone components in *Trichoplusia ni*, predicted from biosynthetic precursors. *J. Chem. Ecol.* *10*:1309–1323 (1984).
11. Golub, M. A. and Weatherston, I. Techniques for extracting and collecting sex pheromones from live insects and from artificial sources. In *Techniques in Pheromone Research.* (Hummel, H. E. and Miller, T. A., eds.). Springer-Verlag, New York, 1984, pp. 223–285.
12. Heath, R. R., Landolt, P. J., Tumlinson, J. H., and Calkins, C. O. A system to collect volatile chemicals emitted by insects. *J. Chem. Ecol.* (in press) 1989.
13. Tumlinson, J. H., Mitchell, E. R., Teal, P. E. A., Heath, R. R., and Mengelkoch, L. J. Sex pheromone of fall armyworm, *Spodoptera frugiperda* (J. E. Smith). Identification of components critical to attraction in the field. *J. Chem. Ecol.* *12*:1909–1926 (1986).
14. Tumlinson, J. H., Mitchell, E. R., Browner, S. M., Mayer, M. S., Green, N., Hines, R., and Lindquist, D. A. *cis*-7-Dodecen-1-ol, a potent inhibitor of the cabbage looper sex pheromone. *Environ. Entomol.* *1*:353–358 (1972).
15. Tumlinson, J. H., Klein, M. G., Doolittle, R. E., Ladd, T. L., and Proveaux, A. T. Identification of the female Japanese beetle

sex pheromone: Inhibition of male response by an enantiomer. *Science 197*:789–792 (1977).

16. Heath, R. R. and Tumlinson, J. H. Techniques for purifying, analyzing and identifying pheromones. In *Techniques in Pheromone Research*. (Hummel, H. E. and Miller, T. A., eds.). Springer-Verlag, New York, 1984, pp. 287–321.
17. Sekul, A. A. and Sparks, A. N. Sex attractant of the fall armyworm moth. *U.S.D.A. Tech. Bull.* 1542, 1976, 6 pp.
18. Mitchell, E. R. Monitoring adult populations of the fall armyworm. *Fla. Entomol. 62*:91–98 (1978).
19. Cardé, R. T., Doane, C. C., Baker, T. C., Iwaki, S., and Murumo, S. Attraction of optically active pheromone for male gypsy moths. *Environ. Entomol. 6*:768–772 (1977).
20. Miller, J. R., Mori, K., and Roelofs, W. L. Gypsy moth field trapping and electroantennogram studies with pheromone enantiomers. *J. Insect Physiol. 23*:1447–1453 (1977).
21. Mori, K. The significance of chirality: Methods for determining absolute configuration and optical purity of pheromones and related compounds. In *Techniques in Pheromone Research*. (Hummel, H. E. and Miller, T. A., eds.). Springer-Verlag, New York, 1984, pp. 323–369.
22. Guss, P. L., Tumlinson, J. H., Sonnet, P. E., and Proveaux, A. T. Identification of a female-produced sex pheromone of the western corn rootworm. *J. Chem. Ecol. 8*:545–556 (1982).
23. Guss, P. L., Sonnet, P. E., Carney, R. L., Branson, T. F., and Tumlinson, J. H. Response of *Diabrotica virgifera virgifera, D.v.zea*, and *D. porraceo* to stereoisomers of 8-methyl-2-decyl propanoate. *J. Chem. Ecol. 10*:1123–1131 (1984).
24. Guss, P. L., Sonnet, P. E., Carney, R. L., Tumlinson, J. H., and Wilkin, P. J. Response of northern corn rootworm, *Diabrotica barberi* Smith and Lawrence, to stereoisomers of 8-methyl-2-decyl propanoate. *J. Chem. Ecol. 11*:21–26 (1985).
25. Krysan, J. L., Wilkin, P. H., Tumlinson, J. H., Sonnet, P. E., Carney, R. L., and Guss, P. L. Responses of *Diabrotica lemniscata* and *D. longicornis* (Coleoptera: Chrysomelidae) to stereoisomers of 8-methyl-2-decyl propanoate and studies on the pheromone of *D. longicornis*. *Ann. Entomol. Soc. Am. 79*:742–746 (1986).
26. Heath, R. R. and Tumlinson, J. H. Correlation of retention times on a liquid crystal capillary column with reported vapor pressures and half-lives of compounds used in pheromone formulations. *J. Chem. Ecol. 12*:2081–2088 (1986).
27. Heath, R. R., Teal, P. E. A., Tumlinson, J. H., and Mengelkoch, L. J. Prediction of release ratios of multicomponent pheromones from rubber septa. *J. Chem. Ecol. 12*:2133–2143 (1986).

28. Chuman, T., Landolt, P. J., Heath, R. R., and Tumlinson, J. H. Isolation, identification, and synthesis of male-produced sex pheromone of papaya fruit fly, *Toxotrypana curvicauda* Gerstaecker (Diptera: Tephritidae). *J. Chem. Ecol. 13*:1979–1992 (1987).
29. Landolt, P. J., Heath, R. R., and Leppla, N. C. (*Z*,*Z*,*Z*)-3,6,9-Eicosatriene and (*Z*,*Z*,*Z*)-3,6,9-heneicosatriene as sex pheromone components of a grass looper, *Mocis disseverans* (Lepidoptera: Noctuidae). *Environ. Entomol. 15*:1272–1274 (1986).
30. Mitchell, E. R. and Heath, R. R. Pheromone trapping system for the velvetbean caterpillar (Lepidoptera: Noctuidae). *J. Econ. Entomol. 79*:289–292 (1986).
31. Chuman, T., Guss, P. L., Doolittle, R. E., McLaughlin, J. R., Krysan, J. L., Schalk, J. M., and Tumlinson, J. H. Identification of female-produced sex pheromone from banded cucumber beetle, *Diabrotica balteata* (Coleoptera: Chrysomelidae). *J. Chem. Ecol. 13*:1601–1616 (1987).

6

Principles of Design of Controlled-Release Formulations

IAIN WEATHERSTON* / Sault Ste. Marie, Ontario, Canada

I. INTRODUCTION

Since the structural elucidation of bombykol, the sex pheromone of the silk worm moth *Bombyx mori*, was completed in 1961 with the total synthesis (1), there has been a plethora of information about many aspects of pheromone research that has been reviewed many times (e.g., 2–11), yet the technology transfer, successful commercialization, and widespread use of these bioactive materials in pest control have, as yet, not been fully realized. In 1977, Boness et al. (12) wrote:

> Even when we have better knowledge of the still sparse principles, it will nonetheless be difficult to reach a final verdict on the value of sex pheromones to crop protection. They are neither the panacea idealistic prophets believed them to be in the first moments of euphoria, nor are they simply curiosities of purely theoretical and scientific interest.

Today, 10 years later, we are still struggling with the question of "the value of pheromones to crop protection."

In the foreseeable future, the acceptance and widespread use of behavior-modifying chemicals in the commercial agricultural arena depends on how well they can be integrated with current pest control methods into crop protection strategies. This, in turn, depends on

---

**Current affiliation*: University of California, Riverside, California.

several factors, among which one of the more important is the development of efficacious and cost-effective controlled-release formulations, about which Campion and Murlis (13) recently reported, "Our experience is that successful promotion of an effective pheromone formulation has involved many people over several years of painstaking work."

Although this report primarily deals with lepidopteran sex pheromones, many of the principles discussed can be applied to other types of behavior-modifying chemicals.

In discussions of the practical use of pheromones, it is generally accepted that there are two use categories: Indirect control—monitoring and Direct control—mass trapping and areawide dissemination. Both the principles of monitoring and mass trapping have already been reported in previous chapters; hence, this overview relates to only aspects of formulation design of materials to be used for areawide dissemination, and these will be discussed for the topics summarized in Table 1. Then, by using six key aspects of formulation design, a comparison is made of the five current commercial pheromone formulations.

II. USE STRATEGIES

A. Disruption

Although the principle of disruption has been envisaged for some time, the feasibility of this strategy was first demonstrated for *Trichoplusia ni*, in 1967, by Gaston et al. (14). The first U.S. Environmental Protection Agency registration was granted to Albany International Corp.

TABLE 1 Aspects of Pheromone Formulation Design

Use strategies	Mode of action	Formulation parameters
Disruption	Adaptation/habituation	Target pest
Attracticide	False-trail following	Type of crop
Bioirritant	Camouflage	Utilization of pheromone Release rate/blend ratio Longevity and stability Application method/rate Economics

in 1978 for a capillary formulation of gossyplure for suppression of the pink bollworm, *Pectinophora gossypiella*, by the disruption of the insect's mating communication system.

By broadcasting controlled-release dispensers throughout a crop, the volatile pheromone permeates through, and above, the plants to be present in a sufficient atmospheric concentration such that the communication between the calling female and any receptive males in the area is disrupted to the detriment of the male's mate-finding ability, with a concomitant reduction in mating.

B. Attracticide

The attracticide strategy, also known as "attract and kill," utilizes the pheromone at the normal application rate in conjunction with a very small amount of a compatible insecticide to increase the robustness of the treatment. It could be said that the insecticide is being used to synergize the pheromone. The small amounts of insecticide associated with the pheromone sources increase control by killing male insects attracted to them. This assumes that the male insect remains at the point source for a sufficient time to absorb a lethal dose. There is also some evidence (15,16) that sublethal doses of insecticide can interfere with the perception of the sex pheromone by affected males; however, data are not available that would relate this to a reduction in mating caused by the male's impaired ability to respond to its conspecific pheromone.

Although cotton growers in the desert Southwest probably had prior field experience, it was not until 1979 that laboratory and replicated field studies were begun under the auspices of the U.S. Department of Agriculture and Albany International Corp. to develop the attracticide strategy.

C. Bioirritant

The bioirritant strategy is described as the use of a behavior-modifying chemical in conjunction with an insecticide in proportions such that the insecticide is synergized. The thesis is that in certain lepidopteran species, adult females can detect their own pheromone (17–22) and, in a few cases, behavioral responses have also been demonstrated (20,23,24); hence, there is a probability of increased movement of both males and females, with an attendant increase in the probability of their contacting the insecticide. It should be noted that, in general, female Lepidoptera do not respond to their own pheromone; however, within some noctuid and tortricid genera, all species investigated exhibit electroantennogram (EAG) responses to to their own pheromones (19). Another possible mechanism envisaged, with certain formulations, is that the insecticide quickly reduces the population, which is then held in check by pheromone disruption (J. Jenkins, personal communication).

III. MODE OF ACTION

Most authorities concede that the mechanisms underlying the behaviors of the adult moths, observed on exposure to an area permeated with synthetic pheromone, are still not really understood (25–27). In general, three modes of action are usually proposed; adaption/habituation, false-trail following, and camouflage.

A. Adaptation/Habituation

Adaptation/habituation are direct neurophysiological effects predicated on the constant exposure of the insect to high-dosage levels of the pheromone, such that there is sensory adaption of the antennal receptors or habituation of the central nervous system, resulting in the male insect being unable to respond to any normal levels of the stimulus. Normal levels of the stimulus are viewed as the amount of pheromone being released by calling virgin female insects.

B. False-Trail Following

The broadcast of discrete point sources of sufficient pheromone strength presents a multiplicity of "false trails" so that insects flying within the treated area are diverted from calling females. This mode of action is very dependent on both point-source strength and the number of dispensers because, as pointed out by Sanders (28), it is the competition of the "hot point sources" with feral calling females that determines the reduction in mate finding. A significant shift in the ratio of sources to feral females in favor of the females will have a negative impact on the reduction of mate finding.

C. Camouflage

Camouflage, like the previous mode, assumes that the males in the treated area are sensorily unaffected by their exposure to the pheromone, and that there is present a sufficiently high background level of applied pheromone that masks or camouflages the natural plumes emitted by the calling females.

Bartell (27) was of the opinion that "...the actual mechanism will be determined largely by the release technique used," while Campion

(26) reported that the mode of action could consist of one or a combination of the mechanisms outlined previously. My thesis is that all three modes could be operative, but one mode may be the "preferred mode" at any given time, dependent on a set of conditions, including release technique, prevailing at that time.

IV. FORMULATION DESIGN PARAMETERS

With both the use strategy and mode of action aspects of formulation design having been defined, attention can now be focused on the third aspect (see Table 1), namely factors affecting the formulation parameters.

A. Target Pest

In 1982, Klassen et al. (29) reported that more than 670 insect pheromones had been identified. In the almost 30 years from the isolation of bombykol (30), through the time of Klassen's report, until now, advancements in both analytical instrumentation and methodology (3) have enabled researchers to show that many of the structural elucidations are incomplete. This resulted in a situation whereby the behavioral researchers were attempting to decipher the complex mating behaviors and the effect of introduced pheromones on these behaviors with incomplete stimulatory systems. In addition, those working toward commercialization of pheromone-based control strategies were also working with incomplete systems, one factor limiting their chances of success.

In many instances, the decision makers in the commercial arena have not appreciated the importance of fully understanding the complexities of mating behaviors; it may well be that a rose is a rose is a rose, but in mating behaviors, belief that an insect is an insect is an insect, is dangerous, false logic. If sufficient behavioral, and general biological information (e.g., number of generations per year, range of host plants, migratory potential, diel periodicity of mating) is not known for the target species, then the potential for success of developing an efficacious formulation is diminished.

B. Type of Crop

The type of crop can affect formulation design in several ways; two examples to illustrate the general principles will suffice at present. Should a control strategy require pheromone applications to be made on, say, vines before leafing, then it would be more logical to use a rope-type formulation, rather than a microcapsule or a liquid flowable material. The second example is in areas in which the cultural practices may present logistical problems in reapplication; then again certain types of formulations may be preferred. This example may be extended to developing countries in which the cultural practices again may dictate the preferred status of certain formulations.

When the type of crop is related to certain economic aspects, there may be cases where the value of a specialized or luxury crop would offset the possible higher costs of a specific pheromone formulation.

C. Utilization of Pheromone

The economics of some of the technical pheromones and our lack of knowledge concerning the fate of pheromones in "biological sinks" dictate that the complete utilization of the pheromone should be an objective in the consideration of a formulation; although this is a worthwhile goal, it is not a practical one.

D. Release Rate/Blend Ratio and Longevity/Stability

These two groups of variables are so closely related that their separate discussion is difficult. Simply stated, the formulation needs to deliver the correct amount of pheromone in the correct component ratio for the correct period. These determinants form the basis of controlled-release formulations, and their control is where most of the major problems are encountered in formulation development, because most pheromones are multicomponent blends. Component ratio would, of course, not be a critical factor if it were possible to utilize one component of the blend to effect an "off-blend overload" and cause disruption by adaptation/habituation or camouflage (27). A similar situation would hold true if it were possible to use an antipheromone or parapheromone. Although reports of successful field experiments have been published (31–37), such techniques have not been commercially developed.

Some of the factors affecting the formulation stability are presented in Table 2. Briefly, isomerization of double bonds can be the result of photochemical or thermal degradation; oxidative decomposition, photochemically or thermally catalyzed, can result in the oxidation of conjugated diene and aldehyde functions; pH changes can result in hydrolysis, enolization, and polymerization. Polymerization can also be caused by metal ions. It should be noted that degradation can take place not only during field use but, also, during the manufacture, packaging, and storage of the formulation. The physical factors of adsorption and desorption can mediate loss of material through irreversible adsorption and catabolism or rerelease at an improper rate.

TABLE 2 Factors Adversely Affecting Stability and Longevity

Chemical	Physical
UV light and/or heat isomerization of double bonds	Adsorption into substrates (leaves, soil, etc.)
UV catalyzed oxidation degradation of aldehydes oxidation of alcohols oxidation of polyenes and conjugated dienes	Desorption from substrates
Effect of pH hydrolysis of esters and epoxides enolization of carbonyl functions trimerization of aldehydes	
Metal ions catalysis of alcohol degradation, trimerization, and polymerization	

E. Application Methods

Application methods influence design for customer acceptance, logistics, and economics. Whether the application is to be by air or by ground rig, any formulation that cannot be applied by existing equipment is disadvantaged. In some quarters specialized application equipment is considered an advantage, for the equipment can be patentable and afford the formulation developer some proprietary protection. On the other hand, among the problems subsequently encountered are the cost of the equipment; modification of aircraft necessitating FAA or equivalent foreign approval, the small number of aircraft or ground rigs that can be fitted out relative to the possible amount of product to be applied, and the various locations of these applications, not to mention the poor acceptance of specialized equipment by aerial and ground applicators.

F. Economics

In 1979, John Siddall (38) was of the opinion that to develop a pheromone formulation would take between 1 and 5 man-years at a cost of between 80,000 and 400,000 dollars; his optimistic estimate was 3.5 man-years and 280,000 dollars. In 1982, Kydonieus and Beroza (39) put the cost at 375,000 dollars and 6 man-years, and they did not take into account needed analytical support. The same authors opined that if a basic formulation technology was developed, then the cost of developing further product formulations based on this technology should drop to about 80,000 dollars per formulation. I concur that after basic technology is in place, subsequent formulation development will be less expensive; however, I believe that in 1988 the initial cost, assuming 3.5 man-years, would be in the order of 500,000 dollars. It is noteworthy that the formulation development estimates from both sources (38,39) are equivalent to 20% of the total product development costs.

V. CURRENT COMMERCIAL FORMULATIONS

A. Formulation Types

The various behavior-modifying chemical formulations have been arbitrarily classified into five types, namely, microcapsules, trilaminates, capillaries, ropes, and liquid flowables. The manufacturers of these formulations, both those in current commercial use and those under development, are given in Table 3.

Microcapsules have been field tested as controlled-release pheromone formulations since the early 1970s, when both a gelatin-based material from NCR (now Eurand America; 40,41) and a nylon microcapsule from Penwalt Corp. (41) were employed against the gypsy moth. A gelatin microcapsule was also tested against the spruce budworm (42) in 1975, whereas, more recently, formulations provided by ICI (43,44) and Bend Research (now Consep; 45) have been tested against the pink bollworm. An excellent review (46) on the development of microcapsule pheromone formulations was published in 1982.

The versatile trilaminate formulations are manufactured by Hercon Laboratories Corp. (formerly Hercon Division, Health-Chem Corp.), and the system has recently been reviewed by Quisumbing and Kydonieus (47). Capillary or hollow-fiber formulations, currently supplied by Scentry Inc., were developed initially by Conrel (Albany International Corp.); these types of formulations were recently reviewed by Swenson and Weatherston (48). Forerunners of the rope type of formulation were tested in Europe (49,50) and in Australia (51) before development and commercialization by Shin-Etsu and Biocontrol Ltd. The liquid flowable formulation, in which the pheromone is attached to particulate material, which in turn is suspended in a liquid medium, is the most recent type of formulation to be commercialized.

TABLE 3 Behavior-Modifying Chemical Formulation Suppliers

Microcapsules	Trilaminates	Capillaries	Ropes	Liquid flowables
		Commercial formulations		
ICI	Hercon Laboratory Corp. (BASF)[a]	Scentry Inc. (Sandoz)[a]	Shin-Etsu (Mitsubishi)[b]	Fermone Inc.
			Biocontrol Ltd.	
		Formulations in development		
3M			3M	Spray Control Systems
Recon[c]			Recon[c]	Dexter Chemical Intl.
Consep				Montedison
				Monterey Chemical

[a]Supplier outside of United States.
[b]Supplier in the United States.
[c]Technology now owned by AgriSense.

Six design factors, namely, use strategy, mode of action, release rate, longevity/stability, application method, and economics have been selected for a discussion of the relative strengths of the current commercial formulations. This comparison is summarized in Table 4; however, a tendency to sum the ratings and obtain an overall comparison should be avoided because the weight attached to each factor for any given formulation can be different.

A note is required here about the design factor—economics; because it is impossible to obtain any accurate financial information from the respective manufacturers, the economic comparisons of the formulations are based on active ingredient (AI) use in the respective pink bollworm formulations. I have used the amount of pheromone to exemplify relative costs, simply because the pheromone is usually the most expensive component of the formulation. Care must be taken when considering such comparisons in examples such as Provesta

TABLE 4 Relative Strengths of the Current Commercial Pheromone Formulations Measured Against Six Design Factors

Factor	Micro-capsule	Tri-laminate	Capillary	Rope	Liquid flowable
Use strategy	++	+++	++	+	+
Mode of action	++	+++	+++	+++	++
Release rate	++	++	+++	++	++
Longevity/stability	++	+++	+++	+++	+++
Application method	+++	+	+	++	+++
Economics	+	++	++	++	++

+++ strong, ++ fair, + weak

(AgriSense) and Shin-Etsu for which the basic producer of the technical active ingredient is also the formulator.

B. Microcapsules

The microcapsule formulations can be used in situations attempting control by the disruption or bioirritant strategy; it cannot be used in an attract-and-kill situation because the applied formulation manifests a miasma and does not possess sufficiently strong point sources. Hence, the modes of action imposed by microcapsules would be those of adaptation/habituation or camouflage (or both).

Ease of manufacture incorporating easily manipulated variables, including capsule size, wall thickness, and wall composition, that affect the release rate, which is claimed to be zero order by the manufacturers, leads to versatile formulations; however, microcapsules tend to retain a rather high percentage of their active ingredients at the end of their targeted use period. Because antioxidants and ultraviolet stabilizers can be incorporated into the capsule wall, the longevity and field stability is good for monounsaturated pheromones.

In current usage, in pink bollworm formulations, microcapsules are being used at an active ingredient rate of 10 g/ha, with three or four applications being used per season (Tables 5 and 6). Of the five formulation types under discussion this is the second largest amount of pheromone utilized, and this, of course, has to be reflected in the economics.

TABLE 5 Comparison of Commercial Formulations Used in Pink Bollworm Control in Ben-Suef and Fayoum Governorates (Egypt) 1985

	Pheromone application				
Treatment	Area (feddan)	Rate/ fed. form.	AI (g)[a]	Applications (no.)	Boll infestation (%)
Microcapsules	3500	200 ml	4.0	4	2.7
Insecticides					2.2
Capillaries	5000	150 g	1.14	4	5.6
Insecticides					3.4
Trilaminates	3000	60 g	3.0	4	3.5
Insecticides					3.5
Liquid flow	120	240 ml	3.72	3	3.12
Liquid flow + insecticide	90	120 ml	1.86	3	1.53
Microcapsules	3500	200 ml	4.00	3	0.90
Insecticides					2.33
No treatment control					7.29

[a]AI, active ingredient.

C. Trilaminates

The trilaminate system can be produced in several forms, for example, sheet, tape, wafer, powder and, the form under discussion, confetti. The confetti can be used for disruption and attracticide strategies; however, the basic system has the flexibility, if formulated as a granule or powder and incorporated into a liquid flowable, to find use in the bioirritant strategy. The release from trilaminates follows first-order kinetics; that is, the release rate is dependent on the amount of material remaining in the vehicle; this could be a disadvantage for longevity, residual pheromone in treatment areas, and economics, depending on the required pheromone threshold and the cost of technical

TABLE 6 Comparison of Commercial Formulations Used for Pink Bollworm Control on the Salauddin Farm (Pakistan) in 1986

	Pheromone application				
Treatment	Area (ha)	Rate/ ha formul.	AI (g)	Applications (no.)	Mating disruption (%)
Microcapsules + insecticides[a]	18.0	500 ml	10.00	3	97.05
Liquid flow + insecticides	1.0	438 ml	6.79	3	88.20
Ropes + insecticides	8.0	1300 tubes	46.80	1	99.20
Ropes + insecticides	10.0	800 tubes	28.80	1	97.90

[a]In all treatments four insecticide applications were used, and the treatments compared to four applications of insecticide alone.

pheromone. Field use stability is a strong point of this type of formulation because it has the ability to protect labile pheromones from environmental conditions that can cause rapid degradation. The various protecting agents (e.g., antioxidants, UV screens) can be incorporated into the outer laminate layers. As with capillary and rope formulations, the trilaminates can affect the male insects in the treated areas by all or any of the three proposed modes of action discussed earlier. The weakest point of this system is that it requires specialized application equipment; this will be discussed in the section dealing with capillaries, because that system also suffers from this serious disadvantage. Ignoring the cost of special equipment and referencing 1985 data from Egypt (see Table 5), where over the season four applications were made, each utilizing 7.5 g/ha (1 feddan ≈ 0.4 ha), could bring technical pheromone costs for trilaminates into line with, or slightly better than, those of the microcapsules.

D. Capillaries

The capillary type of formulation, which contains strong point sources, can be used in either disruption or attracticide strategies, influencing the male moths by any or all of the proposed modes of action. This system has good design flexibility relative to release parameters, longevity, and field stability. Recent modeling studies (54—54; Weatherston and Miller, unpublished) have indicated that further development of this system to give constant release and a high utilization of active material is possible. Longevity is a function of capillary length and can be controlled during manufacture. Initially, this system had a problem in the area of protection of aldehydic and diunsaturated pheromones that was solved by the addition of antioxidants to the liquid charge and the incorporation of UV protectants into the wall material before extrusion of the capillaries.

When a formulation requires specialized application equipment, this is a serious disadvantage. The equipment developed for the capillaries and also for the trilaminates, although affording their developers proprietary protection in the form of patents, subsequently led to the problems outlined in Section IV.E. Once again, if one disregards the equipment and related costs and uses only the cost of the pheromone as a guide to the relative economics, and given 1985 data from Egypt (see Table 5), it can be seen that over the 12-week season four applications were made at the rate of approximately 2.85 g/ha for a total use of less than 12 g—a significantly smaller amount than used by either the microcapsules or the trilaminates.

E. Ropes

The rope formulations are presently aimed primarily at, and restricted to, control by the disruption strategy, although they can achieve this through any or all of the proposed mechanisms. Data for the Shin-Etsu PBW-rope indicates a first-order kinetic-type release, with a first-application-day release of 1.3 mg of gossyplure per dispenser and a mean release, over about 60 days, of 670 μg. This type of formulation, because it contains relatively large amounts of active ingredient per device, compared with a microcapsule or capillary, has the highest longevity of all formulations under discussion. The Isomate M product of Biocontrol Ltd., has a claimed longevity of 3 months per application. In Pakistan, a single application of the Shin-Etsu rope had a use period of 2 months; during this time other plots at the same test site received three applications of liquid flowables and microcapsules (see Table 6). The ropes are hand-applied, and this initially resulted in a great deal of discussion on whether or not this type of application was feasible in developed countries, and whether or not it would be met by customer acceptance. In developing countries in which hand labor is commonly used, or where aerial application technology is not readily available, rope formulations appear well suited. In North America and other developed areas, the type of crop could be an important consideration, the ropes being well-suited to grapes and orchards but not, for example, in certain forestry situations, say, against the spruce budworm or the gypsy moth. Ropes would appear to be the most costly of the five formulations discussed, because in Pakistan (see Table 6) gossyplure was used at the rates of 46.8 and 28.8 g/ha, whereas in a 1985 test conducted in California and Mexico, the active ingredient was used at 78 g/ha with a portion of the area receiving two applications. As can be seen from Table 6, the most effective treatment in the Pakistan test was the rope formulation at the high rate of 46.8 g/ha, which may be assumed to have a 1.6 greater cost factor than the rope formulation at 28.8 g/ha or the microcapsules at 30 g/ha.

Biocontrol Ltd., recommended two applications of Isomate-M for season-long control of the oriental fruit moth at a total cost of 110 dollars per acre, which they claim has a cost equivalency with three insecticide applications.

F. Liquid Flowables

Bioirritant and disruption strategies can be attempted with a liquid flowable type formulation, which must rely on either the adaptation/habituation or camouflage modes of action. Release data, as yet, are not available for many liquid flowable formulations; however, analysis of four Stirrup formulations from Fermone Inc. (Weatherston and Miller, unpublished) indicated that after a brief latency period a fairly constant release (linear regression of release versus time = 0.992) can be achieved. These data also indicated a flexibility in desired longevity but a high percentage retention of active ingredient. Generally, this type of formulation can be applied using conventional spray technology; in fact, like microcapsules, liquid flowables can be applied simultaneously with other materials (e.g., foliar fertilizers) to manifest a savings in application costs. Using the amount of gossyplure as an indicator of relative costs, one sees that in Egypt (see Table 5) 4.65 g/ha, when used in the bioirritant strategy, performed better, as measured by boll damage, than conventional insecticide treatments, although slightly inferior to microcapsules applied at a 2.15 times greater active ingredient rate. In Pakistan (see Table 6), when used at the rate of 6.79 g/ha, it was inferior to microcapsules at a 1.5 times greater rate and ropes at a 1.4 and a 2.4 greater rate, as measured by the percentage of mating disruption.

VI. CONCLUSIONS

As stated in Roelofs (9):

> The design of any practical controlled release device or system will normally represent a set of compromises made to accommodate

release system limitations to uncontrollable variations in environmental conditions as well as circumstances peculiar to a specific insect biology, a particular crop, geographical location, cultural practices and the like.

There are areas that, although they affect a pheromone product, do not directly affect the principles of design of the controlled-release formulations (e.g., patentability, environmental concern, and registration) and, hence, were not discussed. It is hoped that this overview has highlighted some of the "and the like" factors, and their interrelationships that must be considered in the development of behavior-modifying chemical formulations.

What then of the future? The prognosis is good, and with the current surge of activity in formulation development, the "efficacious and cost-effective controlled-release formulations" are beginning to come "on stream"; it is hoped that this will lead to a greater use of behavior-modifying chemicals in plant protection.

REFERENCES

1. Butenandt, A. and Hecker, E. Synthese des Bombykols des Sexual-Lockstoffes des Seidenspinners, und seiner geometrischen Isomeren. *Angew. Chem.* 73:349–353 (1961).
2. Jutsum, A. R. and Gordon, R. F. S. (eds.). *Insect Pheromones in Plant Protection.* John Wiley & Sons, New York, 1989, 300 pp.
3. Hummel, H. E. and Miller, T. A. (eds.). *Techniques in Pheromone Research.* Springer-Verlag, New York, 1984, 464 pp.
4. Kydonieus, A. F. and Beroza, M. (eds.). *Insect Suppression with Controlled Release Pheromone Systems*, Vols. I, II. CRC Press, Boca Raton, 1982, 274 and 311 pp.
5. Leonhardt, B. A. and Beroza, M. (eds.). *Insect Pheromone Technology: Chemistry and Applications.* ACS Symp. Ser. No. 190. American Chemical Society, Washington, D.C., 1982, 260 pp.
6. Mitchell, E. R. (ed.). *Management of Insect Pests with Semiochemicals: Concepts and Practice.* Plenum Press, New York, 1981, 514 pp.
7. Nordlund, D. A., Jones, R. L., and Lewis, W. J. *Semiochemicals: Their Role in Pest Control.* John Wiley & Sons, Chichester, 1981, 306 pp.
8. Ritter, F. J. (ed.). *Chemical Ecology: Odour Communication in Animals.* Elsevier/North-Holland, Amsterdam, 1979, 427 pp.
9. Roelofs, W. L. (ed.). *Establishing Efficacy of Sex Attractants and Disruptants for Insect Control.* Entomological Society of America, College Park, MD, 1979, 97 pp.

10. Beroza, M. (ed.). *Pest Management with Insect Sex Attractants.* ACS Symp. Ser. No. 23. American Chemical Society, Washington, D.C., 1976, 192 pp.
11. Jacobson, M. *Insect Sex Pheromones.* Academic Press, New York, 1972, 382 pp.
12. Boness, M., Eiter, K., and Disselnkötter, H. Studies on sex attractants of Lepidoptera and their use in crop protection. *Pflanzenschutz-Nach. Bayer 30*:213–236 (1977).
13. Campion, D. G. and Murlis, J. Sex pheromones for the control of insect pests in developing countries. *Med. Fac. Landbouww. Rijksuniv. Gent. 50*(2a):203–209 (1985).
14. Gaston, L. K. Shorey, H. H., and Saario, C. A. Insect population control by use of sex pheromones to inhibit orientation between the sexes. *Nature 213*:1155 (1967).
15. Kaissling, K. E. Action of chemicals, including (+) *trans*permethrins and DDT on insect olfactory receptors. In *Proc. Soc. Chem. Ind. Symp. Insect Neurobiology Pesticide Action.* 1982, pp. 351–358.
16. Floyd, J. P. and Crowder, L. A. Sublethal effects of permethrin on pheromone response and mating of male pink bollworms. *J. Econ. Entomol. 74*:634–638 (1981).
17. Cook, B. J. and Shelton, W. D. Antennal responses of the pink bollworm to gossyplure. *Southwest. Entomol. 3*:141–146 (1978).
18. Light, D. M. and Birch, M. C. Electrophysiological basis for the behavioural response of male and female *Trichoplusia ni* to synthetic pheromone. *J. Insect Physiol. 25*:161–167 (1979).
19. Priesner, E. Progress in the analysis of pheromone receptor systems. *Ann. Zool. Ecol. Anim. 11*:533–546 (1979).
20. Palanaswamy, P. and Seabrook, W. D. Behavioral responses of the female eastern spruce budworm *Choristoneura fumiferana* (Lepidoptera, Tortricidae) to the sex pheromone of her own species. *J. Chem. Ecol. 4*:649–655 (1978).
21. Ross, R. J., Palanswamy, P., and Seabrook, W. D. Electroantennograms from spruce budworm moths (*Choristoneura fumiferana*) (Lepidoptera: Tortricidae) of different ages and for various pheromone concentrations. *Can. Entomol. 111*:807–816 (1979).
22. Seabrook, W. D., Linn, C. E., Dyer, L. J., and Shorey, H. H. Comparison of electroantennograms from female and male cabbage looper moths (*Trichoplusia ni*) of different ages and for various pheromone concentrations. *J. Chem. Ecol. 13*:1443–1453 (1987).
23. Mitchell, E. R., Webb, J. R., and Hines, R. W. Capture of male and female cabbage loopers in field traps baited with synthetic sex pheromone. *Environ. Entomol. 1*:525–526 (1972).

24. Birch, M. C. Responses of both sexes of *Trichoplusia ni* (Lepidoptera: Noctuidae) to virgin females and to synthetic pheromone. *Ecol. Entomol.* *2*:99–104 (1977).
25. Bartell, R. J. and Rumbo, E. R. Correlations between electrophysiological and behavioural responses elicited by pheromone. In *Mechanisms of Insect Olfaction.* (Payne, T. L., Birch, M. C., and Kennedy, C. E. J., eds.). Clarendon Press, Oxford, 1986, pp. 169–174.
26. Campion, D. G. Survey of pheromone uses in pest control. In *Techniques in Pheromone Research.* (Hummel, H. E. and Miller, T. A., eds.). Springer-Verlag, New York, 1986, pp. 405–449.
27. Bartell, R. J. Mechanisms of communication disruption by pheromone in the control of Lepidoptera: A review. *Physiol. Entomol.* *7*:353–364 (1982).
28. Sanders, C. J. Disruption of male spruce budworm orientation to calling females in a wind tunnel by synthetic pheromone. *J. Chem. Ecol.* *8*:493–506 (1982).
29. Klassen, W., Ridgway, R. L., and Inscoe, M. Chemical attractants in integrated pest management. In *Insect Suppression with Controlled Release Pheromone Systems*, Vol. I. (Kydonieus, A. F. and Beroza, M., eds.). CRC Press, Boca Raton, 1982, pp. 13–130.
30. Butenandt, A. R., Beckmann, R., Stamm, D., and Hecker, E. Über den Sexuallockstoff des Seidenspinners *Bombyx mori*:Reindarstellung und Konstitution. *Z. Naturforsch.* *14B*:283–284 (1959).
31. Henneberry, T. F., Gillespie, J. M., Bariola, L. A., Flint, H. M., Butler, G. D., Jr., Lingren, P. D., and Kydonieus, A. F. Mating disruption as a means of suppressing pink bollworm. (Lepidoptera: Gelechiidae) and tobacco budworm (Lepidoptera: Noctuidae) populations on cotton. In *Insect Suppression with Controlled Release Pheromone Systems*, Vol. II. (Kydonieus, A. F. and Beroza, M., eds.). CRC Press, Boca Raton, 1982, pp. 75–98.
32. Henneberry, T. J., Bariola, L. A., Flint, H. M., Lingren, P. D., Gillespie, J. M., and Kydonieus, A. F. Pink bollworm and tobacco budworm mating disruption studies on cotton. In *Management of Insect Pests with Semiochemicals: Concepts and Practice.* (Mitchell, E. R., ed.). Plenum Press, New York, 1981, pp. 267–283.
33. Beevor, P. S., Dyck, V. A., and Arida, G. S. Formate pheromone mimics as mating disruptants of the striped rice borer moth *Chilo suppressalis* (Walker). In *Management of Insect Pests with Semiochemicals: Concepts and Practice.* (Mitchell, E. R., ed.). Plenum Press, New York, 1981, pp. 305–311.

34. Tatsuki, S. and Kanno, H. Disruption of sex pheromone communication in *Chilo suppressalis* with pheromones and analogues. In *Management of Insect Pests with Semiochemicals: Concepts and Practice*. (Mitchell, E. R., ed.). Plenum Press, New York, 1981, pp. 313–325.
35. Kanno, H., Hattori, M., Sato, A., Tatsuki, S., Ichiumi, K., Kurihara, M., Fukami, J., Fujimoto, Y., and Tatsuno, T. Disruption of the sex pheromone communication in the rice stem borer moth, *Chilo suppressalis* Walker (Lepidoptera: Pyralidae), with sex pheromone components and their analogues. *Appl. Entomol. Zool. 15*:465–473 (1980).
36. Beevor, P. S. and Campion, D. G. The field use of inhibitory components of lepidopteran sex pheromones and pheromone mimics. In *Chemical Ecology: Odour Communication in Animals*. (Ritter, F. J., ed.). Elsevier/North-Holland, Amsterdam, 1979, pp. 313–325.
37. McLaughlin, J. R., Shorey, H. H., Gaston, L. K., Kaae, R. S., and Stewart, F. D. Sex pheromones of Lepidoptera. XXXI. Disruption of sex pheromone communication in *Pectinophora gossypiella* with hexalure. *Environ. Entomol. 1*:645–650 (1972).
38. Siddall, J. B. Commercial production of insect sex pheromones—problems and prospects. In *Chemical Ecology: Odour Communication in Animals*. (Ritter, F. J., ed.). Elsevier/North-Holland, Amsterdam, 1979, pp. 389–401.
39. Kydonieus, A. F. and Beroza, M. Marketing and economics in use of pheromones for suppression of insect populations. In *Insect Suppression with Controlled Release Pheromone Systems*, Vol. II. (Kydonieus, A. F. and Beroza, M., eds.). CRC Press, Boca Raton, 1982, pp. 187–189.
40. Cameron, E. A. Disparlure: A potential tool for gypsy moth population management. *Bull. Entomol. Soc. Am. 19*:15–19 (1973).
41. Beroza, M., Stevens, L. J., Bierl, B. A., Phillips, F. M., and Tardif, G. R. Pre- and postseason field tests with disparlure, the sex pheromone of the gypsy moth, to prevent mating. *Environ. Entomol. 2*: 1051–1057 (1973).
42. Sanders, C. J. Disruption of sex attraction in the eastern spruce budworm. *Environ. Entomol. 5*:868–872 (1976).
43. Critchley, B. R., Campion, D. G., McVeigh, E. M., McVeigh, L. J., Jutsum, A. R., Gordon, R. F. S., Marrs, G. J., Nasr, E.-S. A., and Hosny, M. M. Microencapsulated pheromones in cotton pest management. *Proc. Br. Crop Prot. Conf. 4A-1*:241–245 (1984).
44. Critchley, B. R., Campion, D. G., McVeigh, L. J., Hunter-Jones, P., Hall, D. R., Cork, A., Nesbitt, B. F., Marrs, G. J.,

Jutsum, A. R., Hosny, M. M., and Nasr, E.-S. A. Control of pink bollworm, *Pectinophora gossypiella* (Saunders) (lepidoptera: Gelechiidae) in Egypt by mating disruption using an aerially applied microencapsulated pheromone formulation. *Bull. Entomol. Res. 73*:289–299 (1983).

45. Flint, H. M. and Merkle, J. R. Studies on disruption of sexual communication in the pink bollworm, *Pectinophora gossypiella* (Saunders) (Lepidoptera: Gelechiidae) with microencapsulated gossyplure or its component *Z,Z*-isomer. *Bull. Entomol. Res. 74*:25–32 (1984).

46. Hall, D. R., Nesbitt, B. F., Marrs, G. J., Green, A. St. J., Campion, D. G., and Critchley, B. R. Development of microencapsulated pheromone formulations. In *Insect Pheromone Technology: Chemistry and Applications.* (Leonhardt, B. A. and Beroza, M., eds.). ACS Symp. Ser. No. 190. American Chemical Society, Washington, D.C., 1982, pp. 131–143.

47. Quisumbing, A. R. and Kydonieus, A. F. Laminated structure dispensers. In *Insect Suppression with Controlled Release Pheromone Systems.* Vol. I. (Kydonieus, A. F. and Beroza, M., eds.). CRC Press, Boca Raton, 1982, pp. 213–235.

48. Swenson, D. W. and Weatherston, I. Hollow-fiber Controlled-release Systems. In *Insect Pheromones in Plant Protection.* (Jutsum, A. R. and Gordon, R. F. S., eds.). John Wiley & Sons, New York, 1989, pp. 173–197.

49. Arn, H., Delley, B., Baggiolini, M., and Charmillot, P. J. Communication disruption with sex attractant for control of the plum moth, *Grapholita funebrana*: A two year field study. *Entomol. Exp. Appl. 19*:139–147 (1976).

50. Charmillot, P. J., Scribante, A., Pont, V., Deriaz, D., and Fournier, C. Technique de confusion contre la tordeuse de la pelure *Adoxophyes orana* F.v.R. (Lep., Tortricidae). I. Influence de la diffusion d'attractif sexuel sur le comportement. *Bull. Soc. Entomol. Suisse 54*:173--190 (1981).

51. Rothschild, G. H. L. A comparison of methods of dispensing synthetic pheromone for the control of oriental fruit moth, *Cydia molesta* (Busck) (Lepidoptera: Tortricidae) in Australia. *Bull. Entomol. Res. 69*:115--127 (1979).

52. Weatherston, I., Miller, D., and Dohse, L. Capillaries as controlled release devices for insect pheromones and other volatile substances—A reevaluation: Part I. Kinetics and development of predictive model for glass capillaries. *J. Chem. Ecol. 11*:953–965 (1985).

53. Weatherston, I., Miller, D., and Lavoie-Dornik, J. Capillaries as controlled release devices for insect pheromones and other volatile substances—A reevaluation: Part II. Predicting release

rates from Celcon and Teflon capillaries. *J. Chem. Ecol. 11*:967–978 (1985).
54. Weatherston, I., Miller, D., and Lavoie-Dornik, J. Commercial hollow fiber pheromone formulations: The degrading effect of sunlight on Celcon fibers causing increased release rates of the active ingredient. *J. Chem. Ecol. 11*:1631–1644 (1985).

7

Dispenser Design and Performance Criteria for Insect Attractants

BARBARA A. LEONHARDT / Agricultural Research Service, U.S. Department of Agriculture, Beltsville, Maryland

ROY T. CUNNINGHAM / Agricultural Research Service, U.S. Department of Agriculture, Hilo, Hawaii

WILLARD A. DICKERSON* / Agricultural Research Service, U.S. Department of Agriculture, Raleigh, North Carolina

VICTOR C. MASTRO / U.S. Department of Agriculture, Otis Air National Guard Base, Massachusetts

RICHARD L. RIDGWAY / Agricultural Research Service, U.S. Department of Agriculture, Beltsville, Maryland

CHARLES P. SCHWALBE / U.S. Department of Agriculture, Otis Air National Guard Base, Massachusetts

I. INTRODUCTION

The nature of insect pheromones and attractants, their chemical identification, and the principles associated with the practical use of these compounds have been discussed in previous chapters. Currently, the general public is becoming more aware of what these attractive chemicals do; commercial traps with baits are now available in stores for a number of insects including Japanese beetle, *Popillia japonica* Newman; gypsy moth, *Lymantria dispar* (Linnaeus); and bagworm, *Thyridopteryx ephemeraeformis* (Haworth).

**Current affiliation:* North Carolina Department of Agriculture, Raleigh, North Carolina

Although the active compound can be deposited on almost any substrate and still attract insects, the highly volatile and chemically unstable compounds should be formulated in polymeric matrices to maintain attractancy for the necessary field lifetime. Many of these polymeric formulations have been used to effectively deliver insect pheromones for field programs (see Chap. 6) and commercial products are now available for many insect species (see Chap. 38). However, established criteria or specifications for field performance of these dispensers have been limited. One usually must rely on the manufacturer for selection of concentration, release rate, and longevity.

There is, or has been, much variation in commercial dispensers. The (Z)-11-hexadecenal content found in pheromone dispensers for *Heliothis armigera* (Hübner) from five sources showed that the amount of this major component ranged from 0.08 to 18 mg; it is reasonable to expect that the attractiveness to *H. armigera* would vary over this range. Similarly, dispensers of *H. virescens* (F.) pheromone from four sources (Table 1) showed variation in the contents and ratios of the two most important components, (Z)-9-tetradecenal and (Z)-11-hexadecenal; the composition of the minor components also varied. Because it is desirable to reproduce as closely as possible the pheromone message emitted by insects to attain consistent capture, it is very important that the following parameters be defined:

1. *Content* of active ingredient (AI): Concentration and purity
2. *Rate* of emission: Effective quantity
3. *Ratio* of components in pheromone blend
4. *Duration* of effectiveness

TABLE 1 Dispensers for Pheromone of *H. virescens*

	Milligrams per dispenser[a]					
Source	Z-9-14:al	14:al	Z-7-16:al	Z-9-16:al	Z-11-16:al	16:al	Z-11-16:ol
A	0.4			0.004	5.5	0.1	
B	0.3			0.3	3.4	0.05	
C	0.6	0.2	0.2	0.3	8.3	0.8	0.6
D	4.0	0.02		0.03	26	0.3	

[a]As determined by gas chromatographic analyses on extracts of the dispensers.

This paper reports the development of these parameters and performance criteria for effective and reliable dispensers for three major insect pests: Mediterranean fruit fly, *Ceratitis capitata* (Wiedemann); boll weevil, *Anthonomus grandis* (Boheman); and gypsy moth, *Lymantria dispar* (L.).

II. DISPENSER AND CRITERIA DEVELOPMENT

A. Mediterranean Fruit Fly

The Mediterranean fruit fly is an established agricultural pest in many parts of the world. A vigilant surveillance (and eradication, when required) program has prevented the permanent establishment of this pest in the continental United States. The potential introduction of this pest through imported produce has necessitated the operation of an array of about 40,000 trimedlure-baited detection traps, principally in California and Florida. In recent years, this early-warning network of traps has detected a number of introductions of *C. capitata*. For example, an incipient infestation was detected in Los Angeles in 1987 and cost 2 million dollars to eradicate.

Trimedlure (TML) is a mixture of eight isomers of the *tert*-butyl esters of 4- and 5-chloro-2-methylcyclohexanecarboxylic acids (1). The four trans isomers, designated A, B_1, B_2, and C, constitute 90% to 95% of the commercially synthesized TML, with the C isomer as not only the major component at 35% to 44% but also the most attractive (2–4). The standard dispenser in these detection traps had been 2 ml of liquid TML on a cotton dental roll; however, this dispenser remains highly effective for only 2 to 4 weeks in normal summer temperatures. The frequent rebaiting of these dispensers is not only costly in terms of labor and lure, but it also can result in contamination of the trap site and the exterior of the trap if the liquid TML is spilled during the rebaiting process. Solid, controlled-release formulations can overcome, or substantially reduce, these difficulties.

Therefore, the objective was to find dispensing systems that would control the rate of release of TML, extend the useful lifetime of the 2-ml dose to at least 6 to 8 weeks, reduce the possibility of contamination and be commercially available. A variety of formulations were evaluated (5–8). The most promising designs were a plastic laminate, a polymeric plug, and a membrane/reservoir. The laminate (Hercon Laboratories, Inc., South Plainfield, New Jersey) consisted of 200-μm outer layers of polyvinyl chloride (PVC) film and a trimedlure-containing middle polymer layer. The deficiency of the laminate is the large size (4.5 × 8 cm) necessary to hold 2 ml of TML and the resulting difficulty in placing the laminate in the Jackson trap (9) used in the detection program.

A second dispenser consisted of a semipermeable membrane covering a microporous pad that contains the reservoir of TML (Bend Research, Bend, Oregon). This formulation was also relatively large (7.5 cm diameter), and a surface film of liquid TML unavoidably caused contamination of the trapper's hands when the protective film covering the dispensing membrane was removed.

The third dispenser is a polymeric plug in which 70% (by weight) TML is held in the Polytrap system (AgriSense, Fresno, California). The plug is solid and, with minimal caution, contamination can be avoided. The plug is small (1.3 cm diameter × 1.7 cm long) and contains 2 ml of liquid TML and just 0.9 g of polymers. The plug fits into a small plastic basket for mounting in the narrow confines of the trap.

The performance of these dispensers was compared in a series of field and chemical tests. Figure 1 illustrates the results of one of these tests (7). The freshly treated cotton dental roll with 2 ml of

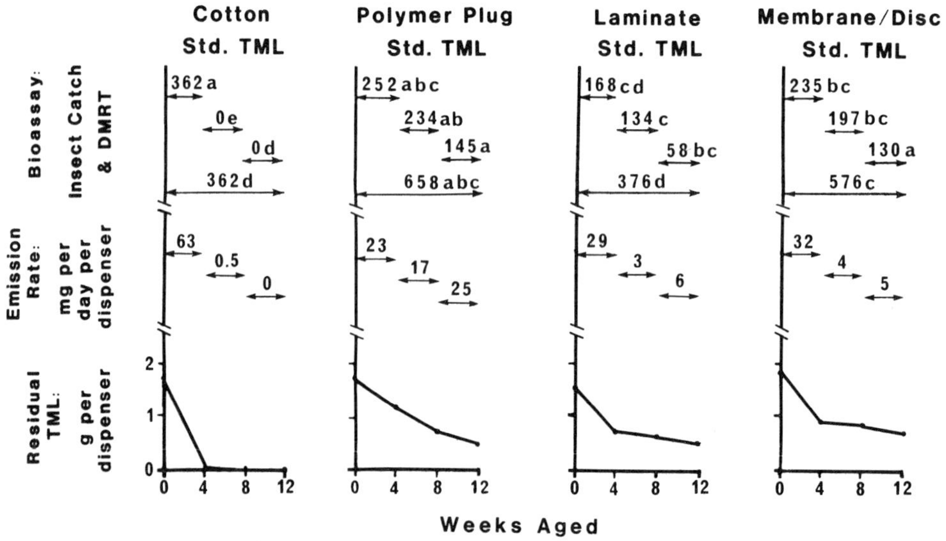

FIGURE 1 Performance of trimedlure dispensers in field and laboratory tests. Insect captures are weighted averages as back calculated from square-root-transformed data, $\bar{x} = (\varepsilon\sqrt{x_i}/n)^2$; for each aging period, averages followed by the same letter are not significantly different [$p = 0.05$; Duncan's (19) new multiple-range test].

TML was used as a reference. The dispensers were aged in California for intervals up to 12 weeks. Aged dispensers were assayed for TML content in Beltsville, Maryland, and bioassayed in Hilo, Hawaii, with released, laboratory-reared Mediterranean fruit flies. The bioassay results showed that the aged cotton wicks were inactive and depleted of TML after 4 weeks. The membrane and laminate continued to emit TML at rates of 3 to 6 mg/day as determined from the residual contents; they also continued to trap relatively large numbers of insects. However, the most consistent performance was given by the polymeric plug; in this test, there was no significant difference in catch between the plug aged for 12 weeks and the reference--a freshly treated cotton dental roll.

The series of tests with released laboratory-reared flies in Hawaii as well as with natural populations of the Mediterranean fruit flies in Egypt and Lebanon (8) indicated that the laminate and the plug often gave similar performance, but the plug had a distinct advantage in handling characteristics. The membrane dispenser was less effective.

In addition to measuring the residual contents to give release rates, a laboratory method was also devised to measure the release rates from aged dispensers by weight loss at a constant temperature of 28°C in an oven through which air flowed at 600 ml/min. Release rates by this procedure, as well as residual TML contents and bioassays, were used in further evaluations of the dispensers.

The following performance criteria were prepared on the basis of the test results for inclusion in the bid solicitation for procurement of dispensers by the Animal and Plant Health Inspection Service (APHIS) for use in the large-scale, cooperative State–Federal detection programs:

1. Initially contain 2 g of TML of specified purity and isomer composition
2. After 1 week of aging (Hawaii), contain $\geq$1.0 g of TML and emit $\geq$0.7 mg/hr at 28°C
3. After 6 weeks of aging (Hawaii), contain $\geq$0.4 g of TML and emit $\geq$0.4 mg/hr at 28°C
4. In bioassay (Hawaii), after 1 and 6 weeks of aging, capture at least 50% as many flies as the reference, freshly treated, cotton wick
5. Handling and storage: No contamination and no deterioration during storage; convenient handling in the field.

In addition to these criteria, standards were set for the liquid TML, based on research results. Because the C isomer in TML is the most attractive, limits were set for its content. The chemical standards are as follows:

≥95% *t*-butyl esters of 4- and 5-chloro-2-methylcyclohexane-1-carboxylic acid
≤0.5% free acid
<2% dehydrohalogenated TML
38–45% *trans* C isomer
≤24% *trans* B_2 isomer
≤10% *cis*-TML (total of four isomers)

Because criteria were now available, APHIS decided to purchase dispensers by competitive bid with a high emphasis on performance rather than price, using the following plan:

Evaluation criteria	Weight (%)
Efficiency	
release	
attractancy	80
operational	
Price	20

For the first time, suppliers were required to prove that their product met the performance criteria for dispensers. APHIS used the competitive bid contract to procure a 2-year supply of 500,000 plug dispensers in August 1986. These dispensers are now used in detection programs in the United States, and similar dispensers are being evaluated elsewhere in the world.

B. Boll Weevil

The aggregation and sex attractant pheromone of the boll weevil is a mixture of four components (10) in the ratio of 30, 40, 15, and 15 for the components I, II, III, and IV, respectively (Fig. 2). This blend (termed grandlure) has been used in a large mass-trapping program by the Southeastern Boll Weevil Eradication Foundation (11; Chap. 26).

The specifications for purchase of grandlure dispensers in 1984 required 10 mg of grandlure per dispenser and a release rate of 0.5 mg/day ± 15% for 2 weeks. This constant or zero-order release rate is a good target but controlled-release dispensers seldom follow this pattern. Although grandlure dispensers were purchased under these specifications, they did not meet the constant emission criterion, and they also did not always give 2 weeks of activity.

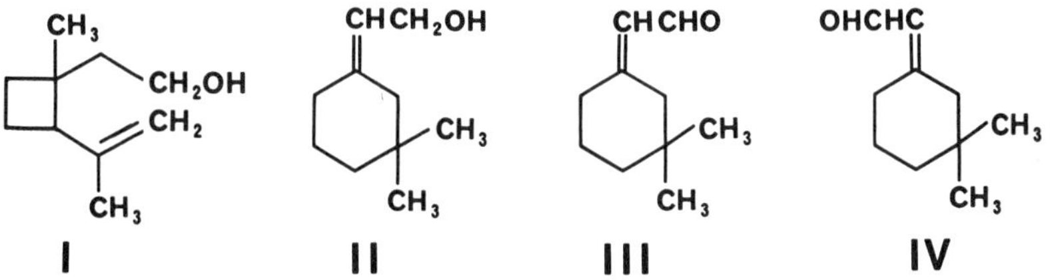

FIGURE 2 Chemical structure of the four grandlure components: (I) (+)-*cis*-2-isopropenyl-1-methylcyclobutaneethanol; (II) (*Z*)-3,3-dimethyl-$\Delta^{1,\beta}$-cyclohexaneethanol; (III) (*Z*)-3,3-dimethyl-$\Delta^{1,\alpha}$-cyclohexaneacetaldehyde; (IV) (*E*)-3,3-dimethyl-$\Delta^{1,\alpha}$-cyclohexaneacetaldehyde.

Most controlled-release dispensers follow first-order kinetics, that is, they release pheromone in proportion to their content. Therefore, as the reservoir of pheromone is depleted, the release rate declines. The problem is to design a dispenser with a conservative initial release rate such that sufficient pheromone is reserved for release for the desired life of the dispenser.

The standard dispenser for grandlure had been a plastic-wrapped cigarette filter (12) in which the 10 mg of pheromone was mixed with polyethylene glycol and other components and then loaded on the filter. The plastic physical barrier and the polyglycol slowed the evaporation of grandlure so that the dispenser remained attractive for at least 2 weeks. This dispenser was subsequently replaced by a plastic laminate dispenser that generally yielded reliable results. However, some batches of these laminates became inactive after just 1 week when used in the Texas High Plains Boll Weevil Program because the high temperatures caused rapid depletion of the grandlure. The effect of temperature on the release rates from two laminate dispensers shows that, at higher temperatures, grandlure is released more rapidly from the dispenser that was used in the Texas program (at right in Fig. 3) than from the other laminate. This is presumably due to differences in the thickness and porosity of the PVC polymer used for the outer layers.

A laboratory method for measuring emission rates from dispensers was devised (13). Candidate dispensers were placed in glass tubes housed in an oven to control temperature. A constant flow of air (100 ml/min) carried the pheromone released from the dispenser onto absorption traps which were subsequently desorbed with solvent for chemical analysis after a 2-hr collection at 28°C.

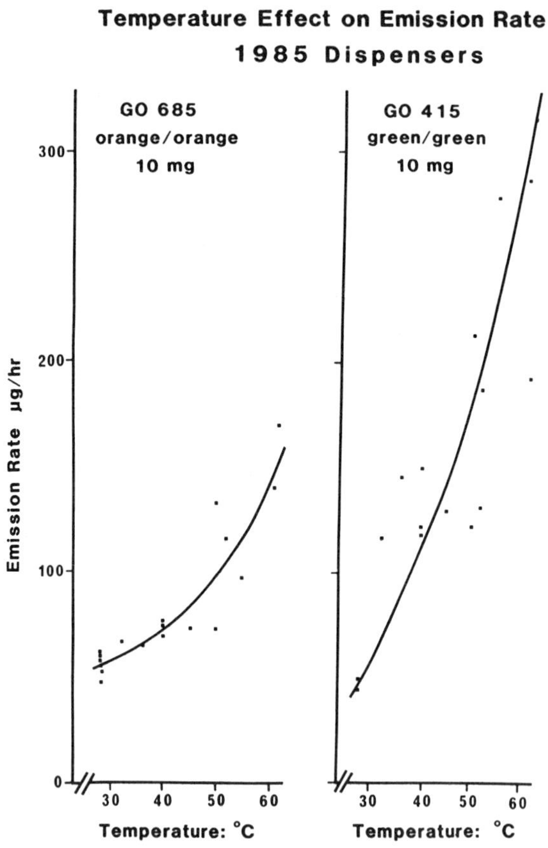

FIGURE 3 Effect of temperature on the emission rate of grandlure from the normal (left) and substitute (right) laminate dispensers.

This method of emission rate determination together with insect captures and residual pheromone content determinations led to a significantly improved laminate for the 1986 tests; this new laminate had outer PVC layers that were double the thickness of the original dispenser (13).

This new laminate and other commercial dispensers were evaluated in 1986 (14), and Table 2 shows the emission rates and the percentage of the total weevil capture for six types of dispensers at aging times of zero and 14 days in the field (Midville, Georgia, in May–June). It is apparent that four of the five commercial dispensers were better in

TABLE 2 Laboratory Measured Emission Rates from Dispensers Nominally Containing 10 mg of Grandlure and Aged 0 and 14 Days in 1986 Field Tests

Treatment[a]	Emission rate[b] (μg/hr) from dispensers aged		Total weevil capture (%) with dispensers aged	
	0 d	14 d	0–2 d	12–14 d
Laminate–Hercon	35	13	9.5	29.8
Membrane–Consep	88	7.1	14.2	20.1
Polymeric rod–AgriSense	18	6.0	17.2	18.0
PVC–Scentry	12	2.2	24.8	21.8
Cigarette filter	21	1.5	22.8	4.0
PVC–Fermone	0.7	0.5	11.5	6.3

[a]Hercon Laboratories, Inc., So. Plainfield, New Jersey; Consep Membranes, Inc., Bend, Oregon; AgriSense, Fresno, California, Scentry, Inc., Buckeye, Arizona; Fermone Chemicals, Inc., Phoenix, Arizona.
[b]Total of four grandlure components.

this test than the cigarette filter, which had been considered to be reliable.

The insect captures, normalized to those with fresh cigarette filters at each aging interval, are shown in Figure 4. The order of treatments is based on the insect captures at 14 days, with the laminate ranking first and the aged cigarette filter last. When the residual content dropped below about 2 mg by day 14, the reservoir of grandlure was considered insufficient to allow for variations in field conditions from one test to another and to assure reliable performance for 14 days.

Considering these results, preliminary performance criteria for grandlure dispensers were proposed as follows:

Content: 9.5 mg of AI initially and about 2 mg of AI after 14 days
Emission rate: equal decline in a regular pattern over 14 days
Bioassay: when aged for 0, 7, and 14 days, catch equal to, or greater than, 50% of that with a fresh cigarette filter

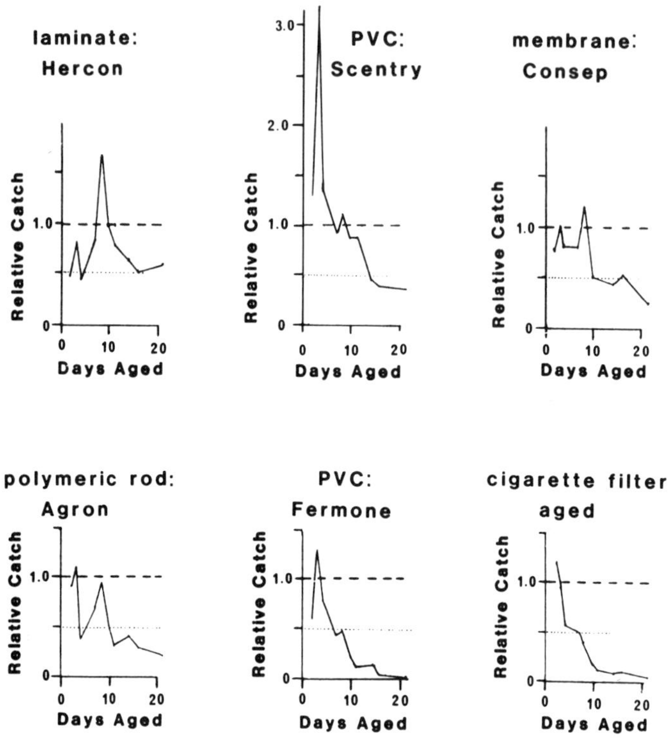

FIGURE 4 Boll weevil captures with commercial dispensers relative to those with reference, fresh cigarette filters. Dashed and dotted horizontal lines indicate equivalency to 100% and 50%, respectively, of the insect capture with the reference filters.

These criteria were included in the November 1987 competitive bid proposal by the Southeastern Boll Weevil Eradication Foundation for purchase of 7.5 million dispensers to be used in 1988 in the expanded eradication program. In addition, standards were proposed and recently used to purchase 88 kg of liquid grandlure.

C. Gypsy Moth

The pheromone of the gypsy moth is (+)-disparlure, (7*R*,8*S*)-*cis*-7,8-epoxy-2-methyloctadecane (15,16). For approximately the last 16 years, disparlure has been used in the State-APHIS cooperative survey program to detect and monitor gypsy moth populations in many parts of the country. The (+) enantiomer of disparlure has been used to bait traps since it became available in the late 1970s. The current dispenser, which is used to bait approximately 200,000 detection traps, is a 3 × 25-mm plastic laminate with 50-μm outer layers and containing 500 μg of (+)-disparlure in the middle polymer layer.

The specifications of this laminate resulted from a series of tests in which the number of insects trapped was used to select the appropriate size, film thickness, and lure concentration, keeping in mind the original high cost and limited availability of the pheromone at that time. These dispensers normally remain attractive throughout the gypsy moth flight season of 4 to 8 weeks. Routinely, to check the performance of each year's lot of dispensers, APHIS subjects the laminates to greenhouse aging for as long as 16 weeks before the insect flight; the elevated temperatures in the greenhouse accelerate the depletion of lure. At the start of gypsy moth flight, the aged dispensers are compared with the unaged laminates in a field bioassay and by chemical analyses (pheromone content and release rate). This procedure assures that year-to-year batches are active. Candidate dispensers from other manufacturers have been evaluated against a standard cotton wick dispenser freshly loaded with 100 μg of (+)-disparlure, and the results have been compared with the performance of the laminate. However, specifications or criteria for the design or performance of suitable dispensers have been lacking. This makes competitive procurement very difficult.

In an effort to develop preliminary performance criteria, the first step was to determine a dose–response relationship, using cotton wicks to deliver a range of doses (Fig. 5). In a series of eight field tests, moth captures were measured with amounts of (+)-disparlure ranging from 1 to 2500 μg/trap (17). The moth captures were normalized to those with 100 to 125 μg on a cotton wick, because this is the quantity that has been used as a reference dose for checking the activity of laminates (18). With amounts less than approximately 20 μg of (+)-disparlure, the moth captures were less than about 80% of the reference. Similarly, the capture generally declined with doses beyond 150 μg.

The emission rates from cotton wicks treated with various doses of (+)-disparlure were measured in the oven apparatus (13) at 35°C and 100 ml/min airflow. The range of doses on cotton wicks (20–600 μg) that yielded consistently high insect captures gave measured release rates of 140 to 600 ng/hr. Batches of the commercial laiminate dispensers from the 1983 to 1986 programs had initial emission rates from 130 to 190 ng/hr under laboratory conditions; this rate was just within the target range. However, after just 2 weeks of greenhouse aging, the emission rates dropped out of the optimum range. The

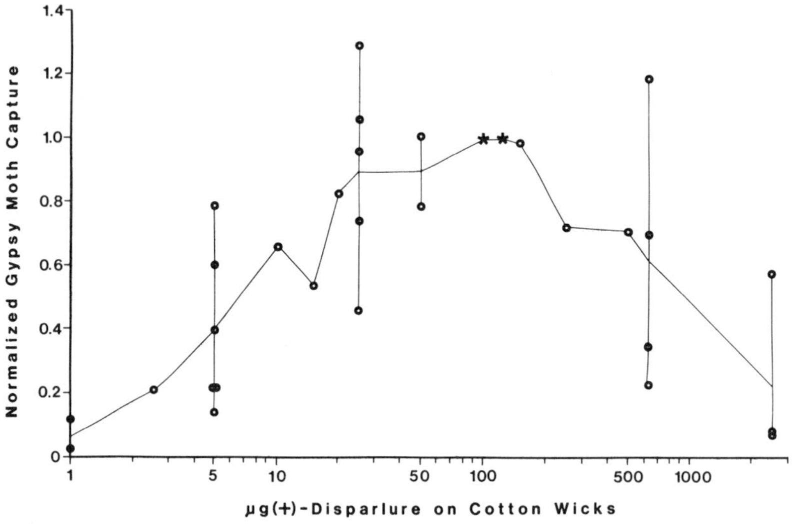

FIGURE 5 Dose–response relationship with gypsy moth capture expressed relative to that with the reference 100- or 125-µg dose on cotton wicks. Vertical lines show variation in results from multiple tests and the stars at 100 and 125 µg denote the reference doses.

emission rates, residual contents, and male captures, each highly correlate with aging times (Fig. 6). The data presented for the 1983 dispenser are representative of similar studies carried out with 1984 to 1986 dispensers. The decrease in pheromone with aging time follows first-order kinetics. The decline in release rate correlates ($r^2 = 0.91$) with the drop in the residual content over time.

Emission rates for all of the 1983 to 1986 laminates correlated with residual contents. Good, although not optimum, insect captures were obtained when the release rate from the laminate dispenser was as low as 50 ng/hr. Therefore, 50 ng/hr was proposed as the low or minimum acceptable threshold emission rate, and a residual content of approximately 130 µg in the laminate is needed to give this emission rate.

The greenhouse-aged laminates were above the threshold of 130 µg content and 50 ng/hr release rate for 5 to 9 weeks. Field-aged dispensers that were retrieved from traps after 11 to 14 weeks in the field (in Michigan, Virginia, West Virginia, Maine, Washington, Wisconsin, and Pennsylvania) had residual contents of about 300 µg and were thus above threshold. This assured high activity throughout

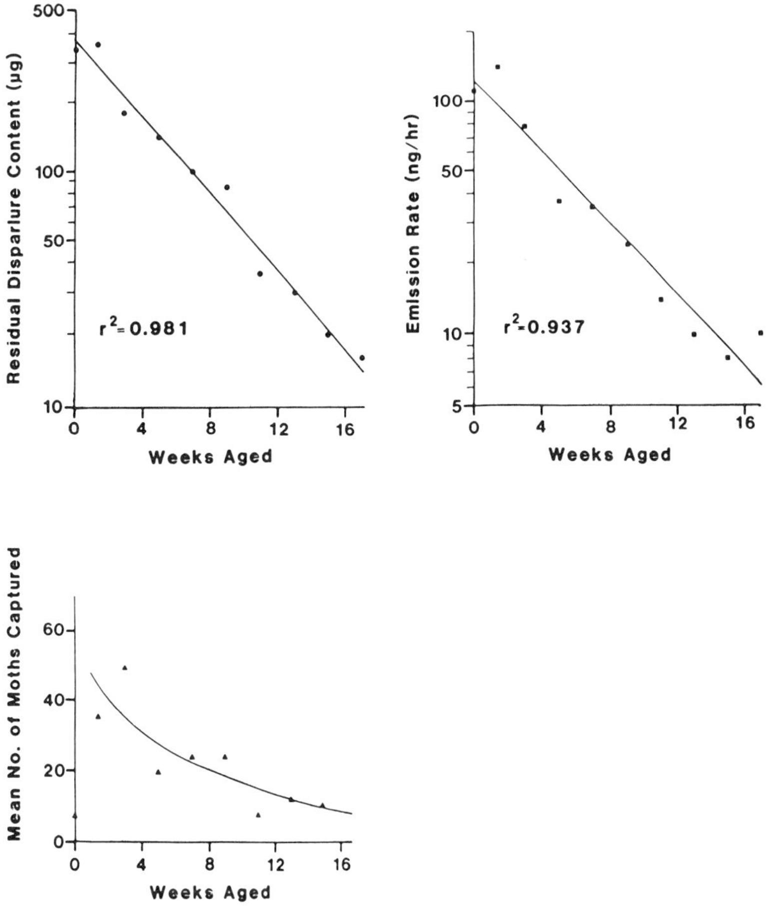

FIGURE 6 Residual disparlure content, emission rate (at 35°C), and gypsy moth capture as a function of duration of greenhouse aging; 1983 laminate dispensers.

the gypsy moth flight. The difference between the two types of aging--greenhouse and field—is primarily the temperature with, of course, the greenhouse temperature being elevated.

The emission rate of the dispensers should fall in the range of 50 to 600 ng/hr for the entire period of their field use. The laminate is a very effective dispenser, but an increase in dose from 500 μg to 1000 μg is suggested to get the initial emission rate well within the optimum range and to assure reliable and consistent performance throughout the flight season. Other commercial dispensers may perform just as effectively as the laminate, and we now have the data to make preliminary judgments and to give manufacturers the targets that they need to design competitive dispensers.

III. CONCLUSIONS

Controlled-release dispensers for the attractants of three major insect pests have been critically examined for many of the following variables:

Active ingredient: purity, amount, ratio of components
Biological assay: effect of aging
Release of active ingredient: rate, composition, residual quantities with aging time
Physical characteristics: size, durability, ease of handling, shelf-life, packaging

On the basis of laboratory and field data, performance criteria are proposed for use in dispenser procurement by the competitive bid process. It is anticipated that these criteria will be used in the decision process.

Table 3 summarizes the planned use of dispensers in 1988 field programs. The dose per dispenser for the three insects ranges from

TABLE 3 Current Dispensers

Dispensers for trapping	Dose (mg) per dispenser	Approx. no. of dispensers for 1988	Status of performance criteria in competitive bids
Mediterranean fruit fly	2000	250,000	Used in 1986
Boll weevil	10	7,500,000	Used in 1987
Gypsy moth	0.5	250,000	Future

0.5 mg to 2000 mg of attractant. Performance criteria for these dispensers are important to assure desired activity, obtain comptetitive prices, and serve as reliable targets for dispenser development by the industry. The Southeastern Boll Weevil Eradication Program is currently the largest user of pheromone dispensers (see Chap. 26).

ACKNOWLEDGMENT

The authors wish to gratefully acknowledge the assistance of R. E. Rice in aging dispensers in California and of D. Y. Sada, T. Urago, and E. C. Paszek in conducting the bioassays. We also wish to thank E. D. DeVilbiss and E. M. Harte for performing the chemical analyses.

DISCLAIMER

This article reports the results of research only. Mention of a proprietary product does not constitute an endorsement or a recommendation for its use by the USDA.

REFERENCES

1. Beroza, M., Green, N., Gertler, S. I., Steiner, L. F., and Miyashita, D. N. Insect attractants. New attractants for the Mediterranean fruit fly. *J. Agric. Food Chem.* *9*:361–365 (1961).
2. McGovern, T. P., Beroza, M., Ohinata, K., Miyashita, D., and Steiner, L. F. Volatility and attractiveness to the Mediterranean fruit fly of trimedlure and its isomers, and a comparison of its volatility with that of seven other insect attractants. *J. Econ. Entomol.* *59*:1450–1455 (1966).
3. McGovern, T. P. and Beroza, M. Structure of the four isomers of the insect attractant trimedlure. *J. Org. Chem.* *31*:1472–1477 (1966).
4. McGovern, T. P., Cunningham, R. T., and Leonhardt, B. A. Attractiveness of *trans*-trimedlure and its four isomers in field tests with the Mediterranean fruit fly (Diptera: Tephritidae). *J. Econ. Entomol.* *80*:617–620 (1987).
5. Leonhardt, B. A., Rice, R. E., Harte, E. M., and Cunningham, R. T. Evaluation of dispensers containing trimedlure, the attractant for the Mediterranean fruit fly (Diptera: Tephritidae). *J. Econ. Entomol.* *77*:744–749 (1984).

6. Rice, R. E., Cunningham, R. T., and Leonhardt, B. A. Weathering and efficacy of trimedlure dispensers for attraction of Mediterranean fruit flies (Diptera: Tephritidae). *J. Econ. Entomol.* 77:750–756 (1984).
7. Leonhardt, B. A., Cunningham, R. T., Rice, R. E., Harte, E. M., and McGovern, T. P. Performance of controlled release formulations of trimedlure to attract the Mediterranean fruit fly, *Ceratitis capitata*. *Entomol. Exp. Appl.* *44*:45–41 (1987).
8. Leonhardt, B. A., Cunningham, R. T., Rice, R. E., Hendrichs, J., and Harte, E. M. Design, effectiveness and performance criteria of dispenser formulations of trimedlure, the attractant of the Mediterranean fruit fly (Diptera: Tephritidae). *J. Econ. Entomol.* (in press).
9. Harris, E. J., Nakagawa, S., and Urago, T. Sticky traps for detection and survey of three tephritids. *J. Econ. Entomol.* *64*:62–65 (1971).
10. Tumlinson, J. H., Hardee, D. D., Gueldner, R. C., Thompson, A. C., Heden, P. A., and Minyard, J. P. Sex pheromones produced by male boll weevils: Isolation, identification and synthesis. *Science* *166*:1010–1012 (1969).
11. Ridgway, R. L., Dickerson, W. A., Brazzel, J. R., Leggett, J. F., Lloyd, E. P., and Planer, F. R. Boll weevil trap captures for treatment thresholds and population assessments. In *Proc. Beltwide Cotton Prod. Res. Conf.* 1985, pp. 138–141.
12. McKibben, G. H., Johnson, W. L., Edwards, R., Kotter, E., Kearny, J. F., Davich, T. B., Lloyd, E. P., and Ganyard, M. C. A polyester-wrapped cigarette filter for dispensing grandlure. *J. Econ. Entomol.* *73*:250–251 (1980).
13. Leonhardt, B. A., Dickerson, W. A., Ridgway, R. L., and DeVilbiss, E. D. Laboratory and field evaluation of controlled release dispensers containing grandlure, the pheromone of the boll weevil (Coleoptera: Curculionidae). *J. Econ. Entomol.* *81*:737–943 (1988).
14. Dickerson, W. A., Leonhardt, B. A., and Ridgway, R. L. Field bioassay and laboratory chemical analysis of controlled-release dispensers for the pheromone of the boll weevil, *Anthonomus grandis* Boheman. In *Proceedings of the 14th International Symposium on Controlled Release of Bioactive Materials.* (Lee, P. I. and Leonhardt, B. A., eds.). Controlled Release Society, Lincolnshire, Ill., 1987, pp. 249--250.
15. Bierl, B. A., Beroza, M., and Collier, C. W. Potent sex attractant of the gypsy moth: Its isolation, identification and synthesis. *Science* *170*:87–89 (1970).

16. Yamada, M., Saito, T., Katagiri, K., Iwaki, S., and Marumo, S. Electroantennogram and behavioral responses of the gypsy moth to enantiomers of disparlure and its *trans* analogues. *J. Insect Physiol.* *22*:755–761 (1976).
17. Leonhardt, B. A., Mastro, V. C., Paszek, E. C., Schwalbe, C. P., and DeVilbiss, E. D. Dependence of gypsy moth (Lepidoptera: Lymantriidae) capture on pheromone dose from cotton and laminate dispensers: Performance criteria. *J. Chem. Ecol.* (in preparation).
18. Schwalbe, C. P. Disparlure-baited traps for survey and detection. In *The Gypsy Moth: Research Toward Integrated Pest Management.* (Doane, C. C. and McManus, M. L., eds.). Washington, D.C., Technical Bull. 1584, U.S. Dept. of Agriculture, 1981, pp. 542–549.
19. Duncan, D. B. Multiple range and multiple F tests. *Biometrics* *11*:1–42 (1955).

8

Olefin Metathesis as an Economical Route to Insect Pheromones

DENNIS S. BANASIAK* and JAMES D. BYERS† / Provesta Corporation, Bartlesville, Oklahoma

I. INTRODUCTION

The preparation of insect pheromone chemicals on a commercial scale is a challenging problem. A wide variety of chemical structures have been identified in the multitude of pheromone compositions now known. These chemicals vary from simple hydrocarbons to complex molecules containing asymmetric centers (Fig. 1; 1). This paper will address the commercial synthesis of olefinic compounds that are found in the pheromones of many economically important pests (Table 1).

Pheromone chemicals, in general, are relatively expensive, often selling for dollars per gram and more. These prices are substantially higher than traditional agricultural active ingredients. Three reasons are important contributors to this high cost. First, pheromone chemicals have typically been prepared by scaling-up laboratory synthetic methods (Fig. 2; 2–4). Unfortunately, these methods often use expensive reagents in multiple-step procedures and are not conducive to large-scale commercial production.

The mode of action for use of pheromones in insect control is indirectly a second contributor to the high cost of pheromone production.

**Current affiliation:* AgriSense, Fresno, California. (Note: AgriSense was formed in 1988 as a joint venture between Provesta [Phillips Petroleum] and Dow Corning.)

†*Current affiliation*: Phillips Petroleum Corporation, Bartlesville, Oklahoma.

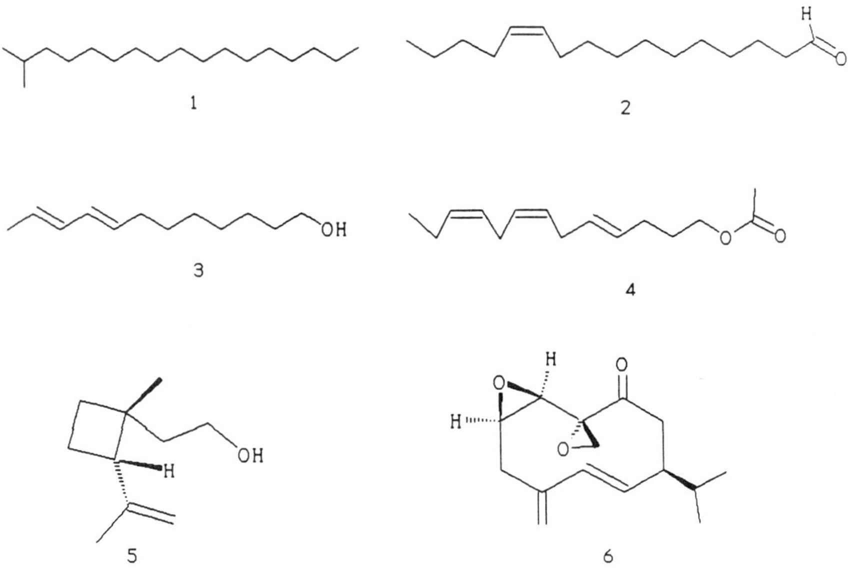

FIGURE 1 Insect pheromone components.

The specificity with which insects respond to pheromones, with each species responding to a different chemical blend, virtually ensures that only small volumes of any single pheromone chemical will be required. Finally, the high degree of purity required to mimic duplicate natural pheromone blends, demands that additional purification steps be included in the manufacture of these chemicals. Thus, pheromone chemicals are, by their very nature, low-volume materials that are manufactured during sophisticated chemistry to exacting specifications. It is not surprising that most pheromones are expensive.

One of the many questions facing the pheromone industry is whether or not pheromone chemicals can be manufactured at prices that allow grower use, while at the same time providing chemical manufacturers and formulators a reasonable rate of return. Until this question is successfully addressed, pheromone technology will never achieve wide-spread use as an insect control method.

II. DISCUSSION

Phillips Petroleum Company and its subsidiary Provesta have addressed the commercial preparation of pheromones by combining inhouse specialty chemical production expertise and proprietary research in olefin synthesis technology. The key to the Provesta/Phillips approach

TABLE 1 Pheromones of Economically Important Pests

"Key" pheromone	Insect common name (*scientific name*)
(*Z*)-11-Hexadecenal	Corn earworm (*Heliothis zea*)
(*Z*)-11-Hexadecenal	Tobacco budworm (*Heliothis virescens*)
(*E*,*E*)-8,10-Dodecadien-1-ol	Codling moth (*Cydia pomonella*)
(*Z*)-8-Dodecen-1-ol acetate	Oriental fruit moth (*Grapholitia molesta*)
(*Z*,*E*)-7,11-hexadecadien-1-ol acetate	Pink bollworm (*Pectinophora gossypiella*)
(*Z*)-9-Tricosene	Housefly (*Musca domestica*)
(*Z*,*E*)-9,12-Tetradecadien-1-ol acetate	Indian meal moth (*Plodia interpunctella*)
(*Z*)-9-Dodecen-1-ol acetate	European grapeberry moth (*Eupoecilia ambiguella*)

is a catalytic reaction known as olefin metathesis. First discivered in the early 1960s, metathesis has grown to significant industrial importance. Metathesis is used commercially to prepare high-purity ethylene and butenes in the Phillips Triolefin Process (5). Fine chemicals such as Neohexane (3,3-dimethyl-2-butene), a key component in the preparation of bicyclic fragrance musks, are also manufactured commercially by metathesis (6).

Metathesis is a transition metal-catalyzed chain reaction in which the carbon–carbon double bonds are fragmented and then recombined. The net result is the synthesis of two new olefins (Fig. 3). This reaction has been applied to the synthesis of pheromones using metathesis catalysts proprietary to Phillips (7). Thus, it is now possible to prepare long-chain olefinic pheromones or pheromone precursors in one or two steps from commercially available materials.

Commercially available materials, such as oleic acid derivatives or 9-decenyl acetate, can be reacted with α-olefins to yield pheromones (Fig. 4). Specifically, the reaction of 9-decenyl acetate with various α-olefins to produce a series of 9-substituted acetates has been examined (Fig. 5). The chain length of the product is determined by the α-olefin chain lengths. Preparation of 9-dodecenyl acetate is one prime example. This product is a pheromone component

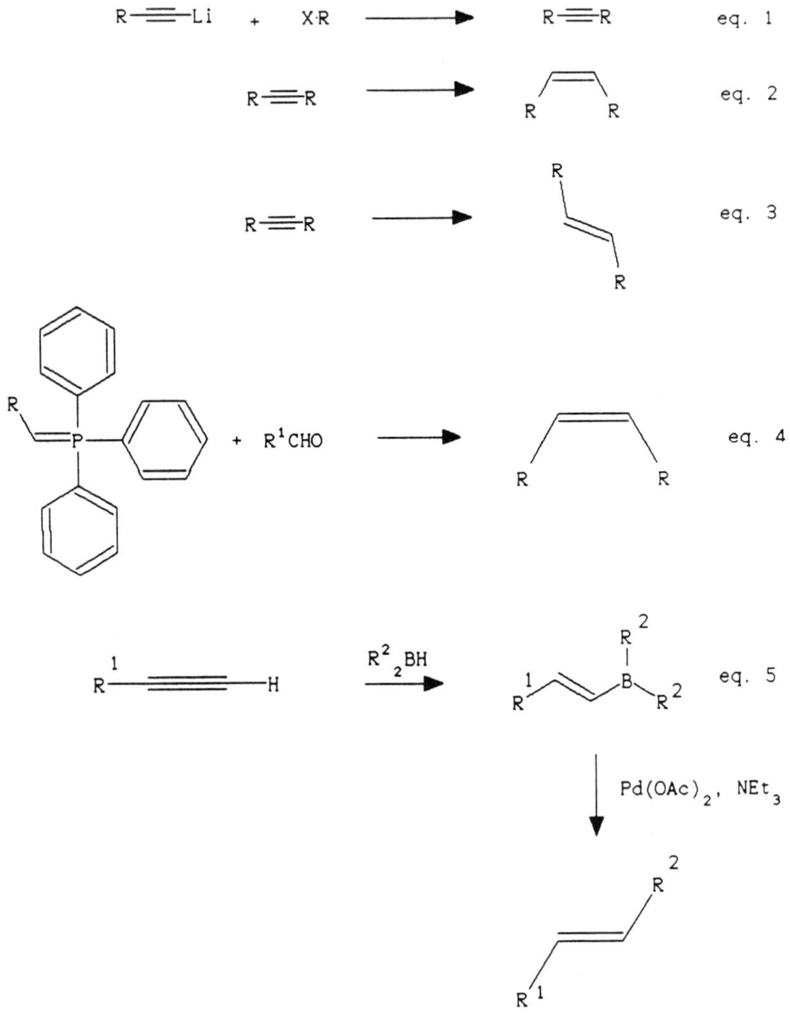

FIGURE 2 General laboratory preparations of pheromones.

of the western pine shoot borer, the Nantucket pine tip moth, the grape berry moth, and the European grape berry moth.

9-Dodecenyl acetate is produced in 21% yield by the reaction of 1-butene with 9-decenyl acetate. The stereochemistry of the olefins

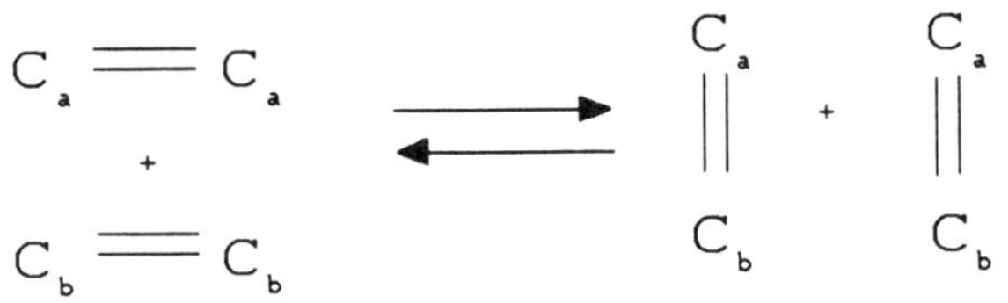

Propylene Self-Metathesis

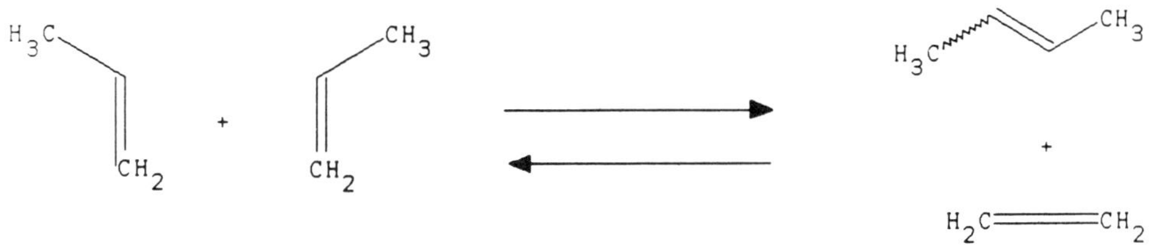

FIGURE 3 Olefin metathesis (disproportionation).

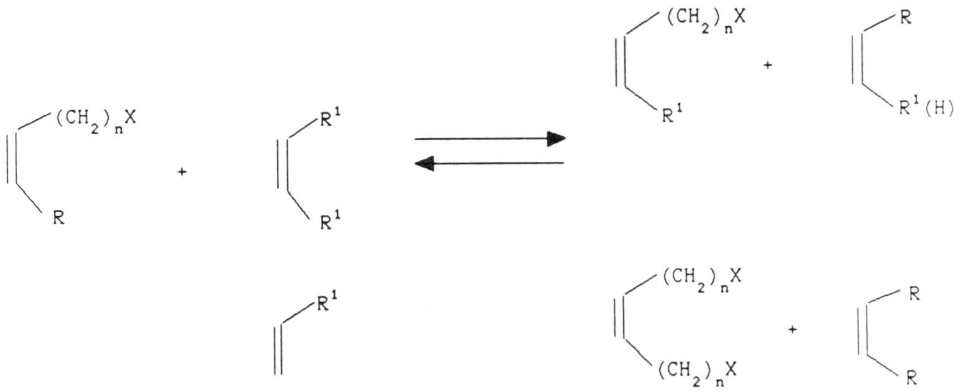

$X = OAc, CO_2R$

R, R^1 = alkyl or H

FIGURE 4 Pheromone synthesis by cross-functional olefin metathesis.

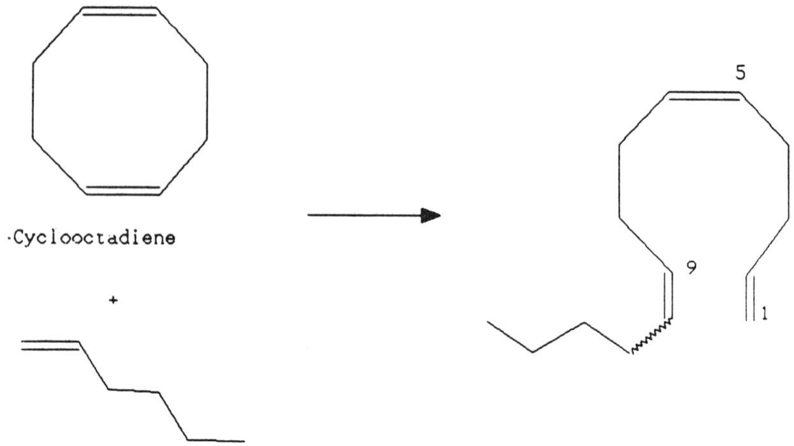

FIGURE 5 Metathesis of 1,5-COD with 1-hexene.

produced in this reaction is a function of reaction conditions and can be varied from an *E*/*Z* ratio of <2:1 to >5:1 (7).

The question of the actual commercial utility of metathesis pheromones depends on whether or not the *E* enriched mixtures can be used in insect control. This is particularly true when *Z* isomers predominate in pheromone mixtures. We have addressed, and continue to address, this question through actual bioassays and field trials. Many of these trials have been quite successful for disruption of insect mating. Specifically, the mating disruption of the western pine shoot borer using a 2:1 *E*/*Z* 9-dodecenyl acetate mixture on 200-acre plots results in greater than 80% damage reduction. Thus, metathesis-produced pheromones, although not always available in the natural ratio, can offer a workable economic alternative to the more expensive procedures.

Metathesis can also be combined with conventional synthetic methodologies to produce pheromones in fewer steps and often more economically. One example is the production of the pink bollworm pheromone gossyplure, a 50:50 mixture of (7*Z*,11*Z*)- and (7*Z*,11*E*)-hexadecadienyl acetate (8). Gossyplure has been the target of numerous academic and industrial syntheses (9–11). As in the synthesis of other pheromones, these gossyplure routes suffer from multiple steps (more than five) and the use of expensive reagents or intermediates.

Recently, Phillips developed a two-pot synthesis of gossyplure utilizing metathesis chemistry to supply the needed *Z*/*E* mixture at

TABLE 2 Metathesis of 9-Decenyl Acetate with Terminal Olefins

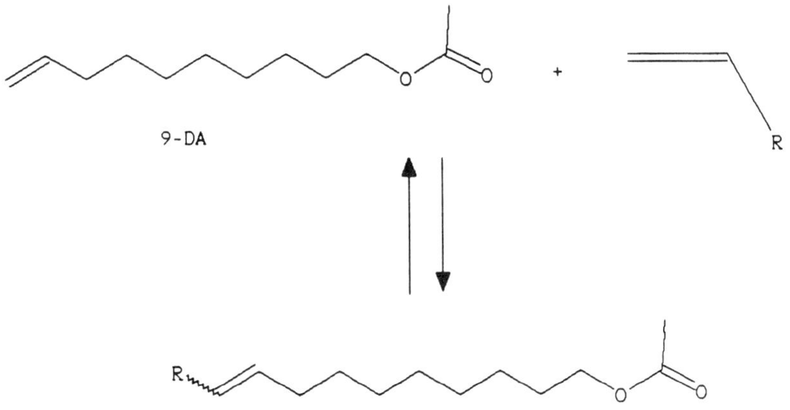

R = 1,2,3,4.6

Olefin	Yield (%)	*E/Z*	Cross-Product
1-$C_3^=$	20	1.3	9-$C_{11}^=$ OAc
1-$C_4^=$	21	1.6	9-$C_{12}^=$ OAc
1-$C_5^=$	14	1.7	9-$C_{13}^=$ OAc
1-$C_6^=$	16	1.7	9-$C_{14}^=$ OAc
1-$C_8^=$	33	1.8	9-$C_{16}^=$ OAc

the C-11 double bond. The synthesis utilized the relatively inexpensive starting materials, 1,5-cyclooctadiene and 1-hexene, to give a C_{14} triene, 1,(5*Z*),(9*E*/*Z*)-tetradecatriene (see Fig. 5). The metathesis reaction gives a molecule with one double bond having all *Z*-(*cis*) stereochemistry (C-5,C-6), one double bond having roughly a 50:50 *E*/*Z* stereochemistry (C-9,C-10), and a vinyl double bond that can be selectively coupled with a C-2 synthon to give gossyplure.

The ring opening of 1,5-cyclooctadiene (COD) with 1-hexene in the presence of a metathesis catalyst leaves 100% cis stereochemistry of one double bond in 1,5-COD untouched. The *cis*-C-5,C-6 double bond of 1,5,9-tetradecatriene can then be translated into the C-7, C-8 double bond of gossyplure, which must have the all-cis configuration. The double bond of 1,5-COD undergoing metathesis produces a vinyl double bond (C-1,C-2) and an internal double (C-9,C-10) bond with a mixed stereochemistry (∿50:50 *E/Z*). The stereochemistry at C-9,C-10 of 1,5,9-tetradecatriene can then be translated into the desired stereochemistry at the C-11,C-12 double bond of gossyplure.

Depending on the metathesis catalyst chosen, the conditions of the reaction, and so forth, the stereochemistry at the C-9,C-10 double bond of 1,5,9-tetradecatriene can be varied from 40:60 *Z/E* to 60:40 *Z/E* (12). Recent studies have shown that the optimum isomer ratio for gossyplure may be 60:40 *Z/E* rather than 50:50 (see Chap. 25).

The metathesis reaction of 1,5-COD with 1-hexene allows the formation of the correct stereochemistry of the two unsaturated sites in gossyplure in one easy step from inexpensive starting materials. Other published procedures require no less than two separate steps to accomplish the same result.

To take advantage of the metathesis chemistry in the preparation of gossyplure, one must be able to selectively react the terminal double bond of 1,5,9-tetradecatriene without affecting the stereochemistry of the other two internal unsaturated centers. Examples of such selectivity are seen with certain metal alkyl and metal hydride reagents (13). 1,5,9-Tetradecatriene can be converted to gossyplure by the addition of a C_2 synthon as shown in Figure 6.

Provesta currently uses this technology to produce gossyplure in the hundreds of kilogram scale in a semicommercial facility dedicated to the production of pheromones.

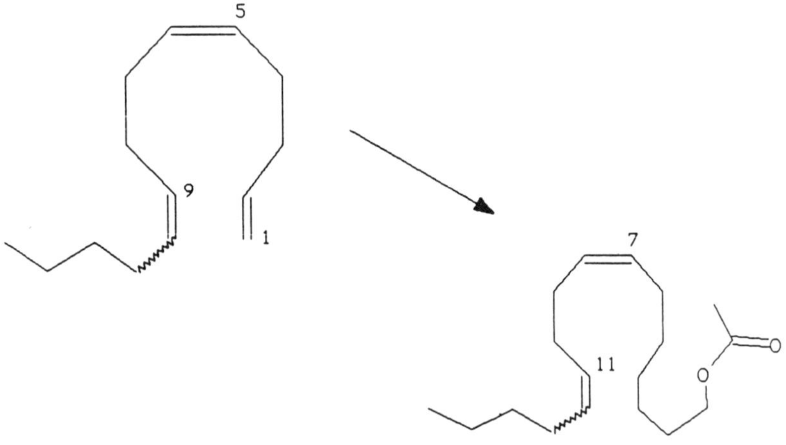

FIGURE 6 Conversion of 1,5,9-TDT to gossyplure.

III. CONCLUSION

The philosophy of Phillips/Provesta of combining their specialty chemical expertise and metathesis technology has lowered the cost of producing pheromone chemicals. We continue to study these and other methods in an attempt to further improve these processes. It is our firm belief that pheromones have a future in insect control. Departures from traditional thinking may be necessary to accomplish this goal. One example of this is the use of off-isomeric mixtures to disrupt certain insect communications. Field trials and bioassays have shown that off-isomeric mixtures can be effective in protecting crops from insect damage.

REFERENCES

1. Inscoe, M. N. Insect attractants, attractant pheromones, and related compounds. In *Insect Suppression with Controlled Release Pheromone Systems*, Vol. II. (Kydonieus, A. F. and Beroza, M., eds.). CRC Press, Boca Raton, 1982, pp. 201–195.
2. Mori, T. The synthesis of insect pheromones. In *The Total Synthesis of Natural Products*, Vol. 5. (ApSimon, J. W., ed.). John Wiley & Sons, New York, 1983, pp. 1–183.2.
3. Rossi, R. Insect Pheromones. 1. Synthesis of achiral components of insect pheromones. *Synthesis ??*:817–836 (1977).
4. Henrick, C. A. The synthesis of insect sex pheromones. *Tetrahedron 33*:1845–1889 (1977).
5. Banks, R. L. Olefins from olefins. *ChemTech 9*:494–501 (1979).
6. Banks, R. L., Banasiak, D. S., Hudson, P. S., and Norell, J. R. Specialty chemicals via olefin metathesis. *J. Mol. Catal. 15*:21–33 (1982).
7. Banasiak, D. S. Insect pheromones from olefin metathesis. *J. Mol. Catal. 28*:107–115 (1985).
8. Banasiak, D. S., Mozdzen, E. C., and Byers, J. D. Preparation of gossyplure. U.S. Patent No. 4,609,498. Sept. 1986.
9. Anderson, R. J. and Henrick, C. A. Stereochemical control in Wittig olefin synthesis: Preparation of the pink bollworm sex pheromone mixture, gossyplure. *J. Am. Chem. Soc. 97*:4327–4334 (1975).
10. Bestman, H. J., Koschatzky, K. H. Stransky, W., and Vostrowsky, O. Pheromone IX. Stereoselektive Synthesen von (*Z*)-7,(*Z*)-11- und (*Z*)-7,(*E*)-11-Hexadecadienylacetat, dem Sexualpheromone von *Pectinophora gossypiella* (Gelechiidae, Lep.). *Tetrahedron Lett.* pp. 353–356 (1976).
11. Ishihara, T. and Yamamoto, A. Novel synthesis of alkynyl halides by a Grignard coupling reaction with *alpha, omega*-

dibromo-1-alkyne: Synthesis of (*Z*,*Z*)- and (*Z*,*E*)-hexadecadienyl acetate: A sex pheromone of pink bollworm (*Pectinophora gossypiella*). *Agric. Biol. Chem.* *48*:211–213 (1984).
12. Byers, J., Banasiak, D. S., Drake, C. A., and Welch, M. W. (unplished results, 1986).
13. Negishi, E.-I. *Organometallics in Organic Synthesis: General Discussion and Organometallics of Main Group Metals in Organic Synthesis*. Vol. 1. John Wiley & Sons, New York, 1980, 532 pp.

9

Commercial Synthesis of Pheromones and Other Attractants

JEFFREY J. SLOCUM* / Orsynex Inc., Clifton, New Jersey

I. INTRODUCTION

Orsynex Incorporated was founded in March 1983, when the Essex Chemical Corporation purchased a chemical manufacturing facility in Columbus, Ohio, which was originally built to produce only pheromones and attractants. Today, we have the technical capability and capacity to produce over 50 such pheromones and attractants. Our experience in manufacturing these products dates back 18 years.

The manufacturing concept of producing only pheromones and attractants could not economically support this facility because most such products do not have markets large enough to justify full-scale commercial production. When the plant was put up for sale, Essex Chemical viewed the facility as a favorable acquisition because:

The plant was located on a 50-acre rural site in Columbus, Ohio, and was environmentally sound.
The facility was relatively new (built and commissioned in 1979–1980), with state-of-the-art equipment.
The facility was FDA-approved for pharmaceuticals and pharmaceutical intermediates manufacturing.
The facility had in place the required permits for pesticide production.

Since the plant acquisition in 1983, Essex Chemical has expanded the facility in several different ways.

The plant site is now 150 acres, and the largest equipment is 4000 gal. Previously, the largest equipment was 1500 gal.

**Current affiliation*: Marlborough Chemicals Inc., Charlotte, North Carolina

Numerous pieces of solids-handling equipment have been installed.
Previously the facility could only manufacture liquid products.
A full-scale research and development laboratory and a quality-control laboratory were constructed immediately after the acquisition.
Other expansions include drum storage facilities, in-process tank farms, and a separate manufacturing facility for pesticides.
Planned expansions include a new warehouse and pilot-plant facility.

Currently, Orsynex sells products in three separate market areas. This strategy has made this facility a sound economic venture. The market areas are agricultural products, pharmaceuticals and intermediates, and industrial specialties each of which is described below.

Agricultural products: contract bulk pesticides and intermediates production for major agricultural companies, proprietary intermediates, pheromones and attractants
Pharmaceuticals and intermediates: contract bulk pharmaceuticals and intermediates production for major drug companies, generic drugs and intermediates using proprietary technology
Industrial specialties: monomers for high-technology polymer applications, cross-linking agents, Grignard reagents, and flame retardants

Being a chemical manufacturer, Orsynex is monitored and regulated by the following governmental agencies and acts:

EPA: Environmental Protection Agency
FIFRA: Federal Insecticide, Fungicide, Rodenticide Act
TSCA: Toxic Substances Control Act
RCRA: Resource Conservation Recovery Act
SARA: Superfund Amendment and Reauthorization Act
OSHA: Occupational Safety and Health Administration
FDA: Food and Drug Administration

The EPA controls and regulates all chemical production in this country apart from pharmaceutical production. The FIFRA controls pesticide regulation and monitors pesticide production.

Under TSCA, all new chemical production is registered with the EPA. Under this act, any new chemical produced in this country in quantities larger than research level must be registered with the EPA through a premanufacturing notification (PMN). Pheromone users need to understand this act, particularly when investigating use of a new pheromone. If the specific product had never been produced on something larger than a research scale, a PMN submittal could add dramatically to the lead time necessary to get the product to market. Also, any intermediate isolated during the manufacturing process may need to be registered through a PMN as well.

RCRA monitors and regulates disposal of hazardous chemical waste. Pheromones can generate both solid and liquid waste. Pheromone users must understand that chemical manufacturers are subject to tremendous liabilities if wastes are not disposed of properly. These liabilities could conceivably be extended to the customer who purchased the product that generated that waste. To limit the risk of liability, pheromone users should obtain their products from reliable, reputable firms.

A major part of SARA is the development of community right-to-know programs. The state of Ohio and the city of Columbus are leaders in getting these programs implemented.

OSHA is concerned with worker safety and standards in the work place. At Orsynex, both chemical and mechanical standards are monitored and met.

The FDA is strictly concerned with the manufacturing of pharmaceuticals and their intermediates. The primary concerns are cleanliness and good manufacturing practices (GMP).

On a local level, the city of Columbus Division of Sewage and Drainage monitors and regulates effluent from the plant site. The fire department regulates chemical storage.

II. PHEROMONE MANUFACTURING

Listed here are chemical structures for the pheromones of some common pests:

Pest	Structure
Gypsy moth	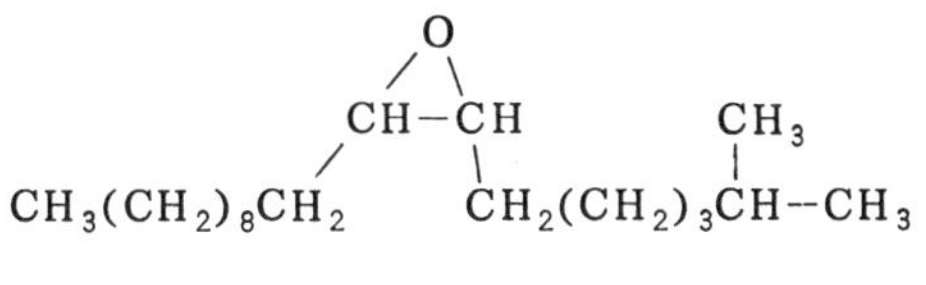 $CH_3(CH_2)_8CH_2\overset{\diagup O \diagdown}{CH-CH}CH_2(CH_2)_3\overset{CH_3}{\underset{|}{CH}}-CH_3$
Greater peach tree borer	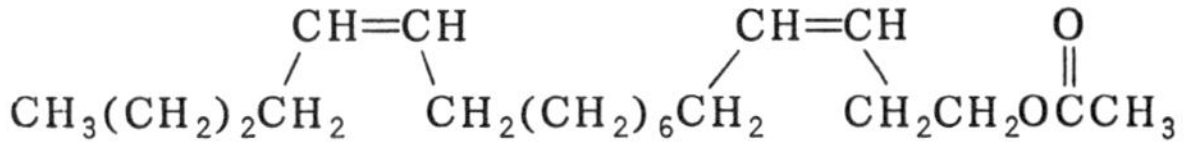 $CH_3(CH_2)_2CH_2\overset{CH=CH}{\diagup \quad \diagdown}CH_2(CH_2)_6CH_2\overset{CH=CH}{\diagup \quad \diagdown}CH_2CH_2O\overset{O}{\overset{\|}{C}}CH_3$
Artichoke plume moth	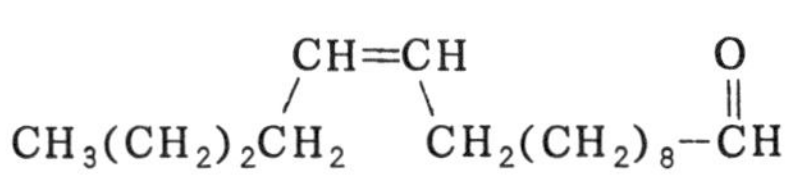 $CH_3(CH_2)_2CH_2\overset{CH=CH}{\diagup \quad \diagdown}CH_2(CH_2)_8-\overset{O}{\overset{\|}{C}}H$

Pheromones are generally complex chemicals. Producing them requires multistep reactions and the use of difficult reagents and solvents. Some examples are sodium amide, lithium amide, sodium acetylide, Grignard reagents, and solvents such as tetrahydrofuran and diethyl ether.

Here is an example of a typical pheromone synthesis:

Typical Pheromone Synthesis

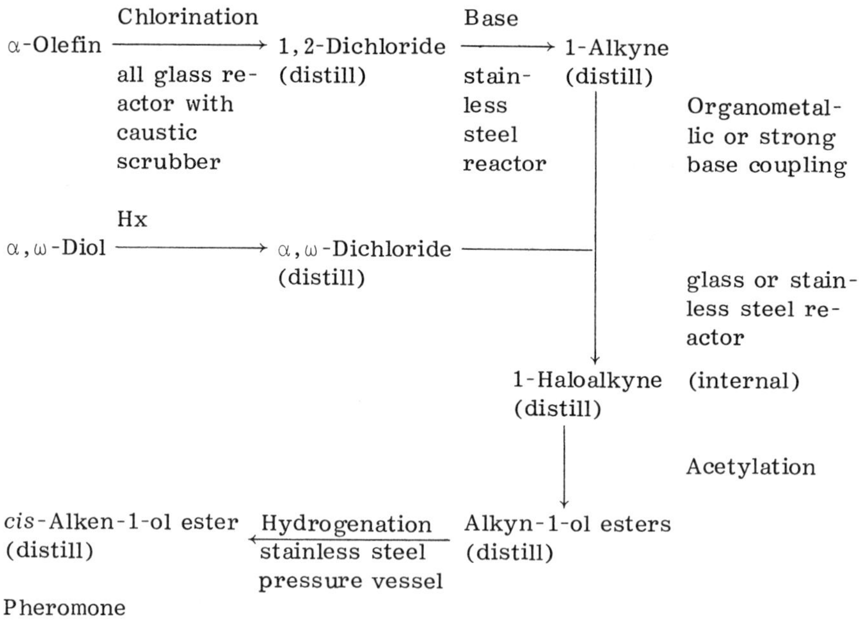

This sequence shows six reaction steps. For four of those reactions, we have identified specific equipment needs. Also, six individual distillations are needed. The figure is a typical example of custom chemical synthesis, in general, as well as pheromone synthesis, in particular. Numerous reactions take place and numerous pieces of equipment are needed. At Orsynex, a wide range of equipment is available for pheromone production. Equipment scale ranges from several hundred gallons to 4000 gal. Material of construction is glass-lined or stainless steel. Some equipment can be operated at 300 psi pressure. A hot-oil system can be used to heat reactors or stills and operates up to 250°C. All the equipment at Orsynex is state of the art and relatively new.

III. MANUFACTURING TIME

The length of time needed to supply a client with a pheromone or chemical depends on three factors: raw material acquisition, actual production time, and equipment scheduling.

Pheromones, being complex chemicals, often require complex raw materials. These raw materials may take several weeks or months to obtain, which will cause an equal delay in starting a pheromone synthesis.

Another factor in manufacturing time is production time. In our typical pheromone synthesis, we identified six reaction steps, and six distillations. Each reaction step may take from several days to a week, and each distillation may take from 1 to 2 days. So for this synthesis, the manufacturing time of 3 weeks to possibly 2 months could be required.

A large part of manufacturing time is equipment scheduling. In the previous paragraph, we identified a manufacturing time from several weeks to 2 months for six reactions and six distillations. That manufacturing time assumed, for each reaction and each distillation, that the specific equipment needed would be available at the appropriate time. When a custom chemical plant produces many products, delays frequently occur from "waiting" for specific equipment to become available.

IV. MANUFACTURING COSTS

At Orsynex, manufacturing costs for a pheromone can be attributed to the five following areas:

Equipment, labor, and overhead
Yields
Volume
Waste charges
Purity

Manufacturing charges for equipment include cost of the equipment, its installation, and its maintenance. Labor refers to the work force involved in operating the equipment. Overhead is a general term that refers to all support functions needed to operate manufacturing equipment. At Orsynex, this includes quality control, research and development, a safety department, an environmental department, management, supervision, and technical service.

Yields play an important part in the manufacturing cost of the pheromone. Here we have identified overall yields for three reaction sequences from raw material A to product E.

Yields	Overall yield (%)
1. A $\xrightarrow{90\%}$ B $\xrightarrow{90\%}$ C $\xrightarrow{90\%}$ D $\xrightarrow{90\%}$ E	65.5
2. A $\xrightarrow{50\%}$ B $\xrightarrow{50\%}$ C $\xrightarrow{50\%}$ D $\xrightarrow{50\%}$ E	6.3
3. A $\xrightarrow{90\%}$ B $\xrightarrow{50\%}$ C $\xrightarrow{10\%}$ D $\xrightarrow{90\%}$ E	4.1

When comparing the overall yields of these three reactions, reaction 3 is 50% more costly than reaction 2 on a fixed-cost basis. Likewise, reaction 2 is 10 times more costly than reaction 1 on a fixed-cost basis. Also, products manufactured by low-yield reactions carry great risk for the manufacturer, particularly when a single batch of the product is produced. The risk comes from off-yield production or batch failures. If either of these happens, the manufacturer will have nothing to sell and will lose the investment in manufacturing time and raw materials. Furthermore, the ability to reprocess off-grade material at any of the process steps can be hampered by scheduling conflicts discussed earlier.

In almost all chemical processes, with few exceptions, the lower the volume produced, the higher the unit cost will be. With lower volume, smaller equipment is used, or perhaps larger equipment is used at a fraction of capacity. Either situation will lead to lower throughput. The lowest unit cost for any pheromone or chemical is obtained when volumes are purchased such that larger equipment is used at full capacity. The larger the volume, the lower the price.

Charges associated with the disposal of chemical waste have become an important part of the manufacturing cost for any chemical or pheromone. In the last year alone, specific costs and charges for disposing of waste have tripled. At Orsynex, an ongoing process development program will limit the generation of chemical waste. The less the amount of waste, the lower the waste disposal charges will be. Waste disposal charges are escalating and are an important part of the manufacturing cost of any pheromone or chemical.

The need for a high-purity product can greatly affect the cost of the chemical. In some cases, several distillations may be needed to obtain a 98% or 99% pure product. Furthermore, the manufacturer may experience low yield. A 50% yield on distillation results in a doubled cost for the product. Low-yield distillations are typical in converting 95% to 97%-pure material to higher-purity products.

V. QUALITY CONTROL

For the pheromone industry, the ultimate test of performance for a chemical is the bioassay. Pheromone users must understand that a chemical manufacturer can only guarantee to produce a chemical to meet chemical specifications. Good chemical manufacturers can do this quite well. They will have inprocess checks for reactions and distillations, and also will assay the final product. When pheromone users supply a chemical manufacturer with complete chemical specifications, they will be sure to obtain a consistently performing product.

VI. CONCLUSION

Pheromone use is becoming an important supplement to conventional pesticides. As this technology grows, pheromone users will feel a greater responsibility to their programs and to the performance of their business. Regulatory constraints, environmental concerns, and worker safety are key parts of safe, reliable chemical production. Pheromone users need to work with companies who stress these factors as well as price and quality.

ACKNOWLEDGMENTS

I gratefully acknowledge the help of Terrence C. Holton, Ph.D., Craig A. Heselton, James A. Cole, Randolph S. Disney, and Larry L. Hilton, Ph.D., all Orsynex employees.

10

The Research, Development, and Application Continuum

B. STAFFAN LINDGREN / Phero Tech Inc., Vancouver, British Columbia, Canada

I. INTRODUCTION

In recent years, increasing emphasis has been placed on so-called applied research, particularly by government funding agencies. Many such agencies in Canada state explicitly that certain funds are intended for research likely to generate some return, whether purely in financial terms or in terms of general socioeconomic impact. For example, the Science Council of British Columbia's application procedures for Assistance Grants for Applied Research state that

> ...the Science Council is concerned with research which involves economic and social issues in B.C. Accordingly the applicants are encouraged, ..., to explain fully the social and economic implications of their research and its potential for producing patentable inventions and innovative technology.

Of their three evaluation criteria, two relate to economic benefits for the Province of British Columbia, whereas the third relates to the quality of the proposed research. Thus, applicants must consider more than just the scientific aspects of their research to qualify for these grants.

Buzz words, such as *integrated pest management* (IPM) and *biotechnology*, reflect increasingly complex approaches to problem-solving in pest management. Semiochemical-based technology is based on integrated biological and chemical research. At the development stage, it requires biological, chemical, and chemical engineering expertise. At the application stage, it demands sound resource management and user cooperation, without losing sight of the intricate biological and chemical systems on which it is based. Semiochemical-based pest management is often directly integrated with other techniques. For example, sex pheromones are used to monitor population trends to justify and time pesticide applications (1), and tree baiting with aggregation pheromones of bark beetles is used to induce attack before silvicultural control by sanitation salvage harvesting (2). Biotechnology may influence semiochemical-based pest management in terms of novel approaches to semiochemical production, such as enzymatic synthesis of optically pure semiochemicals. Integrated pest management, and to some extent biotechnology, then, are essential components of the research, development, and application continuum (henceforth, called the *continuum*). In this chapter, I will discuss some components of the continuum, present some ideas on the philosophy and practice of applied science, and suggest ways in which one can ensure that the process of research, development, and application is truly a continuum, that is, "an *uninterrupted*, ordered sequence" [author's emphasis] (3). Most examples relate to forestry, because I am familiar with that field, but the general discussion should apply to agriculture and other fields as well. My objective is not to present solutions to problems associated with applied research, but to put forward ideas that may stimulate scientists and practitioners involved in semiochemical-based pest management to improve their approach to the continuum.

II. APPLIED RESEARCH: PHILOSOPHY AND PRACTICE

Applied research is defined as: *investigation or experimentation aimed at the discovery and interpretation of facts, ...* [leading to the] *practical application of ... new or revised theories or laws* (3). Basic research aims at *revision of accepted theories or laws in the light of new facts* (3). Although applied research is rarely referred to as *impure* (4), use of the term *pure* for basic research implies that other forms of research are somehow contaminated, presumably by social and political issues (5). Both basic and applied research are regulated by the same scientific principles and utilize the same scientific methods, which guarantee the integrity of the research as long as scientific and ethical principles are followed. By scientific methods in this context, I refer to the

> ...principles and procedures for the systematic pursuit of knowledge involving the recognition and formulation of a problem, the

collection of data through observation and experimentation, and the formulation and testing of hypotheses (3).

Scientific purists argue that social or political involvement violates the assumptions on which scientific research is based (5). They believe that scientific problems, to which science can provide answers, can be distinctly separated from social and political problems, which cannot be dealt with by science, and for which science has no responsibility (5). According to this view, the scientific product is sociopolitically neutral, so that effects and responsibilities are only attributable to the implementors, as opposed to the creators (6). Although such scientific purity undoubtedly exists in some fields of research, it is clear that applied science is, and should be, responding to the socioeconomic and political needs of society. Levy's (6) "mango thesis" states that technology can be likened to a mango fruit, in which the neutral *scientific/technological core* is the pit, the network of *requirements* generated by implementing the core are the fibers that extend from the pit into the flesh of the fruit, and the *values* suggesting how these requirements should be met are analogous to the mango's flesh. Levy (6) argues that because more than just core technology is invariably transferred, scientists must assume some responsibility for the use of their scientific product.

From the perspective of users of new technology, involvement in socioeconomic and political issues by the scientists who produced that technology would appear to be essential. One can argue that scientists, in general, should accept more responsibility than they commonly do in the interest of effective technology transfer (6). However, it is my opinion that too many of us suffer from what I term "the scientific-integrity syndrome"; that is, we hide behind the contention that we lack scientific evidence applicable to use of technology under large-scale, uncontrolled conditions, thereby actually impeding development and technology transfer.

There is a more real side to the scientists' fear of becoming involved in the sociopolitical sphere. Scientists have to guard their integrity, and any association with commercial suppliers may be viewed by peers as unethical. Therefore, some scientists may shun any technology-transfer responsibility they may face, thereby slowing down the implementation of valuable research products.

To a large extent, the scientific-integrity syndrome relates to the cost–benefit of acquiring knowledge. Grossly generalized, we can think of the learning curve as an ascending asymptotic function (Fig. 1), for which the cost of acquiring a unit of knowledge increases drastically as knowledge increases. Application of available scientific knowledge is delayed by the scientific-integrity syndrome sufferer, and more knowledge is acquired at increasing cost; that is, the scientists attempts to go further to the right on the learning curve (see Fig. 1).

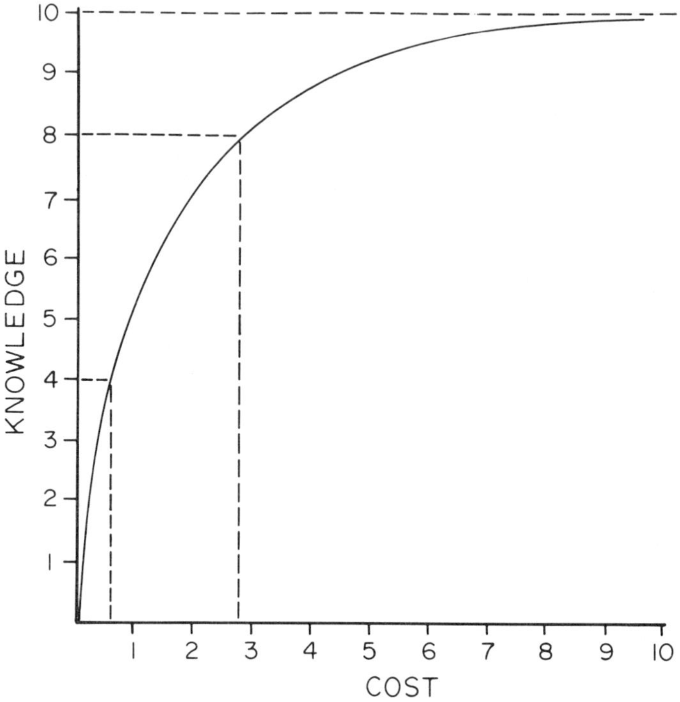

FIGURE 1 Generalized curve illustrating the relationship between the increase in knowledge and its "cost" per unit.

Wood (7) recognized that the development and application of semiochemical-based technology faced some sizable problems. He stated that the lack of resources allocated by industry to applied research in this field was a result of "...high development costs, low market potential because of high specificity, proprietary problems, and the complexity of demonstrating efficiency..." He went on to state that because of the diversity and complexity of these problems "...a much closer collaboration between private and public institutions will be required to develop these compounds for application in pest-management systems."

Funding for large-scale, operational trials is rarely available for scientific purposes. Even when funds are available, it is often virtually impossible to conduct meaningful, replicated trials because of the difficulty of finding large enough areas that are sufficiently homogenous to allow statistically valid comparisons. Furthermore, agricultural interests or resource management concerns often prevent establishment of proper control areas. For example, the reduction of *Ips typographus* (L.) populations in Norway and Sweden after the most extensive semiochemical-based mass trapping program yet undertaken, cannot conclusively be said to have resulted from the trapping program itself because no control areas were established (8). Nevertheless, there seems to be fairly convincing evidence that the mass trapping program aided in preventing damage (9). Similarly, mass trapping of ambrosia beetles at dryland sorting areas, log-boom tie-ups, and sawmills in British Columbia is operational and successful, as judged by company records, but replicated studies, using specific sorting areas as controls, cannot be conducted because of the differences in inventory management, species composition, and weather patterns, all of which affect beetle reproduction and flight patterns (see Chap. 19).

III. DEVELOPMENT AND APPLICATION: THE TECHNOLOGY-TRANSFER PROCESS

For technology transfer to occur, it is essential that all potential users be actively involved at an early stage of the applied research. This involvement was proposed as the primary prerequisite for accomplishing technology transfer by a panel of researchers (10). They further stated that "...open and continuing communication between the originating scientists and the operational personnel, extension specialists, private consultants, industry, and growers are essential in effective transfer of technology from scientists to potential users." Furthermore, scientists "...must also continue to offer technical assistance to programs in the operational stage to assure the misapplication will not result in failure and setback in the use of [a] promising technique" (10).

Campion (1) described the difficulties in developing a mating disruption system for the pink bollworm, *Pectinophora gossypiella* (Saunders). He summarized the important factors leading to the use of the pink bollworm pheromone as the following:

1. A stable pheromone
2. A key pest insect with a narrow food range
3. Limited target species migration
4. Available basic biological knowledge
5. Availability of adequate assessment techniques
6. Availability of formulations and application technology
7. Commercial interest owing to potential market
8. Convincing reasons to alter existing pest management strategy

In addition, it is fairly obvious that the commercial success of the pink bollworm pheromone would not have been possible if cotton had been grown in isolated small fields, rather than over large areas. These points illustrate the multitude of considerations that have to be taken into account to transfer technology from the research stage to the applied stage. Many of the requirements are clearly peripheral in terms of biological sciences. However, if they are not taken into account by the scientist as the research is progressing, the technology transfer may be impeded or even prevented because of incompatibility of the research product with the economic and practical reality of pest management within the scope of existing resource management or agricultural practices. Consequently, input from potential user groups may be crucial when priorities for research direction are established.

A. Questions on Implementation

There are many questions that need to be answered to implement the technology-transfer process. The answers to them may vary considerably depending on the scope of the project, but it may help if they are considered in the planning stages of the research, rather than after the research has been completed.

1. Who Should Transfer Technology?

In general, I agree with Mitchell (10), that the scientist, as producer of the technology, needs to be involved to some extent for the process of technology transfer to proceed smoothly. Many institutions and government agencies now have technology-transfer officers, whose sole responsibility is to make sure that research products are transferred to users. There are some obvious problems with such an arrangement. Because these officers may not be familiar with the research, they usually need input from the principal scientist to gain sufficient familiarity with the product to be able to "sell" it. This means that scientists have to give up their time anyway, whereas quality of the product may suffer owing to the second-hand nature of the information transferred. On the other hand, technology-transfer personnel may help the process by putting the user in contact with the researcher, by coordinating funding applications for applied research, and by securing sources of venture capital for the final development. In some cases, the scope of a project may be so great that technology-transfer personnel must take charge of the technology-transfer process. However, it is important that the scientist be retained as a consultant.

2. *What Should Be Transferred?*

As suggested by Levy's (6) mango thesis and Campion's (1) eight requisite factors for the development of semiochemical-based pest management, much more than the scientific or technological core is transferred. At a recent workshop on technology transfer held at the Southern Forest Insect Work Conference, San Antonio, Texas, one of the participants stated that scientists do not know what they are supposed to transfer. Presumably, this is due to lack of interaction between scientists and users, so that the scientific product fails to meet user needs, or cannot be implemented in the form it was presented to the user. A clear understanding by the scientist of what the user needs would facilitate appropriate decisions on research priorities, and would lead to the best product for the user.

3. *When Should Technology Be Transferred?*

Technology transfer must not be viewed as an event, but as a process. The technology-transfer process should begin as soon as sufficient knowledge has been accumulated to meet the requirements for successful implementation of the technology. It may be difficult to determine what point on the learning curve (see Fig. 1) must be reached before the knowledge accumulated is "sufficient." Depending on the magnitude of the scientific-integrity syndrome, the judgment may vary considerably between different scientists. Moreover, it is never possible to have enough information to provide a perfect application of technology. Rather than delay the technology-transfer process, it is often better to implement technology in such a way that some important information, such as assessment of damage impact and population reduction, may be collected during the operational use of the product. In this manner, rapid technology transfer will be facilitated without sacrificing scientific integrity.

The technology-transfer process does not necessarily end once a product has been accepted by the user. An example is the current development of new application techniques for pink bollworm mating disruption. Slow-release devices for aerial application to disrupt mating of the pink bollworm were developed by several companies in the 1970s and have been successfully used since then (1,12). However, aerial application is not appropriate until the cotton canopy closes, because most of the pheromone dispensers will end up on the ground, where they are ineffective. Because of the need for early control of the pink bollworm, relieving the growers from dependency on pesticides and possible problems with tobacco budworm infestations, a ground applicator system has recently been developed by

Hercon Laboratories Corporation, Health-Chem Corporation, New York, for application of their laminated-flake dispensers (12) to young cotton plants (A. R. Quisumbing, personal communication). This application system may also be more appropriate than aerial application in areas in which the latter may be subject to public opposition owing to association with pesticide sprays, or in certain crops, such as orchards, for which spraying from below the canopy may be more efficient than aerial spraying. Hercon has also developed hand applicators to meet the needs of users in developing countries, where aerial application technology and tractors, which are needed for the ground applicator, are not available (A. R. Quisumbing, personal communication).

4. Where Should Technology Transfer Occur?

Where technology transfer should occur may seem a somewhat ridiculous question, but it is particularly warranted in semiochemical pest management. For example, many insect species respond to more or less well-defined ratios of different structural or optical isomers of particular compounds (13). The optimal ratio may vary between geographic areas. For example, the western avocado leaf roller, *Amorbia cuneana* (Walsingham), responded to a 1:1 ratio of the pheromone (*E*,*Z*)-10,12- and (*E*,*E*)-10,12-tetradecadien-1-ol acetate near San Diego in southern California, but requires a 9:1 ratio if the pheromone is used near Santa Barbara (14,15). Lanier et al. (16) showed that Idaho and California populations of the pine engraver beetle, *Ips pini* (Say), responded to the (*R*)-(-) enantiomer of the aggregation pheromone ipsdienol, whereas New York populations produced the (*S*)-(+)- and (*R*)-(-) enantiomers in a 65:35 ratio, and responded better to racemic ipsdienol than to the (*R*)-(-) enantiomer alone. The western populations were also inhibited by the (*S*)-(+) enantiomer (17). Subsequently, Miller et al. (18) found western Canadian populations that also produced and responded to mixtures of (*S*)-(+)- and (*R*)-(-)-ipsdienol. Consequently, widespread application of a semiochemical technique may not be advisable without local research confirming the applicability of the determined ratios of components. Similar problems may be encountered with threshold levels determined for a particular area. Campion (1, and references therein) described problems of this nature encountered with monitoring threshold levels for the codling moth, *Laspeyresia pomonella* (L.).

5. How Should Technology Be Transferred?

The Expanded Southern Pine Beetle Research and Applications Program (ESPBRAP) used a number of avenues for technology transfer (19). Technology transfer teams were established to be in charge of the process in each of eight different fields. A substantial proportion of the process has been accomplished through publication of fact sheets, by handbooks and articles in applied journals, by production of slide tapes, and by advertising the availability of technology-transfer products in various journals, newsletters, and magazines (19). A more direct approach would involve the use of extension specialists to introduce new technology to the user. Both of these approaches result in some loss of information in the transfer process. When the mass-trapping program for *I. typographus* in Norway was initiated, the problem of setting up the 600,000 traps, used in the first year, was solved by distributing the traps to the individual forest land owners. Instructions on the proper use of the traps were given through pamphlets, through information meetings, and through the press, radio, and television (20).

Whenever the producer of technology can actively partake in the process, particularly by working directly with the user groups, the information flow is unimpeded, and the technology can be adapted to the user's need without sacrificing quality.

6. *To Whom Should Technology Be Transferred?*

The answer to this question may determine the answer to some of the previous questions. For example, technology, such as the aerial application of pheromones for mating disruption of the pink bollworm, cannot be transferred directly to individual farmers if the areas farmed are too small or the farmers do not have direct access to aerial spray equipment. Therefore, government extension specialists and spray contractors would have to be involved. On the other hand, traps for monitoring population levels of an insect, such as the codling moth, can easily be transferred to individual orchard owners. Although it may be difficult for a scientist to assess accurately to whom a particular technology should be transferred, the assignment of research priorities may be more accurate if this question is carefully considered, and the transfer process will benefit as a result.

B. The Pesticide Legacy

A specific problem in the transfer of semiochemical technology is the pesticide legacy. Users have become accustomed to immediate and spectacular population reduction of pest insects whenever a treatment is applied. Semiochemicals will not normally accomplish population reduction in the short-term, and some crop damage may occur, even after treatment. Semiochemicals are often more expensive than pesticides, and it may be difficult to convince users to pay more for what they perceive as less control. Furthermore, registration regulations may impede the use of semiochemicals (7). Even though the situation has improved since 1977, semiochemicals are still classified under the umbrella term *pesticides*, and particularly when used in or near food crops, the registration requirements may prevent their use. An example is the difficulties encountered in implementing the use of the sex pheromone of the grape berry moth, *Endopiza viteana* Clemens, in New York State (see Chap. 15). Successful mating disruption of this species was achieved when the pheromone was formulated in so-called ropes (Shin-Etsu Corp., Japan) and attached to training wires for the grape vines. The dispensers were not in direct contact with the fruits, yet the EPA required expensive residue analysis of the fruits, as well as toxicity tests of the pheromone. Such requirements may make implementation of semiochemicals in small crops virtually impossible. Consequently, agencies that have been established to ensure environmental protection may, in fact, impede development of environmentally safe materials by placing minor-use semiochemicals under the general heading of pesticides, thereby placing them under suspicion in the public eye. Many semiochemicals are structurally very similar (e.g., the position of a double bond may be the only difference between two compounds that are behaviorally active for different species of insects) and, occasionally, it would be logical that this be considered in the registration process. This is not to say that semiochemicals should be exempt from toxicity testing, but they should be regulated by specific guidelines that would encourage their use, rather than discourage it. In the interest of reducing the use of chemical pesticides, governments may have to extend their responsibility by providing grants that would allow toxicity testing to be carried out for semiochemicals or other desirable compounds in cases for which the market would be too limited to allow recovery of the registration costs.

IV. THE CONTINUUM IN PRACTICE: AN EXAMPLE

Research leading up to the commercial implementation of an ambrosia beetle mass-trapping program in British Columbia was initiated in 1967 (21) with the isolation and identification of the pheromones for two species, *Trypodendron lineatum* (Olivier), and *Gnathotrichus sulcatus* LeConte, which are important in coastal British Columbia. Synthetic pheromones in quantities sufficient for preoperational research were available for both species by 1979. A joint research project was initiated between J. H. Borden, Simon Fraser University, and J. A. McLean, the University of British Columbia, and with close and direct cooperation with three forest companies, MacMillan Bloedel, Ltd., British Columbia Forest Products, Ltd., and CIP Inc., Tahsis Pacific Region (then Pacific Logging, Ltd.), including interaction with dryland sorting area and sawmill personnel. After the completion of the research (22,23), two independent companies, Stratford Chemical Developments Ltd. and Pest Management Group Ltd. were brought together through active involvement by the principal scientists in the project. These companies formed PMG/Stratford Projects, Ltd., which later was renamed Phero Tech Inc., and set out to commercialize the research findings. By innovative use of very limited resources, a prototype trap (24) was modified and put into production. Lineatin and (+/-)-sulcatol, the aggregation pheromones for *T. lineatum* and *G. sulcatus*, respectively, were formulated by Consep Membranes Inc. (then Bend Research) in Biolures. Another release device, the bubble cap, was developed through cooperation with the Department of Chemical Engineering, University of British Columbia. This device was more suitable for sulcatol, the optimal release rate of which is relatively high.

As a direct result of their involvement at the research stage, CIP Inc. set up a mass-trapping program run by their pest management specialist, R. H. Heath, and MacMillan Bloedel Ltd. agreed to buy the services of Phero Tech to initiate mass trapping at some of their sorting areas, log-boom tie-ups, and sawmills. Part of the cost was covered by a grant from the National Research Council of Canada. In year 2, MacMillan Bloedel Ltd. paid for all of an expanded program, satisfied that the mass trapping had had an effect on the damage levels. The program was designed as a service, with Phero Tech personnel setting up the traps, collecting the insects, and reporting to the company. The principal scientists in the original research project remained available as consultants and provided advice, even when it was not solicited, throughout the first few years. Details of the fate of this program are given by Borden (Chap. 19).

V. THE SEX PHEROMONE THESIS

Baker and Cardé (25) and Baker et al. (26) first elucidated the importance of using the proper blend of pheromones to elicit the full-response sequence in flying oriental fruit moth males, *Grapholitha molesta* (Busck). Without the correct pheromone components present in certain ratios, the response sequence was interrupted. Other moths respond to mixtures of isomers and components in a similar manner (13) (Fig. 2A).

I would like to suggest that one can think of the continuum as a process analogous to the male moth orienting to a female. Thus, I

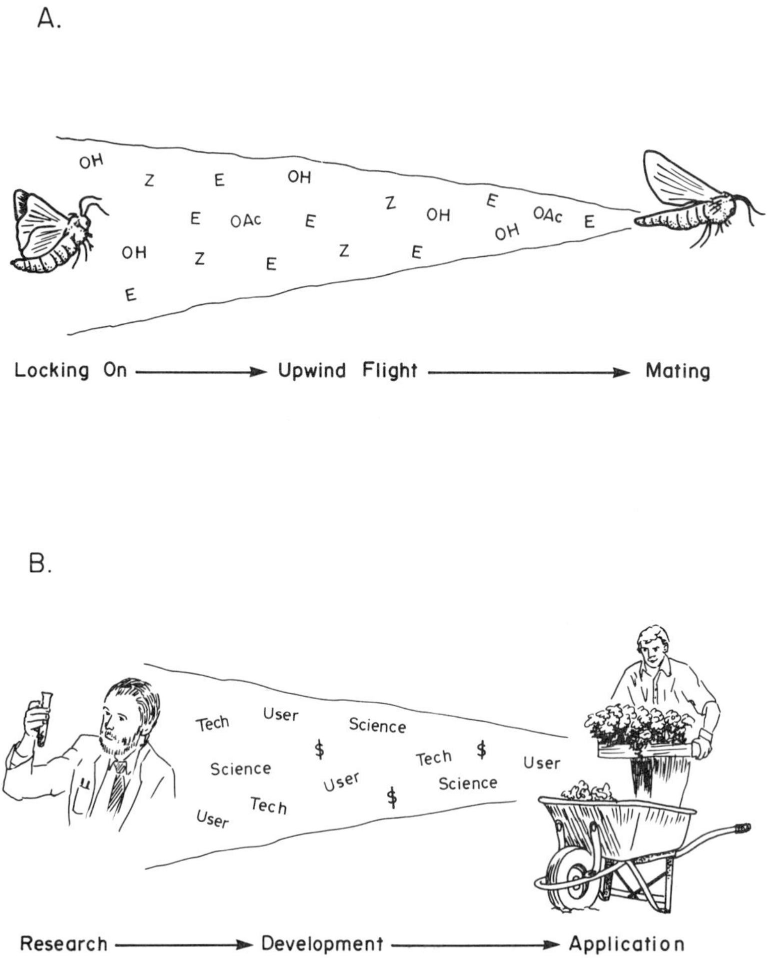

FIGURE 2 Graphic representation of the sex pheromone thesis. (A) Male moth responding to a given ratio of sex pheromone components emitted by a female. (B) Scientist "responding" to information components pertaining to the needs of a user.

put forward *the sex pheromone thesis* as a basis for how the continuum should be approached so that the objectives of a project are met (Fig. 2B). In this thesis, the pheromone components are analogous to information from different sources (i.e., scientific data, economic constraints, technological constraints, and user needs). The upwind flight of the moth is analogous to the progress of the project, and the successful mating with the successful implementation of the product. Some of the components may act at a long range (e.g., the scientific data are necessary to initiate the project) just as some components in the pheromone blend would be necessary to initiate flight and locking-on in the male moth. Other components are necessary only at close range (e.g., the economic constraints and user needs may be required earlier or later in the project, depending on the objectives).

The point is that to progress successfully through the entire continuum, one must stay in touch with the information "plume," that is, one must maintain contact with all of the component sources of the information. Lack of any one component will disrupt the progress, and technology transfer will not be successful.

The sex pheromone thesis describes a linear process in which all the information is contained within a "plume." Obviously, information is not obtained in this fashion nor is the continuum necessarily linear. However, by clearly identifying what components are needed for successful implementation of a product, the continuum can be streamlined and made more cost-efficient.

VI. CONCLUSION

As is clear from the preceding discussion, the approach to the semiochemical research, development, and application continuum may vary among projects. However, the following sources of information would appear to be needed in the "information plume" to make the process as efficient as possible:

1. *Basic scientific data*: basic knowledge of target insect population dynamics; dispersal capability; chemical ecology, including identity of semiochemicals and behavioral responses to them
2. *Applied scientific data*: efficacy of the relevant strategies and tactics
3. *Economic information*: information on value of the crop; damage threshold levels; product development costs, including possible registration costs; current management costs
4. *Technical information*: availability of suitable release devices; trap types; application methods; large-scale syntheses for semiochemicals; stability of semiochemicals
5. *User needs*: current management techniques to determine the best approach for integration of semiochemicals; prophylactic or symptomatic approach to management of target pest species;

acceptable damage thresholds; technology transfer approach, e.g., through service organization or direct to user.

These are just a few of the possible factors to be considered, but they demonstrate that more than biological and chemical research is needed to complete the continuum. By taking factors such as these into account at an early stage, it should be possible to generate cooperation with user groups and to allow the user to become familiar and comfortable with the new technology, thereby streamlining the research, development, and application continuum, and facilitating the efficient transfer of technology.

ACKNOWLEDGMENTS

I thank G. H. Cushon, L. M. Friskie, and Dr. J. H. Borden for valuable criticism of the manuscript.

REFERENCES

1. Campion, D. G. Survey of pheromone uses in pest control. In *Techniques in Pheromone Research*. (Hummel, H. E. and Miller, T. E., eds.). Springer-Verlag, New York, 1984, pp. 405–449.
2. Borden, J. H., Chong, L. J., Pratt, K. E. G., and Gray, D. R. The application of behavior-modifying chemicals to contain infestations of the mountain pine beetle. *For. Chron.* 59:235–239 (1983).
3. Woolf, H. B. (ed.). *Webster's New Collegiate Dictionary*. G & C Merriam Co., Springfield, Mass, 1976.
4. Stark, R. W. and Waters, W. E. Concept and structure of a forest pest management system. In *Integrated Pest Management in Pine-Bark Beetle Ecosystems*. (Waters, W. E., Stark, R. W., and Wood, D. L., eds.). John Wiley & Sons, New York, 1985, pp. 49–60.
5. Nowotny, H. Scientific purity and nuclear danger. The case of risk-assessment. In *The Social Production of Scientific Knowledge*, Vol. I. (Mendelsohn, E., Weingart, P., and Whitley, R., eds.). D. Reidel Publishing Co., Dordrecht, Holland, 1977, pp. 243–264.
6. Levy, E. The responsibility of the scientific and technological enterprise in technology transfers. In *Science, Politics and the Agricultural Revolution in Asia*. (Anderson, R., Brass, P., Levy, E., and Morrison, B., eds.). Westview Press, Boulder, Colorado, 1982, pp. 277–297.

7. Wood, D. L. Manipulation of forest insect pests. In *Chemical Control of Insect Behavior: Theory and Application.* (Shorey, H. H. and McKelvey, J. J., Jr., eds.). John Wiley & Sons, New York, 1977, pp. 369–384.
8. Bakke, A. The utilization of aggregation pheromone for the control of the spruce bark beetle. In *Insect Pheromone Technology: Chemistry and Applications.* (Leonhardt, B. A. and Beroza, M., eds.). American Chemical Society Symposium Series No. 190, Washington, D.C., 1981, pp. 219–229.
9. Lie, R. and Bakke, A. Practical results from the mass trapping of *Ips typographus* in Scandinavia. In *Management of Insect Pests with Semiochemicals. Concepts and Practice.* (Mitchell, E. R., ed.). Plenum Press, New York, 1981, pp. 175–181.
10. Mitchell, E. R. (ed.). *Management of Insect Pests with Semiochemicals. Concepts and Practice.* Plenum Press, New York, 1981, pp. 205–206.
11. Henneberry, T. J., Gillespie, J. M., Bariola, L. A., Flint, H. M., Butler, G. D., Jr., Lingren, P. D., and Kydonieus, A. F. Mating disruption as a means of suppressing pink bollworm (Lepidoptera: Gelechiidae) and tobacco budworm (Lepidoptera: Noctuidae) populations on cotton. In *Insect Suppression with Controlled Release Pheromone Systems*, Vol. II. (Kydonieus, A. F. and Beroza, M., eds.). CRC Press, Boca Raton, 1982, pp. 75–98.
12. Quisumbing, A. R. and Kydonieus, A. F. Laminated structure dispensers. In *Insect Suppression with Controlled Release Pheromone Systems*, Vol. I. (Kydonieus, A. F. and Beroza, M., eds.). CRC Press, Boca Raton, 1982, pp. 213–235.
13. Roelofs, W. L. Pheromone perception in Lepidoptera. In *Neurotoxicology of Insecticides and Pheromones.* (Narahashi, T., ed.). Plenum Press, New York, 1979, pp. 5–25.
14. Hoffman, M. P., McDonough, L. M., and Bailey, J. B. Field test of the sex pheromone of *Amorbia cuneana* (Walsingham) (Lepidoptera: Tortricidae). *Environ. Entomol. 12*:1387–1390 (1983).
15. Bailey, J. B., McDonough, L. M., and Hoffmann, M. P. Western avocado leafroller, *Amorbia cuneana* (Walsingham) Lepidoptera: Tortricidae). Discovery of populations utilizing different ratios of sex pheromone components. *J. Chem. Ecol. 12*:1239–1245 (1986).
16. Lanier, G. N., Claesson, A., Stewart, T., Piston, J. J., and Silverstein, R. M. *Ips pini*: The basis for interpopulational differences in pheromone biology. *J. Chem. Ecol. 6*:677–686 (1980).
17. Birch, M. C., Light, D. M., Wood, D. L., Browne, L. E., Silverstein, R. M., Bergot, B. J., Ohloff, G., West, J. R., and

Young, J. C. Pheromonal attraction and allomonal interruption of *Ips pini* in California by the two enantiomers of ipsdienol. *J. Chem. Ecol.* *6*:703–717 (1980).
18. Miller, D. R., Borden, J. H., and Slessor, K. N. Inter- and intrapopulational variation of the pheromone, ipsdienol, produced by male pine engravers, *Ips pini* (Say) (Coleoptera: Scolytidae). *J. Chem. Ecol.* *15*:233–247 (1989).
19. Hertel, G. D. and Mason, G. N. Technical applications. In *History, Status, and Future Neads for Entomological Research in Southern Forests.* (Payne, T. L., Billings, R. F., Coulson, R. N., and Kulhavy, D. L., eds.). Proc. 10th Anniversary East Texas Forest Entomology Seminar, 1984, pp. 36–39.
20. Bakke, A. and Strand, L. Feromoner og feller som ledd i integrert bekjempelse av granbarkbillen. Noen resultater fra barkbilleaksjonen i Norge i 1979 og 1980. Rapport 5/81 fra Norsk Institutt for Skogforskning, 1981, pp. 1–39. (Norwegian, summary in English).
21. Borden, J. H. and McLean, J. A. Pheromone-based suppression of ambrosia beetles in industrial timber processing areas. In *Management of Insect Pests with Semiochemicals. Concepts and Practice.* (Mitchell, E. R., ed.). Plenum Press, New York, 1981, pp. 133–154.
22. Lindgren, B. S. and Borden, J. H. Survey and mass trapping of ambrosia beetles (Coleoptera: Scolytidae) in timber processing areas on Vancouver Island. *Can. J. For. Res.* *13*:481–493 (1983).
23. Shore, T. L. and McLean, J. A. A survey for the ambrosia beetles *Trypodendron lineatum* and *Gnathotrichus retusus* (Coleoptera: Scolytidae) in a sawmill using pheromone-baited traps. *Can. Entomol.* *117*:49–55 (1985).
24. Lindgren, B. S. A multiple funnel trap for scolytid beetles (Coleoptera). *Can. Entomol.* *115*:299–302 (1983).
25. Baker, T. C. and Cardé, R. T. Analysis of pheromone-mediated behaviors in male *Grapholitha molesta*, the oriental fruit moth (Lepidoptera: Tortricidae). *Environ. Entomol.* *8*:956–968 (1979).
26. Baker, T. C., Meyer, W., and Roelofs, W. L. Sex pheromone dosage and blend specificity of response by oriental fruit moth males. *Entomol. Exp. Appl.* *30*:269–279 (1981).

Part II
Pests of Horticultural Crops

11

Mating Disruption Technique to Control Codling Moth in Western Switzerland

PIERRE-JOSEPH CHARMILLOT / Federal Agricultural Research Station of Changins, Nyon, Switzerland

I. INTRODUCTION

The mating disruption technique consists of diffusing the synthetic sexual attractant of a pest continuously into the orchard that must be protected, so that males are no longer able to locate females to mate with them. This technique has just been registered in Switzerland to control the codling moth, *Cydia pomonella* L. As a specific means of control, it should supplant, where ever it is possible, the classic polyvalent insecticides that, in practice, hinder the introduction of integrated pest management. This paper summarizes trials of the mating disruption technique against this pest in western Switzerland on 275 ha of apple and pear orchards from 1976 to 1987.

II. MATERIALS AND METHODS

A. Attractant and Dispensers

The main component of the pheromone of codling moth is (*E*,*E*)-8,10-dodecadien-1-ol (E8, E10-12:OH), commonly called codlemone (1). Rubber or plastic dispensers containing this substance were hung in the orchards just before the flight period of the codling moth. Their role is to adjust the emission rate while protecting the substance from chemical degradation.

The rubber tubing dispensers used for the trials conducted from 1976 to 1979 were small tubes, 2 mm in internal diameter (ID)

and 4 mm in external diameter (ED). These dispensers were filled with undiluted attractant and sealed at both ends by a metal clip. From 1980 to 1986 the attractant, to which 1% butylated hydroxytoluene (BHT) was added as an antioxidant, was dissolved in ether. The mixture was then sucked into slightly larger rubber tubing (3 mm ID; 5 mm ED) by a peristaltic pump. The amount of attractant was calculated to obtain—after evaporation of the solvent—1.5 mg of codlemone per millimeter. The tubing was then ready to be cut to the desired lengths and stored in the refrigerator until use.

Since 1982, Hercon dispensers have also been tested, cut into pieces of 1 $in.^2$ (6.45 cm^2), each unit containing about 40 mg of codlemone. With both kinds of dispensers, the borders of the orchards were treated by placing one dispenser per tree, 4 to 5 m apart. Larger dispensers were placed inside the orchard, at a greater distance from one another: 40 to 200 sources of attractant per hectare, depending on the size and shape of the orchards.

B. Conditions of Trials

From 1976 to 1987, 48 experiments were conducted in western Switzerland with the rubber dispensers on a total area of 165.6 ha, and from 1982 to 1987, 12 trials were made with Hercon dispensers on 109.6 ha of apple and pear orchards. During that period various experimental factors were modified according to the results obtained in preceding years.

These factors include:

The amount of codlemone used per hectare and its chemical purity
The number of dispensers used per season, as well as their location and density
The isolation and the shape of the orchards
The initial population density of the pest

C. Evaluation Methods Used in Orchards

Pheromone traps were placed in all test orchards as well as in some check orchards to monitor the flight of the codling moths and to measure communication disruption. If necessary, pesticide control treatments were made during the summer based on visual inspection of fruit for larval entrance. At the end of the season a representative preharvest sampling of 2000 to 3000 pieces of fruit made it possible to judge the result of the mating disruption technique. Finally, the capture of diapausing larvae—by 40 corrugated paper bands distributed in the orchard—was a way of estimating the dynamics of hibernating population of codling moth.

D. Emission and Degradation of Attractant

The rate of loss of attractant was followed by weekly weighing of some loaded as well as some unloaded rubber-tubing dispensers. Others were removed regularly and stored in the refrigerator. The residual attractant was later extracted by a solvent; the amount was determined by gas chromatographic (GC) analysis.

III. RESULTS AND DISCUSSION

A. Overall Results of the Trials

Figure 1 sums up the results obtained from 1976 to 1987 with rubber-tubing dispensers. As we noted before, numerous factors were modified during that period to improve the method and to reduce its cost. The first trials were made with high amounts of attractant. The quantity of codlemone was then progressively reduced until 1980, when we used a little more than 10 g/ha. However, that low amount of attractant proved to be insufficient, especially during years favorable to the codling moth. From 1981 to 1984 the amount was progressively raised to 40 g/ha, divided into two applications. At the same time a number of small orchards of less than 3 ha, as well as some insufficiently isolated ones, were abandoned because of excessive damage, especially in the borders. The lower part of Figure 1 sums up the practical results. The upright bars show the trial area considered year after year. The blank sections represent the area that did not need a pesticide treatment. The hatched parts represent the area that needed to be treated because of the damage caused by the summer fruit tortrix moth, *Adoxophyes orana* F.v.R. The black parts represent the area that had to be treated because of the damage caused by the codling moth. It can be observed that from 1983 on, no curative treatment was made against *A. orana*. This is due to the springtime use of an insect growth regulator (IGR)—fenoxycarb—against the hibernating larvae of that pest (2). The reduction in the number of curative treatments against codling moth (since 1984) is a consequence of the increase in the amount of attractant used per hectare. If we consider the total number of trials conducted on 48 orchards (total area, 165.6 ha) in 11 years, using rubber tubing dispensers, we note that 71.9% of that area did not need any curative treatment. The damage caused by *A. orana* justified a curative treatment on 17.2% of the area, whereas that caused by codling moth justified a treatment on 10.9% of the area.

Figure 2 shows the practical results achieved when using Hercon dispensers. From 1982 to 1985 the amount of attractant used was increased from 20 g to 75 g/ha. More than three quarters of

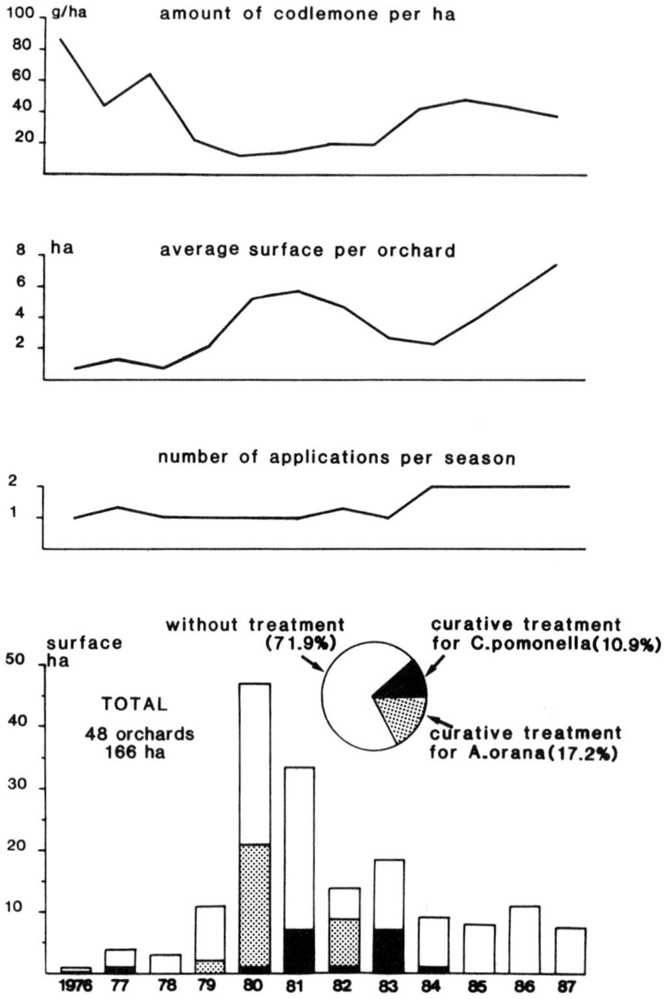

FIGURE 1 Trials summary of the mating disruption technique from 1976 to 1986, using rubber-tubing dispensers. The rectangles in the lower part of the figure show the trial areas for each year. The blank parts represent the areas where no insecticide treatment were made. The hatched parts show the areas where a curative treatment was needed to control *A. orana*. And finally, the dark parts represent the areas where a curative treatment was applied because the tolerance threshold of codling moth was exceeded. The circle shows the same data for the whole duration of trials.

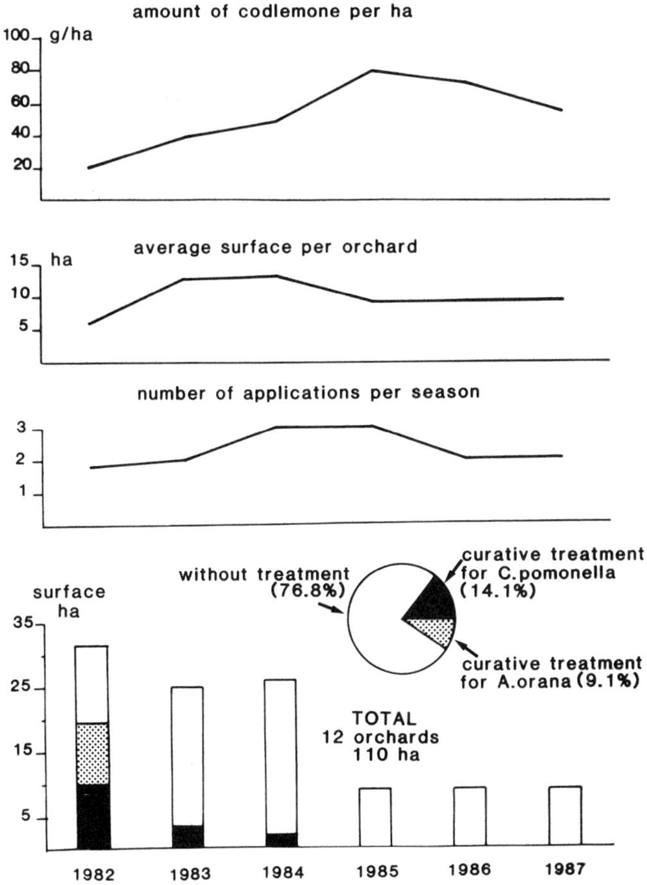

FIGURE 2 Trials summary of the mating disruption technique conducted with Hercon dispensers.

the test area (109.6 ha) did not need any curative treatment and the improvement in the situation paralleled the increase in the amount of attractant used.

B. Sexual Trapping

With both kinds of dispensers, the catches of males in the pheromone traps located in the middle of the trial orchards were always almost completely inhibited, compared with the check traps placed in the same region. Figure 3 serves as an example: it shows the catches made in 1986 in a check trap and in five trial orchards. However, a high reduction of catches is not always a guarantee of successful control, as other researchers have already mentioned (3–5). Conversely, the inhibition of catches may be notably weaker when some traps are placed at particular places, such as the top of the trees or

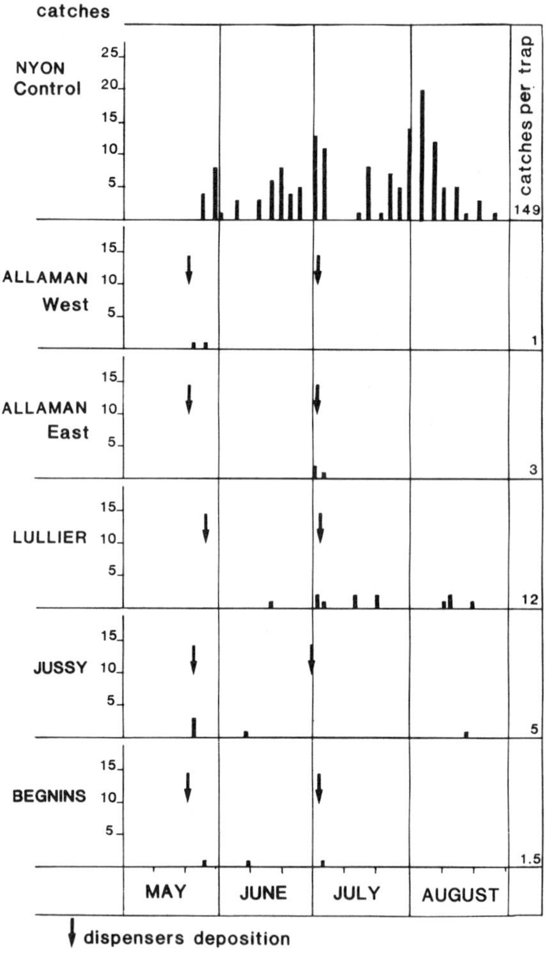

FIGURE 3 Catches of male codling moths in 1986 in a check orchard and in five orchards where the mating disruption technique was tested.

the border of the orchards, but this does not mean that the mating disruption technique has failed (6).

In five trials conducted in 1987 in eastern Switzerland, Mani (personal communication) placed pairs of traps, one at 1.7 m from the ground and the other at the top of a tree. With eight pairs of traps placed inside the trial orchards he caught an average of only 1.6 males per trap when the trap was situated at medium height (1.7 m), but 33.6 moths per trap when it was at the top of the trees. With the 16 pairs of traps placed on the borders of the orchards, he caught an average of 3.4 males at medium height but 45.0 at the top of the trees. Despite the catches made at the top of the trees inside the orchard as well as on its borders, the success of the mating disruption control was not compromised. It must be noted that we had similar results with females tethered in the trial areas and then recovered after a few days and dissected for spermatophores: the reduction of mating was always marked at medium height, but was often much less on the borders or on the top of the trees (4,6,7).

A special experiment on sexual trapping gave some information about the mechanism of mating disruption: sets of five traps baited with capsules loaded with 1, 10, 20, 50, or 100 mg of codlemone were placed in some check orchards as well as in some orchards in which the mating disruption technique was tested. The experiment was repeated three times in 1986 and four times in 1987. Figure 4 shows the relative distribution of catches according to the load of attractant. In the check orchards most males were caught with 1-mg loaded capsules, but the heavily loaded ones were almost ineffective. Conversely, in the orchards where the mating disruption technique was applied, males had difficulties in locating the most weakly loaded capsules, but found the more heavily loaded ones rather easily. It seems then that in the case of codling moth, the mechanisms responsible for mating disruption are based on the camouflage of the females' emissions and on the orientation into false trails rather than on the adaptation of sensory receptors and on the habituation of the central nervous system (8). In the future, it should be possible to provide the trial orchards with two kinds of traps: the first ones—baited with 1 mg of codlemone—should enable us to verify the reduction of catches, and the other, more heavily loaded, should offer the possibility, in spite of air permeation with codlemone, of following changes in the adult population density during the season.

C. Attractant: Quantity and Quality, Number of Applications per Season, Rate of Emission

Figure 1 shows that the first trials, in which rubber tubing dispensers were used, were carried out with high amounts of attractant per

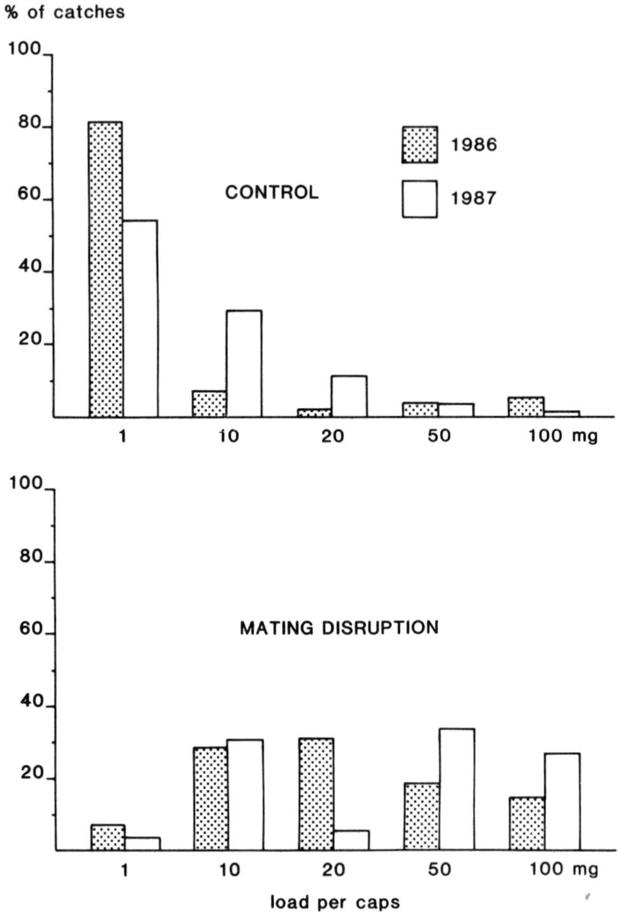

FIGURE 4 Distribution of the catches of codling moth according to the load of attractant in baited traps, in check orchards (top), and in orchards where the mating disruption technique was applied (bottom).

hectare. Then, until 1980, the quantity of codlemone was progressively reduced to about 10 to 15 g/ha to lower the cost of the mating disruption technique. However, some curative pesticidal treatments proved to be indispensable for a relatively high proportion of the trial orchards because the tolerance threshold was exceeded. Even though a small amount of attractant is sufficient during cold years, which are less favorable for codling moths, the population densities nevertheless tend to increase, especially during the years that are more favorable to the pest. Since 1984, we increased the amount of codlemone to about 40 g/ha, divided into two applications. The population densities then declined, so that since 1985 the curative treatments could be avoided. This is especially true with Hercon dispensers, as shown in Figure 2: the progressive increase from 20 g/ha in 1982 to 70 to 80 g/ha (divided into two or three applications) since 1985 has enabled us to avoid curative treatment.

The success of the mating disruption technique depends not only on the amount of attractant, but also on the temporal distribution of the product. We must in fact maintain a minimum concentration of codlemone in the atmosphere during the whole flight period. This requirement can only be achieved by a judicious distribution of the attractant during the season. The amount of attractant that has evaporated can be calculated by a regular weighing of the dispensers exposed in the orchard. In 1985, for example, using Hercon dispensers, we noted that an annual amount of 77.7 g/ha, divided into three applications of 25.9 g/ha made at about 20-day intervals, produced an average diffusion rate of about 10 mg/ha per hour during the beginning of the flight period. The emission rate slowly increased up to 30 to 40 mg/ha per hour until July, after the two new applications and because of the higher temperatures, which accelerated the emission. It then gradually decreased at the end of the season (Fig. 5) as the attractant disappeared. In 1986, an amount of 71.2 g/ha was distributed in two applications, the first one with 47.4 g/ha at the beginning of the flight period and the second one, with 23.8 g/ha, 42 days later. When the first application—made at the beginning of the flight period when lower temperatures usually slow the emission—was followed by a second application, using only a third of the total amount of codlemone, the emission pattern was much more favorable than in 1985, especially at the beginning of the season.

Figure 6 shows the percentage of attractant that remains in the rubber tubing dispensers in the course of the two applications. This percentage was calculated by weighing the dispensers and by GC analysis. The weighing indicates the remaining amount of attractant, as a function of time, but it does not tell by itself whether the substance in question is pure codlemone or if it could be degradation products. The GC analysis, on the other hand, detects nothing but codlemone. The almost overlapping curves indicate that the dispenser is an efficient protection against the degradation of the attractant. The experience of the past few years proved that Hercon dispensers are also an efficacious means of protecting the attractant. The half emission time, that is, the time needed for the evaporation of 50% of of the attractant, averages 45 days with Hercon dispensers. However,

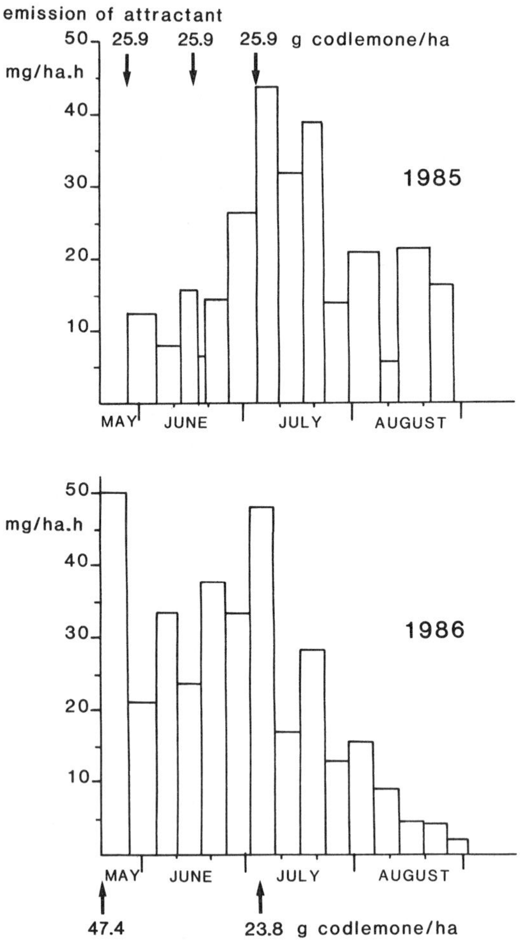

FIGURE 5 Progress of the average hourly emission of attractant in 1985 and 1986: calculated by weighing Hercon dispensers.

it can vary with the temperature and wind conditions. With the rubber tubing dispensers the half emission time is about 36 days, but it can fluctuate from 28 to 42 days according to the year.

The chemical purity of the codlemone impregnated in the rubber tubing dispensers was always at least 99%. In one part of an orchard, however, we compared it with a crude mixture containing about 70% of

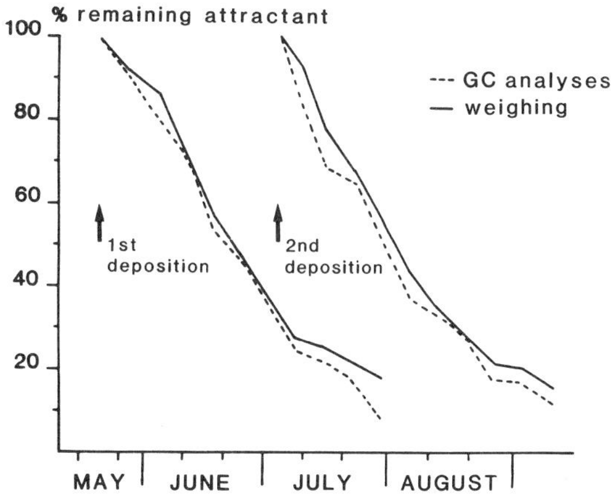

FIGURE 6 Attractant remaining in the rubber tubing dispensers after the two applications of 1986: determined by weighing the dispensers and by GC analysis. The overlapping curves show that the attractant is well protected against degradation.

codlemone (with the amount per hectare increased accordingly so that the amount of active ingredient per hectare would remain identical). During these 3 years of experimentation, the crude product proved to be practically as efficient as pure codlemone. This tends to demonstrate that the unwanted by-products do not reduce the efficacy of the method. However, the same crude product, formulated as baits for trapping, attracted practically no males. These experiments confirm that purity of the attractant is indispensable for sexual trapping but that the mating disruption technique is far less demanding in regard to quality.

D. Isolation, Area and Shape of the Orchards, and Placement of the Dispensers

The male codling moths can move very long distances inside and outside the orchards (9). The migrations of the females are less important; however, they move 50 to more than 100 m inside the orchards and they tend to aggregate on the borders (10,11). It is generally admitted that outside the orchards, isolated, untreated apple trees or orchards with high population density and separated from one another by less than 100 m, constitute real infestation sources (12).

In the trial orchards in which problems appeared, the first attacks always took place on the borders. The proportion of failures decreased greatly after 1984, when most of the orchards of less than 2 to 3 ha or those insufficiently isolated were abandoned (see Fig. 1). As for the placement of the dispensers, about a quarter of the total amount of the attractant was always distributed on the borders; small dispensers were hooked on each tree, that is, at a distance of 4 to 5 m from one another, regardless of the shape and size of the orchards. The interiors of the orchards were always treated with more heavily loaded dispensers, placed further apart. However, it was not possible to get a clear appreciation of what the optimum density of sources should be. Nevertheless, it seems logical to raise the number of dispensers in the small orchards and to reduce it in the larger ones; the load on each dispenser should then be reduced or increased to maintain the same amount of active ingredient per hectare. Satisfactory results were then achieved with 100 sources per hectare inside orchards of 1 to 3 ha and with 40 to 50 sources per hectare inside larger ones. For practical reasons the dispensers were hooked to branches at a height of 1.5 to 2 m from the ground. In a trial in which the dispensers were simply thrown on the ground, the inhibition of catches was satisfactory in the sexual traps, but important damage rapidly appeared. Under these particular conditions the emission rate of the attractant was accelerated and its degradation increased, depleting the dispensers prematurely. Moreover, the concentration of attractant was probably insufficient in the tops of the canopy.

E. Population Density

It is difficult to assess the efficacy of the mating disruption technique. It is indeed impossible to conduct the trials in microplots with checks and repetitions, for the attractant is carried away by the wind and the moths that mate in the check plots move into the whole trial area to lay eggs. However, if the population dynamics in the trial orchards and in the reference orchards are compared year after year, the efficiency of mating disruption technique can be estimated to within 90% on average, with, however, some important variations according to the isolation, the size and shape of the orchards, the climatic factors, and the population densities of the pest (7). With classic insecticide treatments which ensure direct protection of the plant, the efficacy shows little variation with insect density. On the other hand, the efficacy of the mating disruption technique decreases when the densities increase, because the adult behavior is affected.

As an example, Figure 7 sums up the trials conducted in Begnins with Hercon dispensers. No curative treatment was applied in the

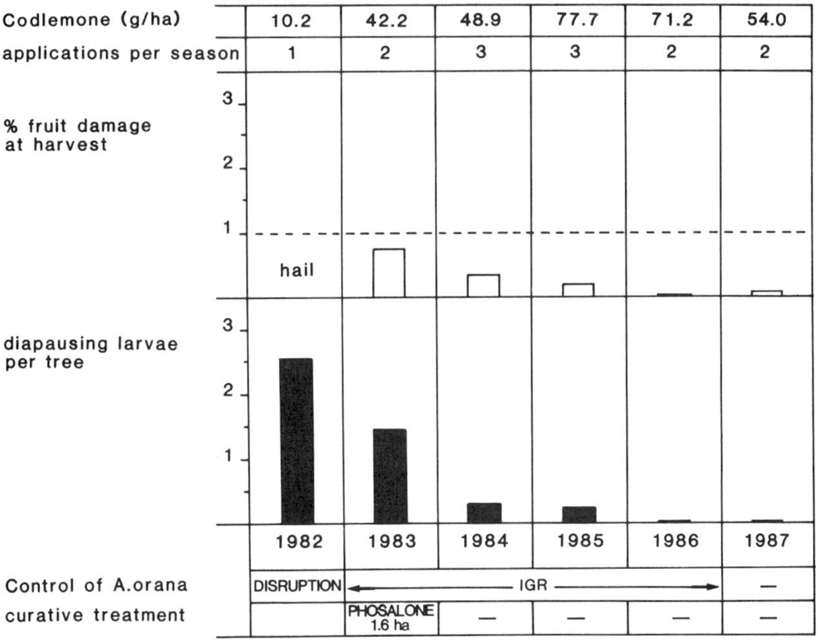

FIGURE 7 Trial summary of the mating disruption technique conducted in Begnins from 1982 to 1987 in an orchard of 9 ha.

summer of 1982, because the harvest was severely damaged by hail. In 1983 a curative treatment was needed to control codling moth on a limited part (1.6 ha) of the orchard where the tolerance threshold was reached. After this, no summer treatment with insecticide was made, the problem with summer fruit tortrix moth was solved by means of fenoxycarb applications each spring. Figure 7 clearly shows that the population density of codling moth, estimated by means of 40 band traps of corrugated cardboard, decreases gradually year after year. In the same way, Figure 8 shows the dynamics of hibernating populations from 1976 to 1985 in a small test orchard (0.1 ha) at the Federal Research Station of Changins. In 1976—an extremely favorable year for the development of codling moth—when no treatment at all was applied, the population reached the record number of 120 diapausing larvae per tree. From 1977 to 1979, mating disruption was conducted by means of rubber-tubing dispensers filled with pure attractant and sealed, having a load corresponding to 71 g/ha. The population

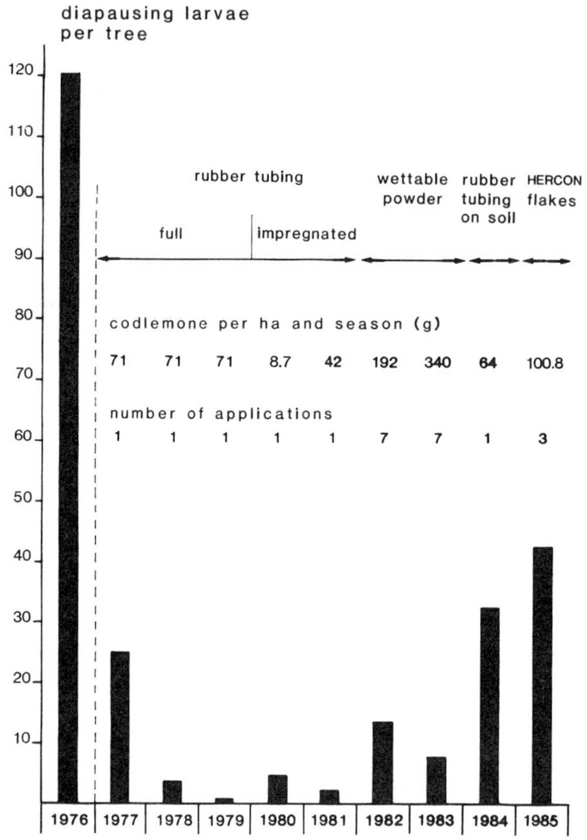

FIGURE 8 Changes in progress of hibernating larvae of codling moths from 1976 to 1985 in a small orchard in Changins where the mating disruption technique was tested.

decreased by about five times during each of these 3 years. In 1980, when the amount of codlemone impregnated in the rubber-tubing dispensers was reduced to give 8.7 g/ha, the population increased by six times, but in 1981, with one application of 42 g/ha, it decreased by half. In 1982 and 1983, a wetable powder containing attractant (codlemone-β-cyclodextrin-complex) was tested by seven applications made during the season by means of a classic spray machine. In spite of an amount of 192 g of codlemone per hectare, the population increased by six times in 1982, because the attractant evaporated much too rapidly. In 1983, the amount of attractant had to be increased up to 340 g/ha to slightly reduce the population. In 1984, when the rubber tubing dispensers were just thrown on the soil, the population increased by four times, in spite of an amount of 64 g attractant per hectare. And finally, in 1985—a very favorable year for codling moth—the population increased somewhat more, despite application of 100 g/ha, because the first of the three applications with Hercon dispensers was made too late. It is obvious that this small area is the typical example of an orchard for which mating disruption technique should not be used in practice because the relatively high proportion of borders and the high initial level of the population favors sudden outbreaks of the pest. It, nevertheless, had the advantage of testing the mating disruption technique under extremely severe conditions. It also shows very clearly that the population decline was always rather slow.

F. Reappearance of Other Pests

During the first years of testing we often encountered difficulties because of the summer fruit tortrix moth, *A. orana*. When the tolerance threshold for that pest was exceeded, broad-spectrum curative treatments had to be applied, thus reducing the possibility of evaluating the efficiency of the disruption technique against the codling moth. The problem was resolved later by using fenoxycarb each spring on the last larval instar of *A. orana*.

Hawthorn leafroller (*Grapholitha janthinana* D.) reappearance has slowly intensified in one of the orchards where it had been possible to avoid all chemical treatments since 1983 (13). The average damage caused by that pest reached 0.76% for the 1986 harvest, with one area exceeding 1.4%. In 1987, a first experiment of mating disruption technique against that insect led to promising results. In other orchards, however, that leafroller did not reappear, although all insecticide treatments had been suppressed for several years. It must be added that when the mating disruption technique against codling moth replaces broad-spectrum chemical treatments, it creates ideal conditions for the introduction of *Typhlodromus* predatory mites (14).

IV. CONCLUSIONS AND RECOMMENDATIONS

In about 10 years, experimentation with the mating disruption technique to control codling moth has progressed to a stage at which practical application may be seriously considered. All problems have certainly not been solved, but the feasibility as well as the limits of this new biotechnical means of control now appear more clearly. The best omen for the future success of this method in the context of integrated pest management is the fact that it is totally innocuous to useful fauna and the environment. The recommendations drawn from the Swiss experiments are supported by the trials conducted either by other research stations (3,4,5,15,16) or by the two firms BASF and Dr MAAG Ltd. in the Swiss process of registration. Because of its mode of action specifically on the behavior of the insect and because of the dispersal capacity of the moth, the practical use of the mating disruption technique is a necessary subject to some topological and ecological restrictions:

The mating disruption technique against codling moth must be exclusively reserved for isolated orchards of no less than 3 ha.

Isolation from the external infestation sources of the codling moth must be at least 100 m.

When the mating disruption technique is applied only on part of a set of juxtaposed orchards, a buffer zone of 40 to 50 m must be kept, where mating disruption technique and classic insecticide treatments should be used concurrently.

The borders of orchards must be protected by small dispensers placed at a maximum distance of 4 to 5 m from one another, but inside the orchard bigger ones can be placed with more space between them.

The mating disruption technique should not be used when the borders are especially irregular or when the fields are too elongated, because the porportion of borders—where matings are more likely to occur—becomes too important in comparison with area.

Difficulties may also appear in crops with heterogeneous crown sizes or with trees of more than 5 m in height. Experience has shown that in the tops of the canopy the attractant is rapidly removed by wind.

The initial populations must not have more than two or three diapausing larvae per tree, so that the tolerance threshold is not exceeded during the first year of application of the mating disruption technique, because insect density decreases gradually with successive generations.

If the half-emission time of the attractant is sufficiently long, the dispensers can be put in place some time before the beginning of the flight period of the codling moth to avoid all mating at that time. The second application should be made about 40 days after the first one.

Finally, it should be noted that the placing of the dispensers—which only takes from half an hour to an hour per hectare, according to the size of the orchard—should not discourage anyone from using the mating disruption technique.

ACKNOWLEDGMENT

We thank the numerous growers for letting us conduct our trials on their orchards and for taking an active part in placing the dispensers.

We thank H. Arn, C. Blaser, B. Bloesch, J. F. Brunner, H. Hoehn, M. N. Inscoe, B. A. Leonhardt, R. L. Ridgway, E. Rothlisberger, and V. Laneve for their invaluable collaboration.

These trials have been supported by BASF from Limburgerhof (RFA). We thank U. Neumann from BASF for his helpful advice.

REFERENCES

1. Roelofs, W., Comeau, A., Hill, A., and Milicevic, G. Sex attractant of the codling moth: Characterization with electroantennogram technique. *Science 174*:297–299 (1971).
2. Charmillot, P. J. and Blaser, C. Le fénoxycarbe, un régulateur de croissance d'insectes homologué contre la tordeuse de la pelure *Adoxophyes orana* F.v.R. *Rev. Suisse Vitic. Arboric. Hortic. 17*:85–92 (1985).
3. Sacco, M. and Pellizzari Scaltriti, G. Prova di lotta contro la *Cydia pomonella* mediante la tecnica della confusione. *Inform. Fitopatol. 11*:51–56 (1983).
4. Mani, E., Schwaller, F., and Riggenbach, W. Bekämpfung des Apfelwicklers (*Cydia pomonella* L.) mit der Verwirrungsmethode in einer Obstanlage im Bündner Rheintal; 1979–81. *Mitt. Schweiz. Entomol. Ges. 57*:341–348 (1984).
5. Moffitt, H. R. and Westigard, P. H. Suppression of the codling moth (Lepidoptera: Tortricidae) population on pears in southern Oregon through mating disruption with sex pheromone. *J. Econ. Entomol.* 77:1513–1519 (1984).
6. Mani, E. Field trials to control codling moth by mating disruption, 1979–85. In Seventh Symp. Integrated Plant Protection in Orchards. (Dickler, E., Blomers, L. H. M., and Minks, A. K., eds.). IOBC/WPRS, 1986, pp. 166–169.
7. Charmillot, P. J. Etude des possibilités d'application de la lutte par la technique de confusion contre le carpocapse *Laspeyresia pomonella* (L.) (*Lep. Tortricidae*). Thèse No 6598. Ecole polytechnique fédérale, Zurich, 1980, 79 pp.
8. Bartell, R. J. Mechanisms of communication disruption by pheromone in control of Lepidoptera: A review. *Physiol. Entomol.* 7:353–364 (1982).
9. Mani, E. and Wildbolz, T. The dispersal of male codling moths (*Laspeyresia pomonella* L.) in the Upper Rhine Valley. *Z. Angew. Entomol. 83*:161–168 (1977).

10. Wildbolz, T. and Baggiolini, M. Über das Mass der Ausbreitung des Apfelwicklers während der Eiablageperiode. *Mitt. Schweiz. Entomol. Ges.* *32*:241–257 (1959).
11. White, L. D., Hutt, R. B., and Butt, B. A. Field dispersal of laboratory-reared fertile female codling moths and population suppression by release of sterile males. *Environ. Entomol.* *2*: 66–69 (1973).
12. Myburgh, A. C., Madsen, H. F., Rust, D. J., and Bosman, I. P. Codling moth (Lepidoptera: Olethreutidae): Sex attractant traps as an adjunct to control programmes. *Proc. I. Congr. Entomol. Soc. South Afr.*, 1975, pp. 99–108.
13. Charmillot, P. J. and Blaser, C. La tordeuse de l'aubépine *Grapholitha janthinana* Dup., un ravageur potentiel de nos vergers de pommiers et pruniers? *Rev. Suisse Vitic. Arboric. Hortic.* *16*:293–296 (1984).
14. Genini, M. and Baillod, M. Introduction de souches résistantes de *Typhlodromus pyri* (Scheuten) et *Amblyseius andersoni* Chant (Acari: Phytoseidae) en vergers de pommiers. *Rev. Suisse Vitic. Arboric. Hortic.* *19*:115–123 (1987).
15. Audemard, H., Charmillot, P. J., and Beauvais, F. Trois ans d'essais de lutte contre le Carpocapse (*Laspeyresia pomonella* L.) par la méthode de confusion des mâles avec une phéromone sexuelle de synthèse. *Ann. Zool. Ecol. Anim.* *11*:641–658 (1979).
16. Rothschild, G. H. L. Suppression of mating in codling moths with synthetic sex pheromone and other compounds. In *Insect Suppression with Controlled Release Pheromone Systems*, Vol. II. (Kydonieus, A. F. and Beroza, M., eds.). CRC Press, Boca Raton, 1982, pp. 117–134.

12

Oriental Fruit Moth in Australia and Canada

RICHARD A. VICKERS / Commonwealth Scientific and Industrial Research Organization, Canberra City, Australia

I. INTRODUCTION

The problems associated with repeated pesticide applications, such as the creation of secondary pests brought about by the destruction of beneficial predators, the increased potential for evolution of resistant pest species, and the dangers pesticides pose to human health are well known. In the search for a more rational approach to pest control, pheromones offered promise in two areas: They could be used as population monitors so that control measures could be more effectively timed, and they could be used directly as a means of control, either by male removal or by mating disruption. Although the pheromone blends of several hundred insect species, many of them pests, have now been identified, attempts to use pheromones as mating disruptants have met with limited success. One notable exception has been the development in Australia of a mating disruption system for the control of *Cydia* (*Grapholita*) *molesta*, the oriental fruit moth (OFM). This chapter summarizes the events leading to the commercialization of the system in Australia. It has subsequently been evaluated in the Northern Hemisphere, and after successful commercial use in the United States, is currently being tested in Canada.

Under Australian conditions, OFM undergoes four to five generations each year and is considered the key insect pest of the stonefruit industry, whose value exceeds 13.4 million dollars (Australian) annually (1). Minor pests include aphids, thrips, and the leafroller caterpillar *Epiphyas postvittana*. A secondary mite problem with

Tetranychus urticae has been created by the repeated use of organophosphates (OPs) such as azinphos methyl and parathion, conventionally applied four to six times a season for OFM control.

II. AUSTRALIAN TRIALS

Identification of the major component of the OFM female pheromone, (Z)-8-dodecenyl acetate (Z8-12:Ac), was made by Roelofs et al. in 1969 (2), and disruption trials were begun in Australia with the compound a year later (3). [The material used in these trials contained as a contaminant 2% to 3% of the trans isomer. This was later shown to be a constituent of the pheromone blend (4)].

Various improvised pheromone dispensers were tested before the final choice of a polyethylene microcentrifuge tube, 400 μl in capacity and manually loaded with 50 μl of pheromone. The first trials measured the ability of males to locate synthetic pheromone sources and, on occasions, caged virgin females, against a background of pheromone released in a variety of ways. This work provided information about the distance over which pheromone released from a single downwind source can influence trap catch (Table 1); the effect of dispenser density and spacing on male catch (Tables 2 and 3, respectively) and the relationship between pheromone release rate and suppression of trap catch (Fig. 1). In these latter trails, Rothschild established

TABLE 1 Influence on Male Captures of Distance Between Microcentrifuge Dispensers and Pheromone Traps

Distance between centrifuge dispenser and trap (cm)	Mean catch/trap (± s.e.)
Control (without dispenser)	28.0 ± 9.6^{a}
60	6.4 ± 3.5^{b}
120	8.6 ± 2.8^{b}

Catches represent total of four successive samplings from five traps; means not followed by same letter are significantly different at $p < 0.05$ (Duncan's M.R. test).
Source: Ref. 3.

TABLE 2 Numbers of Microcentrifuge Dispensers per Tree and Male Catch at Pheromone Traps

No. dispensers/ tree	Mean catch/trap (± SE)	% Disruption
0	16.8 ± 2.2[a]	
0.5	4.7 ± 1.0[b]	71.7
1	4.1 ± 1.3[b]	75.4
2	1.3 ± 0.4[c]	92.3

Catches represent total of three successive samplings from 12 traps; means not followed by same letter are significantly different at $p < 0.05$ (Duncan's M.R. test).
Source: Ref. 3.

TABLE 3 Spacing of Microcentrifuge Dispensers and Male Captures at Pheromone Traps

% Trees with dispensers	Mean area/ dispenser (m^2)	Pheromone release rate/site (μg/hr)	Mean catch/trap (± SE)
0			37.1 ± 2.8[a]
100.0	50	25	9.9 ± 2.3[b]
50.0	100	50	10.0 ± 3.5[b]
25.0	200	100	4.3 ± 1.1[b]
12.5	400	200	7.2 ± 2.1[b]
6.2	800	400	22.0 ± 5.1[c]

Catches represent total of three successive samplings from nine traps; release rate estimated during 3-hour daily flight period; means not followed by same letter are significantly different at $p < 0.05$ (Duncan's M.R. test).
Source: Ref. 3.

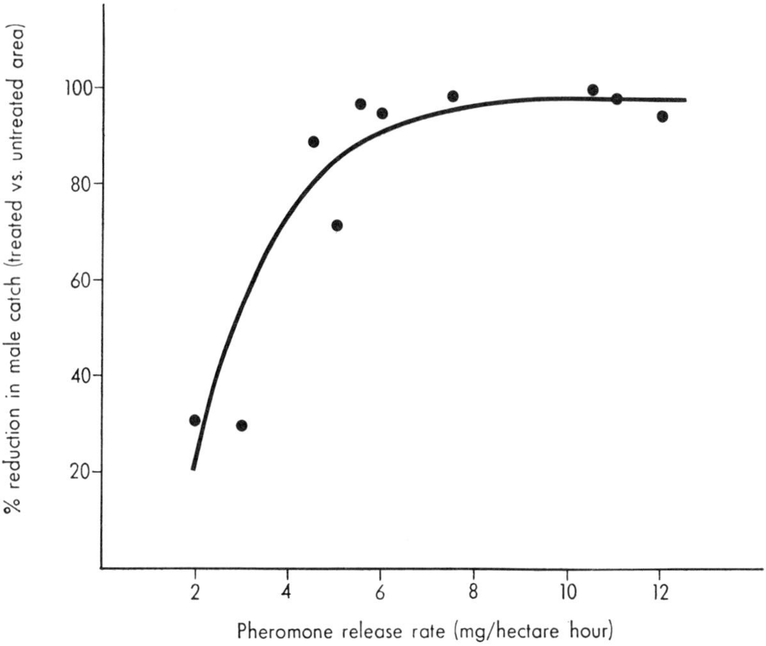

FIGURE 1 Reduction in male captures at pheromone traps in relation to pheromone release rate (from Ref. 3).

that a minimum release rate of 6 mg/ha/per hour was required to reduce trap catch by 90% or more. Pheromone released from dispensers placed at either 1.5 or 3.5 m in trees approximately 3.4 m high was equally effective in disrupting catch.

In 1972, larger quantities of pheromone became commercially available and trials were begun to assess disruption treatments on a larger scale. With a reduced risk of immigrating mated females confusing the trial results, it became feasible to include in the evaluation of the treatments the mating status of females caught at feeding lures and, where season-long protection was attempted, levels of fruit and tip infestation. In all trials, the catches of males at pheromone traps and the proportion of mated females taken at feeding lures (brown sugar solution plus terpinyl acetate) were significantly lower in the treated orchards than they were in the controls. Similarly, levels of fruit and shoot damage were lower in the pheromone-treated area than in the control. The conclusion from these trials was that pheromone released from suitably placed dispensers, at a rate exceeding 6 mg/ha per hour, was as effective as the standard insecticide schedule in controlling OFM (3).

Before proceeding to the next phase of this account it is appropriate to mention a number of other relevant trials conducted by Rothschild during this period. Mating disruption was also attempted with an analogue of Z8-12:Ac, dodecyl acetate (12:Ac); (5). When this compound was released from within traps baited with either synthetic pheromone or virgin females, or from a source 30 cm from the bait, male catch was eliminated. However, 12:Ac released as a background odor over a large area had the opposite effect, and catches were significantly enhanced. The reasons for this result were not clear and Rothschild urged caution in identifying compounds as inhibitors or synergists on the basis of field-trapping results.

Trials concerned with the behavior of males at pheromone sources (6) and the influence of trap design, pheromone loading, and bait composition on male catch (7) were run in conjunction with the early disruption trials. Briefly, these indicated that there was no difference in the period over which males responded to virgin females or synthetic pheromone sources, the onset of which activity being determined by the time of sunset, modified by temperature. Trap design had little influence on total catch. Catches at traps baited with 1 to 1000 μg of Z8-12:Ac were similar but were enhanced by the addition of (*Z*)-8-dodecanol (Z8-12:OH) and dodecanol (12:OH). Both compounds have since been identified as constituents of the pheromone blend (4). The addition of E8-12:Ac at levels greater than 23% reduced trap catches, but at lower levels made no significant difference to catch.

Most of the trials that established the requirements for successful mating disruption of OFM had been completed by the end of the 1974–1975 season. However in his account of that work, Rothschild (3) concluded that "...control of OFM through pheromone treatments is not generally economic at this stage." For this reason and because of a general decline in the peach industry in Australia, further work on the program was suspended.

Under more favorable circumstances, trials were resumed in 1980–1981, when population data were gathered for a trial which was to attempt control of OFM by mating disruption over an entire peach-growing district (8). Pheromone traps were placed in all orchards within two small but discrete districts in southern New South Wales--Stony Point and Stanbridge. Despite considerable differences in the mean catch per trap per day among orchards, there was no significant difference between means for the two districts before the imposition of pheromone treatments. In the following season all Stanbridge orchards were treated with pheromone, whereas those at Stony Point served as controls. In 1982–1983 the Stanbridge orchards and two of the remaining three orchards at Stony Point were treated.

The effectiveness of the treatments was assessed by comparing pheromone trap catches (Table 4), fruit and shoot damage (combined to give an "index of infestation," Table 5), and the mating status of females caught at feeding lures (Table 6). All criteria indicated that the treatments had reduced the population size (as indicated by the pheromone trap and feeding lure captures), levels of infestation, and the mating success of wild females. This latter result was particularly encouraging, as the mating status of wild females is considered to be the most meaningful criterion of mating disruption (9). Records of insecticide application for OFM control in the participating orchards were incomplete. However no more than a single early-season application was made in any of the treated orchards, and most were not sprayed at all.

The district trials provided further evidence that OFM could be successfully controlled by mating disruption. However, it was obvious that for the method to become a commercial proposition the dispenser would have to be modified to allow its automated production.

Dispensers of various kinds, both improvised and commercial, were evaluated by Rothschild between 1974 and 1977 (9). On the basis of information supplied by Rothschild about the physical properties of the microcentrifuge dispensers, the Shin-Etsu Chemical Company of Japan developed a dispenser the production of which could be automated and which incorporated a wire to facilitate application in the field.

TABLE 4 Pheromone Trap Catch (Number of Moths per Trap per Day) for All Orchards Over the Period of the Trial

	Stony Point Orchards				Stanbridge Orchards				
Year	140	162	169	187	842	866	871	891	892
1980–1981	1.35	0.32	1.39	0.31	0.17	0.72	0.43	0.89	0.94
1981–1982	1.61	0.38	1.89	0.52	0.02[a]	0.04[a]	0.02[a]	0.02[a]	0.03[a]
1982–1983	0.01[a]	0.01[a]	4.51		0.00[a]	0.04[a]	0.00[a]	0.00[a]	

[a]Pheromone treatment applied.
Source: Ref. 8.

TABLE 5 Index of Infestation (Combined Percentage Fruit and Shoot Damage Measured at the End of Each Season) in Treated and Untreated Orchards

	Stony Point Orchards				Stanbridge Orchards				
Year	140	162	169	187	842	866	871	891	892
1980–1981	5.4	1.0	10.3	0.5[a]	5.0	3.8	1.4	0.9	2.0[a]
1981–1982	5.7	6.6	7.9	10.4	2.0[b]	0.9[b]	0.3[b]	1.5[b]	1.1[b]
1982–1983	0.3[b]	0.9[b]	11.8		0.5[b]	0.1[b]	0.5[b]	0.0[b]	

[a]Percentage shoot damage only.
[b]Pheromone treatment applied.
Source: Ref. 8.

TABLE 6 Feeding Lure Catch, 1982–1983

	Phermone-treated orchard		Control orchard		Significance level
	%	*n*	%	*n*	
Total catch		219a		1761b	<0.05
Females					
catch	48.4a	106	70.4b	1239	<0.05
mating status					
virgin	76.9b	60	3.4a	11	<0.001
mated once	21.8a	17	93.9b	308	<0.001
multiple mating	1.3a	1	2.7a	9	n.s.

Values followed by the same letter within a row are not significantly different at the indicated level of significance (Duncan's multiple-range test, arc-sine transformation of all % values).
n.s., not significant.
Source: Ref. 8.

An Australian company, Biocontrol Ltd., obtained the Australian marketing rights for the Shin-Etsu dispenser/pheromone combination and the first commercial releases of the product under the name Isomate M were made in time for the 1984–1985 season. The most recent figures available indicate that in the region in which most of the later disruption trials were carried out, about 39% of the area planted with peaches is being protected with Isomate M (J. Slack, personal communication). In the other major Australian peach-growing district, about 10% of the crop is similarly protected (R. Bennett, personal communication). The costs of mating disruption compare favorably with those of conventional control in Australia. In the 1985–1986 season, a typical azinphos methyl/cyhexatin program for the Murrumbidgee Irrigation Area, inclusive of tractor and labor expenses, cost 389 dollars (Aust.)/ha. Control by mating disruption cost 267 dollars (Aust.)/ha (J. Slack, personal communication).

III. CANADIAN TRIALS

The OFM is a worldwide pest of stone fruit. The success of mating disruption as a means of control in Australia encouraged Biocontrol Ltd. to market the method in a number of other countries, one of which was Canada.

The Canadian peach industry, worth 17.8 million dollars (Canadian) in 1985 (D. Pree and M. Trimble, personal communication) is concentrated in British Columbia and Ontario, but only in the latter province is OFM a problem. Here it has three to four generations each season (11) and is part of a complex of pest species that includes the peachtree borers *Synanthedon pictipes* and *S. exitosa* and, sporadically, *Choristoneura rosaceana* and *Anarsia lineatella*. The OFM is traditionally controlled with two to four OP sprays each season (12). These sprays may be more effectively timed by using monitoring traps (11).

Evaluation of Isomate M was begun at Vineland Station, Ontario, during 1987. A 0.6-ha peach block that suffered in excess of 50% OFM damage in 1986 was treated with pheromone. This block was compared with one that received the standard spray program and another that was not treated in any way.

At harvest, fruit in the pheromone-treated, sprayed, and unsprayed blocks sustained 15%, 5%, and 45% OFM damage respectively (P. Kirsch, personal communication). Although 15% is an unacceptable level of damage, it was significantly lower than that sustained in the previous season. Where population densities are high, as they were in this instance, mating disruption might be enhanced by an initial application of insecticides so that the population is reduced to a more appropriate level for control with pheromone.

IV. CONCLUSION

Without knowing how mating disruption works (possible mechanisms were reviewed by Bartell; 13), it is difficult to identify the reasons for the technique succeeding against OFM, whereas it has failed against a range of other pest species. It may have been fortunate that the pheromone used in the early trials contained as a contaminant the trans isomer of the major component, later shown to be a minor constituent of the pheromone blend. If OFM mating is disrupted by the creation of false trails then it may be important to include minor components in the disruption blend (4).

Under Australian conditions, where OFM is the key pest of stone fruit, mating disruption offers a viable alternative to insecticidal control. Concern expressed by occupational safety and health committees for the welfare of spray plant operators handling OPs, community concern about pesticide residues, and the recent banning of a commonly used miticide, cyhexatin, can only help to promote the adoption of OFM control by mating disruption.

Where a complex of pest species exists, as it does on some Canadian peach orchards, a form of integrated pest management in which mating disruption is used to control part of the complex to reduce the reliance on pesticides, may be possible.

The successful development and marketing of a mating disruption system for the control of OFM should provide an incentive for researchers and industry to continue the search for similar systems for the control of other pests.

REFERENCES

1. Cribb, J. (ed.). Canned fruits. In *Australian Agriculture: The Complete Reference on Rural Industry*. National Farmer's Federation, 1987, 291 p.
2. Roelofs, W. L., Comeau, A., and Selle, R. Sex pheromone of the oriental fruit moth. *Nature 224*:723 (1969).
3. Rothschild, G. H. L. Control of the oriental fruit moth *Cydia molesta* (Busck) (Lepidoptera: Tortricidae) with synthetic female pheromone. *Bull. Entomol. Res. 65*:473–490 (1975).
4. Cardé, A. M., Baker, T. C., and Cardé, R. T. Identification of a four-component sex pheromone of the female oriental fruit moth, *Grapholitha molesta* (Lepidoptera: Tortricidae). *J. Chem. Ecol. 5*:423–427 (1979).
5. Rothschild, G. H. L. Problems in defining synergists and inhibitors of the oriental fruit moth pheromone by field experimentation. *Entomol. Exp. Appl. 17*:294–302 (1974).

6. Rothschild, G. H. L. and Minks, A. K. Time of activity of male OFM at pheromone sources in the field. *Environ. Entomol.* *3*:1003–1007 (1974).
7. Rothschild, G. H. L. and Minks, A. K. Some factors influencing the performance of pheromone traps for oriental fruit moth in Australia. *Entomol. Exp. Appl.* *22*:171–182 (1977).
8. Vickers, R. A., Rothschild, G. H. L., and Jones, E. L. Control of the oriental fruit moth, *Cydia molesta* (Busck) (Lepidoptera: Tortricidae), at a district level by mating disruption with synthetic female pheromone. *Bull. Entomol. Res.* *75*:625–634 (1985).
9. Rothschild, G. H. L. Mating disruption of lepidopterous pests: Current status and future prospects. In *Management of Insect Pests with Semio-Chemicals*. Mitchell, E. R., ed.). Plenum Press, New York, 1981, pp. 207–228.
10. Rothschild, G. H. L. A comparison of methods of dispensing synthetic sex pheromone for the control of oriental fruit moth, *Cydia molesta* (Busck) (Lepidoptera: Tortricidae), in Australia. *Bull. Entomol. Res.* *69*:115–127 (1979).
11. Pree, D. J., Herne, D. H. C., Phillips, J. H. H., and Roberts, W. P. Oriental fruit moth. Pest management program for peach series. Factsheet agdex $\frac{212}{624}$. Ministry of Agriculture and Food, Ontario, Canada, 1983.
12. Canadian Ministry of Agriculture and Food. *1986 Fruit Production Recommendations*. Publication No. 360. Ministry of Agriculture and Food, Ontario, Canada, 1986.
13. Bartell, R. J. Mechanisms of communication disruption by pheromone in the control of Lepidoptera. A review. *Physiol. Entomol.* 7:352–364 (1982).

13

Mating Disruption of Oriental Fruit Moth in the United States

RICHARD E. RICE / University of California, Davis, California

PHILIPP KIRSCH / Biocontrol Ltd., Davis, California

I. INTRODUCTION AND EARLY FIELD TRIALS

The oriental fruit moth, *Grapholita molesta* (Busck), is a pest of deciduous fruits throughout most of the world. Its favorite host is quince, but the greatest economic losses from oriental fruit moth (OFM) occur on peaches and nectarines. Oriental fruit moth also attacks apples, pears, plums, apricots, cherries, and almonds. This pest was first detected in the eastern United States in 1913 and, in spite of quarantine regulations, has spread throughout the major fruit-producing areas of North America. It had reached southern California by 1942, and fruit losses were experienced in the San Joaquin Valley by 1954 (1).

There are normally three to four generations of OFM in the northern fruit-production areas of the United States, whereas five to six generations occur in the southern states and in California. Chapman and Lienk (2) have recently reviewed the biology and ecology of the oriental fruit moth.

Control strategies for OFM have normally required the use of insecticides. However, in some areas, such as the East Coast or Midwest where suitable alternate hosts for parasites are available, a hymenopterous parasite, *Macrocentrus ancylivorous* (Rohwer), has been effective in reducing populations of OFM to subeconomic levels (3).

In recent years, the availability of synthetic sex pheromones for OFM has presented an opportunity to use these semiochemicals as alternatives to standard insecticide control programs. Early field trials initiated in 1972 in Australia (4) and in Georgia (5) demonstrated both the problems associated with this control strategy and the potential success of the technique. Improvements in synthetic pheromone composition (6,7) and pheromone dispensers (8) led to increasing successes with the method, culminating in the recent report of successful area-wide control of OFM in Australia (9).

II. CURRENT FIELD TRIALS

A. Western United States

1. *Methods and Procedures*

In 1985, renewed attempts at mating disruption of OFM were initiated in North America in large-scale field trials in California. Trials were continued in 1986 and 1987. These trials were conducted in all of the major peach- and nectarine-growing areas of the San Joaquin Valley, extending from the canning peach districts north of Sacramento and near Stockton, to the primary fresh market production areas between Fresno and Bakersfield. All trials were placed in mature orchards, with plot sizes ranging from 0.8 to 8.0 ha (2.0–20.0 acres). All orchards were watered by flood or furrow irrigation, and with one exception, pheromone-treated blocks were not isolated from adjacent nonpheromone-treated blocks.

The pheromone dispensers used in these trials were manufactured by the Shin-Etsu Chemical Company Ltd., Tokyo, and were provided by an Australian company, Biocontrol Ltd., Warwick, Queensland, Australia. These dispensers each contain a 75-mg blend of (*Z*)-8-dodecenyl acetate (93%), (*E*)-8-dodecenyl acetate (6%), and (*Z*)-8-dodecenol (1%). Dispensers were placed in the upper third of the tree canopy, at heights of 2.5 to 4.0 m (7.5–12.5 ft) and at densities of 1000 dispensers per hectare (400/acre).

The dispensers were designed for a 3-month field life, maintaining a release rate of approximately 30 to 35 μg per dispenser per hour. They were applied twice during each growing season, with first applications made in late February or early March at about the time peach or nectarine trees were in full bloom. Because dispensers must be placed before emergence of the overwintered population, actual timing was determined by the first collection of OFM in pheromone traps. The second application of dispensers was made exactly 90 days after the first, in late May or early June.

Efficacy of mating disruption was measured in pheromone-treated and check orchards using standard wing-type pheromone traps at a density of one trap per hectare or a minimum of two traps in test blocks smaller than 2.0 ha. Terpinyl acetate bait traps were also placed in both pheromone-treated and untreated check blocks to monitor the sex ratio of captured OFM, the mating success of OFM females by dissection and presence of spermatophores, and as indicators of long-term changes in OFM population density. In a few field trials in 1985, counts of infested shoot terminals were also attempted as a measure of mating disruption. This technique is a good indicator of early season mating disruption; however, it was too time-consuming and laborious and was therefore not continued in 1986 and 1987.

As varieties of peaches or nectarines matured in the different test blocks (mid-May to late September), a minimum of 1000 fruit per treatment were examined for the presence of OFM larvae. In test plots involving fresh market cultivars, buckets of fruit picked by field crews from the center one-third of the blocks were randomly selected and the fruit inspected for damage. In plots using canning (cling) peach cultivars, individual fruits were randomly selected and inspected as fruit was dumped into cannery bins by the picking crews. All fruit showing any sign of insect damage were dissected and a determination made of the insect responsible for the observed injury. In most of the cultivar evaluations, at least 2000 fruit were examined in this fashion. Check blocks not treated with pheromones were treated either with standard OFM insecticide programs (10) or were left completely untreated.

In many of the field trials, pheromone traps for the peach twig borer (PTB), *Anarsia lineatella* (Zeller), and the omnivorous leafroller (OLR), *Platynota stultana* (Walsingham), were placed along with the OFM traps to provide population measurements of these pests as well. In several pheromone-treated blocks over the 3 years of trials, peach twig borer populations reached economic levels in the absence of in-season OFM sprays. In these cases, the insecticide endosulfan was used to control twig borer populations, with a minimal impact on the resident OFM population.

2. Results

a. Response to pheromone traps: Dramatic differences were observed in collections of male OFM in pheromone traps in pheromone-treated blocks compared with untreated or insecticide-treated check blocks. Pheromone trap totals at the five mating disruption test trials in 1985 (Table 1) showed high numbers of male moths trapped at all locations in check blocks, whereas pheromone trap collections in the pheromone-treated blocks yielded only three moths for the entire season. Similar results were seen in 1986 (see Table 1).

The inability of pheromone traps to collect male moths (presumably as a result of male disorientation to pheromone point sources) at two test plots at the Kearney Agricultural Center in Parlier is also

TABLE 1 Collections of OFM in Pheromone and Terpinyl Acetate Bait Traps in Pheromone-treated and Untreated (Check) Orchards, California, 1985 and 1986

		Number moths collected			
		Pheromone traps		Bait traps	
Location	Year	Check	Pheromone	Check	Pheromone
Rio Oso	1985[a]	971	1	133	70
Yuba City	1985	1512	0	897	348
Sanger	1985	496	0	34	9
Parlier	1985	1434	0	347	45
Arvin	1985	1184	2		
Dinuba	1986[b]	2378	0	217	75
Exeter	1986	2938	2	1538	636
Parlier	1986	9114	0	6326	301
Arvin	1986	1109	11	333	26

[a]March 1 to September 6, 1985.
[b]March 1 to September 20, 1986.

shown in Figure 1. Pheromone trap collections of OFM in field 24 showed relatively high populations of moths in 1983 and 1984, when this block was untreated with either pheromone or insecticides. This 1.0-ha block of midseason freestone peaches was first treated with pheromone in 1985, and the treatment was repeated in 1986. In both of these years no moths were collected in pheromone traps in this block. In 1987, neither insecticide nor pheromone treatments were applied to the orchard and only two male OFM were collected in two pheromone traps during the season. It should be noted that some of the pheromone dispensers placed in the orchard in 1986 were still in the trees in 1987. Most of them, however, had been removed with normal winter pruning.

The field 21 orchard, which was treated with two insecticide sprays of diazinon in 1984 and again in 1985, also showed OFM collections typical of central California stone fruit orchards (Fig. 2). In the first year of pheromone treatment (from late February to

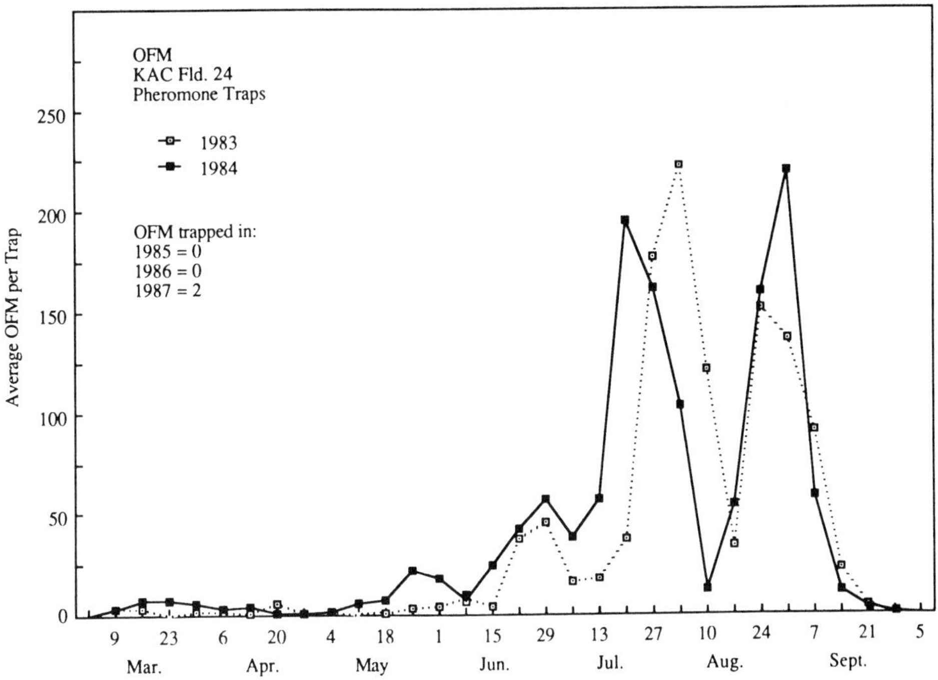

FIGURE 1 Collections of OFM males in pheromone traps in an untreated block of Fay Elberta peaches during two seasons before pheromone treatments (1983–1984), two seasons with pheromone treatments (1985–1986), and one season after pheromone use (1987) (Kearney Agricultural Center, Parlier, Fresno County, California).

mid-October 1986) 17 male OFM were trapped in pheromone traps in this block. The orchard was again treated with pheromone in 1987, and only one moth was trapped in four pheromone traps. These results, which are typical of the pheromone trap collections observed in all of the mating disruption trials conducted in California in 1985 to 1987, indicate that the OFM pheromone quality and quantity used in these trials was sufficient to provide satisfactory disruption of male moth response to point sources of pheromone represented by pheromone traps.

b. *Response to bait traps*: The effects of pheromone treatments on OFM populations was also indicated by collections of male and female moths in terpinyl acetate bait traps. The results of bait trap

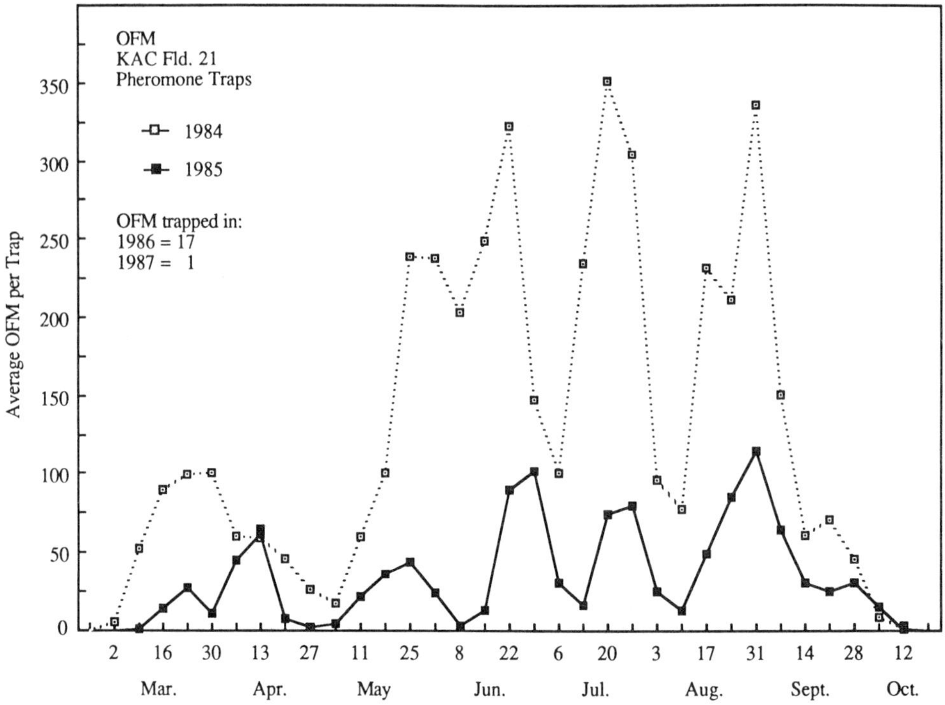

FIGURE 2 Collections of male OFM in pheromone traps in a peach orchard treated with insecticides (one application 1984, two applications 1985) and with synthetic pheromones in 1986 and 1987 (Kearney Agricultural Center, Parlier, California).

collections (see Table 1), from four of the five test locations in 1985 and four locations in 1986 showed total moth captures in the pheromone-treated blocks consistently lower than those in the conventional check blocks. A further example of similar reductions in bait trap collections is shown in one of the 1986 test blocks at Parlier (Fig. 3) in which the pheromone-treated block (field 24) was under the second year of pheromone disruption. Bait trap data from a test block at Parlier (field 21), also in the second year of pheromone treatment in 1987, showed declines in total OFM collections in the pheromone-treated block compared with the untreated check (Fig. 4).

Dissections of female OFM moths collected in bait traps showed consistent reductions in mating success in the pheromone-treated

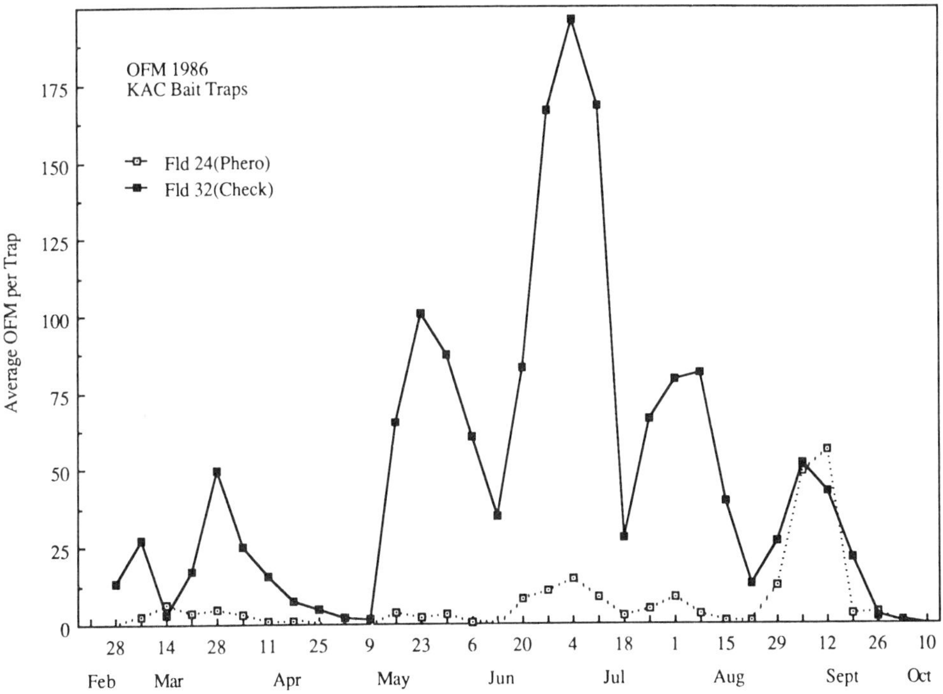

FIGURE 3 Comparison of total (male and female) OFM collections in terpinyl acetate bait traps in a pheromone-treated peach block and an untreated peach check (Kearney Agricultural Center, Parlier, California).

blocks compared with the untreated checks (Table 2). Mated females in the untreated check plots in all cases exceeded 90% (Tables 2 and 3 and Fig. 4). Mating success in pheromone-treated blocks varied considerably, with lower percentages observed in the first two or three flights and a tendency to increase to the 90% or higher level in later flights (see Table 3 and Fig. 4). The reasons for this late-season increase in mating are not known, but may be related to small plot size and migration of mated females (11). Also, the pheromone dispenser has a 90-day field life, so mating disruption resulting from the second seasonal application (in late May) could decrease with declining pheromone release rates in late August and September. However, usually population levels by this time were low (see Fig. 4),

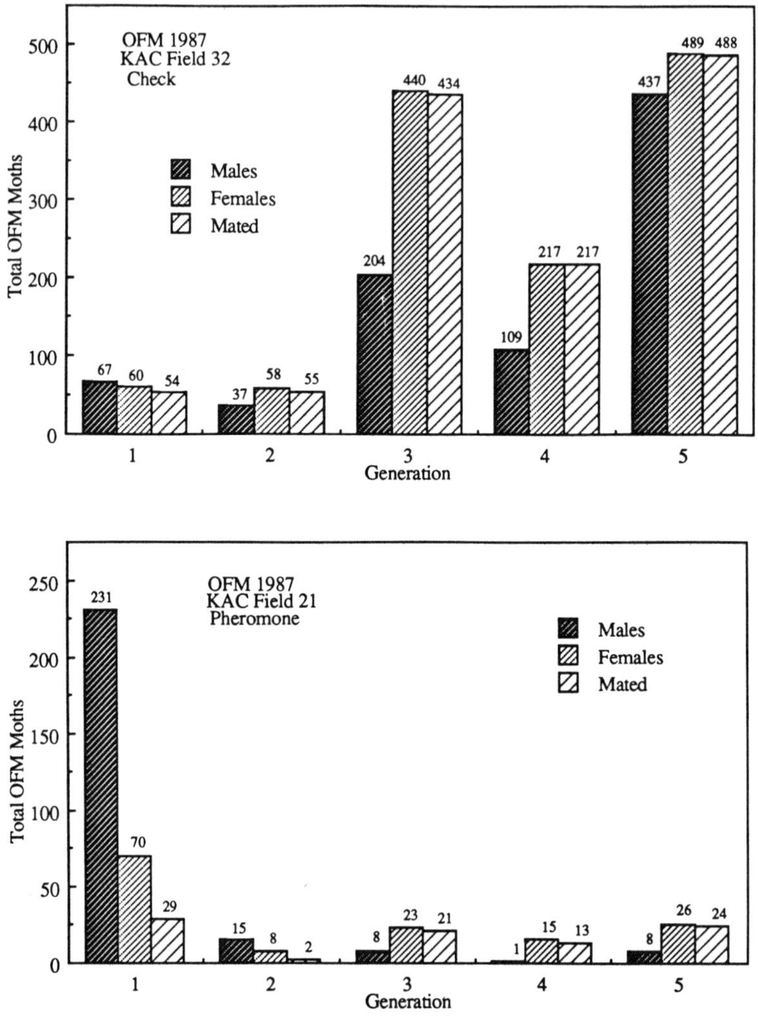

FIGURE 4 Bait-trap collections of male and female OFM and comparison of mated females in a pheromone-treated peach block and an untreated block (Kearney Agricultural Center, Parlier, California).

so that the numbers of viable eggs produced should have been considerably fewer than those in the conventional checks.

Captures of male OFM in bait traps were reduced in pheromone-treated orchards. A mean of the pooled 1985 and 1986 bait trap

TABLE 2 Comparison of Mated OFM Females Trapped in Pheromone-treated and Conventional Check Blocks, California, 1985

	Mated females (%)[a]	
Location	Check	Pheromone
Rio Oso	100.0	47.0
Yuba City	96.9	41.1
Sanger	96.2	77.0
Parlier	96.8	83.7

[a]Collections in terpinyl acetate bait traps, March 1 to September 6, 1985.

collections (Table 4) shows that 44.7% of the total catch in check blocks were males, compared with 27.4% in pheromone-treated blocks. These data correlate with results from Australia (3) and suggest that there could be an increase in mate-searching caused by disruption, with a corresponding decrease in feeding behavior and, therefore, a decreased male response to bait traps.

c. *Fruit damage*: The efficacy of mating disruption with synthetic pheromones in 1985 is indicated in data from five test blocks shown in Table 5. All of the check blocks received at least one insecticide treatment for OFM during the year. The results of these initial trials in 1985 showed that the pheromone treatment invariably achieved control levels equal to, or better than, those measured in the conventional check blocks.

Examination of infested fruit from mating disruption trials in 1986 (Table 6) show a similar pattern of control in the disruption blocks compared with the untreated checks. In those locations where infested fruit levels in the pheromone treatments exceeded those in the checks (e.g., Livingston, Escalon), the checks had been treated specifically for OFM with insecticides. At Yuba City in 1986, the pheromone dispensers were applied 3 weeks late because of winter flooding in the orchard. First-generation mating resulted in a high second-generation population that was treated with one application of insecticide.

The Parlier field 21 location was an example of a situation in which the pheromone, in the first year of application (1986), was apparently

TABLE 3 Comparison of Mated OFM Females in Each Generation Collected in Bait Traps Placed in Pheromone-treated and Untreated Check Peach Orchards in California, 1986[a]

	Untreated checks		Pheromone-treated	
Generation	Collected (no.)	Mated (%)	Collected (no.)	Mated (%)
1	401	90.2	79	84.6
2	1115	91.6	335	61.2
3	2458	95.6	610	76.4
4	1274	97.4	429	87.4
5	515	95.0	368	95.4

[a]Pooled bait trap data from Kingsburg, Parlier, Exeter, Livingston, and Modesto.

TABLE 4 Sex Ratio of OFM Captured at Terpinyl Acetate Bait Traps in Pheromone-Treated and Untreated (Check) Orchards, California, 1985 and 1986

		Moth sex as proportion of total capture[a]			
		Male (%)		Female (%)	
Location	Year	Check	Pheromone	Check	Pheromone
Rio Oso	1985	31.3	22.7	68.9	77.3
Yuba City	1985	69.1	40.6	40.6	59.4
Parlier	1985	43.6	33.3	56.4	66.7
Dinuba	1986	46.5	30.7	53.5	69.3
Exeter	1986	43.2	21.9	56.8	78.1
Parlier	1986	38.0	27.6	62.0	72.4
Arvin	1986	41.1	15.4	58.9	84.6
Mean		44.7	27.4	56.7	72.5

[a]March 1 to September 20.

TABLE 5 Control of OFM Fruit Infestations in Peaches and Nectarines by Mating Disruption with Pheromones, California, 1985

Location and (cultivar)	Treatment	Infested fruit (%)
Rio Oso (Klamt)	Check (2)[a]	0.3
	Pheromone	0.1
Yuba City (Caroline)	Check (3)	0.0
	Pheromone (1-PTB)	0.0
Sanger (Royal Giant)	Check (3)	0.2
	Pheromone	0.1
Parlier (Fay Elberta)	Check (2)	4.0
	Pheromone	0.1
Arvin (Autumn Gem)	Check (1)	0.4
	Pheromone	0.1

[a]Number of insecticide treatments applied during the season for control of OFM or PTB.

unable to entirely overcome the pressure of high OFM populations. In 1985, infested fruit levels in this block exceeded 35%, even with the application of two insecticide sprays for OFM during the season (see Fig. 2). This pheromone-treated block would probably have benefited from an initial insecticide spray directed against the first or second flight of OFM in 1986, similar to the 1986 Yuba City trial, to help reduce the resident population to a level at which the pheromone treatment could achieve satisfactory mating disruption. In contrast, the Parlier field 24 block of Fay Alberta peaches in 1986 showed an extremely low level of fruit infestation compared with the untreated check. This block was treated for the second year in 1986, and started the season with a very low endemic population of OFM.

Table 6 also shows the results of a mating disruption experiment conducted in two small blocks of pineapple quince at Reedley. Quince is reported as the preferred host for OFM. In this location both the check and pheromone-treated blocks were approximately 0.4 ha (1 acre). Both of these blocks had experienced heavy infestations of OFM at harvest over the preceeding 4 to 5 years. Two applications of pheromone, in February and May 1986, reduced fruit infestation

TABLE 6 Fruit Infested by OFM and PTB in Orchards Treated with Pheromones for OFM Mating Disruption, California, 1986

Location and (cultivar)	Treatment	Infested fruit (%)	
		OFM	PTB
Parlier fld. 21	Check	59.3	31.7
(Fay Elberta)	Pheromone	14.2	5.0
Parlier fld. 24	Check	59.3	31.7
(Fay Elberta)	Pheromone	0.9	1.5
Sanger	Check (1)[a]	0.0	0.0
(Royal Giant)	Pheromone (1-PTB)	0.2	0.0
Kingsburg	Check	1.5	0.8
(four cultivars)	Pheromone	0.5	0.2
Arvin	Check (1-PTB)	0.3	0.6
(Autumn Gem)	Pheromone (1-PTB)	0.5	0.3
Reedley	Check (4)	1.7	
(pineapple quince)	Pheromone	0.8	
Exeter	Check (2)	1.2	1.0
(Halford)	Pheromone	1.1	1.7
Livingston	Check (2)	0.0	0.2
(Halford)	Pheromone (1-PTB)	0.9	3.6
Escalon	Check (1)	0.0	0.2
(Monaco)	Pheromone (1-PTB)	0.4	1.9
Lomo	Check (2)	1.2	0.0
(Andross/Tufts)	Pheromone (Andross)	0.3	0.1
	Pheromone (Tufts)	0.3	0.0
Yuba City	Check (4)	0.4	0.1
(Carolyn/Halford)	Pheromone (1)	1.0	0.2
Live Oak	Check (3)	0.5	0.1
(Starn/Corona)	Pheromone	0.1	0.5

[a]Number of insecticide treatments applied during the season for control of OFM or PTB.

to approximately half of that observed in the insecticide-treated check, which received four applications of insecticide during the 1986 season.

Results similar to those observed in 1985 and 1986 were again seen in pheromone-treated blocks in nine field trials in 1987 (Table 7). In all of these plots, OFM damage or infested fruit in the pheromone treatments was lower than or equal to that observed in the untreated checks. The Parlier field 24 block, which had been treated

TABLE 7 Efficacy of Mating Disruption of OFM by Pheromones as Measured by Fruit Infestations, California, 1987

Location and (cultivar)	Treatment	Infested fruit (%)	
		OFM	PTB
Parlier fld. 72 (Fantasia)	Check	1.6	11.6
	Pheromone	0.3	0.3
Parlier fld. 21 (Fay Elberta)	Check	14.9	16.2
	Pheromone	0.0	0.7
Parlier fld. 24 (Fay Elberta)	Check	14.9	
	Pheromone	0.1	
Kingsburg I (Fantasia)	Check	0.2	0.0
	Pheromone[a]	0.0	0.0
Kingsburg II (O'Henry)	Check	0.1	0.0
	Pheromone[a]	0.0	0.1
Kingsburg III (Cal Red)	Check	0.2	0.5
	Pheromone[a]	0.1	0.1
Lomo (Andross)	Pheromone[a]	0.1	0.1
	Pheromone	0.2	0.1
Yuba City (Carolyn/Halford)	Pheromone[a]	0.0	0.0
	Pheromone	0.0	0.0
Live Oak (Corona)	Pheromone	0.1	0.6

[a]Check blocks at these locations in 1986; treated with pheromones in 1987.

with pheromone in 1985 and 1986, was not treated with either pheromone or insecticides in 1987. Still, it sustained only 0.1% OFM-infested fruit, indicating that once OFM populations are reduced to low levels, they may be slow to recover to damaging levels. At the Yuba City location, fruit damage levels exceeded 30% in 1985. The resultant high overwintering population, treated with two applications of pheromone and one insecticide, caused 1.0% fruit damage in 1986 (see Table 6). In 1987, no fruit damage was detected (following only one application of pheromone; see Table 7). As at Parlier, this indicates the potential for long-term OFM population reductions when using pheromones for mating disruption.

These results, from 3 years of field trials for control of OFM in California, have led to the conclusion that mating disruption with synthetic pheromones is a feasible alternative to conventional insecticide sprays for this pest. It should be noted, however, that in several locations, in all 3 years of field trials, it was observed that fruit infestations from the peach twig borer (PTB) and, in some cases, omnivorous leafroller increased in the absence of in-season sprays of insecticide for OFM (see Table 6). This increase in PTB damage was particularly noted in the northern peach-growing districts of the San Joaquin and Sacramento valleys where winter dormant sprays for twig borer control are sometimes not applied, or are applied improperly. In pheromone plots in which dormant sprays were applied correctly (see Table 7, Parlier field 72 and 21), satisfactory control of PTB was achieved without additional in-season sprays.

B. Field Trials in the Eastern United States

In addition to the ongoing field trials with mating disruption for OFM in the western United States, field trials were initiated in 1986 in peaches in Virginia (12) and continued in 1987. In both years of these trials, reduction of fruit and twig damage using pheromones for OFM control has, as in Australia and California, been equivalent to controls using conventional insecticide spray programs. In addition, no males of either OFM or of the lesser appleworm, *Grapholita prunivora* (Walsh), which responds to the same major pheromone components, were collected in pheromone traps placed in pheromone-treated orchards.

III. FIELD IMPLEMENTATION AND USE

A. Benefits

Validation and implementation of the control techniques developed in Australia for OFM, using pheromones for mating disruption, have shown that this technique is equivalent to results achieved with standard insecticide spray programs. In addition to comparable levels of control, there are several other benefits derived from the use of pheromones in comparison with insecticide applications.

The pheromone dispensers used in these trials are easy to apply. Because they are considered essentially nontoxic, they require no special equipment nor do applicators require special training or protective clothing such as spray masks, rubber suits, boots, and the like. There are fewer regulatory restrictions involved with the use of pheromones compared with insecticides. There are no requirements for notices of intent to be filed with local regulatory officials before use of the pheromones. Treated fields are not required to be posted against entry by unauthorized persons. There are no reentry or preharvest intervals that must be observed, nor are there restrictions on amounts of pheromone residue on fruit at harvest.

Use of pheromones is also readily compatible with other cultural and production operations in the orchard. The timing of pheromone applications in February and May is predictable, and once they have been applied other operations such as irrigation, cultivation, thinning, tree propping, or harvest can be scheduled without fear of interference from in-season spray applications. Cultural operations requiring large amounts of hand labor such as thinning for fresh market peach and nectarine varieties can be conducted at the proper time without concern about fruit or foliage residues, as occurs with insecticide sprays.

Use of pheromones for control of OFM enhances preservation of beneficial insects and mites and insect pollinators. It is anticipated that with continued use of pheromones for OFM control, increased levels of biological control of secondary pests, such as the omnivorous leafroller, will be observed over time. Similarly, with decreased use of insecticides for OFM control, there should be fewer increases of phytophagous mite populations which are often observed following applications of organophosphate or pyrethroid insecticides. Biological control of phytophagous mites is becoming increasingly important in deciduous fruit pest management programs, particularly when faced with the loss of effective miticides through regulatory action or resistance (e.g., cyhexatin in California in 1987).

When using pheromones rather than insecticides for OFM control, problems associated with insecticide containers, residual spray mix, or rinse water disposal are also avoided, nor are there requirements for special storage and handling of the pheromone materials as there are with insecticides. As regulatory restrictions on the use of insecticides continue to increase, application of pheromones for insect control may become more appealing even though it may appear that the pheromones are initially more costly in some situations. The benefits from long-term population reduction of OFM (and other pest species) also should not be overlooked. Preliminary observations of orchards under pheromone treatment in California for 2 to 3 years suggest that OFM populations can be reduced to levels so low that only a single early-season application of pheromone may be required to maintain acceptable control. Thus, use of pheromones may possibly be reduced to a point that they are much more competitive with costs of standard pesticide treatments.

B. Limitations

Although pheromones for mating disruption of OFM have been shown to be effective, there are several limitations associated with their use for control of OFM. Some of these problems have been previously noted by other authors, particularly the problem of migration of mated female moths from orchards adjacent to pheromone-treated blocks (9). In California, almonds and plums, and occasionally prunes and apples can produce high populations of OFM, and are often grown adjacent to peach or nectarine orchards.

Pest control advisors in California who have been involved with the implementation of mating disruption for OFM have also pointed out that use of this technique will not lead to a decrease in the amount of orchard examinations or trapping in pheromone-treated blocks. Users of mating disruption technology must continue to use pheromone traps to monitor male moth response to the pheromone point sources and, if possible, also use bait traps to monitor flights of female moths from adjacent orchards and to assess the mated status of trapped females.

As noted in one of the field plots at Parlier in 1986, it may be necessary in some situations to spray out high populations of OFM at the beginning of a pheromone or mating disruption program to reduce the population of moths to a level manageable by the pheromone treatment. This approach should be considered in disruption blocks if pheromone traps collect more than two to three male moths during the emergence of the overwintering population in March or April, or if numerous OFM shoot strikes are observed during the first larval generation. Insecticide sprays would then be directed against the second flight emerging in late May or June.

As also noted in our field trials in California, increases of pests normally considered secondary to a primary pest, such as OFM, may be seen in the absence of in-season sprays. With PTB, it is believed that attention to properly applied dormant sprays usually should adequately control PTB and eliminate the need for additional in-season sprays for this pest. Other tortricid species, however, such as omnivorous leafroller, are not controlled by dormant sprays.

Evaluation of the mating disruption tests in Virginia also emphasized that insecticide applications retain the benefit of controlling other major pests, here catfacing insects, the plum curculio, *Conotrachelus nenuphar* (Herbst), and the Japanese beetle, *Popilia japonica* Newman. It may become necessary, therefore, to use pheromone traps, inspections of green fruit before harvest, or other sampling methods to indicate population trends and the need for additional sprays for such pests. As noted previously, it is hoped that the absence of sprays for OFM will eventually lead to increased biological control of pests such as omnivorous leafroller.

Growers and pest control advisors should also be cautioned that mating disruption for insect control is not a "quick fix" technique. Mating disruption programs must be carefully thought out and prepared. First applications of pheromone must take place in the springtime before the emergence and mating of overwintered moth populations, and additional applications of pheromone must occur before the time that the initial pheromone treatments are depleted and are no longer effective in disruption. Continual assessments of the potential for damage caused by pests moving from adjacent orchards should also be made at each location.

The cost of control is a factor that must be considered by users of mating disruption for control of OFM. In the southern San Joaquin valley where the majority of fresh market peach and nectarine production is located, most growers of these crops spray for OFM once or perhaps only twice each season for OFM control. Two sprays, with perhaps an additional application of miticide, are approximately equivalent in cost to the mating disruption treatments described herein. Costs for two applications of pheromone, or three sprays would be approximately 120 dollars/acre. Thus, on a straight short-term economic analysis, growers of fresh market peaches and nectarines may not consider mating disruption as an acceptable alternative to conventional spray programs. This could change quickly, however, if recommendations for only single applications of pheromone per season can be developed in the future.

Conversely, growers of processing peaches, particularly in the northern San Joaquin and Sacramento valleys, must often apply three to as many as six applications of insecticides per season for acceptable OFM control. In addition, these growers usually apply miticides to control outbreaks of phytophagous mites induced by heavy use of insecticides. Also, the prospect of increasing tolerance to organophosphate insecticides in OFM populations in northern California makes mating disruption even more appealing to growers in that area. Consequently, the cost of pheromones for mating disruption in canning or processing peaches is viewed much more favorably than it presently might be with fresh market peaches or nectarines.

Approximately 750 ha (1800 acres) of peaches and nectarines were treated with synthetic OFM pheromone in California in 1987, following the registration of the Isomate M pheromone dispenser by the Environmental Protection Agency and the California Department of Food and Agriculture in February of that year. It is believed that use of pheromones for mating disruption of OFM will increase as the efficacy and benefits derived from this methodology become better known; as many as 4000 acres (1600 ha) may be treated in California in 1988.

IV. CONCLUSIONS

Field trials in California and Virginia peach orchards with polyethylene dispensers containing synthetic components of the OFM pheromone have shown that pheromone treatments can reduce OFM damage at harvest to levels competitive with standard insecticide control programs at comparable cost. Results suggest that when OFM populations are reduced to very low levels with pheromone treatments, reinfestation occurs slowly. In light of increasing environmental concerns about the use of insecticides, it is anticipated that grower acceptance of the mating disruption technique will increase in coming years.

ACKNOWLEDGMENTS

Sincere appreciation is expressed to W. W. Barnett, W. J. Bentley, D. L. Flaherty, R. A. Jones, and C. V. Weakley for their assistance in conducting the field trials in California.

REFERENCES

1. Summers, F. M. The oriental fruit moth in California. Calif. Agric. Expt. Sta. Circ. 539, 1966, 17 pp.
2. Chapman, P. J. and Lienk, S. E. Tortricid fauna of apple in New York (Lepidoptera: Tortricidae). NY State Agric. Expt. Sta. Special Publ., March 1971, 122 pp.
3. Clausen, C. P., ed. Introduced parasites and predators of arthropod pests and weeds: A world review. *USDA Agric. Hndbk. No. 480*, 1978, 545 pp.
4. Rothschild, G. H. L. Control of oriental fruit moth [*Cydia molesta* (Busck) (Lepidoptera: Tortricidae)] with synthetic female pheromone. *Bull. Entomol. Res. 65*:473–490 (1975).
5. Gentry, C. R., Yonce, C. E., and Bierl-Leonhardt, B. A. Oriental fruit moth: Mating disruption trials with pheromone. In

Insect Suppression with Controlled Release Pheromone Systems, Vol. II. (Kydonieus, A. F. and Beroza, M., eds.). CRC Press, Boca Raton, 1982, pp. 107–115.
6. Roelofs, W. L., Cardé, R. T., and Tette, J. P. Oriental fruit moth attractant synergists. *Environ. Entomol.* *2*:252–254 (1973).
7. Beroza, M., Gentry, C. R., Blythe, J. L., and Muschik, G. M. Isomer content and other factors influencing captures of oriental fruit moth by synthetic pheromone traps. *J. Econ. Entomol.* *66*:1307–1311 (1973).
8. Rothschild, G. H. L. A comparison of methods of dispensing synthetic sex pheromone for the control of oriental fruit moth, *Cydia molesta* (Busck) (Lepidoptera: Tortricidae), in Australia. *Bull. Entomol. Res.* *69*:115–127 (1979).
9. Vickers, R. A., Rothschild, G. H. L., and Jones, E. L. Control of the oriental fruit moth *Cydia molesta* (Busck) (Lepidoptera: Tortricidae) at a district level by mating disruption with synthetic female pheromone. *Bull. Entomol. Res.* *75*:625–634 (1985).
10. Flint, M. L. (ed.). Peach and nectarine pest management guidelines. Univ. of Calif. PMG Publ. 10, 1987, 23 pp.
11. Rothschild, G. H. L. Suppression of mating in codling moths with synthetic sex pheromone and other compounds. In *Insect Suppression with Controlled Release Pheromone Systems*, Vol. II. (Kydonieus, A. F. and Beroza, M., eds.). CRC Press, Boca Raton, 1982, pp. 118–131.
12. Pfeiffer, D. G. and Killian, J. C. Pheromone disruption for the control of oriental fruit moth and lesser peach tree borer. *Proc. 62nd Cumberland-Shenandoah Fruit-Workers Conf.*, 1986, 3 pp.

14

Grape Berry Moth and Grape Vine Moth in Europe

CHARLES DESCOINS / Laboratoire des Médiateurs Chimiques, Saint-Remy-lès-Chevreuse, France

I. INTRODUCTION

The European grape berry moth, *Eupoecilia (Clysia) ambiguella* Hb., and the grape vine moth, *Lobesia (Polychrosis) botrana* Den. and Schiff, are two of the most important moths feeding on the grape plant in Europe.

Eupoecilia ambiguella occurs as a pest, mainly in the northern wine-grape growing areas of Europe, particularly along the Rhine valley, in Switzerland and Austria. It is also present, with *L. botrana*, in some southern vineyards where the climatic conditions are sufficiently humid (as in the Bordeaux region in southwestern France). *Eupoecilia ambiguella* has two generations per year in its distribution area, but a third generation may occur when the summer is particularly warm.

Lobesia botrana is abundant in the Mediterranean area and needs a warm and dry climate. It also has two generations per year, but a third generation is frequent, mainly in the southern limits of its distribution area.

Both species are harmful not only because the larvae directly attack the grapes, but also because they promote growth of a pathogenic fungus, *Botrytis cinerea.*

Accurate determination of the flight periods of the adults and estimation of the population densities of the insects are a prerequisite for the development of integrated control programs (pest management).

In 1917, the French entomologist Feytaud (1) had observed the attraction of *L. botrana* males by females in the Bordeaux vineyards and suggested using this particularity for monitoring purposes. But it was only in 1939 that Götz (2), in Germany, used traps baited with virgin females to lure males for determining the flight periods and the population levels of the insects.

In fact, procedures using virgin females are not easy to develop because such procedures require a permanent rearing of the insects in the laboratory and because the attraction of the females is not always reliable.

Hence, the only monitoring system used in most of the European vineyards, before the discovery of synthetic sex attractants, was food traps. These traps, baited with fermented juices, are not specific; they lure both males and females, but they give good indications on the timing of the flight periods of the adults and also an estimate of the population levels.

II. MONITORING WITH SEX ATTRACTANTS

A. The European Grape Berry Moth

The sex pheromone of the European grape berry moth was identified at the same time in Switzerland by Arn et al. (3) and in France by Saglio et al. (4) about 10 years ago. (*Z*)-9-Dodecenyl acetate (Z9-C12:Ac) has been found to be the major component of the secretion, essential for luring males to a trap. Catches are markedly increased by adding dodecyl acetate (C12:Ac)--a minor component of the secretion or effluvia—with an optimum ratio of Z9-C12:Ac/C12:Ac being 1:1 to 1:5 depending on the initial dose of the unsaturated acetate (5). Given these observations, sex attractants containing 1000 μg of Z9-C12:Ac and 5000 μg of C12:Ac impregnated on rubber septa are used commercially in Europe on a relatively large scale (more than 4000 capsules are sold each year in France) for monitoring purposes. These capsules are effective for 6 weeks in the field, and three capsules per trap per year are enough to cover all the annual activity of the insect and to give a precise knowledge of the beginning of each flight period and of its maximum intensity. Traps are used at a density level of one or two per hectare; depending on the vineyard and its climatic conditions. Unfortunately, it is not yet possible to correlate the number of catches in a trap and potential damage on the crop.

More recently, Arn et al. (6) have reinvestigated the chemical composition of this pheromone and have found, in addition to dodecyl acetate, another saturated acetate, octadecyl acetate (C18:Ac), which behaves as a synergist of the two-component blend—Z9-C12:Ac/C12:Ac (Table 1 illustrates the structures of the compounds). According

TABLE 1 Structures of the Components in the European Grape Berry and Grape Vine Moths' Pheromones

	Relative amount in best lure
Sex Pheromone Blend of *Eupoecilia ambiguella* Females	
 (*Z*)-9-dodecenyl acetate	1
 dodecyl acetate	1
 octadecyl acetate	2
Sex Pheromone Blend of *Lobesia botrana* Females	
 (*E*,*Z*)-7,9-dodecadienyl acetate	20
 (*Z*)-9-dodecenyl acetate	1
 (*E*,*Z*)-7,9-dodecadien-1-ol	4
Parapheromone for *Lobesia botrana* Males	
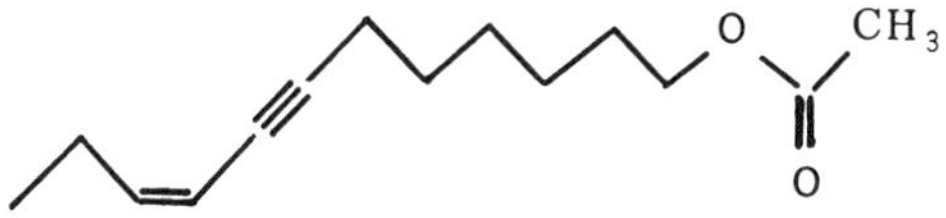 (*Z*)-9-dodecadecen-7-yn-1-yl acetate	

to Arn, a blend of Z9-C12:Ac (100 μg); C12:Ac (100 μg), and C18:Ac (200 μg) would provide the best attractant for *E. ambiguella* yet known, but more field data are required to confirm this proposition. However, it is not always necessary to have an attractant that catches a maximum number of moths. It is often better to have one able to lure a sample of the wild population living the vineyard than to attract insects living outside with subsequent high catches of males and no significant damage on the grapes.

B. The Grape Vine Moth

Early electroantennographic experiments conducted by Roelofs (7) in the United States, suggested that (*E*,*Z*)-7,9-dodecadienyl acetate (E7,Z9-C12:Ac) might be a sex attractant for the males of the grape vine moth.

This structure was confirmed by the biological activity of the synthesized compound and identification of this chemical in the female glands (8).

In fact, this chemical is very efficient by itself for luring males in the field (9), and its steric purity is not a prerequisite for the activity. This is fortunate because it has been shown (10) that the *E*,*Z* dienic system easily isomerizes under sunlight to an equilibrium of the four isomers.

More detailed chemical analysis of female sex gland extracts by Arn (11) revealed the presence of 15 minor components that are straight-chain alcohols and acetates. Two of them (*E*,*Z*)-7,9-dodecadien-1-ol (E7,Z9-C12:OH) and (*Z*)-9-dodecenyl acetate (Z9-C12:Ac) have a synergistic effect on male orientation when they are mixed together (20% and 5%, respectively) with the E7,Z9-C12:Ac at low doses (10 μg). Commercial sex attractants are made up of 1000 or 100 μg of E7,Z9-C12:Ac (95% to 98% steric purity) impregnated on rubber septa, sometimes with anti-UV preparations or antioxidants.

More than 7000 capsules per year are used in France for monitoring purposes. They are effective for 6 weeks (high-dose) or 3 weeks (low-dose) and are used at the density level of three per trap per season.

Low doses are used to get correlations between the trap catches and the density of the native population in the vineyard, whereas high doses are used to detect only the presence of the insect and to determine the timing of its flight periods.

III. DISRUPTION EXPERIMENTS

The grape berry moth and grape vine moth are good candidates for disruption experiments. In fact, there are many vineyards in Europe in which we can find only one of the two species as the major harmful insect.

Both live on high-value crops, mainly vineyards that produce high-quality wines ("appelation contrôlée" in French) for which quality controls are very strict, especially for pesticide residue levels.

Development of a suitable method for disrupting mating behavior of these insects has a double advantage:

Reduction of the population level of the next generation with subsequent reduction of direct damages

Reduction of fungal attacks because of the decrease in the number of the larvae feeding on the grapes.

Cost of a mating disruption treatment must be compared not only with insecticide, but also with fungicide costs.

A. Mating Disruption of the European Grape Berry Moth

The main component of the sex pheromone of this species is (Z)-9-dodecenyl acetate. It is a compound easy to synthesize at low cost, less than 300 DM/kg (150 U.S. dollars).

Most of the experiments of mating disruption were conducted first in Switzerland and later in Germany and Austria.

In 1978, in a vineyard near Geneva, Arn (12) tried to disrupt mating in a small test area (150 × 100 m) by attaching to the wires with staples, at the beginning of the first flight, 4000 pheromone dispensers made by Hercon, each containing 13.2 mg of Z9-C12:Ac. The release rate of this chemical was 500 mg/ha per day for the first generation, but only 50 mg/ha per day for the second.

During all the experiment, catches of males in sex pheromone traps were very low, indicating more than 99% disruption; but, still, 33% of mated females were caught in food traps in the treated area against 51% in the untreated. During the first generation, 4.2% of the grapes were damaged in the treated area and 4.4% during the second, against 9% and 9.8% in the untreated area (control).

Comprehensive experiments were conducted later by Charmillot in vineyards of the Lake Geneva region (13), in the framework of a collaborative program with BASF (West Germany).

Hercon dispensers, 1 in.^{2}, containing 30 to 40 mg of Z9-C12:Ac were used. Dispensers were applied by hand at the beginning of the first flight (May) and also at the beginning of the second flight (July). Dispenser density was calculated to diffuse between 25 to 50 g/ha Z9-C12:Ac for each generation. Treated areas were between 1 to 15 ha. Efficiency of the disruption was evaluated by the following procedures:

Checking of male catches in the sex pheromone traps placed in the treated area

Checking the number of mated tethered females placed in the treated area against the same number in the untreated area
Estimation of the damage on the grapes after each flight on a 100-grape sample taken in various parts of the treated area

As in the first results of 1978, inhibition of male catches in sex pheromone traps was always excellent (between 96.4% and 100%) for all the vineyards and for all the experiments.

Diffusion rates varied from 20 to 40 mg/ha per hour for the first flight and 20 to 100 mg/ha per hour for the second. These differences are due to differing climatic conditions between years and between vineyards. Mating inhibition of tethered females also showed a great variability, but for many vineyards inhibition was between 75% to 90% in the treated area in comparison with the control.

Damage varied between grapes taken within the treated area and those in the first rows of the borders. Results were always better in an isolated vineyard large enough to avoid infestations by mated females from outside. In many cases, reduction of damage was between 70% and 90% for the second generation, but always lower for the first one. This can be explained by the low diffusion rate during the first generation when temperatures were often low. However, threshold levels were higher for the first generation (15 to 30 larvae per 100 grapes) than for the second one (only five larvae per 100 grapes); also, a lower efficiency during the first flight could be expected.

It has also been observed that efficiency was higher when the wild population was not too abundant, mainly during the second generation. Also, it is wise to spray insecticides at the first generation to have a decrease of the population level, and to apply the disruption technique for the second generation. Damage in the border rows can be avoided by using a higher dispenser density on these rows or by treating the borders with insecticides.

These results allowed the German chemical company BASF to obtain registration in 1986 in West Germany for the pheromonal product RAK-1 to control the second generation of the grape berry moth, using the mating disruption technique.

B. Mating Disruption of the Grape Vine Moth

To ensure effective control of the grape moths in most of the European vineyards, we must also control the grape vine moth. The area infested by this insect is much larger than that infested by the grape berry moth. The question is: Why has European research not concentrated more on this economically important species? The main reason for this is that the grape vine moth pheromone was simply not available, at least not in the amounts needed for meaningful field testing. The dienic structure of the pheromone is too complicated for easy industrial synthesis, considering economic factors.

Nevertheless, extensive experiments were conducted by Roehrich in Bordeaux vineyards (southwest France) from 1979 (14) to 1981 (15), using pure E7,Z9-C12:Ac synthesized by INRA (Laboratoire des Médiateures Chimiques).

Three types of dispensers were used:

1. Rubber septa loaded with 7 mg or 50 mg of active material at the density of 5000/ha for the low dose and 650 for the high dose, which corresponds to a total amount of 30 g active material per hectare and an evaporation rate of 10 to 16 mg/ha per hour for 40 days.
2. Hollow fibers (Conrel Albany) sprayed from the ground with a special apparatus and calculated to diffuse 12 mg/ha per hour also for 40 days
3. Strips of hollow fibers attached with staples around the fence wires at a density of 10,000/ha

Treated areas were from 1.5 to 4.5 ha.
Efficiency was evaluated by

Reduction of male catches in traps baited with virgin females or synthetic sex attractant
Reduction of mating of tethered females in the treated area compared with the same number in the untreated area
Evaluation of the sexual status of females collected in food traps both in the treated and untreated areas
Evaluation of the damage on the grapes

Dispensers were placed at the beginning of the second generation (June–July) because it has been demonstrated that the disruption technique is useless for the first generation (April) when the vine has no leaves.

Results were erratic from one year to another and from one vineyard to another.

In all cases, we observed a drastic decrease in the number of males caught in the traps within the treated area, but only a partial decrease in the number of mated females caught in food traps in the treated areas compared with the untreated areas. The number of mated tethered females varied between experiments but was relatively high (20%–30%). Reduction of damage varied from more than 90% to only 40% or even 30% depending on the initial population density. In many cases, it was necessary to apply new dispensers at the beginning of the third generation (August) to prevent excessive damage at vintage. Damage was always more severe on the border, and results were better when the population density was low.

Experiments have now been cancelled because of shortage of active material and because of the difficulty of large-scale synthesis. Nevertheless, BASF has found that the easily available (Z)-9 (yne) 7-C12:Ac had considerable attraction activity in the field (16). Traps baited with loadings of 0.8 to 3 mg of this material on rubber septa attracted approximately the same number of moths as the natural pheromone. First-mating suppression experiments showed considerable reduction in the number of males caught in traps baited with the synthetic pheromone in a treated area impregnated with the eneyne acetate. This does not mean that this compound can be used for disruption, but it is a preliminary step.

IV. CONCLUSION

All the experiments described here clearly demonstrated that it is possible to monitor populations of the two grape moths by using sexual trapping of males. The sex attractants commercially available are selective and efficient and can be used to develop integrated control programs. Unfortunately, they give only qualitative data, and many things remain to be done for quantitative evaluation of risks.

Control of the insect populations by mating disruption is now possible on a commercial basis for the grape berry moth, but it is still difficult to achieve for the grape vine moth because of the high cost of its pheromone. Nevertheless, the possibility of substituting an artificial pheromonal constituent for the true one seems to be promising and must be tested for a better knowledge of its potential as a mating disruptant.

REFERENCES

1. Feytaud, J. A propos de l'attirance des sexes chez les microlépidoptères. Procès-verbaux Soc. Linnéenne Bordeaux, 1–4, 1917.
2. Götz, B. Untersuchungen über die Wirkung den Sexualduftstoffes bei den Traubenwickler *Clysia ambiguella* und *Polychrosis botrana*. *Z. Angew. Entomol.* *26*:306–418 (1939).
3. Arn, H., Rauscher, S., Buser, H.-R., and Roelofs, W. L. Sex pheromone of *Eupoecilia ambiguella*: *cis*-9-Dodecenyl acetate as a major component. *Z. Naturforsch.* *31C*:499–503 (1976).
4. Saglio, P., Descoins, C., Gallois, M., Lettere, M., Jaouen, D., and Mercier, J. Etude de la phéromone sexuelle de la cochylis de la vigne: *Eupocilia (Clysia) ambiguella* Hb. Lépidoptère Tortricoidea Cochylidae. *Ann. Zool. Ecol. Anim.* *9*:553–562 (1977).

5. Rauscher, S., Arn, H., and Guerin, P. Effects of dodecyl acetate and *Z*-10-tridecenyl acetate on attraction of *Eupoecilia ambiguella* to the main sex pheromone component, *Z*-9-dodecenyl acetate. *J. Chem. Ecol. 10*:253–264 (1984).
6. Arn, H., Rauscher, S., Buser, H.-R., and Guerin, P. Sex pheromone of *Eupoecilia ambiguella*: Female analysis and male response to ternary blend. *J. Chem. Ecol. 12*:1417–1429 (1986).
7. Roelofs, W., Kochansky, J., Cardé, R., Arn, H., and Rauscher, S. Sex attractant of the grape vine moth, *Lobesia botrana. Mitt. Schweiz. Entomol. Ges. 46*:71–73 (1973).
8. Buser, H.-R., Rauscher, S., and Arn, H. Sex pheromone of *Lobesia botrana*: *E*7,*Z*9-Dodecadienyl acetate in the female grape vine moth. *Z. Natur. Forsch. 29*:731–748 (1984).
9. Roehrich, R., Carles, J. P., Darrioumerle, Y., Pargade, P., and Lalanne-Cassou, B. Essais en vignoble de phéromones de synthèse pour la capture des mâles de l'Eudemis (*Lobesia botrana* Schiff.). *Ann. Zool. Ecol. Anim. 8*:473–480 (1976).
10. Ideses, R., Klug, J. T., Shani, A., Gothil, F. S., and Gurevitz, E. Sex pheromone of the European grape vine moth, *Lobesia botrana* Schiff. (Lepidoptera: Tortricidae): Synthesis and effect of isomeric purity on biological activity. *J. Chem. Ecol. 8*:195–200 (1982).
11. Guerin, P. In *List of Sex Pheromones of Lepidoptera and Related Attractants*. (Arn, H., Toth, M., and Priesner, E., eds.). International Organization for Biological Control, West Palearctic Regional Section, Paris, France, 1986, p. 47.
12. Arn, H., Rauscher, S., Schmid, A., Jaccard, C., and Bierl-Leonhardt, B. A. Field experiments to develop control of the grape moth, *Eupoecilia ambiguella*, by communication disruption. In *Management of Insect Pests with Semiochemicals*. (Mitchell, E. R., ed.). Plenum Publishing, New York, 1981, pp. 327–338.
13. Charmillot, P.-J., Bloesch, B., Schmid, A., and Neumann, U. Essais de lutte contre la cochylis *Eupoecilia ambiguella* Hbn. par la technique de confusion sexuelle. In *Médiateurs chimiques et Biosystêmatique chez les Lépidoptères*. Valence Décember 13–14, 1985. Colloques INRA (in press).
14. Roehrich, R., Carles, J. P., Tresor, C., and De Vathaire, M. A. Essais de "confusion sexuelle" contre les tordeuses de la grappe, l'Eudémis *Lobesia botrana* Den. et Schiff. et la Cochylis *Eupoecilia ambiguella* Tr. *Ann. Zool. Ecol. Anim. 11*: 654–675 (1979).
15. Roehrich, R. and Carles, J. P. Essais de confusion sexuelle en vignoble centre l'Eudémis de la vigne, *Lobesia botrana* Schiff. In *Les Médiateurs Chimiques Agissant sur le Comportement des Insectes*. Colloques INRA No. 7, 1982, pp. 365–371.

16. Buschmann, E., Seufert, W., and Krieg, W. A new synthetic attractant for the European grape vine moth (*Lobesia botrana*) (*Z*)-9-dodecadecen-7-yn-1-yl acetate. Presented at the Euchem Conference Semiochemicals in Plant and Animal Kingdoms, Angers, October 12–16, 1987.

15

Mating Disruption for Control of Grape Berry Moth in New York Vineyards

TIMOTHY J. DENNEHY, WENDELL L. ROELOFS, EMIL F. TASCHENBERG, and THEODORE N. TAFT / New York State Agricultural Experiment Station, Cornell University, Geneva, New York

I. INTRODUCTION

Although the concept of using pheromones for disruption of orientation of male insects to their mates is well over 25 years old (1), it is only in the past few years that an array of commercially viable pheromone products for disrupting sexual communication of agricultural pests have entered the marketplace. Advancements in the design of pheromone dispensers and reductions in the cost of active ingredient (AI) pheromone are now making pheromone products economically acceptable alternatives to classic insecticidal control of specific major and minor crop pests.

In this paper, we chronicle the efforts expended over the past 20 years that have laid the groundwork for successful disruption of the key insect pest of grape in eastern North America, the grape berry moth (GBM), *Endopiza viteana* Clemens. We then describe field trials conducted in 1985 to 1987 that illustrate successful control of GBM with a commercial pheromone product, the Shin-Etsu pheromone dispenser.

The GBM is an excellent candidate for integrated control tactics such as mating disruption because, in most regions of the Northeast, it is the sole pest of grapes that consistently warrants expenditures on control measures. It is the pest for which insecticide treatments on grapes in the Northeast traditionally have evolved. Growers in the Northeast (inclusive of the Niagara peninsula of Ontario, Canada) typically have applied between one and seven treatments of insecticide to control GBM, the exact number of treatments depending upon location-specific differences in pest severity. Although other pests do occur in vineyards, most notably grape leafhoppers (*Erythroneura* spp.), grape flea beetle (*Altica chalybea* Illiger), rose chafer [*Macrodactylus subspinosus* (F.)], and grape cane girdler and grape cane gallmaker [*Ampeloglypter ater* LeConte and *A. sesostris* (LeConte)], research has shown that in most vineyards these species are periodic in occurrence, generally do not cause economically important damage to healthy vines, and are most effectively treated on an ad hoc basis. A major factor favoring acceptance of techniques like pheromones, that generally control a limited spectrum of target pests, is the apparent absence of multiple key insect pests in the eastern grape system.

Grapes are a relatively high-value commodity. Yet, unlike many fruit commodities, thresholds for allowable damage by GBM are quite high because over 95% of the grapes in the Northeast are destined to be processed into grape products (wine, juice, jelly, or jam). This situation is quite different from that of grapes in Europe where susceptibility of *Vitis vinifera* varieties to *Botrytis* bunch rot mandates that thresholds for similar lepidopteran pests be comparatively low because feeding by larvae is highly correlated with loss of entire clusters to bunch rot. Similarly, in areas of the United States, such as California, where many grapes are destined for fresh fruit or raisin markets, thresholds for allowable levels of damage are but a fraction of that utilized in the Northeast for processed grapes. An emerging incentive for utilization of biorational pest control measures like pheromones is being generated by restrictions on the manner in which pesticides may be used in and around rural settings and limitations on the number and kinds of pesticides that are available to growers in states with aggressive regulatory efforts aimed at limiting pesticide use. This factor may become a driving force in promoting biorationals in New York State, because the average vineyard is less than 2 ha, and a large proportion of vineyards are positioned near dwellings.

Despite these attributes of northeastern grape production that favor development of mating disruption, some important factors have acted in opposition to advancement of this technology. First, regulatory constraints have been a major impediment. All grapes involved in field trials of mating disruption have had to be destroyed, because the products evaluated have not been registered and experimental use permits were not issued. Financial constraints on the amount of grapes for which we could afford to compensate growers placed severe limitations on the size of the trial that could be conducted. Second, because the average size of vineyards in New York is under 2 ha, mating disruption with pheromones must be effective with relatively small treated areas. Having large, contiguous treated areas is out of the question in a large proportion of New York vineyards. Third, and probably most important, wild *Vitis* spp. grow abundantly in the Northeast and serve as an enormous untreated refuge for grape berry moth. Easily half of all vineyards in New York have wild *Vitis* infested with grape berry moth along one or more vineyard borders or short distances from the vineyard. Rothschild (2) pointed out that the main ecological characteristics of a species likely to favor control by mating disruption are that adults (particularly mated females) are nonmigratory and have a narrow host range, both of which would tend to limit invasion of a treated area by mated females. Despite these potential obstacles, the following report will show that the mating disruption technique has been successful in 4 years of field trials and is ready for commercialization in New York State.

II. CHRONOLOGY OF DEVELOPMENT OF MATING DISRUPTION IN NEW YORK VINEYARDS, 1970 TO 1984

A. Background

A GBM pheromone component was identified as (*Z*)-9-dodecenyl acetate (Z9-12:0Ac) (3) and initially was used in a mass trapping program for control of GBM (4). The use of 226 GBM pheromone traps in a 1.1-ha plot in 1971 and in 1972 resulted in the capture of 115 and 300 males, respectively, but fruit damage was lowered to only 5.9% and 7.1%, compared with check plots of 11.3% and 13%, respectively. Although the mass trapping technique had a substantial effect, it did not suppress the pest population to commercially acceptable levels. Because the cost and effort of mass trapping was too high to justify further testing, subsequent research was conducted on the mating disruption technique.

B. Planchet Evaporators

Commercial formulations were not available for the initial tests in 1972 (4), and so the Z9-12:0Ac compound was evaporated from 2.5-cm planchets placed under protective metal covers. The planchets were recharged as needed to maintain continuous evaporation of the pheromone. Tests involving 57 planchets spaced throughout a 0.4-ha plot gave an estimated release rate of 13 mg/ha per hour of pheromone component. No GBM males were captured by synthetic or virgin female traps in the test vineyard, but 62 GBM males were caught by traps in a nearby check vineyard (0.2 ha). Fruit injury by GBM was reduced in the test vineyard from 4.7% in 1971 to 3.8% in 1972, whereas it increased from 6.5% to 9.2% in the check vineyard. This experiment indicated a potential for using the pheromone for control of this pest, but a better formulation definitely was needed to make it commercially feasible.

C. Microencapsulated Formulation

Several microencapsulated formulations were tested in 1974 in small test plots (0.12 ha) for control of GBM (5). A formulation prepared by Pennwalt Corp. (Philadelphia, Pennsylvania) contained approximately 10% by weight of Z9-12:0Ac in polyamide capsules of 30 to 50 μm average diameter. They were diluted with water so that application of 90 l/ha provided 25 g of pheromone per hectare. The formulation was applied to one side of the grape row with a roller pump sprayer driven by the power take-off on a tractor. Also, a 1% wettable powder of Z9-12:0Ac was obtained from Zoecon Corp. (Palo Alto, California) and applied at the rate of 25 g of pheromone per hectare in a similar test. Tests to determine the longevity of a single treatment were conducted by releasing laboratory-reared males in the plots and assessing disruption of male orientation to pheromone traps. In each case, the formulations lost much of their disruptive effect after 8 days. In 1975, the microencapsulated formulation was applied at the rate of 25 g/ha per application in an all-season program of 20 sprays applied at 5- to 7-day intervals. Male orientation to traps was disrupted by 85%, and fruit damage was 87% lower in the treated plot than in the check plot. The requirement of 500 g of pheromone per hectare and excessive applications made this particular formulation undesirable.

D. Hollow-Fiber Dispensers

In 1975, tests were initiated with hollow fibers (Albany International, Conrel, Needham Heights, Massachusetts) filled with the pheromone component (5). Strips containing nine fibers were used, with each fiber containing 270 μg of Z9-12:0Ac to give a release rate of 0.083 μg/hr at 22°C. Tests for male orientation disruption involving the release of 1210 laboratory-reared males over a 10-week period showed excellent shutdown of trap catch with only 2.5 g/ha of pheromone. This gave a calculated release rate of only 0.75 mg/ha per hour. In 1976, mating disruption tests were conducted for season-long control in small plots (0.4 ha) (6). Tests utilizing rates of 50 fibers per vine and 50 fibers on alternate vines, to give pheromone release rates of approximately 2 and 1 mg/ha per hour, respectively, gave excellent results. There was a calculated 99.5% and 97.5% reduction in male orientation to pheromone traps, respectively, for the entire period of adult activity (105 days), and only 0.4% and 0.6% berry damage in the treated plots compared with 1.7% in the check plot. The very low infestation in the check plot made it difficult to determine efficacy, and testing in subsequent years showed that the use of hollow fibers at this prescribed rate would not suppress the damage level below 4%.

E. Rubber Septa Releasers

Under the auspices of a BARD grant, E. F. Taschenberg investigated the possibility of obtaining economically acceptable control of GBM by using higher rates of pheromone release. The tests involved the use of a newly defined pheromone blend consisting of a 9:1 mixture of Z9-12:OAc and Z11-14:OAc. In 1983, tests were conducted with rubber septa dispensers containing 100 mg of the pheromone blend. At a rate of 1 septum per vine (1500 vines per hectare, 150 g pheromone per hectare), there was over 95% reduction in males captured by pheromone traps over a 10-week period. A mating disruption test in small plots (0.4 ha) was conducted in which the rubber septa were applied five times throughout the season at a rate of 1 septum per vine. This program was 100% effective in shutting down male captures in pheromone traps, but the berry injury was 8.4% in the treated plot, compared with 16.4% in the check vineyard. Again, the disruption technique did not lower fruit damage to the desired level.

F. Shin-Etsu Polyethylene Tube Releasers (S-E Tie)

In 1984, in an additional study under the BARD grant, Taschenberg tested the efficacy of Shin-Etsu (S-E) Tube releasers (Type G, 20-cm × 3-mm polyethylene tubes with a fine wire imbedded in the wall) in controlling GBM in vineyards. These S-E ties (containing approximately 88 mg of a blend of 10 parts Z9-12:OAc and 1 part Z11-14:OAc) were applied three times throughout the season at a rate of two ties per vine for each application. Release rate studies in the field showed that each tie emitted the blend at a rate of approximately 0.035 mg/hr. This provided a minimum of 105 mg/ha an hour in the disruption test area. This high release rate was 100% effective in shutting down male captures in traps in the treated plot, whereas traps in the check area caught 211 GBM males. More important, however, was the finding of fewer than 3% injured berries per cluster in the treated area, compared with 22.7% damage in the check area. The ease of tying this formulation to the vines combined with its initial success in reducing GBM damage to a tolerable level provided the impetus to continue the testing of this commercial formulation.

III. TRIALS WITH THE SHIN-ETSU PHEROMONE DISPENSER 1985–1987

A. Materials and Methods

1. *Vineyard Sites*

All trials were conducted within the grape belt that extends along Lake Erie in New York. The vineyards contained mature "Concord" vines, *Vitis labrusca* L., planted at a density of 1512 vines per hectare. The approximate sizes of the vineyards were as follows: Francis, 12 ha; Christy, 20 ha; Militello, 20 ha: DeGolier, 30 ha. The disruption plots (including control treatments) at each of these locations were spaced throughout the vineyard to provide minimal interference between treatments (Fig. 1). Each treatment comprised a rectangle encompassing 600 vines (20 rows × 30 vines per row), an area of about 0.4 ha. Growers at these vineyard sites typically apply two to four treatments of insecticide per year to control GBM.

To prevent fungal pathogens, which are serious pests of grape in the Northeast, from damaging grapes in the disruption plots and thereby complicating assessments of GBM, it was necessary to adhere to a conventional fungicide treatment program. During 1985 to 1987, growers applied three to five fungicide treatments to the vineyards. These applications were made from mid-May to mid-August, and the materials used for each treatment were one of the following: captan (Orthocide), triadimefon (Bayleton 50 WP), mancozeb (Manzate 200), maneb (Dithane M22: Manzate 200), and copper and lime. These materials were applied at recommended rates (7) and did not present a problem for the mating disruption experiments, because at each vineyard all vines (including the mating disruption control blocks) received the same fungicide treatments.

Owing to extremely high percentages of damaged fruit in 1987 in the control block and the block treated at the rate of 1000 ties per hectare, these plots were treated with insecticide on August 20. One application of carbaryl (Sevin 50% WP) was applied at the rate of 0.37 kg of active ingredient per hectare, using a conventional airblast sprayer (Windmill Spraymaster, GerVan Co., Modesto, California).

2. *Pheromone Traps in Plots*

Three Pherocon 1C wing-type pheromone traps (Zoecon Corporation, Palo Alto, California) were placed along a line transecting each treatment block. One trap was placed midway in the plot, and the other two traps were placed approximately 18 m in either direction of the center trap along the transect line. A 10-mm diameter rubber septum (Fisher Scientific, No. 14-126AA) loaded with 110 μg of a blend consisting of 10:1 mixture of Z9-12:OAc and Z11-14:OAc was placed in each trap. Septa were replaced monthly. During 1985 and 1986,

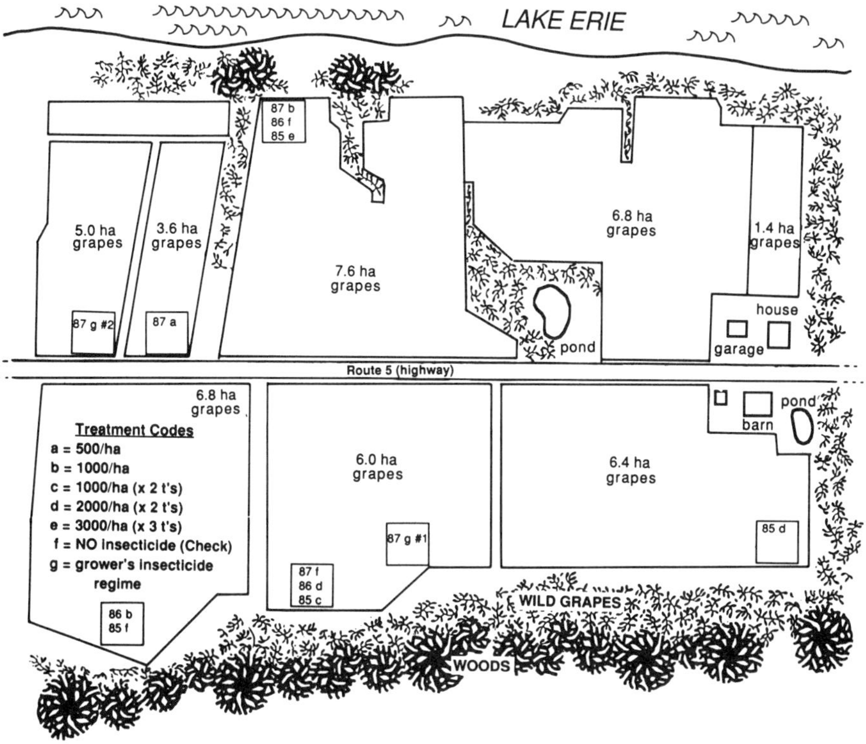

FIGURE 1 Map of the DeGolier vineyard location indicating placement of plots in 1985–1987. Note the location of wooded areas and clumps of trees and wild *V. riparia*, which serves as an untreated refuge for GBM. Treated blocks are designated as squares containing the treatment codes. Treatment codes are given on the figure and the years 1985–1987 are coded as 85, 86, and 87.

the pheromone traps were inspected biweekly for trapped moths. In 1987, the traps were checked weekly.

3. *Shin-Etsu Pheromone Product*

The pheromone product was obtained from Mitsubishi Corporation, Plant Protection and Intermediate Department, Tokyo, Japan. This product, described previously, is manufactured by Shin-Etsu Chemical

Company, Ltd., Tokyo, Japan. Initial treatments were placed in the field about May 15. In the cases where two applications were evaluated in the same year (1986–1987), the second application was applied on July 15. When three treatments per season were evaluated (1985), the second treatment was placed in the plots about July 1, and the third treatment on August 12. The pheromone dispensers were placed on the top wire of the grape trellis, a height of approximately 1.3 to 1.6 m. The ties were evenly spaced throughout the treated blocks; spacing along the trellis was varied according to the rate of ties used per hectare.

4. *Evaluation of Fruit Damage*

Evaluation of GBM damage in the mating disruption trials commenced each year on July 15 and was continued until harvest (late September or early October). During this period counts were made at 3-week intervals in each of the 0.4-ha blocks. Each evaluation involved inspecting a total of 10 clusters per vine on each of 20 vines. The 20 vines on which clusters were observed were selected from a rectangle delineated by five vines by four rows. Each time that a block was sampled, a randomized starting point within the 0.4-ha block was predetermined and this starting point became the lower left-hand corner of a five-vine by four-row rectangle in which all 20 vines were inspected for GBM. The randomization procedure was constructed by using the random number generator on Minitab (8); values were assigned to each possible starting position within the block, taking into account a buffer of one row around the perimeter of the entire block.

Within each of the randomly chosen 20-vine areas, 10 clusters per vine were nondestructively sampled and counts were made of the total berries per cluster that had any signs of feeding by GBM larvae. No statistical randomization procedure was used for selecting the clusters sampled from each vine, but clusters were chosen from all regions of the vine canopy. To allow computation of the percentage of damaged berries within each mating disruption treatment, the number of berries per cluster was assessed for 10 clusters on each of 20 vines, using the same sampling methodology employed for estimating larval damage. Berry counts were made on July 15 and again at the end of the season. The percentage of damaged berries was computed each time that the plots were sampled. Binomial confidence intervals for the percentage of damaged berries were computed, as were standard error (S.E.) values generated from the mean infestation rates of the 200 clusters (20 vines × 10 clusters per vine) observed in each plot on each sampling. It was observed that 95% binomial confidence intervals (9) were slightly larger than 95% confidence intervals generated from standard errors of mean cluster infestation, so we used only the former in depicting variances in percentage of damaged berries.

5. Weight Loss of Shin-Etsu Ties

Weight loss of a group of 50 S-E ties was recorded throughout the 1986 season. The 50 dispensers were suspended on a wooden rack in such a manner that they could be removed and replaced (using forceps) without touching or bending the ties. The rack of pheromone dispensers was placed at the height of the top trellis wire within a Concord vineyard at the Cornell University Vineyard Laboratory, Fredonia, New York. The ties were placed in the vineyard in May 15, 1986. On 6- to 10-day intervals throughout the growing season, the rack of dispensers was brought into the laboratory adjacent to the vineyard and the 50 ties were individually weighed. After weighing, the ties were promptly returned to the vineyard. Weights were obtained using a Mettler P160 balance. Mean (±SEM) weight loss per tie per day was computed. Proportional loss of product from the dispensers was estimated based on an initial loading of 0.088 g for each tie.

B. Results and Discussion

1. 1985 Trials

a. Pheromone trap-catch: No male GBM were captured in any of the pheromone-treated plots in 1985. A total of 266 male moths were captured in the untreated plot.

b. Fruit damage: Evaluations of the Shin-Etsu (S-E) ties conducted in 1985 at the DeGolier vineyard (see Fig. 1) revealed very acceptable levels of control of GBM (Fig. 2). When larval damage was first monitored on July 15, 1985, there were already biologically and statistically significant differences between the three plots treated with pheromone and the untreated control. These differences were sustained throughout the season (see Fig. 2). By harvest time (mid-September), the untreated (control) plot contained over 16% berry damage, whereas the pheromone-treated plots averaged close to, or below, 1% GBM-damaged berries. Although the pheromone-treated plots were well below the USDA threshold of 2% damage by the season's end, a pronounced increase in infestation rates was observed during the final sampling period (September 9). It is unlikely that this increase reflects reductions in active pheromone released by the ties during the final weeks of the season, because the increase was observed even in the 3000-×-3 treatment in which ties were placed in the vineyard at monthly intervals during May, June, and July. The block treated with the 3000-×-3 rate contained ties which were placed

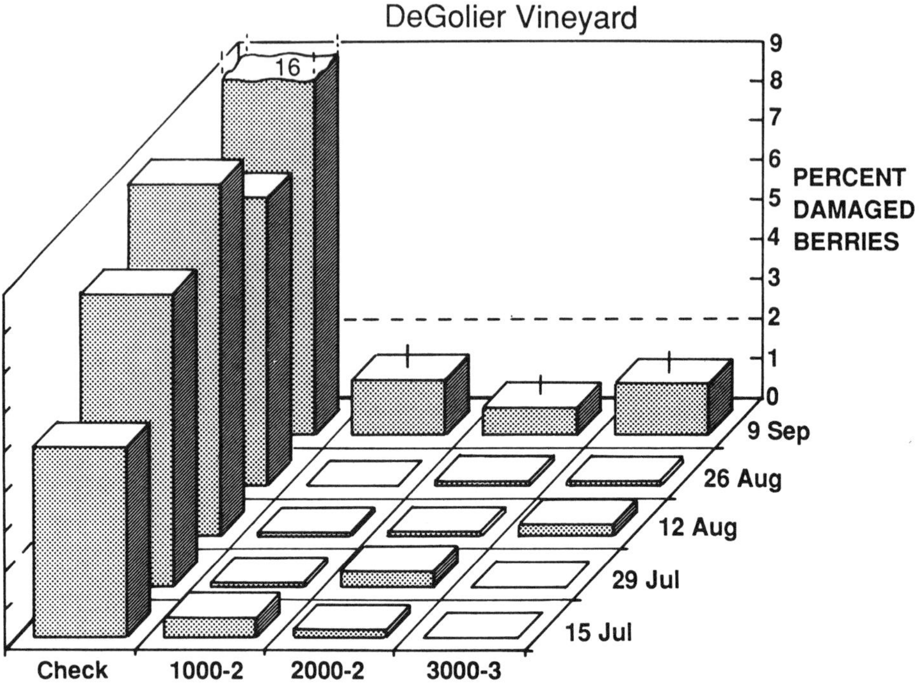

FIGURE 2 DeGolier vineyard mating disruption trials conducted in 1985. Shin-Etsu pheromone ties were placed in vineyards two (×2) or three (×3) times per season at rates of 1000 to 3000 ties per hectare. The percentage of berries damaged by GBM (95% binomial confidence intervals) is given. Two-percent-damaged berries is the industry-established threshold for GBM damage of processed grapes. The GBM densities in the grape belt were average or slightly above average in 1985.

in the vineyard on August 12, less than 30 days before the ultimate sampling on September 9.

All three rates of pheromone evaluated in 1985 were equally effective in suppressing GBM damage. This was despite the fact that over four times as much pheromone was dispensed in the 3000-×-3 tie treatment as in the 1000-×-2 plots. On the basis of the relative levels of damage observed throughout New York vineyards, we characterized overall GBM populations in 1985 as being average or somewhat above average. Our trial indicated that the S-E pheromone product could provide economically acceptable suppression of GBM in processing grapes during years of average pest densities. Indications were that use of rates of less than two treatments of 1000 ties per hectare should be investigated.

2. *1986 Trials*

a. *Pheromone trap-catch*: No male GBM were captured during the entire season in pheromone traps in the treated blocks (400 × 1 and 200 × 1) at the Francis vineyard. A total of 90 males were trapped in the untreated block. A total of 66 males were trapped in the check block at the DeGolier vineyard. Treated blocks at DeGolier (800 × 2 and 400 × 1) rendered no trapped males until the first week of September. At that time, two males were caught in the 800 × 2 block, and one moth was caught in the 400 × 1 block. The significance of these late-season catches is unclear. It seems unlikely that the result is explained by low release rates from the pheromone ties, because two males were captured in the 800 × 2 treatment. This block had 400 new ties placed in it on July 15, about 6 to 7 weeks before the males were trapped.

b. *Weight loss of S-E pheromone ties*: Shin-Etsu pheromone ties lost between 0 and 1 mg/day (Fig. 3) during the 1986 season. Slight weight gains were recorded on three occasions and were attributed to moisture during rainy periods. On the basis of the loading rate of 0.088 g per tie reported by Shin-Etsu, ties had lost approximately 33, 49, 61, and 69% of the volatile product by 30, 60, 90, and 120 days, respectively. Average weight loss per day was 0.91, 0.50, 0.39, and 0.25 mg/day for the periods of days 1–30, 31–60, 61–90, and 91–120, respectively. Because release rates are influenced by temperature, it should be emphasized that the 1986 season was wetter and cooler than average.

c. *Fruit damage*: After a relatively warm spring, unusually wet conditions prevailed in western New York throughout the summer of 1986. As a result of the favorable spring conditions, GBM populations began developing earlier than normal and then were subjected to cold, wet weather. At two of the four vineyard sites where the S-E ties were evaluated in 1986, the control (untreated) blocks had no greater GBM damage than those treated with pheromone, and all of these blocks had levels of GBM damage that were well below the economic injury level of 2% damaged berries. However, despite the generally lower levels of GBM damage observed in 1986, trials conducted at the DeGolier and Francis vineyards resulted in good suppression of GBM in treated plots relative to control plots (Fig. 4).

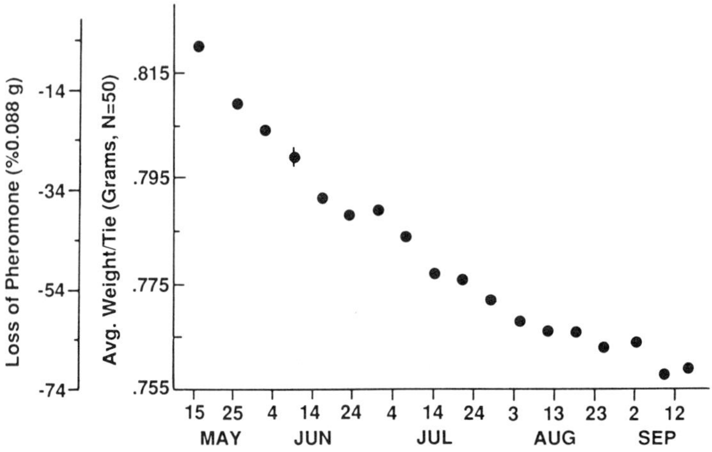

FIGURE 3 Weight loss of Shin-Etsu pheromone ties in 1986. Weight was recorded on 7- to 10-day intervals. Percentage weight loss (±2 SEM) is based on a sample of 50 ties and an average loading of 0.088 g/tie.

Noteworthy is the result that greater percentages of damaged berries were detected at the DeGolier and Francis vineyards on July 15 than on August 4. This finding is attributed to abscission of small berries damaged by GBM during the early season and to persistent rainfall that thwarted GBM development during late July and August. By season's end the untreated blocks at both the DeGolier and Francis vineyards had greater than 3% damaged berries, whereas all pheromone-treated blocks sustained less than 1% damage, well below the industry threshold of 2% damaged berries.

The DeGolier block treated in 1986 with 1000 ties per hectare was the same block that in 1985 was the check (see Fig. 1) block and that at the end of the 1985 season had more than 9% damaged berries. Although it must be emphasized that 1986 was a season of below-average pressure from GBM, it merits mention that this plot that had over 9% damage at the end of the 1985 season, had less than 1% damage when treated with 1000 ties per hectare in 1986.

In 1986, higher rates of pheromone resulted in small, but statistically significant improvements in suppression of GBM relative to lower rates (see Fig. 4). In September the 2000 × 2 rate at DeGolier resulted in 0.20% (±0.13%) damaged berries, whereas the 1000 × 1 rate had 0.90% (±0.29%) damaged berries. Similarly, at the Francis

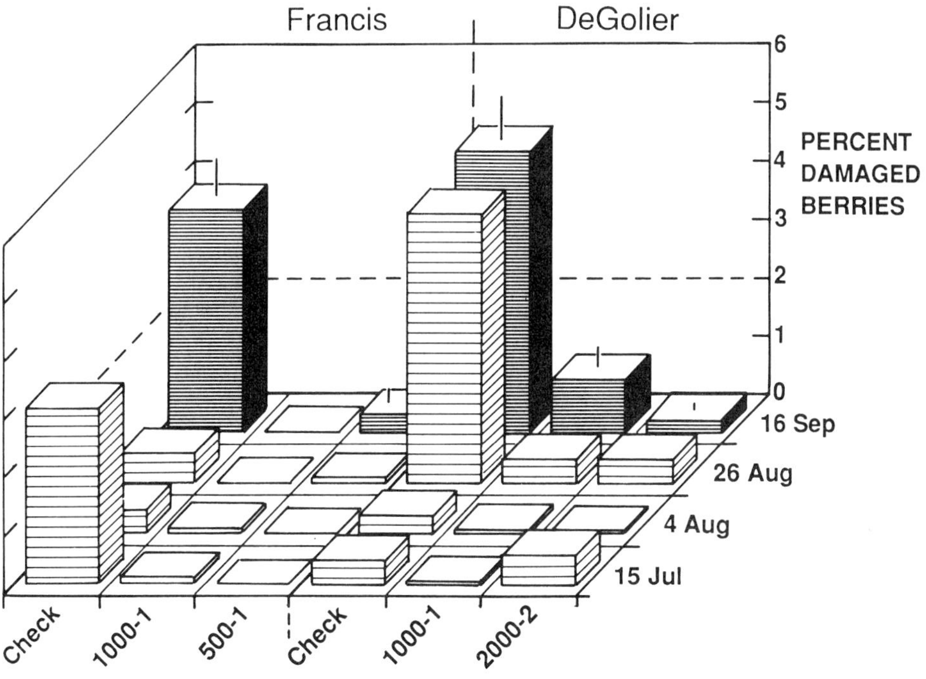

FIGURE 4 Mating disruption trials conducted in 1986 at the DeGolier and Grancis vineyard sites. Shin-Etsu pheromone ties were placed in vineyards at rates of 500 to 2000 ties per hectare on May 15. The block receiving 2000 ties per hectare received a second treatment of 2000 ties on July 15. The percentage of berries damaged by GBM (95% binomial confidence intervals) is given. Two-percent-damaged berries is the industry-established threshold for GBM damage of processed grapes. The GBM densities in the grape belt were below normal in 1986.

vineyard the 1000 × 1 block had no detectable (0±0) damage in September, whereas the 500 × 1 block had 0.30% (±0.17%) damaged berries. These differences do not appear to be important, because all treated plots were well below the 2% damage threshold. However, such seemingly small differences in rate response may translate into larger differences during seasons of greater GBM pressure. The 1986 trials showed that during years of below-average damage from GBM, rates as low as 500 ties per hectare could provide economically acceptable control.

3. 1987 Trials

a. Pheromone trap-catch: Surprisingly large numbers of males were caught within the pheromone-treated plots in 1987 at the DeGolier and Christy vineyards. These high male trap-catches were correlated with unusually high densities of GBM throughout New York vineyards in 1987. Field personnel reported that the intensity of GBM damage observed at specific vineyards in 1987 was estimated to occur once every 10 to 20 years. High trap counts in pheromone plots did not, however, necessarily reflect concomitantly high levels of berry damage in respective plots. At the DeGolier location 71, 20, and 88 males were trapped in the 500/ha, 1000/ha, and control plot, respectively. The first males were trapped on July 16 for the 500/ha rate, July 23 for the 1000/ha rate, and a full month earlier, June 11, for the control block. Seventy-nine to eighty-three percent of the DeGolier males were trapped after August 15.

At the Christy location 16, 7, and 76 males were trapped in the 500/ha, 1000/ha, and control plot, respectively. The first males were trapped on July 16 in the 500/ha rate, July 30 in the 1000/ha rate, and, as with DeGolier, over a full month earlier (June 3) in the control block. Unlike the DeGolier result, trap-catch in the 500/ha plot (total, 16) was sharply lower than that of the control plot (total, 76). Most males were trapped after August 15.

A total of 50 males were captured in the check plot at the Militello vineyard. Yet only one male was captured in each of the pheromone-treated plots. The single males were captured in the pheromone-treated plots on August 19 (500 ties per hectare plot) and September 11 (1000 ties per hectare plot). The earliest males were captured in the untreated plot over 2 months earlier, on June 11. The relative number of males caught in the check plot at the Militello vineyard was high (total, 50) but not as high as observed at DeGolier and Christy.

b. Fruit damage: Grape berry moth infestations in many New York vineyards reached unusually high levels in 1987. At the DeGolier test site, populations of the magnitude observed in 1987 had not been experienced for over 10 years (T. N. Taft, personal communication). Populations at the Christy and Militello locations also were higher than usual, although not the extreme observed at the DeGolier vineyard (Fig. 5).

Control during the 1987 season in pheromone-treated blocks at the Christy and Militello vineyards was comparable with that observed in blocks treated by the growers with conventional insecticides (see Fig. 5). At harvest (early September), plots treated with rates of 500 and 1000 ties per hectare resulted in 0.34% (±0.18) and 0.30% (±0.16%) damaged berries, respectively, at the Christy vineyard.

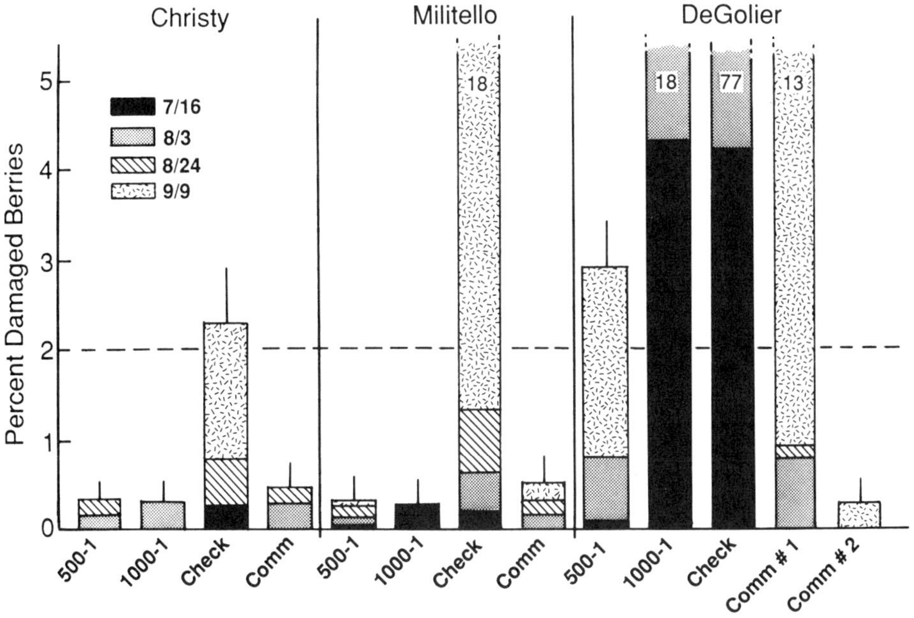

FIGURE 5 Mating disruption trials conducted at the Christy, Militello, and DeGolier vineyards in 1987. Shin-Etsu pheromone ties were placed in vineyards at rates of 500 to 1000 ties per hectare on May 15. The percentage of berries damaged (95% binomial confidence intervals) by GBM was recorded. Two-percent-damaged berries is the industry-established threshold for GBM damage of processed grapes. The GBM densities in the grape belt were unusually high in 1987. Insecticide-treated blocks are denoted as Comm.

A Christy vineyard block in which conventional insecticides were used had 0.44% (±0.20%) damaged berries, and the untreated block at this location had 2.3% (±0.46%) damaged berries.

Trends in suppression of GBM in 1987 at the Militello vineyard were similar to that of the Christy vineyard (see Fig. 5). At harvest, plots treated with 500 and 1000 ties per hectare had 0.30% (±0.20) and 0.26% (±0.16%) fruit damage, whereas plots treated by the grower with conventional insecticides had 0.49% (±0.21%). The untreated plot at the Militello site had 18% (±1.2%) damaged berries. The Militello and Christy plots supported our hypothesis that average infestations of GBM could be controlled with rates of as low as 500 S-E ties per hectare.

Densities of GBM observed in 1987 at the DeGolier location greatly exceeded those recorded in 1985 and 1986. Untreated plots at this site had 16% (±1.1%) and 4.8% (±0.66%) damage in 1985 and 1986, respectively, whereas in 1987 the untreated plot had 77% (±1.3%) damaged berries (see Figs. 2, 4, and 5). In 1987, a plot treated by DeGolier with four applications of conventional insecticides was located east of the untreated plot (see Fig. 1: insecticide plot 1). This insecticide-treated plot had 13% (±1.0%) damaged berries, indicating that four treatments of insecticide failed to provide economically acceptable control of GBM during the 1987 season in areas of high GBM densities. Both the insecticide-treated plot and the check plot were in close proximity to wooded areas that harbor wild *V. riparia* and serve as an untreated refuge for GBM. Equally as unacceptable as the insecticide-treated plot was the suppression of GBM in the plot treated with 1000 ties per hectare in 1987. This plot, located on the shoreline cliffs of Lake Erie (see Fig. 1) and between two clumps of trees heavily laden with wild *V. riparia* infested with GBM, had 18% (±1.2%) damaged berries at harvest.

The DeGolier vineyard plot treated in 1987 with 500 ties per hectare was located toward the interior of the 44-ha vineyard and further from untreated sources of GBM than was the other pheromone-treated plot or the control plot. Nonetheless, the plot treated with 500 ties per hectare incurred 2.9% (±0.52%) damaged berries: an amount close to, but exceeding the threshold of 2% damage. A second insecticide-treated plot located to the west of the 500-tie per hectare plot (Fig. 1) was evaluated when it was observed that the interior of the vineyard had lower overall GBM pressure than did the perimeter of the vineyard. This plot, which received four insecticide treatments, had 0.27% (±0.16%) damaged berries. Therefore, in the interior of the DeGolier vineyard, the insecticide treatments provided better control than did the 500/ha rate of pheromone ties.

Although male trap-catch at DeGolier was nearly as high in the 500/ha plot as in the control plot, less than 3% berry damage was recorded at harvest in the former block, whereas the latter block had over 77% damage (see Fig. 5). Contrasting male trap-catch and berry damage at the DeGolier and Christy locations serves to underscore the weakness of attempting to relate male trap-catch with the amount of damage that will occur in the vineyard. Although very similar numbers of males were trapped in the control plots at these two locations, the control plot at Christy ended up with just over 2% damaged berries, whereas the DeGolier control had over 77% damage. Similar examples are found by contrasting the trap-catch and berry damage in the pheromone-treated plots at these two locations.

IV. CONCLUSIONS

The 1987 DeGolier trials pushed both insecticide and pheromone treatments beyond their abilities to provide acceptable suppression of populations in areas of intensive pressure by GBM. This result does not denote a major shortcoming of mating disruption; rather, it indicates that higher rates of pheromone may be needed when extraordinarily high densities of GBM are being combated. During 3 years of field evaluations, in only one instance, in the DeGolier vineyard in 1987, did the Shin-Etsu pheromone dispenser fail to provide economically acceptable control of GBM. In the one instance where the S-E ties failed, infestation rates were unusually high, and plots receiving conventional insecticide also had unacceptable levels of GBM damage. It appears as though single treatments of 500 to 1000 ties per hectare may be a viable, environmentally safe alternative to the insecticide treatments that are typically applied to vineyards in New York for control of GBM.

ACKNOWLEDGMENT

The authors acknowledge the technical assistance of H. Crowe, C. Cummings, and J. Hahn of the Cornell University Vineyard Laboratory, Fredonia, New York. We thank B. Francis, B. Militello, and P. Christy for allowing us to conduct tests in their vineyards and G. DeGolier for providing vineyards and assistance throughout the course of these studies. This research was funded in part by Mitsubishi Corporation and the Cornell University Integrated Pest Management Program.

REFERENCES

1. Henneberry, T. J., Bariola, L. A., Flint, H. M., Lingren, P. D., Gillespie, J. M., and Kydonieus, A. F. Pink bollworm and tobacco budworm mating disruption studies on cotton. In *Management of Insect Pests with Semiochemicals: Concepts and Practice.* (Mitchell, E. R., ed.). Plenum Press, New York, 1981, pp. 267–283.
2. Rothschild, G. H. L. Mating disruption of lepidopterous pests: Current status and future prospects. In *Management of Insect Pests with Semiochemicals: Concepts and Practice.* (Mitchell, E. R., ed.). Plenum Press, New York, 1981, pp. 207–228.
3. Roelofs, W. L., Tette, J. P., Taschenberg, E. F., and Comeau, A. Sex pheromone of the grape berry moth: Identification by

classical and electro-antennogram methods and field tests. *J. Insect Physiol.* *17*:2235–2243 (1971).
4. Taschenberg, E. F., Cardé, R. T., and Roelofs, W. L. Sex pheromone mass trapping and mating disruption for control of redbanded leafroller and grape berry moths in vineyards. *Environ. Entomol.* *3*:239–242 (1974).
5. Taschenberg, E. F. and Roelofs, W. L. Pheromone communication disruption of the grape berry moth with microencapsulated and hollow fiber systems. *Environ. Entomol.* *5*:688–691 (1976).
6. Taschenberg, E. F. and Roelofs, W. L. Mating disruption of the grape berry moth, *Paralobesia viteana*, with pheromone released from hollow fibers. *Environ. Entomol.* *6*:761–763 (1977).
7. Wilcox, W. F. and Agnello, A. M. *Grape Pest Control Guide*. Cornell Cooperative Extension, Ithaca, 1987, 10 pp.
8. Minitab. 1985. *Minitab Reference Manual*. Release 5.1. October, 1985. Minitab, State College, Penn. 1985.
9. Daniel, W. W. *Applied Nonparametric Statistics*. Houghton Mifflin Co., Boston, 1978, 503 pp.

16

Peachtree Borer and Lesser Peachtree Borer Control in the United States

J. WENDELL SNOW / Agricultural Research Service, U.S. Department of Agriculture, Byron, Georgia

I. INTRODUCTION

The use of atmospheric permeation with pheromones to disrupt mating has been investigated for many species (1). The concept was first tested in 1974 (2), when it was determined that hexalure in a cotton field reduced damage caused by the pink bollworm to a level comparable with that achieved with insecticides. With the discovery (3) of (*Z*,*Z*)-3,13-octadecadienyl acetate (ZZA) and (*E*,*Z*)-3,13-octadecadienyl acetate (EZA) as attractants for the peachtree borer (PTB), *Synanthedon exitiosa* (Say), and lesser peachtree borer (LPTB), *S. pictipes* (Grote and Robinson), respectively, research with these species was conducted by several researchers (4–7) to determine the feasibility of the male disruption technique. The selection of clearwing moths for this research was based principally on their economic significance and the powerful, long-range nature of their pheromone. Also, both species were day fliers, which facilitated their study. The data from this research are the subject of this chapter.

II. ECONOMIC AND BIOLOGICAL IMPORTANCE

A. The Peachtree Borer

The PTB is the most economically important member of the family Sesiidae, and many papers have been written about its importance, distribution, seasonal population trends, control, and such (5–8). It is a native American species that existed on wild species of *Prunus* until the introduction of cultivated cherry, almond, plum, and peach, which have become the primary hosts. It is distributed throughout most of the United States and parts of Canada and occurs in several color forms (9). Its larval attack is confined to the base of the tree and to the roots at or below the ground surface. Masses of gum mixed with brownish frass exude at and below the soil surface. The winter is passed as larvae, and in the spring larval development is completed, pupation occurs, and the adults begin to emerge in central Georgia during early May (Fig. 1). The emergence is not significant until August and September, when about 80% of the adults emerge. Barry et al. (10) reported that a 96:4 blend of ZZA to EZA was the best formulation for attraction, and this blend was used in the reported studies unless otherwise stated.

B. Lesser Peachtree Borer

The LPTB is probably the second most important sesiid species but its distribution is confined to the eastern and central United States, where it attacks the same tree species as the PTB. However, it usually confines its attack to the above-ground parts of the tree. At the base of an attacked tree, both PTB and LPTB larvae can be found but all larvae above this point are LPTB (8). I have noted

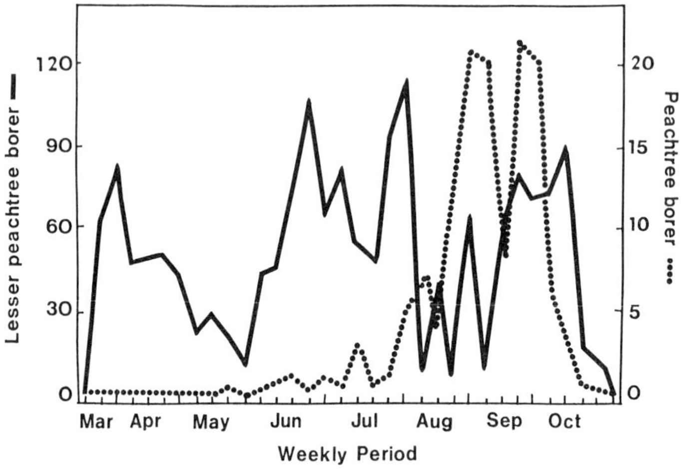

FIGURE 1 Seasonal capture of the PTB and LPTB in pheromone traps operated in peach orchards in central Georgia.

infestations as high as 10 m from the ground on black cherry where two limbs had rubbed together and caused a wound. For the LPTB, masses of frass-laden gum also exude from the trunk and limbs of infested trees, especially at sites where injuries have occurred. The winter is passed as larvae and these begin to feed and pupate in the early spring and start to emerge by late March or early April (11). Figure 1 also shows a typical seasonal curve for the species. The LPTB is much more abundant in the Southeast than the PTB and usually three distinct generations occur each year (see Fig. 1). The species is of particular importance in the Northeast because of its association with *Cytospora* canker (12).

C. Current Control Methods

Methods to control the PTB have been varied and many. The first recommendation before chemicals were available was to dig the larvae out with a sharp knife or probe them with a flexible wire. This was ineffective and only a few of the larvae could be reached, but it was the only method available. Later an application of para-dichlorobenzene (moth ball crystals) around the trunk of the tree during the fall was recommended. After application, the crystals were covered with soil to contain the fumes (8). This was later replaced by the use of ethylene dichloride emulsion that could be sprayed or poured on and around the tree trunk. This treatment was very effective and, until recently, was still used for yard trees and small orchards. There were no early direct control recommendations for the LPTB, but control of catfacing insects also provided some control of this species.

Current insecticide recommendations for both species are the use of an application of Thiodan immediately after harvest, followed by a Lorsban application after peak emergence between August 1 and September 1, applied to the larger limbs and trunk. If only one application is used, the Lorsban application is usually recommended. Depending upon the actual rate applied per acre, Lorsban costs between 11.25 dollars and 22.50 dollars/acre, whereas Thiodan costs between 12 dollars and 21 dollars/acre. Thus, the average cost to control borers ranges from a low of 11.25 dollars/acre to a high of 43.50 dollars/acre. The average cost is about 20 dollars/acre. The effectiveness of chemical control for both pests is often poor because the timing and precision of applications are very critical.

III. PHEROMONE RESEARCH

A. Early Research

The attractiveness of males to females of the PTB was first reported by Smith (13) in 1898, Cory (14) in 1913, and Gossard and King (15) in 1918. Smith (13) also described a brush of hairs at the base of the last abdominal segment of the females which he considered to be the source of the sex pheromone. Jacklin et al. (16) found females to be most attractive to males during their second and third day of life. Maximum attraction was at about 11:00 AM EST. This information aided greatly in the eventual isolation, identification and synthesis of both the LPTB and the PTB pheromone.

Following this discovery, McLaughlin et al. (5) were the first to report good results using the permeation method of control. They concluded that the sexual communication of both the PTB and the LPTB could be disrupted by permeation of the atmosphere with their respective pheromones. They also concluded that the two isomers seemed equally effective against both species. Disruption was greatest when the pheromone was evaporated from the tops of peach trees. On the basis of these data, these and later researchers concluded erroneously that either isomer would be effective for both species. The ZZA was the isomer of choice because of its lower cost. This research was followed by new tests by Gentry et al. (7) and Yonce (6), and their results were good when "control" was measured by reduction of numbers of males captured or when treatment was initiated in young orchards before they became infested. However, in all of these cases, economic control was not recorded or did not occur over long periods.

Following these studies, attempts were made in the early 1980s to commercialize the ZZA pheromone by various companies, but these attempts failed, principally because insufficient biological data had been collected and the tests had been of insufficient size to obtain the needed data. In addition, several conclusions reached in earlier tests led companies and researchers in the wrong direction. In the remainder of this paper we will concentrate on these erroneous conclusions and also on large-scale testing and the current status of the technology.

B. Orchard Tests

In 1984, we began experiments in total orchard situations and under treatment regimens that were economically feasible and aimed at partitioning the levels of control obtained for both the PTB and LPTB (17). All tests were conducted using Conrel hollow-fiber dispensers, Hercon vinyl-laminated dispensers or Shin-Etsu rope dispensers that were prepared commercially to provide the appropriate isomeric mixture and dosage rate. The evaporation rates of Hercon and Shin-Etsu dispensers, loaded with 43 and 68 mg of pheromone per dispenser, respectively, are shown in Figure 2. These dispensers were placed in a peach orchard and three were removed and stored in a refrigerator every 2 weeks for a 16-week period. Evaporation rates were determined by gas chromatographic techniques. Note that evaporation was

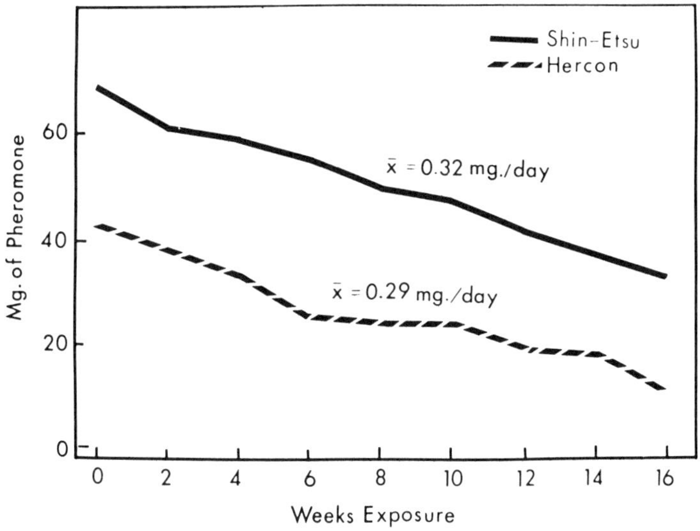

FIGURE 2 The evaporation of ZZA from Shin-Etsu (0 wek = 68 mg) and Hercon (0 wk = 43 mg) dispensers hung on a peach tree.

linear over time, and that the average release rates were 0.29 and 0.32 mg/day for the Hercon and Shin-Etsu dispensers, respectively. Our studies were also conducted to confirm the control aspects of the earlier work and to include immigration and specific mating information that would provide an understanding of the insects' behavior in a permeated area (18). Only the ZZA isomer was tested in the earlier research in the belief that it would control both species, but in later, previously unreported tests both isomers were included. In the first tests, the Hercon dispensers loaded with 43 mg of pheromone per dispenser were used. Experiments were conducted in three different blocks of peach trees located in central Georgia; each tree (in two blocks) was treated with one dispenser per tree. In one treatment area the wild hosts (plums and black cherries) within 1.5 km of the area were treated, but they were not treated in the other area. This scheme was reversed in the second year. Male PTB and LPTB populations were monitored using Pherocon 1C sticky traps baited with their respective pheromone. A single control orchard was maintained for the two treatment orchards.

In 1984, mating activity was determined by using mating tables. Tables were operated both within the orchards and outside the orchards in wild host areas (within the 1.5-km radius area). Female

PTB were collected from orchards with insect nets. The LPTB females fly too fast to be collected in this way. Where appropriate, a two-way analysis of variance ($p < 0.05$) or χ^2 contingency table analysis ($p <$ 0.01) was used to analyze the data.

Area M

Treatment area M consisted of six orchards comprising 44.8 ha that were moderately infested with PTB and LPTB. (The "Area M" designation represents moderate infestation.) We treated the entire 44.8 ha in 1984 and in 1985 we treated the same 44.8 ha, but we also treated the wild host areas with 1.5 km of the central orchard. The nearest untreated peach orchard was 20 ha and located about 6.0 km away.

Area H

Treatment area H consisted of seven orchards, but all were young and lightly infested or noninfested, except for a 19.2-ha orchard that was heavily infested with both species. ("Area H" represents heavy infestation.) During 1984, we treated the entire area including the wild host areas within a 1.5 km radius area but, in 1985, the wild host areas were excluded.

Data collected from these areas for the PTB are shown in Figure 3 and Table 1. No males were captured after treatment with pheromones on August 1 to 10 (see Fig. 3). Table 1 shows data from mating tables. More females were placed in some locations than in others because data from certain locations were more critical and because the number of available females was limited. In area M, the moderately infested orchard where the adjacent wild hosts were not treated, no mating occurred in the orchard but 23% of the females mated in the wild host area within the 1.5-km radius area. In area H, the heavily infested orchard, 2% of the exposed PTB females mated in the orchard, but none mated on the tables in the wild host area (which was treated) within the 1.5-km radius area. Females on mating tables in the control orchard were 84% mated.

Figure 4 shows the data on PTB females collected with sweepnets. In 1984, 70 females were collected in the control orchard and 8.6% of these were virgins. In area M, 70% of all collected females (28 of 40) were virgins, and in area H, 52% (69 of 132) were virgins. In 1985, in area M (which in this year had the wild hosts treated) 76% of the collected females (45 of 59) were virgins, and in area H, in which the wild hosts were not treated, 79% (31 of 39) of the collected females were virgins. The control orchard had 12% virgins (19 of 160). In general, the populations in both of the treated orchards were much lower during 1985, and the collection of females was difficult.

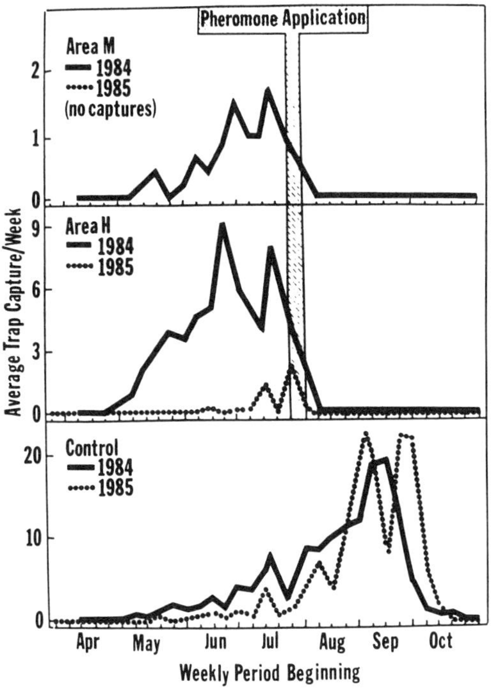

FIGURE 3 The capture of male PTB in moniotor traps in the 3 test areas in 1984 and 1985.

Control of the PTB was almost equal both years in area M regardless of whether or not the wild host area was treated. In neither year was mating observed in this area on mating tables (1984) or by observations (1984–1985). However, in area H the control level varied (52% and 79%) during the 2 years. Although the reduction in mating during 1984 was low, it was more significant than first perceived because of the many mortality factors that come into play against eggs, larvae, and pupae. The good level of control and difficulty in finding adults the following year (1985) is evidence of the effect that the 1984 treatment had in area H. After 1985, it was impossible to conduct tests for the PTB in these orchards because the population had been reduced to near zero.

In these tests, based on mating table and trap captures, the ZZA isomer was successful in reducing and delaying, but not preventing,

TABLE 1 Collection of Mated Pairs of PTB from Mating Tables Operated In and Around Test Areas, 1984

Mating table location	Days data collected[a]	Females[b] Unmated	Mated	% mated	Separation of counts by χ^2
		Area M			
Within	10	100	0	0	a
Outside	8	60	18	23	b
		Area H			
Within	10	100	2	2	a
Outside	8	43	0	0	a
		Control			
Within	9	8	42	84	c

[a]A day's data are considered one table per treatment with five to ten wing-clipped females.
[b]Accumulated totals in each category. Totals are significant at ($p < 0.01$).

mating of LPTB. This was proved by the collection of mating pairs on mating tables and by the collection of wild mating pairs in area H. Although mating was reduced, I believe that mating eventually occurred and no real control resulted. This was established in this and other tests where damage estimates of LPTB showed that treatment with ZZA did not reduce damage levels.

C. Pheromone Requirement

It became evident in these early large-scale tests that whereas control of the PTB was being obtained with the ZZA isomer, damage was still being caused by the LPTB. As a result several tests were conducted to determine if the EZA isomer (its own pheromone) was necessary for LPTB control. Table 2 shows the capture of males in traps that were baited with various ratios of EZA to ZZA. The test was replicated five times, with the total dosage being held constant at 100 μg/dispenser. Note that the LPTB was highly attracted to only the pure isomer of EZA. However, for the PTB a 92.5:7.5 blend of ZZA/EZA was the most attractive. This data confirms the increased attraction of the PTB to blends of ZZA and EZA reported earlier by Barry et al.

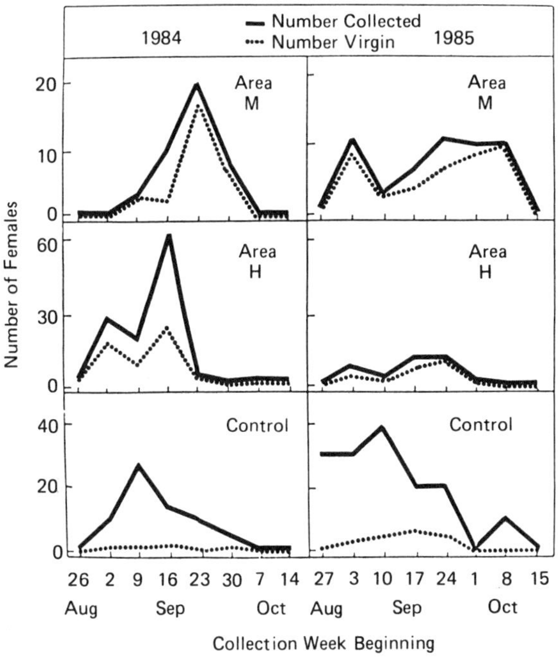

FIGURE 4 The collection and mating status of wild females collected from the treatment areas in 1984 and 1985.

(10). However, if more than 15% EZA was included with the ZZA, the capture of PTB males was greatly reduced.

In another test (Table 3), dispensers loaded with ZZA, EZA, 50:50 blend of EZA/ZZA, 75:25 blend of EZA/ZZA, and alternating placements of EZA and ZZA dispensers on every other tree were evaluated. The test was replicated four times in a randomized complete block design and the plot size was six trees by six trees (total 36 trees). Evaluation criteria were based on captures of males in two monitor traps that were placed in the center of the 36-tree blocks. One trap was baited with EZA and one was baited with ZZA. The dispensers were Conrel hollow-fiber dispensers and the total dosage was held constant at 17 mg/dispenser. The only pheromone that successfully prevented capture of LPTB was the pure EZA. This formulation was equally successful when used alone or when it was alternated with ZZA on separate trees. For the PTB, all blends except the pure EZA significantly reduced capture. We concluded from these data that, for best results, the species' own pheromone must be used for disruption. In the last

TABLE 2 Effect of Various Concentrations of EZA and ZZA on Captures of LPTB and PTB

% of isomer[a]		Mean no. males/trap[b]	
EZA	ZZA	LPTB	PTB
0	100.0	1.8 a	17.5 b
2.5	97.5	8.3 a	31.5 bc
7.5	92.5	6.0 a	35.0 c
10.0	90.0	1.8 a	22.5 bc
15.0	85.0	4.3 a	23.3 bc
30.0	70.0	2.5 a	3.0 a
50.0	50.0	1.0 a	0 a
75.0	25.0	0 a	0 a
100.0	0	100.8 b	0 a
Blank		0 a	0 a

[a]Total of 100 μg attractant per dispenser.
[b]Means followed by the same letter are not significantly different at the 5% level of confidence based on Duncan's multiple range test.

few years, excellent results have been obtained when EZA has been used for LPTB control (J. W. Snow, unpublished data). These later tests have all been conducted with the Shin-Etsu dispenser.

As the LPTB is active during the entire season and the PTB does not have significant populations until August--September, different strategies must be developed for their control. For the LPTB, one dispenser of EZA per tree is required sometime in late March or before flight activity begins. An 80-mg dispenser is required because of the long period that the dispenser must operate. This EZA dispenser will also provide some disruption of PTB during June and July when only a few adults are in the orchard. Then on August 1 or shortly before the principal flight of the PTB, an application of ZZA (40 mg) is required. We place the dispensers on the opposite sides of the trees so they are not in direct competition with one another. This means that on a seasonal basis approximately 8 g of EZA and 4 g of ZZA are required (assuming 100 trees per acre) for borer control.

TABLE 3 Effects of Treatments with EZA, ZZA and Blends of the Two Isomers on Captures of LPTB and PTB in Pheromone Traps. Conrel Dispensers Loaded with 17 mg/dispenser

% of each pheromone		Mean catch/trap[a]	
EZ	ZZ	LPTB	PTB
0	0	54.6 c	48.8 c
100	0	4.5 a	11.0 b
0	100	38.0 c	1.7 a
50	50	11.3 ab	2.8 a
75	25	15.8 b	0.8 a
50	50[b]	4.3 a	.3 a

[a]Means followed by the same letter are not significantly different at the 5% level of confidence based on Duncan's multiple range test.
[b]Standard EZA+ZZA dispensers alternated (every other tree), 18 dispensers each.

However, the dispensers from one season carry over into the next season and actually contribute to control in the following season.

The effect of the pheromone, except in the highest population situations, is to totally prevent mating within an orchard. The population already within the field causes no future damage because mating is prevented, thus the egg–larvae–adult cycle is broken. However, no control is obtained on the mated females that are migrating into the orchard, and this damage must be sustained. Fortunately, their progeny cause no harm because their mating is prevented once they reach the adult stage. These migrating females also have a reduced biotic potential because of their age and previous oviposition. Their contribution is further lessened when natural egg and larval mortality factors are considered. However, with the PTB and LPTB and all pheromone control systems, the migrating mated females are unaffected by treatment. If sufficient females come into the orchard, then economic control is not possible. In our situation, migration usually is not sufficient to overcome economic control. In most orchards in the southeast, adjacent infected orchards which could change this situation are owned by the same grower, or by a grower who would also use pheromones as a control.

IV. CONCLUSION

Tests with both the LPTB and PTB show that effective control can be obtained by using each insect's pheromone. Usually, the treatment of adjacent wild plums and black cherries is not necessary because they do not contribute enough insects to overcome the effects of control. However, we do believe that all infested orchards within an area must be treated if they contribute significant numbers of mated females. The following is established from these data: (a) one dispenser per tree of EZA gives season-long protection against LPTB and also suppresses the low PTB population during early spring through midsummer, (b) one application of ZZA is required in August for PTB, (c) control with pheromones is as good or usually better than with insecticides, and (d) the pheromone is likely to be cost-competitive with insecticides.

REFERENCES

1. Beroza, M. Control of the gypsy moth and other insects with behavior-controlling chemicals. In *Pest Management with Insect Sex Attractants*. (Beroza, M., ed.). American Chemical Society, Washington, D.C., 1976, pp. 99–118.
2. Shorey, H. H., Kaae, R. S., and Gaston, L. K. Sex pheromones of Lepidoptera. Development of a method for pheromonal control of *Pectinophora gossypiella* in cotton. *J. Econ. Entomol.* *67*:347–350 (1974).
3. Tumlinson, J. H., Yonce, C. E., Doolittle, R. E., Heath, R. R., Gentry, C. R., and Mitchell, E. R. Sex pheromones and reproductive isolation of the lesser peachtree borer and the peachtree borer. *Science* 185:614–616 (1974).
4. McLaughlin, J. R. Disruption of mating communication in peach borers. *Agric. Res. Results Northeast. Ser.* *6*:27--31 (1979).
5. McLaughlin, J. R., Doolittle, R. E., Gentry, C. R., Mitchell, E. R., and Tumlinson, J. H. Response to pheromone traps and disruption of pheromone communication in the lesser peachtree borer and the peachtree borer. *J. Chem. Ecol.* *2*:73–81 (1976).
6. Yonce, C. E. Mating disruption of the lesser peachtree borer, *Synanthedon pictipes* (Grote and Robinson), and the peachtree borer, *S. exitiosa* (Say), with a hollow fiber formulation. *Misc. Publ. Entomol. Soc. Am.* *12*:21–29 (1981).
7. Gentry, C. R., Bierl-Leonhardt, B. A., McLaughlin, J. R., and Plimmer, J. R. Air permeation tests with (*Z*,*Z*)-3,13-octadecadien-1-ol acetate for reduction in trap catch of peachtree and lesser peachtree borer moths. *J. Chem. Ecol.* *7*:575–582 (1980).

8. Metcalf, C. L., Flint, W. P., and Metcalf, R. L. The peachtree borer and lesser peachtree borer. In *Destructive and Useful Insects*. McGraw-Hill Book Company, New York, 1962, pp. 757–761.
9. Engelhardt, G. P. The North American clear-wing moths of the family Aegeriidae. *Smithsonian Inst. US Natl. Mus. Bull. 190*: 1–122 (1946).
10. Barry, M. W., Nielsen, D. G., Purrington, F. F., and Tumlinson, J. H. Attractancy of pheromone blends of male peachtree borer, *Synanthedon exitiosa* (Say). *Environ. Entomol. 7*:1–3 (1978).
11. Snow, J. W., Eichlin, T. D., and Tumlinson, J. H. Seasonal captures of clearwing moths (Sesiidae) in traps baited with various formulations of 3,13-octadecadienyl acetate and alcohol. *J. Agric. Entomol. 2*:73–84 (1985).
12. Swift, F. C. *Cytospora* canker and the establishment of lesser peachtree borer (Lepidoptera: Sesiidae) in peach trees. *J. Econ. Entomol. 79*:537–540 (1986).
13. Smith, J. B. Notes on some structural peculiarities of *Sanninoidea exitiosa* (Say). *Entomol. News 9*:114–115 (1898).
14. Cory, E. N. The peachtree borer, *Sanninoidea exitiosa* (Say). *Univ. Md. Exp. Sta. Bull. 176*:181–218 (1913).
15. Gossard, J. A. and King, J. L. The peachtree borer, *Sanninoidea exitiosa* (Say). *Ohio Agric. Exp. Sta. Bull. 329*:56–87 (1918).
16. Jacklin, S. W., Yonce, C. E., and Hollow, J. P. The attractiveness of female to male peachtree borers. *J. Econ. Entomol. 60*: 1291-1293 (1967).
17. Snow, J. W., Gentry, C. R., and Novak, M. Behavior and control of the peachtree borer and lesser peachtree borer (Lepidoptera: Sesiidae) in peach orchards permeated with (*Z*,*Z*)-3,13-octadecadien-1-ol acetate. *J. Econ. Entomol. 78*:140–196 (1985).
18. Snow, J. W. and Gentry, C. R. The controlled release of (*Z*,*Z*)-3,13-octadecadienyl acetate and its effect on the peachtree borer. *Proc. 13th Int. Symp. Controlled Release Bioactive Materials 13*:124–125 (1986).

17

The Male Lures of Tephritid Fruit Flies

ROY T. CUNNINGHAM / Agricultural Research Service, U.S. Department of Agriculture, Hilo, Hawaii

RICHARD M. KOBAYASHI / Animal and Plant Health Inspection Service, U.S. Department of Agriculture, Honolulu, Hawaii

DORIS H. MIYASHITA / Agricultural Research Service, U.S. Department of Agriculture, Hilo, Hawaii

I. INTRODUCTION

There is a peculiar class of attractants which, for lack of a more precise definition, are called *male lures*. Male lures are a biological conundrum that seems to be confined to the Tephritidae of the Old World tropics and subtropics, Australia, and Oceania. This class of attractants has not been found to occur in the temperate zone Tephritidae, such as *Rhagoletis*, nor in the larger *Anastrepha* complex of the neotropics. However, where the phenomenon does occur, in species around the *Ceratitis* complex and the *Dacus* complex, male lures are of paramount importance. Male lures provide the basis for an extensive detection trap array and, in a few cases, for an efficient eradication system that is largely unnoticed and unknown to the general public and to most entomologists.

On the other hand, pheromones have, as of yet, failed to achieve any practical utility in tephritid fruit fly programs. There is another class of attractants, however, the proteinaceous food baits (1) and the associated derivatives of protein degradation, that are generalized tephritid attractants. These food baits are short-range attractants, but they are used routinely in detection trap arrays, field control programs, and with spectacular success in certain eradication programs (2,3). We will not discuss either food baits or pheromone lures but, rather, will concentrate upon the extensive, but little-known programs utilizing the male lures.

A. History

With the introduction of the oriental fruit fly into Hawaii in the 1940s and its explosive population growth, a large-scale research program (4) was mounted in Hawaii to investigate the three species of pest tephritids that occurred there, the oriental fruit fly, *Dacus dorsalis* Hendel; the Mediterranean fruit fly, *Ceratitis capitata* (Wiedemann); and the melon fly, *Dacus cucurbitae* Coquillett. Among the many lines of research that were started was a search for attractants. This was a cooperative effort between the U.S. Department of Agriculture (USDA) Fruit Fly Laboratory in Hawaii and the USDA Chemical Laboratories in Beltsville, Maryland. Thousands of compounds were screened (5). In a relatively few years a powerful male lure was available for each of the three species; methyl eugenol for the oriental fruit fly (6); trimedlure for the medfly (7); and cue-lure for the melon fly (8; Fig. 1). No one set out to find male lures and, in fact, there was considerable disappointment that equally attractive lures for the females, the sex that does the damage, were never found. Surprisingly, the first and most efficient male lure, methyl eugenol, was discovered serendipitously and identified by Howlett in 1912 in India (6,9). This discovery lay unexploited for decades until Loren Steiner rediscovered its efficacy on *Dacus dorsalis* (10).

B. Nature of Male Lures

These male lures are definitely not pheromones in that they are not produced by or found in the insects themselves. In fact, they defy easy classification, and their role in the biology of the insect is not understood. Each of these male lures is essentially unattractive for the other two species in Hawaii. However, with further research in other areas of the world, it was found that each of those three male lures acts as an attractant for three separate groups of tephritid fruit flies (11–14). Drew (11) found that the division of response to these lures, in the species he studied, followed recognizable taxonomic divisions—that is, allied species respond to the same lure with no significant cross-attraction to the other male lures (12).

Conversely, the early work showed that for any given species there was not just a single male lure but, rather, a group of attractive compounds of various degrees of efficacy (5). Later research has confirmed this multiplicity of male lures (15–17). These tests,

(a)

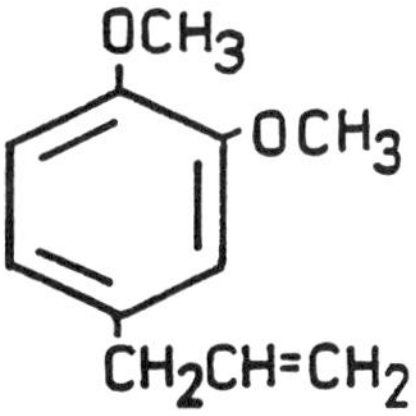

(b)

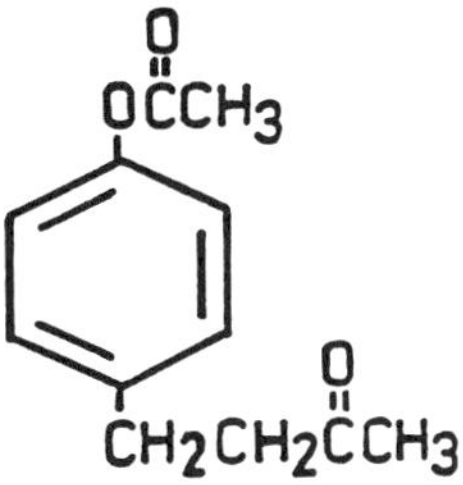

(c)

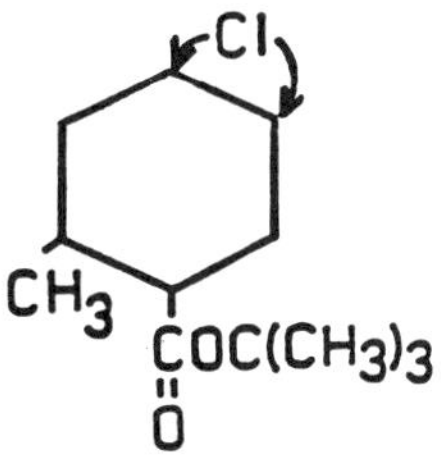

FIGURE 1 Male lures for tephritid fruit flies. (a) Methyl eugenol, (b) cue-lure, (c) trimedlure.

for the most part, were done in small cages or room-sized olfactometers. In recent years, T. P. McGovern and I have explored this phenomenon further through field tests of hundreds of compounds (unpublished). Although many compounds that were found to be attractive in the close confines of an olfactometer were functionally nonattractive in field tests, we found many new compounds to take their place. We have found a number of other male lures of varying degrees of attraction for each of the three pest species in Hawaii. Usually, these male lures were analogues of the recognized male lures, but, occasionally, quite strong attraction was found with unrelated compounds. However, at the same time that there is a relatively broad range of attractive compounds, as opposed to the lepidopteran sex pheromones, there is also high specificity--little cross-over attraction to another species group—and acutely tuned structural sensitivity. This structural sensitivity is best illustrated by the response of the Mediterranean fruit fly male to the structural isomers of trimedlure (18,19) and their enantiomers (20).

II. DETECTION PROGRAMS

Irrespective of our lack of easy classification and understanding of the biological basis for male lures, they have played a decisive role in keeping continental United States free of these important pests. None of these male-lure-responding species of tropical fruit flies has become established, in spite of frequent introductions. In a large measure, this is because of the use of the male lures in detection trap arrays and in an eradication system. In the last two decades there has been a great increase in the amount of air traffic between Hawaii and the continental United States and, furthermore, between Southeast Asia and Oceania and United States. The resident immigrant population from Southeast Asia has also greatly increased. This increased movement and displacement of people has greatly increased the risk of exotic fruit fly introductions because, when people move, their food moves with them. This is best illustrated by the record of the number of passengers boarding aircraft at Hawaiian airports who were found to be carrying fruit that was prohibited for entry into the continental United States by agricultural quarantines (Table 1). This represents a typical pattern of interceptions. The number of fruit fly specimens found in these fruits is the result of a short visual examination alone, and it would naturally underestimate the true number of forms, because eggs and first instar larvae are difficult to find.

California has had the most frequent problem with exotic fruit fly introductions, with the oriental fruit fly being the most commonly introduced. The record in Table 2 of oriental fruit fly discoveries in California over the last 27 years shows a trend toward an increase in frequency that corresponds with the increase in the movement of people. These data in Table 2 were collected from an array of detection traps run throughout the population centers of the state. Oriental fruit flies have been detected in eight counties of California, from Santa Clara by San Francisco Bay in the north to San Diego near the Mexican border in the south and in every year but two of the last 22 years.

The fruit and vegetables industry in California is a 14 billion dollar/year business. In the midst of the medfly invasion of California in 1981, it was estimated that the establishment of the medfly would

TABLE 1 Passengers Intercepted with Contraband Fruit[a] Boarding at Hawaiian Airports

	1984											
Month	J	F	M	A	M	J	J	A	S	O	N	D
No. passengers with fruit	1002	1815	1969	1785	3284	2683	4232	3449	2607	2158	1571	1361
No. larvae in fruit	40	13	8	29	64	112	104	146	103	36	21	14

[a]22 species of host fruits.
From unpublished report, USDA-APHIS-PPQ.

TABLE 2 *Dacus dorsalis* Detections in California

Year	60	61	62	63	64	65	66	67	68	69	70	71	72	73	74
Detected	X						X	X	X	X	X	X	X	X	X
Year	75	76	77	78	79	80	81	82	83	84	85	86	87		
Detected	X	X	X			X	X	X	X	X	X	X	X		
No. flies detected	3	35	12	0	0	13	9	9	14	97	123	29	52		

From unpublished report, California Dept. Food & Agriculture.

cost California at least 260 million dollars/year in crop losses alone (unpublished report presented to State Board of Food and Agriculture, August 6, 1981, by C. E. Hess). Furthermore, because most of these species of fruit flies attack the fruit as it begins to ripen, it would be difficult to incorporate insecticidal control without dislocating many current pest management regimens. This is to say nothing of the increased use of pesticides by, and loss of fruit to, the many dooryard fruit tree growers and gardeners in California. Establishment of these pests would indeed have many negative ramifications.

Because of the magnitude of the potential losses, the California Department of Food and Agriculture invests several millions of dollars each year in running a statewide trapping detection program aimed at tephritid fruit flies. Similar programs are run in Florida through the cooperative efforts of the Florida Department of Agriculture and USDA/APHIS (Animal and Plant Health Inspection Service). Some detection traps are also run in the other southern tier states. The primary tools for the detection of the paleotropical species are the three male lures deployed in Jackson traps (21), a tent-shaped trap with a sticky insert for a floor and a cotton wick for the lure dispenser.

In 1987, 15,000 traps baited with methyl eugenol, 33,000 traps baited with trimedlure, and 7000 traps baited with cue-lure were deployed in the populated area of California.

Traps baited with trimedlure for the Mediterranean fruit fly are deployed at the rate of five traps per square mile ($2/km^2$) in the Los Angeles basin throughout the year. Oriental fruit fly traps with methyl eugenol are deployed at the lower rate of two traps per square mile ($0.8/km^2$), both because the lure is relatively more powerful and because the fly itself is more mobile and, therefore, more likely to encounter a trap. This lower trapping rate also reflects our great confidence in the male annihilation eradication system that is discussed later.

Table 3 gives the history of captures of all male–lure-responding species captured in California in the last 8 years. The first three listed species are those that occur in Hawaii. *Dacus tryoni*, the Queensland fruit fly, is a cue–lure-responding species from Australia. *Dacus zonatus* responds to methyl eugenol and is a serious pest in Pakistan and India, among other areas. *Dacus correctus* is another methyl eugenol-responding species from Southeast Asia. *Dacus scutellatus* is a cue–lure-responding species endemic to Japan and Taiwan. *Dacus bivattatus*, the pumpkin fruit fly, is a cue–lure responding species from Africa. Finally, an unidentified species of *Dacus* was collected in a methyl eugenol trap in 1987. All of the other species on this list are fruit-infesting forms, so the presumption is

TABLE 3 Numbers of Male–Lure-Responding Tephritid Fruit Fly Species Captured in California

Year	1980	1981	1982	1983	1984	1985	1986	1987
Species								
Ceratitis capitata	186	201	2	0	2	0	2	43
Dacus cucurbitae	0	0	0	0	0	2	2	7
D. dorsalis	13	9	9	14	97	123	29	52
D. tryoni	0	0	0	0	0	1	0	0
D. zonatus	0	0	0	0	2	0	0	6
D. correctus	0	0	0	0	0	0	3	1
D. scutellatus	0	0	0	0	0	0	0	1
D. bivittatus	0	0	0	0	0	0	0	1
D. sp.	0	0	0	0	0	0	0	1

(From unpublished report, California Dept. Food & Agriculture.)

that this new species is also frugivorous. The introduction of these exotic fruit flies into the continental United States is a problem of movement of people, and all of the introductions into California have been found in population centers. The Los Angeles basin has the largest concentration of people on California, and it also has the most frequent incursions of fruit flies. The actual sources of these particular introductions into California are not known. Probably not all or even most of these introductions were from passenger-carried fruit. There is increasing evidence that much contraband fruit is being mailed or shipped into California.

With the constant influx of these exotic fruit flies, the key to keeping them from establishing themselves in the southern parts of the continental United States is early detection—detection before the population has spread to an unmanageable extent. However, it is not within the realm of practicality to run sufficient traps to have a high probability of detecting the founder population (22). It is, however, within the realm of economic possibility to run trapping arrays of sufficient density to have a very great likelihood of detecting the succeeding generations before they become unmanageably large in numbers and extent—before the population has spread over more than a few square kilometers. This is the principle upon which the current detection programs are predicated.

III. TRAPPING PROTOCOLS

When an exotic fruit fly is detected, the most urgent question then becomes: "Are There More?" This question is addressed by the rapid deployment of a much more intense trapping grid around the original find, according to a predetermined protocol. The protocol for response to a Mediterranean fruit fly find, for example, calls for the installation of 1700 additional traps baited with trimedlure, among other things, in an area of 81 mi^2 (210 km^2) within 48 hr of the initial find. One hundred traps are placed in the center square mile (2.59 km^2); 50 traps in each of the 6 mi^2 surrounding the center (15 km^2); and so on, out to a 4.5-mi radius (7 km) away from the find. If no further captures are made and fruit inspection fails to show larvae, these added traps are run for three hypothetical generations (day-degree determined) (23) and then removed if no further flies are found.

These protocols were first formalized in 1969 through conferences among California Department of Food and Agriculture, California County Agricultural Commissioners, USDA Animal and Plant Health Inspection Service, and USDA Agricultural Research Service personnel. The were based on research data, historical experience, and probabilistic estimates (otherwise known as educated guesses). They are subject to revision as historical experience of their use dictates but, in the main, they have stood the test of time well.

If more flies are found in the area through the intensive survey, an eradication program may be started as outlined in the protocols, depending upon the temporal and spatial pattern of the discoveries. Fortunately, over 50% of all the fly finds in California are single finds indicating that the introduced population was too sparsely scattered after reaching sexual maturity to find mates or that some other adversity snuffed them out.

IV. ERADICATION SYSTEMS

A. In Hawaii and the Marianas

One of the male lures, methyl eugenol, apart from its vital importance in the detection of the oriental fruit fly and related species, has been the basis of one of the most efficient insect eradication systems in use today—the so-called male annihilation system. This is largely the development of L. F. Steiner who, working in the ARS laboratory in

Hawaii in the 1950s, recognized the potential for the utilization of this powerful male lure in an eradication system (24). Steiner and his colleagues first demonstrated the complete eradication of the oriental fruit fly on the island of Rota in the Marianas (25). They used 5% by weight of an insecticide in methyl eugenol, which was used to saturate fiberboard squares about 5 cm on a side holding about 24 g of the mixture. These fiberboard squares were thrown out of an aircraft over uninhabited areas at given rates and at given reapplication times to achieve eradication in about 6 months.

Subsequently, under Steiner's direction in Hawaii, we started working on thickener–extender agents that could be used in sprayable formulations with sufficient residual life to be a practical system. Thickener–extenders were found (26,27), and we are currently using a third-generation one (28). In aerial applications to field plots, several square miles in extent, we found that the male population could be reduced over 98% using as little as 5 lb of formulation per square mile (0.9 kg/km^2) in a single application (29). Very high levels of male kill are necessary to push fertility into a downward trend by male annihilation.

B. In California

From these and other lines of research (30) our laboratory was able to recommend by 1970 an easily applied eradication system for use in urban California against the introductions of the oriental fruit fly (31). The program in California consists of applications of approximately 70% methyl eugenol with 10% insecticide and 20% of the thickening agent. This gives a formulation with a custardlike consistency that is applied as 5-ml spots to telephone poles and roadside trees 6 ft or more above the ground. The attractive spots are applied at the rate of 600 spots per square mile (230/km^2) throughout the treatment zone. A treatment zone is a minimum of 9 mi^2 around each find (23 km^2), that is, a 1.5-mile (2.4-km) radius out from each find. The treatments are applied every 2 weeks for two hypothetical generations past the last fly-find date—usually about six applications over a 3-month period. This treatment is not a registered use but, rather, it is applied under an emergency exemption.

Because of the dense grid of streets in these urban–suburban areas, the attractive lure spots can be distributed rapidly in an even spatial pattern to achieve good coverage. The current system utilizes 2-man crews riding in pickup trucks. The passenger seat crewman has a pistol-grip spray gun connected to a pressurized spray tank carried in the bed of the truck. This system was adapted from portable tree-marking paint sprayers used in timber cruising. The tank pressure is recharged periodically through a portable electric air compressor that is run off the truck electrical system. The nozzle delivers a solid jet stream that is capable of traveling several meters. As the truck is driven slowly down the curbside of the streets the passenger-seat crewman gives a short burst on the poles and tree trunks in passing. With this system two such two-man crews can treat all of the streets in the 9-mi^2 (23-km^2) treatment zone readily on a 2-week reapplication schedule.

Over the last 17 years this male annihilation system has been used dozens of times over an accumulative area of hundreds of square miles with never a failure to achieve eradication. The year 1987 was an especially severe year for California. There were seven separate male annihilation programs directed against the oriental fruit fly, the northernmost being in San Jose and the southernmost in San Clemente. The largest male annihilation program in 1987, however, was directed against *D. zonatus* using the same technique and same methyl eugenol formulation as that used for the oriental fruit fly. In that program, 25 mi^2 were placed under treatment after six flies were found scattered in an area just to the north of the Los Angeles airport in September.

C. In Japan

The largest eradication by male annihilation to date, however, has been carried out in Japan by Ushio et al. (32) and Koyama et al. (33). Steiner's technology was adapted for their situation to achieve complete eradication of the oriental fruit fly from their southern archipelago—the Ryukus—from Amami in the north to Okinawa in the south. This national eradication program demonstrates the possibilities that exist for eradicating other isolated established pest populations on a large scale by male annihilation.

Grobstein (34), writing on the relationship of science to public policy said, "We would be remiss to withhold what is useful because it is not perfect." The male lure phenomenon, although still very imperfectly understood, has indeed proved to be useful.

V. CONCLUSIONS

Although pheromones have not yet proved to have practical applications in the management of tephritid fruit flies, the so-called male lures have great utility in programs against certain species from the Paleotropics and Oceania. These male lures are used in large-scale detection-trapping arrays in the continental United States, where there is a continuous influx of accidental introductions of these species through movement of infested fruit past agricultural quarantine barriers. Early detection of these exotic species by means of these lures has enabled agricultural officials to eradicate incipient populations. Male annihilation, an outstanding eradication technique utilizing methyl eugenol, has frequently been used successfully to eradicate introductions of the oriental fruit fly. Despite our imperfect understanding of the action of these male lures, their usefulness in programs against specific tephritid species has been clearly demonstrated.

REFERENCES

1. Steiner, L. F. Fruit fly control in Hawaii with poison-bait sprays containing protein hydrolysate. *J. Econ. Entomol.* *45*:838–843 (1953).
2. Ayers, E. L. The two medfly eradication programs in Florida. *Proc. Fla. State Hort. Soc.* *70*:67–69 (1957).
3. Jackson, D. S. and Lee, B. G. Medfly in California 1980–1982. *Bull. Entomol. Soc. Am.* *31*:29–37 (1985).
4. Carter, W. Insects on fruit. The oriental fruit fly. In *The 1952 Yearbook of Agriculture.* (Steffered, A., ed.). U.S. Dept. of Agric., Washington, D.C., 1952, pp. 551–559.
5. Beroza, M. and Green, N. Materials tested as insect attractants. *Agric. Handbook 39.* U.S. Dept. of Agric., Washington, D.C., 1963, pp. 148.
6. Howlett, F. M. Chemical reactions of fruit flies. *Bull. Entomol. Res.* *6*:297–305 (1915).
7. Beroza, M., Green, N., Gertler, S. I., Steiner, L. F., and Miyashita, D. H. New attractants for the Mediterranean fruit fly. *J. Agric. Food Chem.* *9*:361–365 (1961).
8. Alexander, B. H., Beroza, M., Oda, T. A., Steiner, L. F., Miyashita, D. H., and Mitchell, W. C. The development of male melon fly attractants. *J. Agric. Food Chem.* *10*:270–276 (1962).
9. Howlett, F. M. The effect of oil of citronella on two species of *Dacus*. *Trans. Entomol. Soc. Lond.* *60*(Part II):412–418 (1912).
10. Steiner, L. F. Lures for *Dacus dorsalis*. *Hawaiian Entomol. Soc. Proc.* *14*:204 (1951).
11. Drew, R. A. I. The responses of fruit fly species (Diptera: Tephritidae) in the South Pacafic area to male attractants. *J. Aust. Entomol. Soc.* *13*:276–270 (1974).
12. Drew. R. A. I. and Hooper, G. H. S. The responses of fruit fly species (Diptera: Tephritidae) in Australia to various attractants. *J. Aust. Entomol. Soc.* *20*:201–205 (1981).

13. Hancock, D. L. New species and records of African Dacinae (Diptera: Tephritidae). *Arnoldia Zimbabwe 9*:299–314 (1985).
14. Hancock, D. L. Two new species of African Ceratitinae (Diptera: Tephritidae). *Arnoldia Zimbabwe 9*:291–297 (1985).
15. Valega, T. M. and Beroza, M. Structure-activity relationships of some attractants of the Mediterranean fruit fly. *J. Econ. Entomol. 60*:341–347 (1967).
16. Valega, T. M., McGovern, T. P., Beroza, M., Miyashita, D. H., and Steiner, L. F. Candidate attractants for the control of the Mediterranean fruit fly. *J. Econ. Entomol. 60*:835–844 (1967).
17. Metcalf, R. L., Mitchell, W. C., and Metcalf, E. R. Olfactory receptors in the melon fly, *Dacus cucurbitae* and the oriental fruit fly, *Dacus dorsalis*. *Proc. Natl. Acad. Sci. USA 80*:3143–3147 (1983).
18. McGovern, T. P., Cunningham, R. T., and Leonhardt, B. A. *cis*-Trimedlure: Attraction for the Mediterranean fruit fly (Diptera: Tephritidae). *J. Econ. Entomol. 79*:98–102 (1986).
19. McGovern, T. P., Cunningham, R. T., and Leonhardt, B. A. Attractiveness of *trans*-trimedlure and its four isomers in field tests with the Mediterranean fruit fly (Diptera: Tephritidae). *J. Econ. Entomol. 80*:617–620 (1987).
20. Sonnet, P. E., McGovern, T. P., and Cunningham, R. T. Enantiomers of the biologically active components of the insect attractant trimedlure. *J. Org. Chem. 49*:4639–4643 (1987).
21. Harris, E. J., Nakagawa, S., and Urago, T. Sticky traps for detection and survey of three tephritids. *J. Econ. Entomol. 64*:62–65 (1971).
22. Cunningham, R. T. and Couey, H. M. Mediterranean fruit fly (Diptera: Tephritidae): Distance/response curves to trimedlure to measure trapping efficiency. *Environ. Entomol. 15*:71–74 (1986).
23. Tassen, R. L., Hagen, K. S., Cheng, A., Palmer, T. K., Feliciano, G., and Blough, T. L. Mediterranean fruit fly life cycle estimations for the California eradication program. In *Fruit Flies of Economic Importance*. (Cavalloro, R., ed.). Balkema, Rotterdam, 1983, pp. 564–570.
24. Steiner, L. F. Methyl eugenol as an attractant for oriental fruit fly. *J. Econ. Entomol. 45*:241–248 (1952).
25. Steiner, L. F., Mitchell, W. C., Harris, E. J., Koyama, T. T., and Fujimoto, M. S. Oriental fruit fly eradication by male annihilation. *J. Econ. Entomol. 58*:961–964 (1965).
26. Hart, W. G., Steiner, L. F., Cunningham, R. T., Nakagawa, S., and Faris, G. Glycerides of lard as an extender for cue-lure, medlure, and methyl eugenol in formulations for programs of male annihilation. *J. Econ. Entomol. 59*:1395–1400 (1966).

27. Ohinata, K., Steiner, L. F., and Cunningham, R. T. Thixcin E as an extender of poisoned male lures used to control fruit flies in Hawaii. *J. Econ. Entomol.* *64*:1250--1252 (1971).
28. Cunningham, R. T. and Suda, D. Y. Male annihilation of the oriental fruit fly, *Dacus dorsalis* Hendel (Diptera: Tephritidae): A new thickener and extender for methyl eugenol formulations. *J. Econ. Entomol.* *78*:503--504 (1985).
29. Cunningham, R. T., Steiner, L. F., and Ohinata, K. Field tests of thickened sprays of methyl eugenol potentially useful in male-annihilation programs against oriental fruit flies. *J. Econ. Entomol.* *65*:556--559 (1972).
30. Cunningham, R. T., Chambers, D. L., and Forbes, A. G. Oriental fruit fly: Thickened formulations of methyl eugenol in spot applications for male annihilation. *J. Econ. Entomol.* *68*: 861--863 (1975).
31. Chambers, D. L., Cunningham, R. T., Lichty, R. W., and Thrailkill, R. B. Pest control by attractants: A case study demonstrating economy, specificity, and environmental acceptability. *BioScience* *24*:150--152 (1974).
32. Ushio, S., Yoshioka, K., Nakasu, K., and Waki, K. Eradication of the oriental fruit fly from Amami Islands by male annihilation (Diptera: Tephritidae). *Jp. J. Appl. Entomol. Zool.* *26*:1--9 (1982). (In Japanese with English summary.)
33. Koyama, J., Teruya, T., and Tanaka, K. Eradication of the oriental fruit fly (Diptera: Tephritidae) from the Okinawa Islands by male annihilation. *J. Econ. Entomol.* 77:468--472 (1984).
34. Grobstein, C. Advise and Dissent Section. *Science* *83*:18 (1983).

18

Development and Commercial Application of Sex Pheromone for Control of the Tomato Pinworm

JACK W. JENKINS / Scentry Inc., Buckeye, Arizona

CHARLES C. DOANE / Scentry Inc., Buckeye, Arizona

DAVID J. SCHUSTER / Gulf Coast Research and Education Center, University of Florida, Bradenton, Florida

JOHN R. McLAUGHLIN / Agricultural Research Service, U.S. Department of Agriculture, Gainesville, Florida

MANUEL J. JIMENEZ / University of California, Cooperative Extension Service, Community Civic Center, Visalia, California

I. INTRODUCTION

The tomato pinworm, *Keiferia lycopersicella* (Walsingham), is a primary pest of tomatoes in North America. The insect occurs in many tomato-growing areas but causes greatest damage in Mexico, southern California, southern Texas, and Florida. The pest has also been reported in Haiti, Cuba, Central America, and Peru (1). Larvae are leafminers during early instars and leafrollers as they increase in size. Foliar injury is of little economic importance although severe infestations can suppress growth of younger plants. Larger larvae also enter tomatoes below the fruit calyx, predisposing them to decay. Traditional methods of controlling tomato pinworm (TPW) involve repeated applications of broad-spectrum insecticides, often in mixtures aimed at control of other pests as well; however, TPW larvae within the plant tissue are shielded from most insecticides, reducing their effectiveness. Insecticides recommended for TPW control have caused outbreaks of secondary pests such as the vegetable leafminer, *Liriomyza sativae* Blanchard (2–4) and mites (N. C. Toscano, personal communication). Broad-spectrum insecticides also kill beneficial fauna important in the control of *Spodoptera exigua* (Hübner) and *Heliothis zea* (Boddie). Furthermore, use of pesticides has resulted in above-tolerance residues on fresh market tomatoes (5). Researchers suspect that high levels of insecticide resistance have developed in many TPW populations. Alternatives to the traditional insecticidal approach for TPW control are urgently needed.

II. FORMULATION DESCRIPTION

Development of a system using sex pheromone for control of the tomato pinworm was initiated in 1979 soon after identification of the TPW pheromone as a 96:4 mixture of (*E*)- and (*Z*)-4-tridecenyl acetate (W. L. Roelofs, unpublished) and description of adult sex pheromone biology of the adult (6). The TPW was chosen as a candidate for product development because of previous success with another gelechiid, the pink bollworm, *Pectinophora gossypiella* (Saunders) (PBW) (7). As with PBW, development centered on use of hollow fibers as controlled-release devices. The end-sealed hollow-fiber dispenser emits controlled amounts of active ingredient through its single open end. Emission rates and longevity can be varied by adjusting configuration and density per unit area (8,9). Fibers can be applied singly or in groups, either on adhesive-backed tape or in flowable polybutene stickers. Application may be made with specialized equipment or simply by hand.

III. SYSTEM DEVELOPMENT

A. Release Rates

A prerequisite for evaluating mating disruption was to develop an attractive lure and trap for use in monitoring and determining pest density of TPW adults (10). Early research with hollow-fiber dispensers revealed an inverse relationship between pheromone release rate and numbers of moths captured in traps (11). A single hollow fiber attracted significantly more moths than 5-, 7-, 9-, 10-, 30-, or 100-fiber lures. Additional dose–response testing conducted in 1982 also demonstrated a clear upper threshold of male response to the sex pheromone (Scentry, unpublished data). Dispensers designed with two fibers caught 35% to 52% more moths than six-fiber dispensers and 67% to 88% more moths than 18-fiber dispensers. Since then two-fiber dispensers have been used as a standard lure for monitoring with traps. Active life of this lure is approximately 4 weeks.

In the laboratory, release rates of (*E*)-4-tridecenyl acetate in hollow fibers were measured by meniscus regression. Analyses indicated rates were 2.72 μg/day and 5.93 μg/day at 72°F and 100°F, respectively. The same fiber configuration was loaded with 96:4 (*E*)- and (*Z*)-4-tridecenyl acetate and analyzed for emission rate by gas chromatography. Fibers were mixed with polybutene adhesive used in field application and exposed in the field for 0, 7, 14, 21, and 28 days. Average daily temperature during this period was 70.6 ± 7°F. Residual pheromone was extracted and quantitated by gas chromatography (fused silica Carbowax 20M column). One hundred fibers were analyzed from each age class. Results indicated a mean release rate of 3.7 μg/fiber per day after the initial burst of pheromone on days 1 and 2.

B. Traps

The trap design most effective for capturing TPW was determined by Wyman in 1979 (11). A wing-style trap was six times more effective in capturing TPW moths than either Delta-style or mineral oil traps. Superiority of the wing trap was attributed to ease of access to the sticky surface and emission of pheromone from the trap. The horizontal orientation of the wing-style trap is more efficient for capturing these moths as they alight. Research in California showed a close correlation between trap captures of TPW, larvae in foliage, and percentage of infested fruit in untreated fields (12). Trap-monitoring guidelines and economic thresholds have been established for TPW on pole tomatoes in southern California (13,14).

C. Mating Disruption

The first replicated mating disruption trial of TPW sex pheromone in hollow-fiber formulation was conducted in Florida during 1979. Hollow-fiber dispensers in tape formulation were hand-stapled to tomato stakes in 2.8-ha plots at approximately 1235 dispensers per hectare. Each dispenser contained 50 fibers, producing a release rate of 40 g AI/ha. Adjacent 2.8-ha plots were used as controls. Treatments were replicated three times. Trap captures in treated plots remained low after application (Fig. 1), but larval infestations failed to develop in either treatment. Both the pheromone and control plots were treated equally with conventional insecticides during the period.

In 1980, an unreplicated mating disruption trial was conducted in Mexico's Culiacan valley under supervision of Sanidad Vegetal. Two applications using hollow-fiber dispensers in the tape formulation with 50 fibers per tape were hand-placed on tomato stakes at 312 dispensers per hectare (10 g AI/ha) in the first application and 625 dispensers

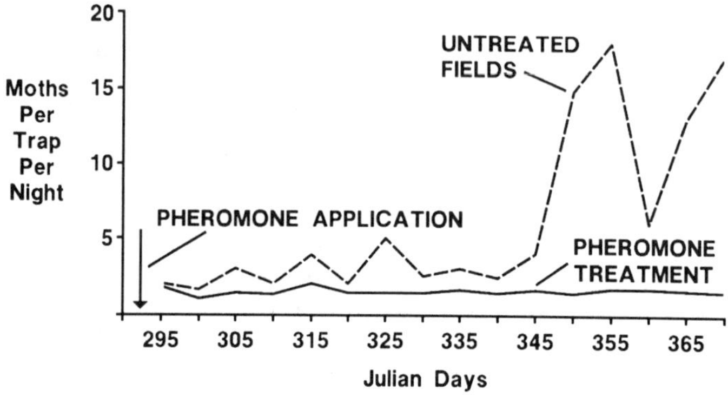

FIGURE 1 Trap captures of male TPW in pheromone-treated and untreated fields, Florida, 1979.

per hectare (20 g AI/ha) 20 days later. The 4-ha treated field was compared with an adjacent field that remained completely untreated. Traps baited with standard two-fiber lures were used to monitor male populations. Trap counts were greatly reduced in the pheromone-treated field (Fig. 2). Samples of 100 tomatoes collected at random on four dates were examined for TPW infestations. The mean TPW infestation in the treated field was 2% compared with 34% in the untreated field (Table 1).

Mating tables were first used in Florida in 1980 to measure disruption. Three sets of paired blocks, approximately 0.1-ha each, were established and half were treated with the hollow-fiber tape dispensers at 10 g AI/ha. All fields received insecticides at the discretion of the pest control advisor. A single mating table was placed in the center of each plot at canopy height. Virgin TPW females, 3 days old, were placed on mating tables on bouquets of excised tomato foliage in vials of water. Depending upon availability, three to seven females were placed on the table each day. Females were collected after the daily mating period and examined for presence of spermatophores. Results from four dates indicated only 4.2% of females were mated in pheromone-treated plots, compared with 52.6% in the check plots (Table 2).

During 1981 in Florida, plots treated with three formulations of hollow fibers applied at 10 g AI/ha were compared with an untreated check. Treatments were polyethylene fibers on sticky tape (PET), Celconese plastic fibers on sticky tape (Celcon), and Celconese fibers applied with polybutene sticker. Treatments were replicated

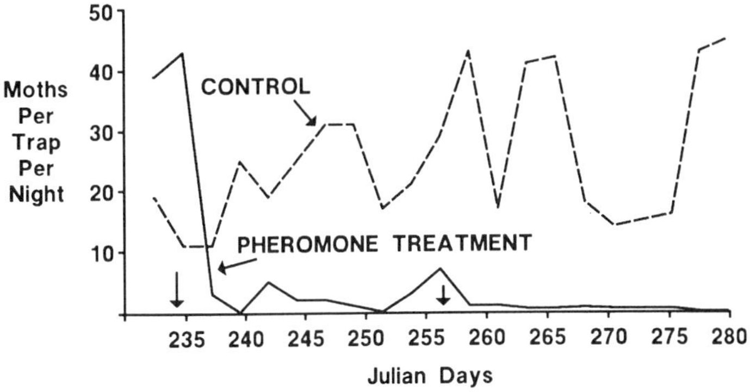

FIGURE 2 Trap captures of male TPW in pheromone-treated and untreated fields. The two vertical arrows indicate dates of pheromone application, Culiacan, Mexico, 1980.

TABLE 1 Average Percentage of Fruit Infested with TPW in Pheromone-Treated and Untreated Plots, Culiacan, Mexico, 1980

	% Infestation			
Treatment	9/24	10/2	10/4	10/6
Pheromone	2	3	1	2
Untreated	40	37	28	33

TABLE 2 Mating Frequency of Virgin TPW Females on Mating Tables in Plots Treated with Pheromone and Control Plots, Florida, 1980

	Number of females		
Treatment	Mated	Unmated	% mated
Pheromone	1	23	4.2
Control	10	9	52.6

four times on 0.4-ha plots. Point-source pattern was consistent between treatments—36 fibers per site, 1300 sites per hectare. Trap counts and mating tables were used to measure efficacy. Test plots were monitored twice weekly for 8 weeks before application, and treatments were assigned to plots so preapplication populations did not differ statistically.

The trap captures are summarized in Table 3. Counts did not differ significantly among the three pheromone formulation treatments, but all treatments were significantly lower than catches in the untreated check. There was a 98.7% trap-catch reduction in plots treated with PET lures, followed by Celcon lure plots at 98.3%, and Celcon fiber plots at 96.1%. In no pheromone treatment did average trap count exceed 1.5 moths per trap a week during 7 weeks of posttreatment monitoring.

Mating table data correlated well with trap data in check plots, $r = 0.88$. Peak trap pressure came 5 weeks after treatment, which coincided with peak mating table activity, as seen in Table 4. Mating table data did not differ significantly among the three pheromone formulation treatments, but all treatments had significantly less mating than in the untreated check.

In 1981 a large demonstration was supervised by Sanidad Vegetal in Culiacan, Mexico. Fields comprising 600 ha of staked tomatoes treated with pheromone were compared with 180 ha of fields treated only with conventional insecticides. Applications began during the

TABLE 3 Average Number of Male TPW Males Captured Per Trap Per Week in Pheromone-Treated and Untreated Fields, Florida, 1981

	Number of weeks after application							
Treatment	1	2	3	4	5	6	7	Total[a]
PET tape	0	0	0.2	0	0.7	0.5	0.2	1.7[b]
Celcon tape	0.2	0	0	0.2	1.0	0	0.7	2.2[b]
Celcon fiber	0	0.2	0.2	1.2	1.5	1.0	0.7	5.0[b]
Control	2.0	2.7	6.0	28.2	60.5	18.7	10.0	128.2[a]

[a]Numbers within the column followed by the same letter are not significantly different ($p > 0.05$; Duncan's multiple range test, 1955).

TABLE 4 Percentage of Mated TPW Females after Placement of Virgin Females on Mating Tables in Pheromone-Treated and Control Tomato Fields, Florida, 1981

Treatment	Number of weeks after application						Total[a]
	2	3	4	5	6	7	
PET tape	0	0	11.1	17.0	58.9	25.0	17.3[b]
Celcon tape	3.6	0	62.5	9.7	46.8	33.3	19.1[b]
Celcon fiber	10.3	41.7	33.3	52.8	38.0	68.7	36.8[b]
Control	58.9	72.9	92.8	100.0	83.6	79.2	78.2[a]

[a]Means within the column followed by the same letter are not significantly different ($p > 0.05$; Duncan's multiple range test, 1955).

period of highest TPW pressure in tomatoes in Culiacan. Pheromone in Celcon hollow fibers was mixed in sticker and applied by hand at 10 g AI/ha. The distribution pattern was 36 fibers per point and 1300 points per hectare. A mean of 2.6 pheromone applications were made in tomato fields during the demonstration. Pest populations were monitored in traps before and after treatment. All conventional insecticide applications in both pheromone-treated fields and check fields were recorded. Fewer conventional insecticide applications for all insect pests were required in pheromone-treated plots (Table 5).

TABLE 5 Results from 600 ha of Tomatoes Treated with TPW Pheromone in Hollow Fibers Compared with 180 ha Treated with Conventional Insecticide, Culiacan, Mexico, 1981

Treatment	Average number of applications		Average % infested fruit
	Pheromone	Insecticide	
Pheromone	2.6	5.8	1.2
Control	0	12.3	5.4

Each pheromone application replaced 2.5 conventional insecticide applications. Throughout the test the highest level of infested fruit in pheromone-treated fields was lower than levels in pesticide-treated check fields.

The relationship between point-source strength and number of point sources was examined for TPW in 1980 by evaluating two distribution patterns and application rates of the hollow-fiber system (15). Pinworm populations were monitored by adult capture in traps and larval infestation in foliage and fruit. Within the test range, distribution patterns did not result in significantly different larval counts; however, all rates and patterns resulted in significant reduction of moth captures compared with the untreated control. The lack of difference in larval infestations between treatments and the control was apparently due to poor isolation and to immigration of mated females from adjacent plots.

Mating disruption was further evaluated in California pole tomatoes in 1981, when the hollow-fiber formulation was applied at an average density of six fibers per site with approximately 5500 sites per hectare for an application rate of 10 g AI/ha (15). Fibers were mixed with sticker and applied by hand to tomato stakes at canopy level. Applications were made once each month during August, September, and October. Three plots, 16-ha each, were treated with pheromone. All plots, including corresponding controls, were treated with mixtures of insecticides for beet armyworm (BAW) control. The TPW populations were monitored by traps and plant inspection.

Moth captures were greatly suppressed in pheromone-treated plots compared with check plots (Table 6). Larval infestations and the percentage of fruit infested were generally reduced in pheromone-treated

TABLE 6 Average Trap Captures and Larval Infestation of TPW in Pheromone-Treated and Control Tomato Fields, California, 1981

Treatment	Moths/trap/night	Mean no./meter of row		Mean % infested fruit
		Mines	Larvae	
Pheromone	0.6[b]	8.0[b]	1.6[b]	0.6[a]
Control	38.6	9.9	2.3	0.8

[a]Means are significantly different (matched pair comparison) from corresponding control: $p < 0.1$.
[b]$p < 0.01$.

plots as well, although not as dramatically as trap captures. The TPW infestations were low because of insecticide applied for BAW control.

IV. ECONOMICS

In his 1981 trial, Van Steenwyk (15) estimated hand application cost about 80 dollars/ha, not including the cost of the pheromone-filled fibers. It was concluded, at that time, that although pheromone treatments suppressed TPW populations, the degree of control did not justify the expense of the system. In normal commercial programs, cost for hand application is far less. A single laborer can treat 1 ha/hr using fibers in sticker. Application cost is less than 5 dollars/ha in the United States and less than 30 cents/ha in Mexico.

High costs of labor, land, and water in southern California have resulted in a gradual shift of fresh market tomato production to Mexico. Tomatoes are Mexico's largest export crop, with revenues in the hundred's of millions dollars annually (16). The Culiacan valley of Mexico produces as much as 50% of all fresh market tomatoes consumed in the United States each winter. Growers are highly motivated to protect this market through economic production of clean tomatoes. In 1979 the USDA-FDA stopped large quantities of tomatoes from importation from Mexico because of pesticide residues above tolerance (5). This action caused considerable economic loss to Mexican growers. Alternative methods of pest control, especially nontoxic alternatives, such as pheromones, are therefore desirable.

V. PRODUCT DESCRIPTION

The TPW mating disruption system development efforts culminated in 1982 with full EPA registration of Attract'n Kill Tomato Pinworm (A&K TPW). This is the only sex pheromone product currently registered for control of TPW. Efforts to develop a similar product are in progress at Shin-Etsu Chemical Company.

The agent A&K TPW is registered in Mexico, Florida, and California. The system is most widely used in the Culiacan valley and San Quintin regions of Mexico. Monthly applications normally begin no later than 2 weeks after transplanting. Pheromone traps and larval sampling are used to determine population levels and product longevity. Typically, trap captures will average fewer than one moth per trap a night for 3 to 4 weeks after an application. Warm temperatures increase the release rate and decrease product longevity. When captures increase to an average of five moths per trap a night, it is recommended that the pheromone be reapplied.

The A&K TPW is applied by hand directly to plant foliage or to stakes supporting the tomatoes and does not come into contact with the tomato fruit. No special equipment is needed to apply the mixture. Clusters of fibers and adhesive are applied at canopy level.

VI. INTEGRATED PEST MANAGEMENT

A recent demonstration by University of California researchers showed that TPW pheromone products applied to cherry tomatoes provided effective control (17). Native beneficial insects were conserved and augmented by release of *Trichogramma pretiosum* Riley. Except for a single application of *Bacillus thuringiensis* on one field treated with A&K TPW, no insecticides were needed for any insect pest during the season. In comparison, these same fields had required weekly insecticide applications the previous year. Furthermore, TPW populations in nearby fields in 1987 were not controlled, despite weekly applications of the standard insecticides. In fact, TPW infestations were so high on some insecticide-treated fields that tomatoes were rejected by packers. During 1987, the average percentage of infested fruit in pheromone-treated fields was less than 5%, compared with more than 20% in the same fields in 1986 under traditional insecticide protection.

VII. CONCLUSION

Concern for pesticide residues on tomatoes and difficulty in controlling the TPW economically with available conventional insecticides have created a dilemma for tomato growers. Experiments have proved that controlled-release formulations of TPW sex pheromone provide an effective control of this pest through mating disruption. One system, A&K TPW is registered for control of tomato pinworm, both in the United States and in Mexico. This system can be integrated with other pest management practices, such as narrow-spectrum insecticides and biological and cultural control methods, to further reduce populations of primary and secondary pest species.

REFERENCES

1. Lin, S. Y. H. and Trumble, J. T. Bibliography of the tomato pinworm, *Keiferia lycopersicella* (Walsingham) (Lepidoptera: Gelechiidae). *Bibliogr. Entomol. Soc. Am. 1*:65–74 (1983).

2. Oatman, E. R. and Kennedy, G. C. Methomyl induced outbreak of *Liriomyza sativae* on tomato. *J. Econ. Entomol. 69*: 667–668 (1976).
3. Johnson, M. W., Oatman, E. R., and Wyman, J. A. Effects of insecticides on populations of the vegetable leafminer and associated parasites on fall pole tomatoes. *J. Econ. Entomol. 73*:67–71 (1980).
4. Oatman, E. R., Van Steenwyk, R. A., and Trumble, J. T. Insect pest management of fresh market tomatoes. In *California Fresh Market Tomato Research and Education Programs, 1981–82.* (Pasateri, F. P., ed.), 1981, pp. 95–110.
5. Food and Drug Administration. Results of FDA Surveillance of Mexican Produce for Pesticide Residues. (Compliance program 7305.008. Fiscal year, 1979.) Bureau of Foods and Office of Associate Comm. for Reg. Aff., FDA Administration. April 23, 1979.
6. McLaughlin, J. R., Antonio, A. Q., Poe, S. L., and Minnick, D. R. Sex pheromone biology of the adult tomato pinworm, *Keiferia lycopersicella* (Walsingham). *Fla. Entomol. 62*:34–41 (1979).
7. Doane, C. C. and Brooks, T. W. Research and development of pheromones for insect control with emphasis on the pink bollworm. In *Management of Insect Pests With Semiochemicals.* (Mitchell, E. R., ed.). Plenum Press, New York, 1980, pp. 285–303.
8. Ashare, E., Brooks, T. W., and Swenson, D. W. Controlled release from hollow fibers. In *Controlled Release Polymeric Formulations.* (Paul, D. R. and Harris, F. W., eds.). ACS Symp. Ser. No. 33, American Chemical Society, Washington, D.C., 1976, pp. 273–282.
9. Brooks, T. W., Ashare, E., and Swenson, D. W. Hollow fibers as controlled vapor release devices. In *Textile and Paper Chemistry and Technology.* (Arthur, J. C., ed.). ACS Symp. Ser. No. 49, American Chemical Society, Washington, D.C., 1977, pp. 116–126.
10. Roelofs, W. E., Brooks, T. W., Burkholder, W. E., Cardé, R. T., Chambers, D. L., Shorey, H. H., Wood, D. L., Ludvik, G. F., and Russel, P. L. *Establishing Efficacy of Sex Attractants and Disruptants for Insect Control.* Entomological Society of America, College Park, Md., 1979, 97 pp.
11. Wyman, J. A. Effect of trap design and sex attractant release on tomato pinworm catches. *J. Econ. Entomol. 72*:865–868 (1979).
12. Van Steenwyk, R. A., Oatman, E. R., and Toscano, N. C. Optimal timing of pesticide applications for tomato pinworm through

a pheromone monitoring program in commercial fall fresh market tomatoes. *Calif. Dept. Food Agric. Final Rep.*, 1979--1980.

13. University of California. Tomato pinworm, *Keiferia lycopersicella.* In *Integrated Pest Management for Tomatoes.* Univ. Calif. Statewide IPM Project. Div. Agric. Sci. Publ. 3274. 1985, pp. 57–59.
14. Van Steenwyk, R. A., Oatman, E. R., and Wyman, J. A. Density treatment level for tomato pinworm (Lepidoptera: Gelechiidae) based on pheromone trap catches. *J. Econ. Entomol.* *76*:440–445 (1983).
15. Van Steenwyk, R. A. and Oatman, E. R. Mating disruption of tomato pinworm (Lepidoptera: Gelechiidae) as measured by pheromone trap, foliage, and fruit infestation. *J. Econ. Entomol.* *76*:80–84 (1983).
16. Buckley, K. C., VanSickle, J. J., Bredahl, M. E., Belibasis, E., and Gutierrez, N. Florida and Mexico competition for the winter fresh vegetable market. *USDA ESA Agric. Econ. Rept. 556.* 1986, 101 pp.
17. Jiménez, M. J., Toscano, N. C., Flaherty, D. L., Ilic, P., Zalom, F. G., and Kido, K. Controlling tomato pinworm by mating disruption. *Calif. Agric.* *42*:10--12 (1988).

Part III
Forest Insect Pests

19

Use of Semiochemicals to Manage Coniferous Tree Pests in Western Canada

JOHN H. BORDEN / Simon Fraser University, Burnaby, British Columbia, Canada

I. INTRODUCTION

In Western Canadian forests, semiochemical-based pest management programs have been operationally implemented against two major targets: ambrosia beetles, principally the striped ambrosia beetle, *Trypodendron lineatum* (Olivier); and the mountain pine beetle, *Dendroctonus ponderosae* Hopkins (Table 1). In addition, three lepidopteran pests have been subjected to semiochemical-based monitoring on an operational basis (see Table 1), whereas semiochemicals for other insect pests are in various stages of research and development (Table 2; see Chap. 20 for similar data on forest insect pests in the western United States). Operational application of semiochemicals in 1987 required an investment by government and industry of over 850,000 dollars. One company, Phero Tech Inc., Vancouver, British Columbia, has been predominant in the development of semiochemical-based, pest management services and the sale of traps and baits. For ambrosia beetles, integrated pest management (IPM) programs that are based primarily on semiochemicals have replaced the use of chemical pesticides. For the mountain pine beetle, semiochemicals have also been incorporated into an IPM approach, facilitating the effectiveness of intensive silvicultural operations and allowing the judicious and effective use of chemical pesticides to occur.

The development and implementation of these exciting new programs is described in this chapter.

TABLE 1 Semiochemical Products in Operational Use by 1987 in Western

Order, common name, and scientific name	Semiochemical product and supplier
Coleoptera	
Mountain pine beetle, *Dendroctonus ponderosae*	Multiple funnel traps and semiochemicals (myrcene, *trans*-verbenol, *exo*-brevicomin) supplied by Phero Tech Inc., Vancouver, B.C.
	Tree baits (devices with myrcene, *trans*-verbenol, *exo*-brevocomin) supplied by Phero Tech Inc., Vancouver, B.C.
Striped ambrosia beetle, *Trypodendron lineatum*, and *Gnathotrichus sulcatus*	Multiple funnel traps and semiochemicals (lineatin, sulcatol, ethanol) supplied by Phero Tech Inc., Vancouver, B.C.
Lepodoptera	
European pine shoot moth, *Ryacionia buoliana*	Sex pheromone lures and wing traps supplied by Phero Tech Inc., Vancouver, B.C.
Cranberry girdler, *Chrysoteuchia topiaria*	Sex pheromone lures and wing traps supplied by Phero Tech Inc., Vancouver, B.C.

Canadian Coniferous Forests

Use	Estimated annual costs, including purchased materials, transportation, labor, and overhead (30%) (Canadian dollars)
Used by Prince Rupert and Prince George Forest Regions in B.C. to assess spread of infestations and to determine peak flight periods, during which "no-haul" restrictions can be imposed on companies logging infested trees.	1,750
Integrated with silvicultural control throughout western Canada to contain and concentration infestations and to mop up infestations after sanitation–salvage logging or single tree treatments.	600,000
Baited traps and baited log bundles positioned around timber-processing or holding areas to assess risk and to intercept beetles before they can attack vulnerable timber	250,000
Approx. 100 traps distributed each year among 10 conifer nurseries in B.C. for detection program to meet regulatory stipulations. Chemical sprays applied if any moths caught. Outside zone of establishment, traps placed in mature pines to ensure that no moths occur on nursery site.	2,150
Approx. 100 traps distributed each year among five conifer nurseries	1,800

(continued)

TABLE 1 (Cont.)

Order, common name, and scientific name	Semiochemical product and supplier
[Cranberry girdler, *Chrysoteuchia topiaria*]	
Douglas-fir tussock moth, *Orgyia pseudotsugata*	Sex pheromone lures and delta traps prepared and manufactured by Pacific Forestry Centre, Victoria, B.C.

II. AMBROSIA BEETLES

A. The Problem

In the late 1960s, direct control of ambrosia beetles on the British Columbia coast ground to a halt. In previous years, first DDT, then lindane, and finally methyl trithion had been sprayed from the ground and the air onto decked logs, log booms, and unseasoned lumber to control these pests (1,2).

But resistance to the use of chemical pesticides by labor, as well as increasing environmental concern, eventually resulted in the termination of all insecticide applications. Yet these pests, *T. lineatum*, *Gnathotrichus sulcatus* (Le Conte), and *G. retusus* (Le Conte), continued to infest the sapwood of felled timber and freshly milled, green lumber. The "pin-holes" or tunnels, stained dark by the beetles' ambrosia fungus, caused degrade of lumber and plywood, and the rejection of lumber for export. They also necessitated an unacceptable investment in resorting and remanufacturing to eliminate infected or damaged wood (1,2). This impact now exceeds 63 million dollars (Canadian) per year, *every year* (3).

Use	Estimated annual costs, including purchased materials, transportation, labor, and overhead (30%) (Canadian dollars)
in B.C. for survey and detection. If mean catch per trap $\geq$ 10 moths per week, chemical sprays applied in the fall to protect seedlings from girdling.	
Seven traps placed in each of 19 permanent sampling sites each year in fully operational program to monitor high-hazard areas to predict upcoming infestations.	2,850

Despite host–habitat management efforts designed to reduce the inventory and increase the processing speed of vulnerable timber during beetle flight periods (1,2), the beetles continue to be a major problem. This problem is particularly acute in dryland sorting areas (dryland sorts) where logs trucked from the field are sorted by species and grade and held for transport to the mill (1). In these field "warehouses," *T. lineatum*, which overwinter in the surrounding forest, mass-attack logs in the spring. A little later, *Gnathotrichus* spp, after emerging from stored logs, debris, or newly arrived logs in which they overwinter, also attack fresh logs. Even though timber brought into a dryland sort in the spring might already have been attacked in the field, the resident beetles in the sorting area greatly intensify the infestation (4). The risk subsides somewhat in early summer, but in August, the main flight-and-attack period of *G. sulcatus* begins, extending the period of vulnerability in the sort to 8 months, from March through October (1). Often, logs in the sort turn white from boring dust produced by the beetles intent on raising their broods in the succulent sapwood.

TABLE 2 Status in 1987 of Research and Development for Semiochemicals of Forest Insect Pests in Western Canadian Coniferous Forests

Order, common name (if applicable), and scientific name[a]	Semiochemical(s) under investigation	Potential use	Progress in research and development continuum		Research institution[b]
			Re-search	Develop-ment	
Coleoptera					
western balsam bark beetle, *Dryocoetes confusus*	*exo*-Brevicomin	Containment and concentration	X	X	CERG, SFU
	endo-Brevicomin	Protection of felled timber and high-hazard stands	X		
Douglas-fir beetle, *Dendroctonus pseudotsugae*	3-Methylcyclo-hex-2-en-1-one (MCH)	Protection of felled timber		X	Phero Tech
mountain pine beetle, *Dendroctonus ponderosae*	Verbenone	Protection of high-hazard stands, dispersion of populations from infested stands	X	X	Phero Tech; CERG, SFU
spruce beetle, *Dendroctonus rufipennis*	Frontalin, α-pinene	Containment and concentration		X	BCFS; NFC, CFS; U. CALGARY

	3-Methylcyclo-hex-2-en-1-one (MCH)	Protection of felled timber		X	Phero Tech
Gnathotrichus retusus	(*S*)-(+)-Sulcatol (retusol)	Mass trapping		X	Phero Tech
Lepidoptera					
western blackheaded budworm, *Acleris gloverana*	(*E*)-11-Tetra-decenal	Detection		X	PFC, CFS
black army cutworm, *Actebia fennica*	(*Z*)-7-Dodecenyl acetate and (*Z*)-11-tetradecenyl acetate	Detection and monitoring		X	PFC, CFS
Choristoneura spp.	(*E*)-11- and (*Z*)-11-Tetradecenal, (*E*)-11- and (*Z*)-11-tetradecenyl acetate, and (*E*)-11-tetradecenol	Detection, species distribution		X	PFC, CFS
"two-year cycle spruce budworm," *Choristoneura biennis*	Unknown sex pheromone	Detection and monitoring	X		CERG, SFU; PFC, CFS
spruce budworm, *Choristoneura fumiferana*	(*E*)-11- and (*Z*)-11-Tetradecenal	Detection and monitoring		X	NFC, CFS

(continued)

TABLE 2 (Cont.)

Order, common name (if applicable), and scientific name[a]	Semiochemical(s) under investigation	Potential use	Progress in research and development continuum		Research institution[b]
			Re-search	Develop-ment	
[Lepidoptera]					
larch casebearer, *Coleophora laricella*	(*Z*)-5-Decenol	Detection		X	PFC, CFS
western pine shoot borer, *Eucosma sonomana*	(*Z*)-9- and (*E*)-9-Dodecenyl acetate	Detection		X	PFC, CFS
Douglas-fir tussock moth, *Orgyia pseudotsugata*	(*Z*)-6-Heneicosen-11-one	Disruption of mating		X	USFS
various sesiids	Unknown sex pheromones	Detection, species distribution	X		PFC, CFS

Vespamima novaroensis	Unknown sex pheromones	Detection	X	PFC, CFS
"larch bud moth," *Zeiraphera improbana*	(*E*)-11- and (*Z*)-9-Tetradecenyl acetate	Detection of two biotypes	X	PFC, CFS
Diptera				
"Douglas-fir cone gall midge" *Contarinia oregonensis*	Unknown sex pheromone	Detection and monitoring in seed orchards	X	CERG, SFU; PFC, CFS

[a]Common names in quotation marks are not approved by the Entomological Society of America (75).
[b]Institutions as follows: CERG, SFU = Chemical Ecology Research Group, Simon Fraser University; BCFS = B. C. Forest Service; Phero Tech = Phero Tech Inc., Vancouver, B. C.; PFC, CFS = Pacific Forestry Centre, Canadian Forestry Service; NFC, CFS = Northern Forestry Center, Canadian Forestry Service; USFS = U. S. Forest Service.

Ambrosia beetles are also a problem in some sawmills. *Gnathotrichus sulcatus* is a severe threat to green lumber, which it attacks freely, and in which it can successfully reproduce, constituting a significant threat to lumber-importing countries (5).

B. Development of Semiochemicals for Pest Management

In the early 1960s, research by J. A. Rudinsky and G. E. Daterman of Oregon State University and J. A. Chapman of the Canadian Forestry Service led to the surprising (at the time) conclusion that mass attack by *T. lineatum* was mediated by a female-produced pheromone (6–8). Thus, the scarce natural resource of windthrown or broken trees could be found and fully exploited by the majority of the population (9). This discovery was followed by the demonstration that similar mass attack by *Gnathotrichus* spp. is mediated by male-produced, aggregation pheromones (10, 11).

By 1967, collaborative research by entomologists at Simon Fraser University (SFU) and R. M. Silverstein's group of chemists at the Stanford Research Institute (later at the New York State College of Environmental Science and Forestry) had begun. The objective was to identify the critical semiochemicals for each of the major species of ambrosia beetle. This research soon involved two more chemists, K. N. Slessor and A. C. Oehlschlager of SFU.

For *T. lineatum*, the research culminated in the identification, synthesis, and field testing of lineatin (Fig. 1), a single aggregation pheromone for this species (12,13). *Gnathotrichus sulcatus* was shown to utilize sulcatol (see Fig. 1), again as a single aggregation pheromone (14).

Both lineatin and sulcatol are optically active, that is, they have two enantiomers (or mirror images). *Trypodendrom lineatum* was

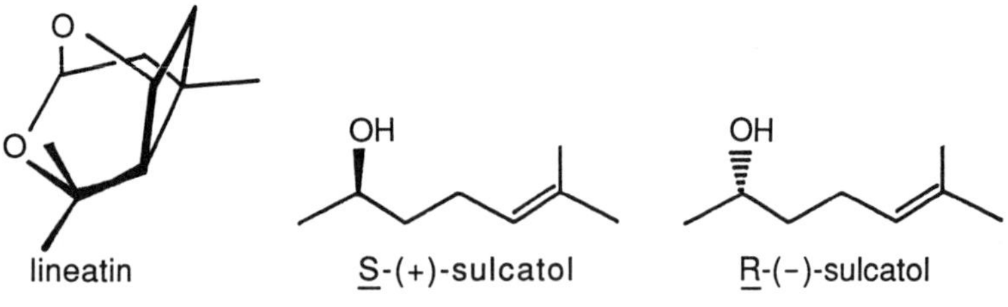

FIGURE 1 Aggregation pheromones for the three main species of ambrosia beetles in western Canada.

shown to use the (+) enantiomer of lineatin, whereas the (-) enantiomer is inert (15). Thus, the racemic compound, a 50:50 mixture of both enantiomers can be used operationally. On the other hand, *G. sulcatus* requires both enantiomers of sulcatol, which act synergistically to attract responding beetles (16). Finally, it was shown that *G. retusus* utilizes *S*-(+)-sulcatol (see Fig. 1), and is repelled by the antipode (17,18). This adaptation ensures species specificity, but demands the availability of *S*-(+)-sulcatol in pest management programs that incorporate *G. retusus*.

Aggregation pheromones alone, however, are only part of the semiochemical message. Perceptive research by Cade et al. (19) and Moeck (20,21) had shown that ethanol, produced by anaerobic metabolism in cut logs (22), acts as a primary attractant (i.e., a host-produced kairomone) for ambrosia beetles. Subsequently, ethanol was shown to act synergistically with sulcatol to create an enhanced attraction and a balanced sex ratio of captured *Gnathotrichus* spp. (23, 24). For *T. lineatum*, recent studies have shown conclusively that ethanol also acts synergistically with lineatin (B. S. Lindgren, personal communication). Uncertainty has persisted over whether or not a second host compound, α-pinene, is active as a synergist.

Lineatin and sulcatol were first deployed experimentally in glass vials, capillary tubes, or hollow-fiber dispensers that released volatiles at a more or less constant rate because of their fixed aperture size. Ethanol was originally released from plastic bottles. In a procedure developed by Phero Tech, it is now released from closed polyvinylchloride tubes, with the chemical passing through the plastic at a fairly constant rate. In operational programs, lineatin is released from Hercon Luretape dispensers. Sulcatol is released from bubble-cap dispensers designed by A. Meisen and V. Lee of the University of British Columbia. They provide fairly constant release rates through a rate-controlling membrane covering a pheromone reservoir. Operational release rates per trap used by Phero Tech, based on laboratory determinations at 20°C for the three pertinent semiochemicals are: lineatin, 0.4 mg/24 hr; sulcatol, 2.0 mg/24 hr; and ethanol, 60.0 mg/24 hr.

The final requirement for an operational, mass-trapping program was an effective trap. The first traps used were modifications of Browne's (25) cylinder and vane traps, made of wire-mesh screen covered with Stikem Special on which beetles responding to attractive baits adhered. These traps, although quite effective in capturing beetles, were subject to various problems. These included inactivation because of dust and debris adhering to the sticky surface, fragility or instability in conflict with wind or heavy equipment, and the necessity of laboriously counting beetles on sample portions of the trap, removing them mechanically, or separating them in solvent for volumetric determination of their numbers. Tests of Norwegian drainpipe traps (26) indicated that they could readily capture beetles, but for *T. lineatum*, not at rates competitive with sticky traps (27); moreover, they were bulky and cumbersome to handle.

The ultimate solution was to use the newly invented, multiple-funnel (Lindgren) trap (28), a series of vertically aligned black, plastic funnels above a collecting receptacle. Baits are suspended inside the open column in the middle of the trap. These traps have several advantages. In common with drainpipe traps, they provide a prominent vertical silhouette to which the beetles orient, and eliminate sticky material. However, they are more efficient than drainpipe traps and are easier to use because they are collapsible for transport, and can be hung from ropes, wires, or from metal bars driven into the ground. Because only the top end is fixed, they move with the wind and return to their original vertical position when the wind subsides. Dust does not impede the effectiveness of the trap, although large debris, such as leaves and spider webs occasionally reduce their trapping power. Finally, beetles can be captured in great numbers. Originally water, with a detergent and microbial inhibitor added, was used in the collecting receptacle to capture and kill the beetles. However, it was later found that beetles could be captured dry. Thus, collecting and volumetrically counting the captured beetles is greatly facilitated. In field trials over short periods the Lindgren traps, although efficient, captured fewer *T. lineatum* than sticky vane traps (27). This deficiency is compensated for by the greater trapping longevity of the multiple-funnel traps and, in operational use, simply by increasing the numbers of traps deployed.

C. Operational Research

The availability of only (±)-sulcatol, but not lineatin in 1974 dictated that the first operational research program be directed against *G. sulcatus* in sawmills, where it is by far the predominant pest. Perhaps not fully realizing the historical significance of the moment, a courageous new PhD student, J. A. McLean undertook the pioneering challenge of mass trapping *G. sulcatus* in the immense lumber storage yards of the Chemainus sawmill on Vancouver Island. After first running a baited trap survey of the mill site for areas of highest beetle activity (29), McLean turned to investigating the hypothesis that *G. sulcatus* populations could be suppressed by mass trapping with sulcatol-baited traps (in this case, sticky screen traps; 30).

He worked on the assumption that the total *G. sulcatus* population in the sawmill was represented by those that were caught in traps or on semiochemical-baited trap-piles of suitably-aged lumber, plus those that attacked other lumber piles in the mill site. Any not in those categories were of no consequence. Therefore, a weekly survey for new attacks on piles of green lumber was made, and the efficiency of the mass trapping program was calculated using the following formula:

$$\%\ \text{Suppression} = \frac{\text{Number of beetles caught on traps and trap loads} \times 100}{\begin{array}{c}\text{Number of beetles}\\ \text{caught on traps}\\ \text{and trap loads}\end{array} + \begin{array}{c}\text{Estimated number of}\\ \text{beetles attacking}\\ \text{lumber in the mill site}\end{array}}$$

By using this formula, McLean and Borden (31) achieved an overall efficiency of 65.1% for the entire 1976 season. During the midsummer, when the population was low, the efficacy approached 100%. However, during the peak flight periods in the spring and late summer, the efficacy decreased, apparently as beetles bypassed the traps and set up attraction centers of their own on freshly attacked lumber. This problem can be eliminated, in part, during peak flight periods by the integration of mass trapping with inventory control, in which infested logs are neither brought to nor stored at the mill, and in which rough-cut green lumber is either processed rapidly into finished lumber and is subjected to kiln-drying, or is removed from the mill site as rapidly as possible. The interaction between the sawmill and an adjacent booming ground was later investigated by Shore (32), who surveyed the entire location for ambrosia beetles and expanded the mass-trapping program to embrace *T. lineatum* and *G. retusus*.

The availability of lineatin by 1979 (33) allowed research on mass-trapping to be extended into dryland sorts. In 1981, another PhD student, B. S. Lindgren mounted a concerted mass-trapping program in the Canadian Pacific Forest Products Ltd. dryland sort in Sooke, British Columbia (34). The primary target was *T. lineatum*, with the trapping of *G. sulcatus* as a secondary goal. The size of the overwintering population of *T. lineatum* on the forest floor in a 60–m–wide band covering 10 ha around the sort was estimated from the recovery of beetles from samples of the litter and duff. Low and high estimates were based on extending the area in which maximal densities occurred around the base of trees either 0.45 or 0.90 m from the root collar.

The strategy of the program was to intercept emergent beetles as they flew into the sort by trapping them in multiple-funnel, drain-pipe, and sticky vane traps placed around the sort perimeter, with the traps concentrated mainly in margins having the highest overwintering populations. These traps were supplemented by semiochemical-baited bundles of appropriately aged, second-growth, trap-logs strategically placed around the sort margin following attack by ambrosia beetles. Before emergence of the brood beetles, these trap-logs were reduced to chips for making pulp, killing all beetles therein.

Taking into account the high and low estimates of overwintering populations, the 2.26 million *T. lineatum* captured in 1981 represented from 44% to 77% of the target population (34).

D. Operational Programs

In 1982, R. H. Heath, Pest Management Forester for Canadian Pacific Forest Products assumed responsibility for the semiochemical-based management of ambrosia beetles in the Sooke dryland sort. The company purchased traps and baits from Phero Tech (then P.M.G. Stratford Ltd.), and because of Heath's expertise, had the technical competence and professional commitment necessary to make the program a success. In 1983, the fifth year after research-oriented trapping had begun on the sort, Heath reported that virtually no beetles broke through the trap barrier to attack logs in the sort. The success of the program was aided by rigorous inventory control. The program is run today by R. G. Fraser, who has expanded its scope to include three dryland sorts, two booming grounds, and three sawmills with adjacent booming grounds. It is a rare example of a continuing semiochemical-based pest management program run by an enlightened forest products company.

In 1977, the year after completion of J. A. McLean's research at the Chemainus sawmill and booming grounds, the trapping program was taken over by MacMillan-Bloedel Ltd. until further research was initiated in 1979 by T. L. Shore. In 1983 to 1984, the mill was closed, but in 1985 a new mill was reopened. In 1982, and from 1985 on, the mass-trapping program at the mill has been run by Phero Tech.

By far the largest annual, commercial program for the mass trapping of any insect was initiated by S. E. Burke, P. Mumby, and M. G. Banfield of Phero Tech in 1982. Since 1984 it has been aided by a new, commercial synthesis for lineatin (35). By 1987, over 50 timber-processing areas in British Columbia, as well as in Washington, Oregon, and Alaska were utilizing Phero Tech's trapping technology (36). Thirty-seven areas are under pest management contract to the company, and 20 additional areas are managed independently, with baits, traps, or consulting purchased from Phero Tech.

A fundamental precept of Phero Tech's program was introduced by the incumbent specialists in ambrosia beetle management, S. H. Krannitz and E. Stokkink. This precept is the company's insistence that a new customer purchase the services of the company's professionals along with the purchase of traps and baits. Previously, users of baits and traps had suffered from a lack of comprehension of the ambrosia beetle problem and the techniques and dedication required to solve it. Misuse of the tools often decreased the success of the pest management program and created an unjustifiably bad image of semiochemical-based management.

There are four possible components of an IPM program for ambrosia beetles in a timber-processing area (1,37): (a) host-habitat management, i.e., cultural control through manipulation of log inventories in the woods as well as in processing areas; (b) suppression of beetle populations by using semiochemicals to intercept and kill host-seeking beetles; (c) product protection by the use of such agents as toxic insecticides and repellents; and (d) alteration of the overwintering habitat for *T. lineatum*, e.g., by increasing the distance between the processing area and the surrounding forest. Phero Tech's program has incorporated primarily the first two components.

Fundamental to such an IPM program is the hypothesis that (as proposed for *T. lineatum*) the size of ambrosia beetle populations is determined by the amount of host-habitat available, rather than by such factors as natural enemies or the physical environment (2). When there are fewer hosts available, there will be fewer beetles in the next generation, although there will be some compensation for reduced habitat, by increased attack densities in the available habitats. When the population of beetles is thus reduced as part of an IPM program, the prospect of success of the mass-trapping component is greatly enhanced.

In the first year of a contract, a timber-processing area is surrounded by survey traps to detect the areas of greatest beetle activity. Thereafter, mass trapping supplemented by trap-logs, is conducted, with emphasis on these "hot" areas. Typically a timber-processing area with a 1000-m perimeter will have about 100 to 200 traps for *T. lineatum* plus 10 to 15 bundles of trap-logs. In early spring, there would also be 15 to 18 traps for *G. sulcatus*. (Trapping for *G. retusus* is still under development.) From midsummer on, the ratio of *G. sulcatus*/*T. lineatum* traps would be increased up to 50:50, and the new ratio would be maintained until the *G. sulcatus* flight finished in late October.

Phero Tech personnel collect captured beetles, replenish spent baits, and repair and replace broken traps every 5 weeks. The opportunity to talk with the customers' employees on their "turf" is never wasted. Such conversations serve to keep working personnel informed and to satisfy their curiosity, but also involve them personally in the process so that they become watchdogs over the program. In addition, these informal encounters are taken as opportunities to advise on such procedures as inventory control, sanitation, and overwintering habitat management that will minimize attack on vulnerable logs as well as the size of future beetle populations.

The numbers of beetles captured are rapidly determined by volumetric measure according to species. These counts, as well as a written assessment of the problem and recommendations for integrated management, are provided to the customer within 5 to 10 working days of the collection date. The most critical part of the report, however, is a map of the processing area (Fig. 2). This map indicates each trap site and by color code the magnitude of the catch. Highest catches are designated by "shocking pink," a color that the customer soon learns to dread. This color message is perhaps the single most important stimulus for the customer to realize that a problem exists and to do something about it.

Quantitative evaluation of every semiochemical-based management operation, as well as an entire IPM program is virtually impossible. For example, the dangers from heavy equipment and rapidly moving logs often preclude entry into a dryland sort to inspect vulnerable timber for attacks. Rather, confidence is placed in large part on the original research programs that demonstrated efficacy of the strategy (30,31,34). To some extent, the numbers of beetles captured can be related to the number of logs that could be potentially attacked at given densities, but this measure does not take into account the benefits achieved through inventory control, i.e., host–habitat management. Therefore, a large component of the overall evaluation process, to date, has been the visual observation of uninfested logs by Phero Tech and sort personnel, who remember the days of white sawdust dripping from attacked logs. There has been a decrease in complaints from sawmill divisions who used to inherit huge ambrosia beetle problems passed on from the woods, the dryland sort, or the booming ground. And when the pest management program has been terminated (sometimes by a doubting customer), there have been striking examples of renewed impact of the beetles. Ultimately, evaluations of degrade in the sawmill may be integrated with data from the woods and in processing areas to provide a realistic simulation model of ambrosia beetle infestation dynamics and impact. The efficacy of an IPM program could then be evaluated in terms of its eventual benefit to the final product.

III. THE MOUNTAIN PINE BEETLE

A. The Problem

In 1977, a small infestation of the mountain pine beetle was observed in lodgepole pine in the remote western reaches of the Chilcotin Plateau in central British Columbia. It was dismissed by the zone forester for that region by the memorable question, "where can it go?" He had forgotten the massive outbreaks of the beetle that had devastated vast areas of the province in the 1930s (38).

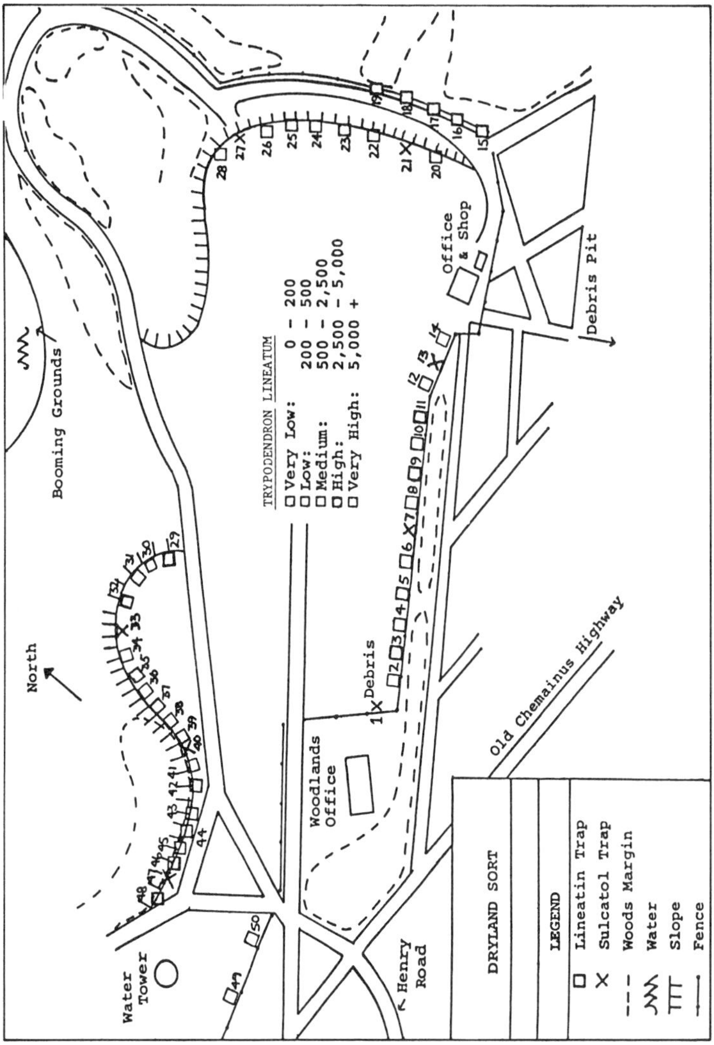

FIGURE 2 Working map of representative dryland sorting area on Vancouver Island, British Columbia, showing trap locations (numbered squares and *x*s) which are color-coded in periodic reports according to numbers of beetles captured at each location.

By 1986, *D. ponderosae* from this and a few other infestation centers had killed most of the mature pine on the Chilcotin plateau. From 1972 to 1985, it killed an estimated 195.7 million pines in British Columbia alone (39–41). As each cubic meter of harvested wood will generate 150 to 200 dollars of economic activity (1978 to 1982 figures in Canadian dollars; 42,43), and as an average lodgepole pine has a volume of 0.49 m^3 (40,41), the potential economic impact before salvage logging of the current infestation for the 14 years in question is 14.4 to 19.6 billion dollars. As an estimated 80% of the infested trees were logged within 4 years, but with an estimated mean value of 90% of a healthy tree, the real economic impact is reduced to 4.1 to 5.4 billion dollars. Similar infestations killed millions of pines in western Alberta. An infestation in Cypress Hills Park on the Alberta–Saskatchewan border threatened the beauty and recreational value of this priceless resource, an "island" left untouched in the middle of the prairie by the last glaciation.

These huge infestations are thought to be a consequence of the vast areas of mature and overmature, even-aged lodgepole pine monoculture that followed widespread wildfires in the latter half of the last century. Many of these were apparently set by prospectors to expose mineral deposits. These stands were then preserved intact by the loggers' preference for other species and by assiduous control of wildfires. And when the outbreaks started, Nature blessed them with a series of mild winters that resulted in very low overwintering mortality. When man woke up, he had on his hands, a series of roaring outbreaks, larger than any in recorded history.

When management efforts were finally directed against the beetle, the first strategy was to remove as many beetles as possible from the forest by sanitation-salvage clearcutting (45,46). This tactic was supplemented by single-tree disposal (i.e., cutting, piling, and burning of small groups of individual trees). But both of these tactics constitute "beetle chasing," that is, response to established infestations, whatever their size. The possibility then arose that if there were some way to manipulate the attacks so that they occurred when and where one wanted them to, infestations might at least be reined in, if not wiped out altogether. Thus, the stage was set in 1984 for the injection of semiochemicals into IPM of the mountain pine beetle.

B. Development of Semiochemicals for Pest Management

Research on semiochemical-based communication by *D. ponderosae* in British Columbia had begun in 1974. However, "Pondelure" (Fig. 3), a mixture of the host tree monoterpene α-pinene and its oxidation product the female-produced aggregation pheromone *trans*-verbenol, failed to attract significant numbers of beetles in lodgepole pine

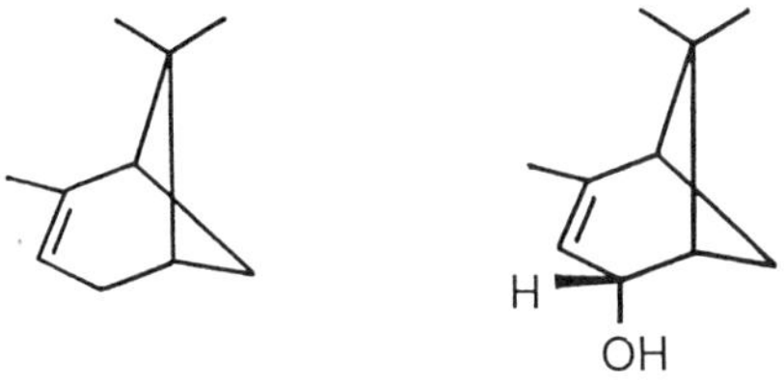

α-pinene trans-verbenol

FIGURE 3 "Pondelure," a blend of the host tree monoterpene α-pinene and the female-produced aggregation pheromone, *trans*-verbenol. This formulation was used effectively in operational research on the mountain pine beetle in the United States.

forests in British Columbia (47), despite its attractiveness in ponderosa and white pine forests in the United States (48,49). In British Columbia, myrcene replaced α-pinene as the most effective synergist with *trans*-verbenol for the attraction of males (47,50), a result in an agreement with those of Billings et al. (51) in the U. S. Pacific Northwest.

Attention then turned to *exo*-brevicomin, which is released by males when they join females that have initiated attack on a new tree (52,53). When added to a blend of myrcene and *trans*-verbenol, *exo*-brevicomin served to balance the sex ratio by differentially attracting females (47). Additional research showed that *exo*-brevicomin and another male-produced compound, frontalin, act as multifunctional pheromones, enhancing attraction at low concentrations and serving as antiaggregation pheromones at high concentrations (54). A final compound, verbenone, produced by microorganisms associated with the beetles (55), also serves as an antiaggregation pheromone (54,56). Borden and co-workers (54) present a conceptual model describing the perceived role of each of the above compounds in five phases of attack on a new tree.

Although even more compounds than those just described may be involved in semiochemical communication by *D. ponderosae*, a blend of myrcene, *trans*-verbenol, and *exo*-brevicomin, affectionately called "Mountain Do" (Fig. 4), has proved to be a most effective pest management tool. Remember that it consists of a female-produced pheromone *trans*-verbenol, that attracts mostly males; a male-produced pheromone, *exo*-brevicomin, that attracts mainly females; and the host tree monoterpene, myrcene, that synergizes the activity of both pheromones. Mountain Do was demonstrated to be a

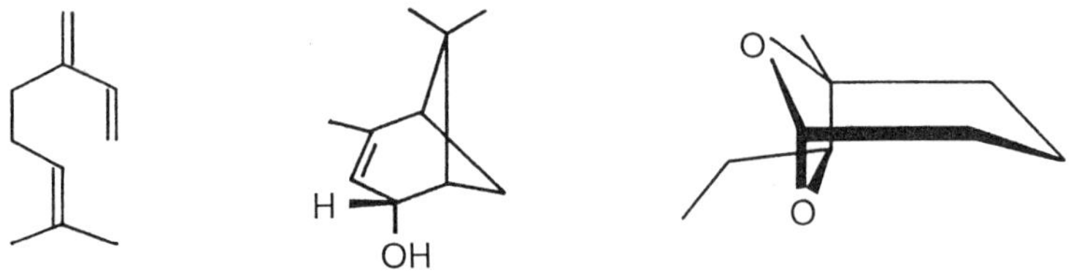

myrcene trans-verbenol exo-brevicomin

FIGURE 4 "Mountain Do," a blend of the host tree monoterpene, myrcene, the aggregation pheromone, *trans*-verbenol, and the multifunctional pheromone, *exo*-brevicomin, that is used effectively to manipulate populations of the mountain pine beetle in western Canada.

potent stimulus for inducing attack on and around baited trees (50) and a moderately good stimulus for attracting beetles to baited Norwegian drainpipe traps (47) and Lindgren traps (54). Several years of research have resulted in the delineation of effective release rates for the components of Mountain Do as tree baits as follows: myrcene, 20 mg/24 hr; *trans*-verbenol, 1 mg/24 hr; and *exo*-brevicomin, 0.2 mg/24 hr. For use in traps, the *exo*-brevicomin release rate must be reduced by at least tenfold.

A critical step in the development of Mountain Do was the design of a cheap, durable, "idiot-proof," bait receptacle for affixing to trees. Makeshift bait-holders, such as those made of perforated aluminum, film cans (57), or wooden blocks (50), were too cumbersome to be used on a large scale. Prototype receptacles invented by L. J. Chong utilized foldable, aluminum crosses that could be compactly stacked for transport (50,58,59). A cross was nailed to a tree and folded into a receptacle by bending the bottom up and the side arms of the cross over it. Bait dispensers were placed in the receptacle, and the top was folded over the opening to provide a sun and rain shield. Although effective, the stacked, unfolded receptacles and the required nails and hammer were heavy and cumbersome. The operational successor to this device (60) was invented by M. G. Banfield. It consisted of a waterproofed cardboard envelope that already contained the bait dispensers (2 closed 1.9-ml polyethylene tubes for myrcene; a 1.9-ml polyethylene tube for *trans*-verbenol, which required opening; and a 1.0-mm ID capillary tube, which needed to be broken, for *exo*-brevicomin). The envelope was stapled to a tree and the flap was folded into a slot on the outside to protect against sun and rain. When affixed to the north side of a tree, this device had an effective life of >100 days, leaving considerable time for operational treatments before beetle flight.

C. Operational Research

Four potential operational uses for Mountain Do were the subject of research and development programs from 1983 to 1987: population monitoring, containment and concentration, postlogging mop-up, and the eradication of spot infestations.

1. Monitoring

Monitoring for the location of most *D. ponderosae* infestations does not require semiochemicals. Rather, it is conventionally done from midsummer to early fall by aerial reconnaissance for red trees killed by beetles the previous year. This survey is followed up by ground probes (46) to locate freshly attacked green trees, that can be subjected to one or more treatments, possibly involving the use of semiochemicals.

However, semiochemicals have been tested for determining the occurrence of peak flights of the beetle (61). This knowledge is beneficial in instituting "no-haul" restrictions on loggers so that infested logs are not transported out of an attacked forest when the beetles are emerging and flying. Thus, new infestations are prevented at sites where trucks stop or break down, or around mill sites where the infested logs are stored before manufacture. Companies that have mill sites in or near a town have learned the hard way that they can become instantly unpopular if infestations in pines that beautify the town can be attributed to escaped beetles from a mill site.

Using strategically placed, multiple-funnel traps, Stock (61) monitored *D. ponderosae* populations in the Prince Rupert Region of British Columbia. During a 9-week trapping period from June 15 to August 16, 1983, only 16 days occurred in which there was sufficient flight to constitute a perceived problem. On a calendar basis, no-haul restrictions would have been imposed for a continuous 11-week period. By using the daily trap-catch data from the monitoring program in 1983, this period could have been reduced to 4 to 5 weeks, with considerable savings returned to the companies concerned. The reduced restrictions, of course, would be dependent upon rapid processing of logs arriving at the mill site so that any resident beetles would be killed in the debarking process.

2. Containment and Concentration

Under the pressure of mountain pine beetle outbreaks, one of the most frustrating situations encountered by forest pest management staff in the past was the necessity to neglect certain infestations for extended periods while attention was focused on others. The beetles in these neglected infestations were allowed to disperse and attack new trees, often far from their original infestation. Thus, new outbreaks flared up continuously, while others were exterminated. If, however, there were a mechanism whereby infestations could be contained, and possibly concentrated, allowing them to intensify without expanding, then their eventual excision by clearcutting would be much more effective and limited in size. Moreover, the chances of new infestation centers flaring up would be greatly reduced. It was with this prospect in mind that my research group began to investigate the hypothesis that semiochemical tree-baits could be used to contain and concentrate *D. ponderosae* infestations.

Research was set up in several areas of British Columbia in collaboration with the B. C. Forest Service, an arrangement that later facilitated rapid technology transfer. The plan was to bait trees at 50-m centers with Mountain Do, a cost-effective tactic requiring only four baits per hectare (58). With this layout, no beetle attempting to fly out of a stand could avoid passing within 25 m of a baited tree. It would thus have an optimal chance of perceiving the bogus message that the baited tree was a "good" host, already undergoing the early phases of mass attack. Dispersing beetles would orient to this tree and attack it, supplementing the bait with their own pheromones. When the baited tree was full, the beetles' antiaggregation pheromones would divert the attack toward trees around it (62,63).

The results of operational research upheld the hypothesis that infestations could be contained and concentrated. Attacked trees were significantly more concentrated around baited trees than around marked but unbaited trees in control stands (58). In two studies (58,64), the ratios of newly mass-attacked, green trees/red trees attacked the previous year (G/R ratios) were significantly higher in baited than in control stands, suggesting that potentially dispersive beetles had been retained in the baited blocks and possibly that beetles had been attracted into the baited blocks. [Mass attack was arbitrarily set at 31.25 attacks per square meter, a very conservative estimate of the minimum attack rate at which the beetles could kill an attacked tree. By comparison, 40 attacks per square meter were required to kill trees in the northern Rocky Mountain area of the United States (65).]

The hypothesis was upheld by Gray and Borden (66) who compared attack intensification in three zones, a well-infested zone that would be chosen for semiochemical baiting, and in concentric 50- and 100-m wide zones around the infested zone. To obtain a more true measure of actual attack intensification than simple ratios of green/red trees, they also used a G/R ratio and an attack intensification ratio (AIR) that were weighted by attack density and diameter of newly attacked trees.

The results indicated that in the three stands in which the central zone was baited, there were significantly greater ratios in the baited zone than in the concentric zones surrounding them. In two control stands, however, the attack intensification ratios were significantly higher in the first concentric zone than in either the control or the outer zone. Thus, the baiting regimen inhibited dispersal of beetles away from the infestation epicenter and contained and concentrated beetles within the baited zone.

For large stands (e.g., >10 ha), it does not matter if beetles rattle around in the central area. Thus, as supported by Heath's (64) data, it is sufficient to bait a double line of trees around the perimeter of a large infestation with the lines offset so that there is a baited tree for every 25 m of perimeter. Dispersing beetles would then be retained as they attempted to fly through the perimeter barrier.

After a beetle attack, it is recommended that baited stands be logged in the ensuing 10 to 11 months before the brood from infested trees emerge. For moderate infestations, it may be possible to bait a second year. Even for very lightly infested areas, prelogging baiting was found to be cost-effective if it avoided the minimum 20 dollars per tree cost of removing two infested trees per hectare outside of the baited and logged area (67).

During the course of research and development, trial and error has demonstrated that even the grandest strategy can be sabotaged by lack of attention to apparently trivial, tactical details. This experience has led to the development of six rules for a successful baiting program:

1. Bait the north side of a tree to avoid excessive evaporation of semiochemicals from dispensers exposed to direct sunlight.
2. When baiting, check to make sure that the baited tree is alive and that it is a pine.
3. Bait the largest tree possible, even if it requires going off-line 10 to 15 m to do so, as the beetles often will not attack small-diameter (<20 cm dbh) trees, and if they do are unable to create an intense source of attraction to supplement the bait.
4. Do not kneel, sit, squat, or lie down when baiting a tree. Mountain pine beetles apparently fly preferentially 2 to 5 m above ground. Therefore, the bait should be affixed as high as possible so that flying beetles will intersect the odor plume from baited trees.
5. Avoid baiting in brush higher than can be reached from the ground. The beetles fly above the understory canopy where their path is unencumbered and seldom will detect an odor plume that descends into the understory vegetation.
6. Read the instructions. Trees with sealed bait dispensers are not very attractive.

3. *Postlogging Mop-Up*

Semiochemical baiting can be used to mop up mountain pine beetles after a sanitation–salvage logging operation. To use a medical analogy, it is equivalent to chemotherapy applied after a malignant neoplasm has been surgically removed.

Investigation of this tactic occurred in an infestation in Manning Park, British Columbia (59). By 1982, this infestation reached a magnitude of between 5000 and 6000 huge green-attacked trees. Many of these trees were of sufficient size to generate 5000 to 10,000 brood beetles each. A selection cut was carried out, taking only the attacked trees. Their stumps were peeled to kill all possible beetles. To monitor the effectiveness of the sanitation-salvage cut, we worked with the park personnel to bait 50 trees well dispersed within the logged area. Clearly the logging operation had not removed all of the beetles from the park. Forty-eight of the 50 baited trees were attacked, as were 80 trees within 12 m of them, representing 128 attacked trees on an area of 2.2 ha. A further 86 attacked trees were found on the remaining 63 ha. All of these trees were cut and burned (60).

The next year 10 trees were baited, and 29 were attacked (68). Again, these were cut and burned. The process was repeated in 1984, with three trees baited, attacked, and cut and burned (69). No infestation recurred. While immigrating beetles could create a new infestation, we had successfully mopped up the old one. An IPM approach of intensive survey, radical surgery, chemotherapy, resurvey, minor surgery, and repeated resurvey and treatment had solved the problem.

4. *Eradication of Spot Infestations*

Large infestations of *D. ponderosae* usually arise from small infestations that expand, multiply, and eventually coalesce (44). In British Columbia, disposal of these infestations on a single tree basis (i.e., cutting, bucking, piling, and burning of infested trees) has been practiced routinely and effectively. However, felling and burning must cease at the advent of fire season in mid- to late-spring. If some other way were found to eradicate these infestations, single-tree disposal could continue until the emergence of brood beetles in late July through August. Moreover, because felling and burning is time-consuming and expensive, any cheaper and faster alternative methods of postattack treatment would be welcome. These objectives have been met through two tactics that are now developed to the operational stage.

I refer to the first tactic as *Cut, Spray, Bait, and Treat*. In collaboration with T. E. Lacey of the B. C. Forest Service, a noncontrolled operational trial was set up in July 1984 in four spot infestations totaling 62 red trees at Similkameen Falls, British Columbia. It utilized a combination of treatments judged to have an optimal probability of success in reducing the green/red ratio to zero.

A seven-man helicopter-borne "Rap-Attack" crew (usually employed in fire fighting) was used because of the steep, relatively inaccessible terrain in which the spots occurred. The crew rappelled from a helicopter into the treatment zones, and water for mixing of pesticide was lowered to the ground in 318 liter plastic tanks for two sites that had no natural water supply. The crew felled the 62 trees previously identified by probe crews as containing living brood beetles. Other B. C. Forest Service staff, utilizing backpack sprayers sprayed each of the felled trees completely around the bole to attack height with 2% Sevin SL in water, a treatment shown to be virtually 100% effective in killing emergent brood beetles (70). For the four spots, a total of 37 surrounding green trees were baited with Mountain Do, an approximate ratio of one baited tree to every two felled and sprayed trees.

On August 28, a three-man crew surveyed the four treatment areas. Nine mass-attacked trees and 37 lightly attacked trees were found, all of which were treated by the hack and squirt technique with one-half strength monosodium methanearsonite (MSMA), a treatment double that known to cause 100% mortality of brood beetles (71).

A follow-up cruise in October indicated that the trial was very successful (Table 3). Although it was not a controlled experiment, data on the ratios of mass-attacked green trees/red trees from 20 untreated spot infestations throughout British Columbia in 1983 and 1984 indicated that a green/red ratio of 2:1 could be expected. The actual ratio of 0.2:1 was far below that, and very encouraging. In addition, the maximum attack height dropped to a mean of 4.4 m on the mass-attacked green trees, as opposed to a mean height of 12.6 m on the red trees attacked the previous year, further indicating a sharp reduction in numbers of attacking beetles. In three of the four spots, there was only one mass-attacked, green tree, while nine occurred in the other spot. Eight of the nine trees that were mass attacked by August 28 (one was missed), as well as 36 other, lightly attacked trees, were treated with MSMA. A successful attack would not have occurred in trees that were not mass-attacked (65), but it may be easier for forestry crews to treat all but the most poorly attacked trees, rather than to make a decision about the attack success on each tree. Three trees were mass-attacked after the MSMA treatment on August 28, indicating that two MSMA treatments might be required in some instances.

Two galleries in each of 23 MSMA-treated trees were checked on October 15. Seven had live larvae in one of two galleries, and one

TABLE 3 Results of *Cut, Spray, Bait, and Treat* Trial in 1984 at Similkameen Falls, British Columbia

Criterion assessed	Assessment per infestation				All infestations combined
	1	2	3	4	
No. of trees felled and sprayed	12	14	17	19	62
No. of trees baited	8	7	9	13	37
No. of baited trees attacked	8	7	8	10	33
No. of baited trees mass-attacked	1	6	1	1	9
No. of unbaited trees attacked	0	5	6	8	19
No. of unbaited trees mass-attacked	0	3	0	0	3
Total trees attacked	8	12	14	18	55
Total trees mass-attacked	1	9	1	1	12
Ratio of mass-attacked, green/red trees	0.1	0.6	0.1	0.1	0.2

[a]Infested trees in four small infestations were felled and treated with Sevin, green trees were semiochemical-baited and newly attacked trees were treated with MSMA.

had live larvae in two galleries, i.e., 17.4% of 46 galleries had live larvae. In contrast, six of six (100%) of the galleries in the three trees that were mass attacked after August 28 had larger live larvae in longer galleries than in the galleries in MSMA-treated trees. As in previous trials with arsenicals (72,73), a final assessment of the effectiveness of MSMA should be made just before emergence in the summer.

Costs for the project were as follows:

Manpower, 176.5 man-hours @ $12.18/man-hour	$2150.00
Helicopter time, 3.2 hr @ $345/hr	1104.00
Chain saws, 6 for 1 day @ $21.25 ea.	127.50
Fuel, 1236.5 L. @ $0.44/L	544.00
Sevin SL, est. 15 L @ $7.10/L	106.50
MSMA, est. 1 L @ $4.40/L	4.40
Vehicles, 2 for 1 day @ $40.00 ea.	80.00
	$4116.00

Despite this apparently large sum, the 66.40 dollar cost per tree is considerably below the estimated cost of 75.00 dollars per tree for single-tree disposal by the fell-and-burn technique (P. M. Hall, personal communication). Undoubtedly the cost would decline markedly if helicopters were not needed, and as well-practiced crews became more efficient, reducing the time required per tree.

A second operational trial for eradicating small infestations employs a procedure that I call *Treat, Bait, and Treat*. It is being conducted by T. E. Lacey in the Merritt Timber Supply Area in southern British Columbia. When new infestations are encountered shortly after attack has occurred, all attacked trees are treated with MSMA to kill any developing brood beetles.

In 1986 and 1987, approximately 12,000 and 20,000 newly infested trees, respectively, were treated with MSMA. Treatments extended through mid-October in 1986, well beyond the 3-week period after attack in which MSMA is 100% effective in assessments in the fall (73). However, in a previous trial, MSMA applied 6 to 8 weeks after attack reduced the surviving brood in the spring to 24% of that in untreated control trees (74). In 1987, trees were assessed prior to MSMA treatment by checking the lengths of the egg galleries. If they exceeded 10.0 cm, the trees were set aside for fell-and-burn treatment later in the fall, thus restricting the MSMA treatment to trees in which the symbiotic fungus would not yet be well established and in which there was a good probability of MSMA being translocated (71).

In each treatment site, trees immediately surrounding those treated with MSMA were semiochemical-baited the following spring. If these trees or those surrounding them were attacked, they were treated postattack with MSMA. Preliminary assessment of the 1986 treatment areas indicates very little spill-over into the semiochemical-baited, mop-up trees and suggests that this tactic will be very effective.

D. Operational Programs

Only two forest regions in British Columbia have adopted semiochemical-baited, multiple-funnel traps to monitor mountain pine beetle infestations and to impose no-haul restrictions on logging companies (see Table 1). By far the greatest use of semiochemicals has been to integrate tree-baiting with regionwide silvicultural controls in British Columbia and Alberta. In late summer, fall, and spring, Forest Service or industrial personnel probe infested and high-hazard forests for *D. ponderosae* attack, concentrating on areas in which aerial photography or sketch mapping has disclosed the presence of red trees attacked the previous year. Moderate to large-sized infestations are scheduled for immediate logging and small infestations for single-tree disposal by fell-and-burn, cut-and-spray, or MSMA treatment. Trees on the peripheries of these infestations are semiochemical-baited to mop up populations of beetles that were missed during logging or single-tree treatments. After beetle flight in the next year, these trees and those surrounding them are checked for attack, and infested trees either logged (if economically accessible) or treated to kill the beetles, increasingly with MSMA.

If logging cannot be done immediately, future cut blocks are semiochemical-baited the following spring to contain and concentrate infestations before logging during the next 10 months. Depending on available green trees and logging priorities, this procedure may sometimes be repeated for a second year before the stand is logged.

Often, the population of emerging beetles in a baited stand is smaller than anticipated, for example, because of severe winter kill or other unknown mortality factors. Here, only the baited trees and one or two trees adjacent to them may be attacked. The beetles in the mass-attacked trees may then be killed with MSMA. These stands are then removed from logging priority, and the remaining trees are preserved alive for later, more regulated harvesting.

IV. CONCLUSION

I have described the development and operation of two successful programs in which semiochemicals are used in the management of bark and timber beetles in western Canadian coniferous forests. Several factors were critical to the rapid implementation and success of these two programs. Coupled with technological capability not available to previous researchers, there was a fair amount of luck. For instance, the single-compound pheromone messages used by ambrosia beetles eliminated the necessity to unravel how the beetles use a complex pheromone blend. For both ambrosia beetles and the mountain pine beetle, the users of the IPM programs had no other solutions to the problem. Thus, there was ready acceptance of new technology. There was also no requirement for registration of semiochemicals as pesticides in Canada, as long as they were released from devices such as traps and tree baits. After considerable argument and deliberation, the U. S. Environmental Protection Agency has now followed suit, and as of 1987 has exempted from registration pheromones emitted from both traps and tree baits. Finally, most of the initial technology transfer in both examples was done by the primary researchers. Thus semiochemical-based management passed from one year to the next out of operational research and into full, operational use.

In addition to the applied use of semiochemicals for scolytids, sex pheromones are fully operational for monitoring the populations of two lepidopteran pests of coniferous tree nurseries and one forest defoliator (see Table 1). Numerous other semiochemicals are in the research and development process (see Table 2), some of them very near to being fully operational.

As for the case histories described, the fanciful expectations of some futurists that semiochemicals would replace broad-spectrum toxicants in virtually annihilating damaging populations have proved to be unfounded. Thus, one cannot expect a "magic, semiochemical bullet" to solve a pest problem. However, one can expect many more semiochemicals to be integrated into the pest management strategies and tactics for many more forest pests.

ACKNOWLEDGMENTS

I thank J. A. Carlson, H. F. Cerezke, R. L. Chan, G. E. Daterman, D. Doidge, A. J. Dupilka, T. Ebata, R. G. Fraser, P. M. Hall, R. H. Heath, D. Heppner, R. S. Hodgkinson, T. E. Lacey, B. S. Lindgren, L. E. Maclauchlan, J. A. McLean, E. V. Morris, S. H. Krannitz, S. Salom, G. M. Shrimpton, H. Spahan, E. Stokkink, and G. A. Van Sickle for advice, information, or review of the manuscript.

REFERENCES

1. Borden, J. H. and McLean, J. A. Pheromone-based suppression of ambrosia beetles in industrial timber processing areas. In *Management of Insect Pests with Semiochemicals.* (Mitchell, E. R., ed.). Plenum Press, New York, 1981, pp. 133--154.
2. Borden, J. H. The striped ambrosia beetle, *Trypodendron lineatum* (Olivier). In *Forest Insect Oubreaks: Patterns, Causes and Management Strategies.* (Berryman, A. A., ed.). Plenum Press, New York, 1988, pp. 579–596.
3. McLean, J. A. Ambrosia beetles: A multimillion dollar degrade problem of sawlogs in coastal British Columbia. *For. Chron.* *61*:295--298 (1985).
4. Gray, D. R. and Borden, J. H. Ambrosia beetle attack on logs before and after processing through a dryland sorting area. *For. Chron.* *61*:299–302 (1985).
5. McLean, J. A. and Borden, J. H. *Gnathotrichus sulcatus* attack and breeding in freshly sawn lumber. *J. Econ. Entomol.* *68*:605–606 (1975).
6. Rudinsky, J. A. and Daterman, G. E. Field studies on flight patterns and olfactory responses of ambrosia beetles in Douglas-fir forests of western Oregon. *Can. Entomol.* *96*:1339–1352 (1964).
7. Rudinsky, J. A. and Daterman, G. E. Response of the ambrosia beetle *Trypodendron lineatum* (Olivier) to a female-produced pheromone. *Z. Angew. Entomol.* *54*:300–303 (1964).

8. Chapman, J. A. The effect of attack by the ambrosia beetle *Trypodendron lineatum* (Olivier) on log attractiveness. *Can. Entomol.* *98*:50–59 (1966).
9. Atkins, M. D. Behavioral variation among scolytids in relation to their habitat. *Can. Entomol.* *98*:285–288 (1966).
10. Borden, J. H. and Stokkink, E. Laboratory investigation of secondary attraction in *Gnathotrichus sulcatus* (Coleoptera: Scolytidae). *Can. J. Zool.* *51*:469–473 (1973).
11. Borden, J. H. and McLean, J. A. Secondary attraction in *Gnathotrichus retusus* and cross attraction of *G. sulcatus* (Coleoptera: Scolytidae). *J. Chem. Ecol.* *5*:79–88 (1979).
12. MacConnell, J. G., Borden, J. H., Silverstein, R. M., and Stokkink, E. Isolation and tentative identification of lineatin, a pheromone from the frass of *Trypodendron lineatum* (Coleoptera: Scolytidae). *J. Chem. Ecol.* *3*:549–561 (1977).
13. Borden, J. H., Handley, J. R., Johnston, B. D., MacConnell, J. G., Silverstein, R. M., Slessor, K. N., Swigar, A. A., and Wong, D. T. W. Synthesis and field testing of 4,6,6-lineatin, the aggregation pheromone of *Trypodendron lineatum* (Coleoptera: Scolytidae). *J. Chem. Ecol.* *5*:681–689 (1979).
14. Byrne, K. J., Swigar, A. A., Silverstein, R. M., Borden, J. H., and Stokkink, E. Sulcatol: Population aggregation pheromone in the scolytid beetle, *Gnathotrichus sulcatus*. *J. Insect Physiol.* *20*:1895–1900 (1974).
15. Borden, J. H., Oehlschlager, A. C., Slessor, K. N., Chong, L., and Pierce, H. D., Jr. Field tests of isomers of lineatin, the aggregation pheromone of *Trypodendron lineatum* (Coleoptera: Scolytidae). *Can. Entomol.* *112*:107–109 (1980).
16. Borden, J. H., Chong, L., McLean, J. A., Slessor, K. N., and Mori, K. *Gnathotrichus sulcatus*: Synergistic response to enantiomers of the aggregation pheromone, sulcatol. *Science* *192*:894–896 (1976).
17. Borden, J. H., Handley, J. R., McLean, J. A., Silverstein, R. M., Chong, L., Slessor, K. N., Johnston, B. D., and Schuler, H. R. Enantiomer-based specificity in pheromone communication by two sympatric *Gnathotrichus* species (Coleoptera: Scolytidae). *J. Chem. Ecol.* *6*:445–456 (1980).
18. Borden, J. H., Chong, L., Slessor, K. N., Oehlschlager, A. C., Pierce, H. D., Jr., and Lindgren, B. S. Allelochemic activity of aggregation pheromones between three sympatric species of ambrosia beetles. *Can. Entomol.* *113*:557–563 (1981).
19. Cade, S. C., Hrutfiord, B. F., and Gara, R. I. Identification of a primary attractant for *Gnathotrichus sulcatus* isolated from western hemlock logs. *J. Econ. Entomol.* *63*:1014–1015 (1970).

20. Moeck, H. A. Ethanol as the primary attractant for the ambrosia beetle *Trypodendron lineatum* (Coleoptera: Scolytidae). *Can. Entomol.* *102*:985–995 (1970).
21. Moeck, H. A. Field test of ethanol as a scolytid attractant. *Environ. Can., Bi-Mon. Res. Notes* *27*(2):11–12 (1971).
22. Graham, K. Anaerobic induction of primary chemical attractancy for ambrosia beetles. *Can. J. Zool.* *46*:905–908 (1968).
23. McLean, J. A. and Borden, J. H. Attack by *Gnathotrichus sulcatus* (Coleoptera: Scolytidae) on stumps and felled trees baited with sulcatol and ethanol. *Can. Entomol.* *109*:675–686 (1977).
24. Borden, J. H., Lindgren, B. S., and Chong, L. Ethanol and α-pinene as synergists for the aggregation pheromones of two *Gnathotrichus* species. *Can. J. For. Res.* *10*:290–292 (1980).
25. Browne, L. E. A trapping system for the western pine beetle using attractive pheromones. *J. Chem. Ecol.* *4*:261–275 (1978).
26. Bakke, A. and Saether, T. Granbarkbillen kan fanges i rorfeller. *Skogeieren* *65*(11):10 (1978).
27. Lindgren, B. S., Borden, J. H., Chong, L., Friskie, L. M., and Orr, D. B. Factors influencing the efficiency of pheromone-baited traps for three species of ambrosia beetles (Coleoptera: Scolytidae). *Can. Entomol.* *115*:303–313 (1983).
28. Lindgren, B. S. A multiple funnel trap for scolytid beetles (Coleoptera). *Can. Entomol.* *115*:299–302 (1983).
29. McLean, J. A. and Borden, J. H. Survey for *Gnathotrichus sulcatus* (Coleoptera: Scolytidae) in a commercial sawmill with the pheromone, sulcatol. *Can. J. For. Res.* *5*:586–591 (1975).
30. McLean, J. A. and Borden, J. H. Suppression of *Gnathotrichus sulcatus* with sulcatol-baited traps in a commercial sawmill and notes on the occurrence of *G. retusus* and *Trypodendron lineatum*. *Can. J. For. Res.* *7*:348–356 (1977).
31. McLean, J. A. and Borden, J. H. An operational pheromone-based suppression program for an ambrosia beetle, *Gnathotrichus sulcatus*, in a commercial sawmill. *J. Econ. Entomol.* *72*:165–172 (1979).
32. Shore, T. L. A pheromone-mediated mass-trapping program for three species of ambrosia beetle in a commercial sawmill. PhD Thesis, University of British Columbia, Vancouver, B. C., 1982.
33. Slessor, K. N., Oehlschlager, A. C., Johnston, B. D., Pierce, H. D., Jr., Wickremsinghe, L. K. G., and Grewal, S. Lineatin–a regioselective synthesis and resolution yielding the chiral pheromone of *Trypodendron lineatum*. *J. Org. Chem.* *45*:2290–2297 (1980).

34. Lindgren, B. S. and Borden, J. H. Survey and mass trapping of ambrosia beetles (Coleoptera: Scolytidae) in timber processing areas on Vancouver Island. *Can. J. For. Res. 13*:481–493 (1983).
35. Johnston, B. D., Slessor, K. N., and Oehlschlager, A. C. A synthesis of (±)-lineatin by 2+2 cycloaddition of dichloroketene with a cyclic allyl ether. *J. Org. Chem. 50*:114–117 (1985).
36. Phero Tech's Lindgren funnel trap. A versatile aid to forest insect management. Phero Tech Inc., Vancouver, B. C., undated.
37. Gray, D. R. Ambrosia beetles (Coleoptera: Scolytidae) in a Vancouver Island dryland sort: Their damage and proposed control. M.P.M. Professional Paper, Simon Fraser University, Burnaby, B. C., 1985.
38. Van Sickle, G. A. The mountain pine beetle situation in Canada. In *Proceedings of the Joint Canada/USA Workshop on Mountain Pine Beetle Related Problems in Western North America.* (Shrimpton, D. M., ed.). Environ. Can., Can. For. Serv. Pap. No. BC-X-230, pp. 13–15.
39. Kondo, E. S. and Taylor, R. G. *Forest Insect and Disease Conditions in Canada 1984.* Can. For. Serv., Ottawa, 1985.
40. Wood, C. S. and Van Sickle, G. A. Forest insect and disease conditions. British Columbia and Yukon, 1985. Can. For. Serv. Inf. Rep. BC-X-277, 1986.
41. Wood, C. S. and Van Sickle, G. A. Forest Insect and disease conditions. British Columbia and Yukon 1986. Can. For. Serv. Inf. Rep. BC-X-287, 1987.
42. Jones, J. R. Economic benefits of timber and productive forest land in British Columbia. *For. Chron. 63*:112–118 (1987).
43. Morrison, P. Value of forest land overestimated. *For. Chron. 63*:307 (1987).
44. Safranyik, L., Shrimpton, D. M., and Whitney, H. S. Management of lodgepole pine to reduce losses from the mountain pine beetle. Can. For. Serv., Pac. For. Res. Cen., For. Tech. Rep. 1, 1974.
45. McMullen, L. H., Safranyik, L., and Linton, D. A. Suppression of mountain pine beetle infestations in lodgepole pine forests. Can. For. Serv., Pac. For. Cen., Inf. Rep. BC-X-276, 1986.
46. British Columbia Ministry of Forests and Lands. *Pest Management.* Protection Manual, Vol. II. Victoria, B. C., 1987.
47. Conn, J. E., Borden, J. H., Scott, B. E., Friskie, L. M., Pierce, H. D., Jr., and Oehlschlager, A. C. Semiochemicals for the mountain pine beetle, *Dendroctonus ponderosae* (Coleoptera: Scolytidae) in British Columbia, field trapping studies. *Can. J. For. Res. 13*:320–324 (1983).

48. Pitman, G. B. and Vité, J. P. Aggregation behavior of *Dendroctonus ponderosae* (Coleoptera: Scolytidae) in response to chemical messengers. *Can. Entomol.* *101*:143–149 (1969).
49. Pitman, G. B. *trans*-Verbenol and alpha pinene: Their utility in manipulation of the mountain pine beetle. *J. Econ. Entomol.* *64*:246–340 (1971).
50. Borden, J. H., Conn, J. E., Friskie, L. M., Scott, B. E., Chong, L. J., Pierce, H. D., Jr., and Oehlschlager, A. C. Semiochemicals for the mountain pine beetle, *Dendroctonus ponderosae* (Coleoptera: Scolytidae), in British Columbia: Baited tree studies. *Can. J. For. Res.* *13*:325–333 (1983).
51. Billings, R. F., Gara, R. I., and Hrutfiord, B. F. Influence of ponderosa pine resin volatiles on the response of *Dendroctonus ponderosae* to synthetic *trans*-verbenol. *Environ. Entomol.* *5*:171–179 (1976).
52. Rudinsky, J. A., Morgan, M. E., Libbey, L. M., and Putnam, T. B. Antiaggregative rivalry pheromone of the mountain pine beetle, and a new arrestant of the southern pine beetle. *Environ. Entomol.* *3*:90–98 (1974).
53. Libbey, L. M., Ryker, L. C., and Yandell, K. L. Laboratory and field studies of volatiles released by *Dendroctonus ponderosae* Hopkins (Coleoptera: Scolytidae). *Z. Angew. Entomol.* *100*:381–392 (1985).
54. Borden, J. H., Ryker, L. C., Chong, L. J., Pierce, H. D., Jr., Johnston, B. D., and Oehlschlager, A. C. Response of the mountain pine beetle, *Dendroctonus ponderosae* Hopkins (Coleoptera: Scolytidae) to five semiochemicals in British Columbia lodgepole pine forests. *Can. J. For. Res.* *17*:118–128 (1987).
55. Hunt, D. W. A. Production and regulation of oxygenated terpene pheromones in the bark beetles *Dendroctonus ponderosae* Hopkins and *Ips paraconfusus* Lanier. PhD Thesis, Simon Fraser University, Burnaby, B. C., Canada, 1987.
56. Ryker, L. C. and Yandell, K. L. Effect of verbenone on aggregation of *Dendroctonus ponderosae* Hopkins (Coleoptera: Scolytidae) to synthetic attractant. *Z. Angew. Entomol.* *96*:452–459 (1983).
57. Furniss, M. M., Daterman, G. E., Kline, L. N., McGregor, M. D., Trostle, G. C., Pettinger, L. F., and Rudinsky, J. A. Effectiveness of the Douglas-fir beetle antiaggregative pheromone methylcyclohexenone at three concentrations and spacings around felled host trees. *Can. Entomol.* *106*:381–392 (1974).
58. Borden, J. H., Chong, L. J., Pratt, K. E. G., and Gray, D. R. The application of behaviour-modifying chemicals to contain infestations of the mountain pine beetle, *Dendroctonus ponderosae*. *For. Chron.* *59*:235–239 (1983).

59. Borden, J. H., Chong, L. J., and Fuchs, M. C. Application of semiochemicals in postlogging manipulation of the mountain pine beetle, *Dendroctonus ponderosae* (Coleoptera: Scolytidae). *J. Econ. Entomol. 76*:1428–1432 (1983).
60. Mountain pine beetle management with tree baits. Tech. Bull., Phero Tech, Inc., Vancouver, B. C., undated.
61. Stock, A. J. Use of pheromone baited Lindgren funnel traps for monitoring mountain pine beetle flights. B. C. For. Serv. Int. Rep. PM-PR-2, 1984.
62. Geiszler, D. R. and Gara, R. I. Mountain pine beetle attack dynamics in lodgepole pine. In *Theory and Practice of Mountain Pine Beetle Management in Lodgepole Pine Forests*. (Berryman, A. A., Amman, G. D., and Stark, R. W., eds.). University of Idaho, Moscow, Idaho, USDA, Forest Service, Washington, D. C., 1978, pp. 182–187.
63. Geiszler, D. R., Gallucci, V. F., and Gara, R. I. Modelling the dynamics of mountain pine beetle aggregation in a lodgepole pine stand. *Oecologia 46*:244–253 (1980).
64. Heath, D. Assessment of operational pheromone-based containment programs for mountain pine beetle control in the Cariboo Region. B. C. For. Serv. Int. Rep. PM-C-1, 1986.
65. Raffa, K. F. and Berryman, A. A. The role of host plant resistance in the colonization behavior and ecology of bark beetles (Coleoptera: Scolytidae). *Ecol. Monogr. 53*:27–49 (1983).
66. Gray, D. R. and Borden, J. H. Containment and concentration of infestations of the mountain pine beetle by semiochemical baiting: Validation by assessment of baited and surrounding zones. *J. Econ. Entomol.* (in press).
67. Borden, J. H., Chong, L. J., and Lacey, T. E. Pre-logging baiting with semiochemicals for the mountain pine beetle, *Dendroctonus ponderosae*, in high hazard stands of lodgepole pine. *For. Chron. 62*:20–23 (1986).
68. Wood, R. O. Mountain pine beetle in Manning Park. A post-control survey. Can. For. Serv., Pac. For. Res. Cen., Pest Rep., September, 1983.
69. Wood, R. O. Mountain pine beetle in Manning park. A post-control survey. Can. For. Serv., Pac. For. Res. Cen., Pest Rep., September, 1984.
70. Fuchs, M. G. and Borden, J. H. Pre-emergence insecticide applications for control of the mountain pine beetle, *Dendroctonus ponderosae* (Coleoptera: Scolytidae). *J. Entomol. Soc. BC 82*: 25–28 (1985).
71. Maclauchlan, L. E., Borden, J. H., D'Auria, J. M., and Wheeler, L. A. Distribution of arsenic in MSMA-treated lodgepole pines

infested by the mountain pine beetle, *Dendroctonus ponderosae* (Coleoptera: Scolytidae), and its relationship to beetle mortality. *J. Econ. Entomol.* *81*:274–280 (1988).

72. Chansler, J. F., Cahill, D. B., and Stevens, R. E. Cacodylic acid field tested for control of mountain pine beetles in ponderosa pine. USDA, For. Serv., Res. Note RM-161, 1970.
73. Dyer, E. D. A. and Hall, P. M. Timing cacodylic acid treatments to kill mountain pine beetles infesting lodgepole pine. *Envir. Can. Can. For. Serv. Bi-Mon. Res. Notes* *35*:13 (1979).
74. Bristow, W. Time span for the application of M.S.M.A to control mountain pine beetle. Northwood Pulp and Timber, Ltd., Prince George, B.C., unpublished report, 1985.
75. Werner, F. G. (Chm.). *Common Names of Insects and Related Organisms.* Entomol. Soc. Am., College Park, Md, 1982.

20

Pheromones for Managing Coniferous Tree Pests in the United States, with Special Reference to the Western Pine Shoot Borer

GARY E. DATERMAN / Forest Service, U.S. Department of Agriculture, Corvallis, Oregon

I. INTRODUCTION

Synthetic pheromones are now available for most major forest defoliators, bark beetles, ambrosia beetles, and various seed pests, nursery pests, sawflies, and shoot borers. At first glance, therefore, the title of this chapter would appear to cover an excessively broad topic. However, when one is restricted to operational use of pheromones, there are still relatively few examples of pheromone application in forest pest management. The featured example in this chapter will be the case history of pheromone development for managing the western pine shoot borer, *Eucosma sonomana* Kearfott. Biology of the insect will be reviewed, and the damage caused by the species described. Research and development of the pheromone and its current and prospective use will be outlined. Other examples of pheromone applications will include use of an antiaggregation pheromone for preventing population buildups of Douglas-fir beetle, *Dendroctonus pseudotsugae* Hopkins, and a pheromone-baited trapping system for monitoring population increases of the Douglas-fir tussock moth, *Orgyia pseudotsugata* (McDunnough).

Field testing of other forest insect pheromones has also taken place, and some of these compounds are used operationally for applications such as detection surveys to determine presence or absence of an introduced pest, and timing of seasonal flights for synchronizing spray treatments. In addition, Borden (Chap. 19) reports outstanding progress in the development of pheromone-based pest management programs for the mountain pine beetle, *Dendroctonus ponderosae* Hopkins, and the striped ambrosia beetle, *Trypodendron lineatum* (Olivier). Although developed in Canada, use of these bark beetle programs is expanding rapidly in the western United States.

II. THE WESTERN PINE SHOOT BORER

The western pine shoot borer (WPSB) infests ponderosa pine, *Pinus ponderosa* Laws., throughout its range in the western United States (1). Other common host trees include lodgepole pine, *P. contorta* Dougl., and jeffrey pine, *P. jeffreyi* Grev. The species was not identified as a pest until about 1970 (2,3). Considering that it is distributed throughout the western pine zone and is a native species, this is somewhat surprising. The delay in identifying the problem was, no doubt, a result of the insect's characteristic low-density populations and the subtle damage caused by its inconspicuous attacks on leader shoots.

A. Biology

The western pine shoot borer overwinters as a pupa in the litter beneath host pines. Adult moths emerge early in the spring with flight lasting 4 to 6 weeks. Depending on geographic location and local weather patterns, flight and oviposition may occur from February into May. Eggs are deposited on the buds, and larvae bore directly into the bud. The larvae then mine the pith of the elongating new shoots. Rarely is more than one larva found per infested shoot, and the insect consistently chooses the terminal or leader buds over the much more numerous lateral branches (4). As larval development nears completion, the larva exits the shoot and moves to the ground to pupate. A single generation develops per year. The WPSB is a persistent pest because it characteristically occurs year after year in the same stand of young trees. A moderate to high infestation may consist of only several hundred moths per hectare, but because of the insect's preference for terminal shoots, this density is enough to cause damage to 50% or more of the trees in a plantation.

B. Tree Injury

Evidence of WPSB infested trees is not obvious to the untrained observer. Except for the occasional shoots that are actually killed, the typical infested shoot appears green and healthy (Fig. 1). However, it has shortened needles on the distal portion, and has an increased density of needles per unit of shoot length (2,4).

Mining by WPSB larvae in the pith of growing terminal shoots causes a stunting of shoot elongation and a corresponding 25% loss in height growth (2,4). Some large lateral shoots may also be infested, but the feeding that occurs on lateral branches is of little economic consequence. The insect clearly prefers the dominant leader shoots, which is the characteristic that causes problems for management. Besides stunting of height growth, larval feeding in terminal shoots can cause forked stems and other distortions in tree form (3). Such distortions occur when feeding actually kills the terminal shoot or when the terminal is overtopped by one or more lateral branches.

Trees appear most susceptible to damage between the heights of 1.5 and 15 m. Even-aged stands and plantations with 50% or higher rates of infestation are not uncommon (2,4), and because each infested tree loses 25% or more height growth, such stands lose 12% or more in height growth per year. Stoszek's (2) impact analysis showed a 15% to 23% annual loss in height growth for plantations with 40% to 70% rates of infestation. Such impacts are sufficient to cause substantial economic losses, particularly for high-value, intensively-managed plantations.

C. Pheromone Research and Development

When the WPSB was identified as a problem, no methods for control were known (2,3). Development of some nontraditional control technique appeared necessary, and synthetic pheromones seemed particularly appropriate because of the pest's characteristically low-population densities. Consequently, a joint research and development project between the U. S. Forest Service and The Weyerhaeuser Company was begun in 1976 to chemically identify the sex attractant pheromone and develop it for curbing the damage.

1. *Pheromone Chemistry*

Low densities of field populations and lack of a laboratory colony of WPSB were distinct disadvantages for studying the insect's pheromone chemistry. Identification efforts included screening available synthetic chemicals as potential attractants and attempting to collect a sufficient supply of insects for chemical analysis. The screening was implemented by baiting adhesive traps with candidate synthetic compounds formulated in polyvinyl chloride pellets (5). The field screening yielded positive results when lures contained (*Z*)-9-dodecenyl acetate and small quantities of (*E*)-9-dodecenyl acetate. No males were captured by any of the other 12- and 14-carbon compounds screened as potential attractants (6). Additional field tests with different combinations of the (*Z*)- and (*E*)-9-dodecenyl acetates were very effective if the lure preparations contained between 95:5

Uninfested

FIGURE 1 Ponderosa pine terminal shoot infested with western pine shoot borer (right) and uninfested terminal (left). Note shortened needles and height of infested terminal compared with height of the longest lateral shoot.

and 50:50 ratios of (*Z*)- and (*E*)-9-dodecenyl acetates (6). Live female moths were difficult to obtain, but gas chromatographic analysis of a small number indicated the *Z* and *E* isomers of 9-dodecenyl acetate were present in solvent rinses of female abdomens at a ratio of about 3:1 (6). Traps baited with synthetic lures containing 75% *Z* and 25% *E* isomers of 9-dodecenyl acetate and emitting these chemicals at 28 ng/hr (at 23°C) captured nine times more males than traps baited with live females (6). The attractiveness to males of synthetic blends, particularly those of 70:30 and 80:20 ratios of (*Z*)- and (*E*)-9-dodecenyl acetates, plus the evidence that these compounds are produced by female moths, led to the qualified conclusion that they constitute the insect's sex pheromone (6).

2. *Field Testing of Mating Disruption*

Early objectives in developing the mating disruption approach to control were to evaluate dosage effects, ground and aerial application

Infested

techniques, and available controlled-release formulations. All field testing was done on isolated or semi-isolated plots to minimize the possibility of fertile females flying into treated plots from nearby untreated areas because egg laying by such females would confound assessment of mating-disruption efficacy.

a. Aerial application of hollow-fiber formulation: In 1978, an aerial test was conducted in southern Oregon using the hollow-fiber formulation that is currently marketed by Scentry Inc. Three treated plots of 4 to 9 ha were isolated on three sides by cleared brushfields. Each plot received 15 g/ha of the synthetic sex attractant. By various evaluation criteria, the treatment and formulation were successful. During the 80-day WPSB flight period, 75% to 80% of the chemical was discharged from the fibers (7). Only one male moth was captured in evaluation traps in the treated plots, whereas 234 moths were captured in the untreated plots. This difference suggests nearly total disruption of male-to-female orientation. Evaluation showed a 67% decrease in damage to terminal shoots in the treated plots. Some of the damage that did occur could have resulted from mated females flying into the plot from untreated locations or by a few females that could have emerged and mated before treatment (7). We cannot discount the possibility that some or all of the damage was caused by

WPSB mating and oviposition that occurred within the treated plots in spite of the mating disruption treatments.

The results of this first aerial test were particularly important in showing that damage caused by WPSB in pine plantations could be substantially reduced by aerial applications of synthetic sex attractant. Also, the test was noteworthy because it was the first demonstration in forestry of controlling pest damage by mating disruption with a commercially available treatment (7).

b. *Dose—response evaluation*: A second formulation for aerial treatment was applied in 1980 at three different rates of synthetic disruptant chemical to plots of young ponderosa pines in northern California (8). The formulation was a laminated plastic flake available from the Hercon Division of the Health Chem Corp. This material was applied with an acrylic sticker that held the flakes to the foliage. Blocks of 30- to 35-ha plots were treated with 0, 0.2, 2, and 20 g/ha of synthetic pheromone. Each treatment was replicated three times. Results of this test clearly demonstrated that increased dosage of the disruption chemical reduced the numbers of infested terminal shoots (Fig. 2). Dosages of 2 g/ha or less were insufficient for practical

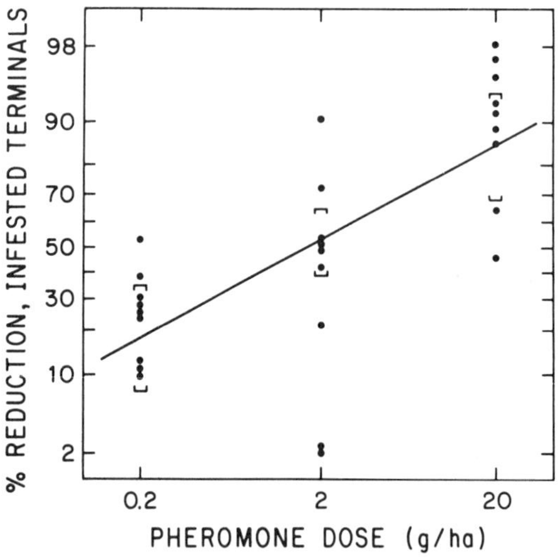

FIGURE 2 Mating disruption dosage response of western pine shoot borer with aerial treatment of Hercon flake formulation (from Ref. 8).

use (8). The performance of the formulation was excellent, with some 80% of the active ingredient dispensed during an 8-week period that spanned the insect's seasonal flight (8).

c. Manual application of point-source releasers: Small plots of 15-year-old, 2- to 4-m tall ponderosa pines were treated with 3- × 50-mm strips of polyvinyl chloride that contained 10% pheromone formulated at a 4:1 mixture of *Z* and *E* isomers of 9-dodecenyl acetate. The strips were wired to trees in 5 × 5- and 10 × 10-m grids, with each of these treatments replicated three times on isolated or semi-isolated plots that varied from 0.25 to 5 ha. The 400 releasers per hectare, spaced at 5-m intervals, had a maximum delivery of 14 g/ha, and the 100 releasers per hectare, at 10-m intervals, contained 3.5 g/ha. Actual release was determined by gas chromatographic analysis of check releasers, and showed 10.2 g/ha for the 5-m spacing and 2.5 g/ha for the 10-m spacing over the 10-week period. Comparisons of formulated strips that remained in the laboratory with field samples showed a loss of pheromone from field samples that could not be accounted for by release of pheromone. Degradation caused by exposure to sunlight was suggested as the reason for this additional loss (9).

Both spacing treatments were highly effective in disrupting male moth orientation to females, as measured by male response to monitoring traps baited with synthetic attractant. No moths were caught by traps in treated plots, but 32 moths per trap were captured in untreated check plots (9). Damage to terminal shoots was also markedly reduced by both treatments (Table 1). Reductions of 83% to 84% fewer injuries to terminal shoots were recorded for the treated areas compared with the untreated checks (9). This experiment was significant in demonstrating that manual application of point-source releasers could be used successfully to reduce damage. The experiment also provided dose—response information for successful use of mating disruption for this species.

Subsequent experiments with wider spacings of manually applied disruptant releasers indicated considerable flexibility in this approach to application (10—12). On small, 4-ha plots, for example, Daterman et al. (10) found no significant differences in treatment effects, as measured by captures in traps baited with synthetic pheromone, when a standard dose of 3.5 g/ha synthetic attractant was distributed as a mating disruptant on grid intervals ranging from 10 to 100 m (Table 2). At these same grid-interval treatments, the numbers of releasers per hectare varied from 1 (100-m grid) to 100 (10-m grid). The dosage per hectare was held constant by varying the quantity of disruptant releaser at each releaser station. Although the number of moths captured at the widest spacing was slightly higher than at the other treatments, the difference was not statistically significant (10).

TABLE 1 Effect of Mating Disruption on Damage by Western Pine Shoot Borer in Pine Plantations, Bly, Oregon, 1978

Treatment (g/ha)	Damaged terminals/ha 1977	Damaged terminals/ha 1978	Control (%)
0	154	199	-
3.5	150	31	83
14.0	159	32	84.2

Source: Ref. 9.

3. *Commercial Development*

Because of the success of initial mating-disruption trials, interest in product registration and commercial development was high in 1980 and 1981. Consequently, plans for field tests began to be shaped by practical considerations such as treatment costs, frequency of treatments needed, techniques for large-plot applications, and requirements for regulatory agency registration.

TABLE 2 Effect of Mating Disruption with Widely Space Releasers on Capture of Western Pine Shoot Borer in Baited Traps

Spacing (m)	Treatment releasers No./ha	Treatment releasers Size (cm)	Moths captured per trap[a]
Check	0	-	28.9 a
100	1	500	3.2 b
50	4	125	1.8 b
33	9	57	1.9 b
20	25	20	2.4 b
10	100	5	1.6 b

[a]Means followed by the same letter are not significantly different at the $p = 0.05$ level.
Source: From Ref. 10.

a. Large-plot treatments with hollow fibers: The primary objective of this test was to simulate operational use of the mating disruption technique on a large area at a dosage of 10 g/ha (4 g/acre). Three plots ranging in size from 200 to 250 ha were treated with synthetic pheromone formulated in hollow fibers (8). Plots were in pine plantations 3 to 6 m in height in southern Oregon near Bly, Keno, and Chiloquin. Untreated plantations, at least 200 m from the treated plots, were used as checks. Results of this trial were excellent (Table 3), with an average damage reduction of 76% (range 68%–81%). Monitoring traps placed inside treated areas captured no male moths, whereas averages of 70 to 131 moths per trap were captured in the untreated check plots, a further indication of treatment effectiveness. The results of this large-scale treatment were consistent with the outcome of prior aerial applications; they indicated that dosages of 10 to 20 g/ha yielded practical control.

Performance of the controlled-release formulation was effective over the 10-week period required for the three southern Oregon locations (8). The pheromone remaining in the hollow fibers at the end of the flight season was about 20% of the amount formulated, which corresponds very closely to data derived from monitoring the pheromone expended from the Hercon laminated flake releasers (Fig. 3).

b. Large-plot ground applications: In 1980 and 1981, plots of up to 20 ha were treated manually at grid intervals of 20, 25, and 50 m with two controlled-release formulations. A total of 225 ha were treated by hand application in three areas in central Oregon in 1980 and 1981 (12).

TABLE 3 Terminal Pine Shoots Infested with WPSB After Treatment of Large Plots with 10 g/ha Pheromone in Hollow Fibers

Location (Oregon)	Untreated 1978	Untreated 1979	Treated 1978	Treated 1979	Control (%)
Keno	45	70	36	11	80.8
Chiloquin	29	63	26	12	78.4
Bly	28	38	33	14	68.1
X ± SD	34.2 ± 9.6	57.0 ± 16.9	31.5 ± 5.0	12.3 ± 1.9	75.8 ± 6.8

Source: From Ref. 8.

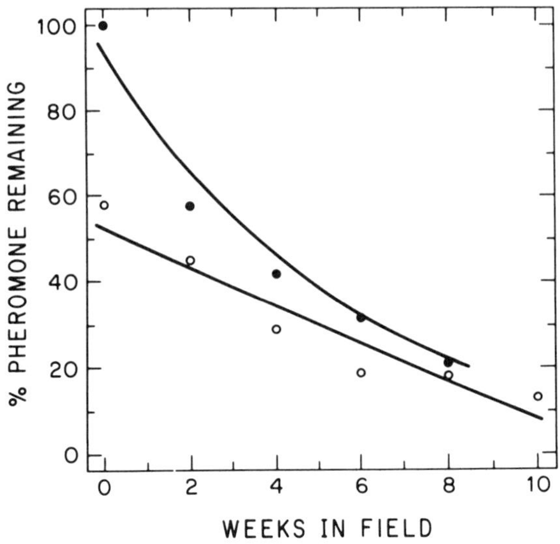

FIGURE 3 Patterns of pheromone release from Hercon flakes (closed circles), and Conrel hollow fibers (open circles) (from Ref. 11).

Treatment formulations consisted of the standard polyvinyl chloride (PVC) containing 10% pheromone by weight, or the commercially available Hercon Luretape, which is a laminated plastic strip (HLT) manufactured by the Hercon Division of Health Chem Corporation. Pheromone dosage was low at 3.5 to 5 g/ha in the interest of minimizing costs. Records were kept to compare times necessary to treat at different grid spacings.

There was a significant reduction in damage to terminals for all treated plots, although the 45% to 62% levels of reduction were generally less than had been realized in most of the earlier tests on smaller plots. One HLT application at 25-m spacing and 3.5-g/ha dosage did average 73% damage reduction. Possibly these results are as good as can be expected at these spacings at such low dosages. Also, at least part of the reduced success may have been caused by photodegradation of the pheromone formulated in PVC (12). For operational applications, we concluded that PVC should not be used until protection against photodegradation is provided. For use of HLT releasers, we would not recommend treatments of less than 5 g/ha nor grid-releaser intervals of more than 25-m. Experience from these and earlier tests (9,12) showed that, at 10-,

25-, and 50-m spacings, two applicators can treat 3, 8, and 16 ha/hr, respectively.

c. *Need for retreatment*: Pine plantations require many years to reach cutting age, and the question arises of how often treatment would be needed for adequate protection. Because WPSB populations are characteristically low, it was postulated that treatment effects might carry over into future years (13). A series of plots treated by mating disruption in southern Oregon and northern California were monitored for 3 years after treatment to determine population recovery rates. The results suggested that, for sustained protection, treatments will have to be repeated at 2- to 3-year intervals for most locations (13). Recovery rates probably depend on the degree of isolation from surrounding infested pine stands and how much infestation remains after initial treatment.

Figure 4 illustrates WPSB recovery in a location that had low infestation densities of about 7% to 18% before treatment in 1978. After

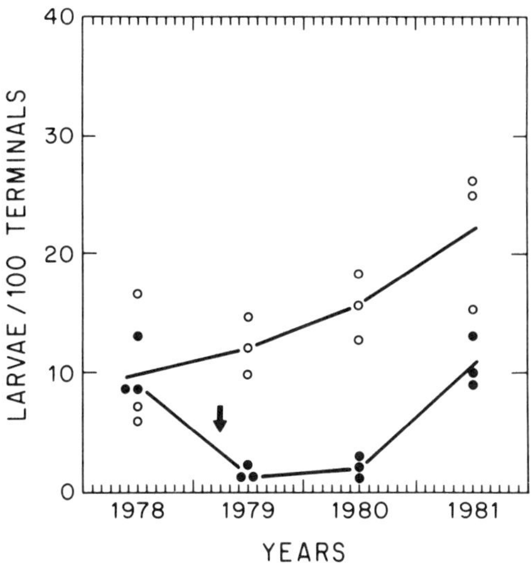

FIGURE 4 Western pine shoot borer damage before (1978) and after (1979–1981) a mating disruption treatment with 20 g/ha pheromone in Hercon flakes. (Arrow indicates time of pheromone application, solid dots indicate treated plots and clear dots represent untreated check plots) (from Ref. 13).

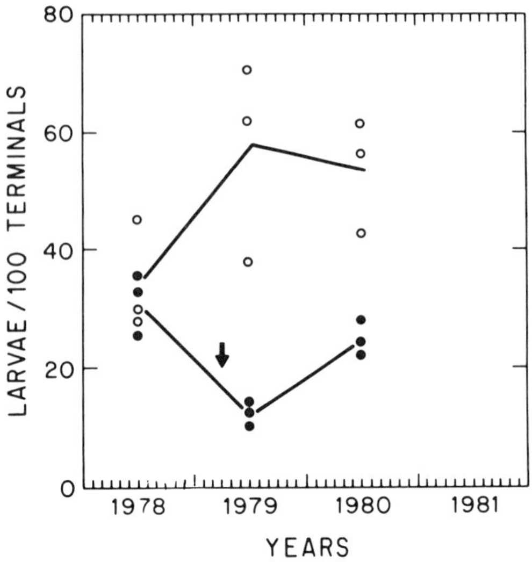

FIGURE 5 Western pine shoot borer damage at higher insect densities before (1978) and after (1979–1980) a mating-disruption treatment with 20 g/ha pheromone in Hercon flakes. (Arrow indicates time of pheromone application; solid dots indicate treated plots and clear dots represent untreated check plots) (from Ref. 13).

treatment, populations remained very low for 2 years, but in the third year (1981) they had recovered to about 10% of available shoots. Although still numerically lower than infestation rates in nearby check plots, the infestations were no longer statistically different from the untreated check areas. In such locations, retreatment could have been recommended for 1982, or possibly 1983. Figure 5 illustrates recovery in a location that had higher pretreatment populations of 25% to 45%. In 1979, larval populations in all treated plots had uniformly decreased to about 15%, but by the following year they were above 20% and would likely have required retreatment in 1981. Any mated females flying into these areas from surrounding untreated pine stands would clearly influence recovery. The authors also make the point that, for sites of unusually high value, annual treatment by mating disruption may be desirable (13).

A persistent, recurring pest such as WPSB can be expected to impose its 12% annual loss in height growth through much of the life of a pine plantation. If the plantation is on an 80-year rotation cycle, for example, and its trees suffer WPSB damage for 40 to 50 years of that time, the loss in height could be equivalent to a 16-ft sawlog. This estimate is based on a loss of 10 cm of height growth annually. It points out a need for more precise evaluation of damage caused by WPSB and the frequency of mating disruption treatments that would be cost-effective for particular areas.

d. Releaser formulation and probable mode of action: The success of mating disruption by all aerial and ground formulations showed considerable flexibility in what is required in a suitable mating disruption dispenser. An experiment with evaluation traps that have a range of bait strengths, however, indicated some limitations on dispenser designs. Traps with PVC-formulated lures, containing pheromone concentrations ranging from 0.001% to 10.0%, were placed in mating disruption treatment areas and untreated check plots (14). Figure 6 illustrates how the increasing lure strength captured increasing numbers of males in the untreated check areas, and how the lower strength lures were disrupted by the treatment and resulted in no captures in the treated locations. The higher strength lures, however, did capture moths in the treated plots, probably because the 1% and 10% lures emitted pheromone signals that rivaled

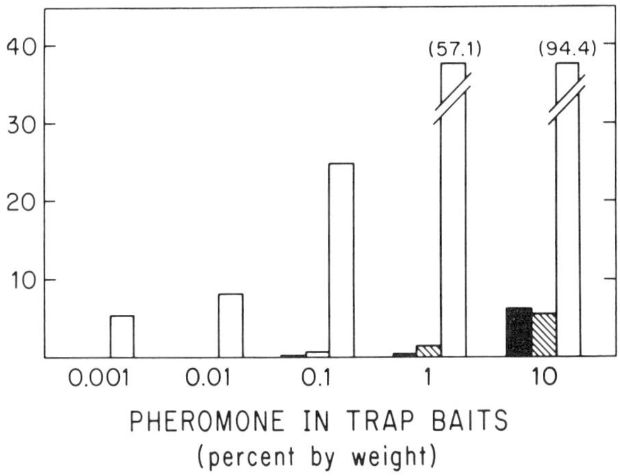

FIGURE 6 Response of WPSB males to traps baited with different strength lures, and placed in untreated check plots (open area) and plots treated with 10 g/ha pheromone (hatched area) or 5 g/ha pheromone (solid area) (from Ref. 14).

the release rates of pheromone from the disruption releasers. The 10% pheromone lure captured as many moths in the treated plots as did the 0.001% and 0.01% lures in the untreated check plots (14).

These results support the hypothesis that mating disruption results from a false–trail-following mode of action. Even in the center of areas treated with disruption releasers, the trap lure signal was strong enough to be detected over the disruption-releaser plumes. The omnipresence of pheromone odor *per se* may not be adequate to achieve mating disruption. Rather, the key factor seems to be the relative strength of the odor signals produced by female moths (simulated by trap lures in this example) versus the strength of the disruption releasers. Consequently, disruption releasers should be designed with substantially stronger release characteristics than the release rates of female moths, and this relative potency must be maintained throughout the insects' mating season. In other words, a dispensing system using very high numbers of releaser particles may be less effective if individual particle emissions are weak relative to the natural pheromone emissions of female moths.

e. Precision of pheromone blend: Beginning in 1983, mating disruption treatments in high-value genetic improvement plantations of ponderosa pine have taken place in Idaho and Montana. The treatment objective was to reduce injury caused by WPSB, thus clarifying genetic evaluations for growth characteristics and heritability. In 1984, an inexpensive blend of the *Z* and *E* isomers of 9-dodecenyl acetate was compared with the standard 4:1 mixture of *Z*/*E*. The Phillips Petroleum Co. produced this blend by a low-cost bulk synthesis procedure, but the product has a *Z*/*E* ratio of 2:3 rather than the 4:1 standard. In a replicated comparison of both blends at 18 g/ha dispensed from HLT releasers, no significant differences in treatment efficacy were found (J. E. Dewey et al., 1986, unpublished report). The degree of success as measured by reduced terminal shoot damage was in the 60% to 65% range and would likely have been higher except for the proximity of some treated plots to surrounding untreated, infested stands.

The success of the inexpensive, off-blend mating disruption product was somewhat surprising. Recent studies (15,16) have demonstrated that optimal responses to synthetic lepidopteran pheromones occur with precise synthetic blends of the same components released by the insects. This would suggest that the synthetic blend used for mating disruption should match the blend emitted by the female moths. However, the objective of mating disruption is to block or suppress male-to-female attraction, rather than to induce or optimize it. Consequently, an adequate amount of the principal components of the sex pheromone, even if they are present in inappropriate ratios, may be adequate to disrupt male-to-female orientation. In the case of WPSB, there are two additional factors to consider in this context. First, the male moths respond to a relatively wide range of *Z/E* isomer blends (6), with captures by lures containing 1:1 *Z/E* ratios often approaching those of the standard 4:1 ratio. Secondly, the imprecise blend has been used only at the relatively high dosage of 18 g/ha, whereas the precise blend formulations have been successful at 3.5 to 20 g/ha.

f. Importance of plot isolation: From the outset, efforts to develop the mating disruption method for control of WPSB have aimed at isolated or semi-isolated plots to minimize the possibility of fertile females flying into the treated areas from adjacent untreated locations. This practice was based on the assumption that female immigration could pose problems of treatment assessment and reduction in efficacy. Four years of mating disruption treatments in genetic improvement plantations in Idaho and Montana showed that immigration of females from outside treatment boundaries could reduce efficacy. The plantations varied from almost total isolation to being generally surrounded by other pine stands. Treatment evaluations revealed the highest efficacy occurred in the more isolated plantations, and one plantation in close proximity to surrounding infested areas was eventually dropped from further treatment plans because of the reinvasion problem (Dewey et al., 1986, unpublished report).

The significance of the observations on plot isolation is clear. If areas to be treated are surrounded by infested areas, successful crop protection is questionable. In such circumstances, treatment of buffer zones around the periphery of treated plots may be helpful (17). Most effective, certainly, would be areawide treatments that would encompass large units of land, including that supporting plantings of marginal value. This approach would increase costs but would assure a greater degree of protection to the high-value plantings.

g. Commercial registration: Because of the high interest in use of mating disruption for control of WPSB in 1980 and 1981, plus the general success of field trials (Table 4), two companies pursued registration with the Environmental Protection Agency (EPA) for commercial use of their pheromone formulations. The companies were Health Chem Corporation, New York, New York with their Hercon laminated plastic flake formulation, and Albany International Corp., for the hollow-fiber formulation (now marketed by Scentry Inc., Buckeye, Arizona). The U. S. Forest Service's PNW Research Station provided both companies with laboratory toxicity data on the pheromone blend. These basic toxicity tests included oral LD_{50} for rats, dermal LD_{50} for rabbits, acute inhalation study (rats), eye irritation test

TABLE 4 Efficacy of WPSB Mating Disruption Treatments with Various Formulations at Different Dosages

Formulation and application	Area treated (ha)	Dose (g/ha)	Damage reduction (%)
PVC strips (10 × 10 m)	8	3.5	83
Hollow fibers (aerial)	20	15	67
Hollow fibers (aerial)	700	10	76
Hollow fibers (aerial)	634	5	75
Hercon flakes (aerial)	100	20	88
HLT (25 × 25 m)	48	3.5	73
HLT (10 × 10 m)	300	18	39–85

(rabbits), and primary skin irritation (rabbits). Although the results of these tests were insufficient in themselves to secure registration, they likely provided assistance. *Such assistance may be necessary to encourage the commercialization of pheromone products for pest species that are bound to be considered as minor marketing opportunities.* In 1982, Health Chem Corporation was successful in registering their Hercon laminated plastic flake as the product "Disrupt," and Albany International registered their hollow-fiber formulation as "NoMate Shootgard." Both products were called western pine shoot borer "suppressants" on their registration labels.

4. Current Use and Future Prospects

The research-and-development phase of developing WPSB pheromone into an effective control agent moved quickly and successfully to completion. Following commercial registration, however, interest has lagged. Only 125 ha/year have been treated to suppress damage on genetic-improvement plantations, plus a few hundred hectares per year of Weyerhaeuser Co. plantations in southern Oregon. At best this use is minor in comparison with the potential on the hundreds of thousands of hectares of ponderosa pine plantations in the West. Lack of interest in operational treatment apparently has two causes. First, the economic recession that began in the United States in 1981, at about the same time the mating disruption formulations were registered, led to a general reduction of expenditures on forest lands. Commercial use of WPSB suppression formulations was simply one of the first casualties of the faltering economy.

Second, the damage caused by this insect is poorly understood. Most infested trees remain green, and effects of feeding are subtle. Sower and Shorb (4) emphasized the difficulty of accurately assessing injury because of the insects' preference for infesting the larger, faster-growing buds and shoots. Sower and Mitchell (18) point out that cursory evaluations of infested plantations can lead to the conclusion that WPSB causes little damage. Western pine shoot borer appears to be an "equalizer" of growth characteristics because its preference for the faster-growing shoots can obscure differences between faster- and slower-growing trees. Thus, an infested, fast-growing tree may lose a substantial percentage of its potential growth to WPSB injury but still measure the same, or even greater, growth for the year of infestation than an adjacent uninfested but slow-growing tree (4).

The key to future use of the mating-disruption technique against WPSB is increased awareness of the injury caused by this pest. Studies by Sower and Shorb (4) and Sower and Mitchell (18) have helped clarify this issue, and ongoing research (T. W. Koerber and C. Williams, personal communication) is expected to further clarify the impact question and raise awareness for the additional growth potential of pine plantations infested by WPSB. Cost of treatment will remain an issue, but if frequency of treatment can be reduced to once every 2 to 3 years (13), it would encourage use of WPSB control for pine plantation management. Reduced pheromone and formulation costs would also encourage treatment.

Future prospects for increased use of mating-disruption treatments are good. With increased awareness of the damage actually caused by the inconspicuous injuries from WPSB feeding, forest land managers will likely be motivated to suppress the insect's activity to retain the growth potential that otherwise would be lost.

III. THE DOUGLAS-FIR BEETLE

The Douglas-fir beetle, *Dendroctonus pseudotsugae*, is a major pest of Douglas fir, *Pseudotsuga menziesii* (Mirb.) Franco, throughout the tree's natural range in western North America. Logging debris and trees felled by storms are prime breeding material, and infestations of large quantities of such host material can lead to population buildups of this insect. The resulting outbreak populations of beetles can then attack and kill living Douglas fir trees, sometimes leading to catastrophic losses. Removal of susceptible host trees is the most effective means of preventing population buildups, but inaccessibility and other considerations sometimes prevent such action. Other means of preventing population buildup may then be needed (19), such as the use of the antiaggregative pheromone, 3-methyl-2-cyclohexen-1-one (MCH) to prevent colonization of host trees (20).

A. Pheromone-Mediated Behavior

Like many bark beetles, the Douglas-fir beetle has an aggregation pheromone that is emitted by the first female beetle to enter a suitable tree or log. The ensuing plume of aggregation pheromone draws other beetles to the host material, thus causing more infestations and pheromone production. In this way, the log or tree is quickly colonized. As the tree is colonized by more beetles, an antiaggregant (MCH) is produced. The antiaggregant serves to terminate attraction by the aggregation pheromone and, thereby, prevents overcolonization and competition for food and space by the beetle progeny (20).

B. MCH Field Tests

Early field tests with adhesive traps baited with lures of synthetic aggregation pheromones and female-infested log sections demonstrated that attraction of Douglas-fir beetles to such baits could be nullified by the addition of the antiaggregation pheromone (MCH; 21). In large-scale field experiments, colonization of entire trees was prevented by point-source releasers of MCH (19). A granular controlled-release formulation of MCH was then developed and applied by aircraft to plots as large as 36 ha (20). The result of this test was excellent, with a composite decrease in attacks of 96% on treated plots (20).

C. Future Prospects for MCH Treatment

The U.S. Forest Service is currently working with the Environmental Protection Agency to register MCH for operational treatment of Douglas-fir beetle to prevent population buildups. This registration will likely be well received because benefit/cost estimates associated with outbreaks of this species and their prevention by MCH have been determined to be as high as 25:1 (M. D. McGregor, personal communication).

This high ratio is primarily because of the major damage a bark beetle outbreak can cause, once the population buildup has occurred.

IV. DOUGLAS-FIR TUSSOCK MOTH

Populations of the Douglas-fir tussock moth, *Orgyia pseudotsugata*, can undergo rapid changes in population densities and within a few years increase from low, innocuous numbers to an outbreak capable of causing severe defoliation (22). The primary hosts of this western defoliator are Douglas fir, grand fir, *Abies grandis* (Dougl.) Lindl., and white fir, *A. concolor* (Gord. and Glend.) Lindl. Because of the insect's abrupt population changes and the vast areas of susceptible forests, a monitoring technique to warn against impending outbreaks is needed.

A. Monitoring with Pheromone-Baited Traps

A monitoring system was devised that uses inexpensive triangular-shaped, adhesive-lined traps constructed from paper milk cartons (23). These traps are baited with low-strength lures formulated at 0.001% of the Douglas-fir tussock moth pheromone in 3 × 5-mm pellets of PVC. The traps are placed in groups of five in plots in potentially susceptible forests throughout five western states. Traps are placed in the field before the onset of seasonal flight and retrieved after flight has ceased. The lure strength is made deliberately weak to avoid problems with trap saturation that commonly occur in adhesive traps with a limited capture surface.

Figure 7 illustrates the approach to monitoring tussock moth populations with this trap and lure design. In this diagram of a population continuum for tussock moth, the monitoring target is the area between the "high-" and "suboutbreak-" population densities. The weak pheromone lure is calibrated to capture few moths at populations below the high stage because the insect poses no management problems at the lower-population densities. When trap captures indicate populations in the high or suboutbreak ranges, 1 or more years of intensive monitoring in that area can be conducted before an outbreak occurs (23,24). As populations increase into the high, suboutbreak, and outbreak phases, the traps become saturated with trapped moths. Before or at this point, however, managers are alerted that a tussock moth population buildup has occurred.

During a 1-week trapping period in a location with a moderate population of tussock moths, traps baited with the standard weak-strength monitoring lure (0.001% pheromone in lure) captured an average of 0.6 males per trap, whereas traps baited with lures approximating the attractive strength of a female (0.1% pheromone in

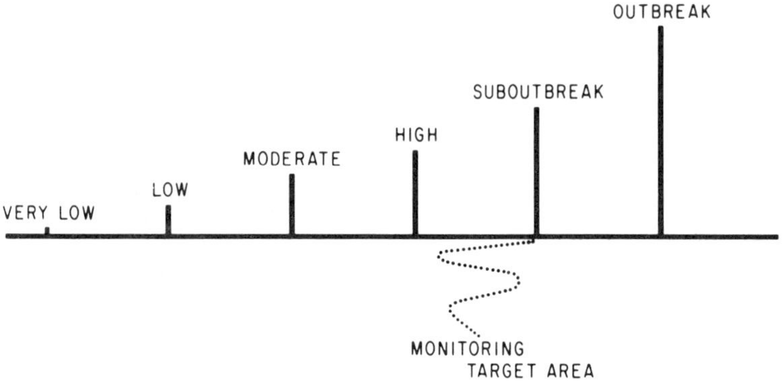

FIGURE 7 Douglas-fir tussock moth population continuum indicating population target for monitoring. Low-strength lure is calibrated to avoid or minimize captures at lower populations.

lure) captured an average of 37 males per trap (G. Daterman, unpublished). Low to moderate populations of tussock moth characteristically show no evidence of defoliation. At such population densities, visible activity or evidence of any tussock moth life stage is extremely rare, other than by luring adult males to pheromone-baited traps. In such locations, the monitoring traps would capture few moths, which would signal managers that no outbreak is impending for that general area. Conversely, average trap captures exceeding 25 moths signal managers to pay closer attention to those areas for possible surges in tussock moth activity. The trapping system is a first-stage monitoring method, which serves to alert managers to the comparatively few locations in which populations may be building up to potential outbreaks. This triggers other, more intensive monitoring for egg mass and larval densities in those areas.

This monitoring system has been operational in five western states for several years. It has successfully detected several population surges that resulted in outbreaks or near-outbreaks. Table 5 summarizes trapping results for several years for the operational application of this method. It appears that the trigger threshold of 25 moths per trap is probably set too low. Too many areas showed potential high or suboutbreak populations, but did not develop visible defoliation. Hundreds of other locations were correctly categorized as low-population areas, however, which permitted managers to focus attention on the comparatively few locations with potential buildups.

A similar approach for monitoring tussock moth has been applied in British Columbia. The primary differences in application are the

TABLE 5 Douglas-Fir Tussock Moth Monitoring with Pheromone-Baited Traps in Five Western States (1982–1985)

Year	Total plots	No. plots with <25 moths/trap	No. suboutbreak or outbreak areas[a]
1982	908	50	4[b]
1983	887	45	0
1984	727	40	7
1985	728	68	11[b]

[a]Suboutbreaks designated by finds of fresh egg masses in 30-min timed searches; outbreaks designated by evidence of visible defoliation.
[b]Outbreaks occurred in one or more areas.

use of a larger series of traps per plot and the incorporation of a series of bait strengths (25). The objective is still to provide early warning against population buildup but also to provide additional information on low populations, which is accomplished by the use of stronger lures in combination with traps containing weak lures. Shepherd et al. (25) concluded that traps baited with 0.01% lures provide the most consistent captures. They reported poor correlations of trap captures with egg mass or larval densities in the next generation. Shepherd et al. (25) also reported that 25 moths per trap provided a useful threshold for indicating a possibility of defoliation in those stands. These authors point out the importance of following numerical trends in trap captures over a series of years.

V. OTHER CONIFER PESTS

Considerable field testing to develop pheromones of other coniferous pests has taken place in recent years. Some of these pheromones are in operational use in the United States, and others are nearing that degree of development. Some outstanding examples include monitoring of coneworms of southern pines, *Dioryctria* spp., for timing of control treatments (26,27); suppression of damage caused by mountain pine beetle, *Dendroctonus ponderosae* Hopkins, and striped ambrosia beetle, *Trypodendron lineatum* (Olivier) (see Chap 19);

detection-trapping for new infestations of the European pine shoot moth, *Rhyacionia buoliana* (Schiff.) (5,28,29); monitoring the Nantucket pine tip moth, *R. frustrana* (Comstock), for optimal timing of chemical treatments (30–32); and detection and monitoring of the cranberry girdler, *Chrysoteuchia topiaria* (Zeller), for pest management in Douglas fir nurseries (33). Table 6 includes additional conifer pests for which pest management use of pheromones is operational or nearly operational.

VI. CONCLUSIONS

Several examples of pheromones are in operational or near-operational use for management of forest insect pests in the United States. The case history of western pine shoot borer pheromone development, in particular, emphasizes some experiences and circumstances that may be useful to the development of other insect semiochemicals.

TABLE 6 Coniferous Forest Pests in the United States for Which Use of Pheromones in Management Is Operational or Near-Operational

Pest species	Application
Douglas-fir bark beetle	Suppression
Spruce beetle	Suppression
Mountain pine beetle	Suppression
Western pine shoot borer	Suppression
Ambrosia beetles	Suppression
Southern pine beetle	Monitoring
Douglas-fir tussock moth	Monitoring/suppression
European pine shoot moth	Monitoring
Nantucket pine tip moth	Monitoring
Cranberry girdler (nursery pest)	Monitoring
Coneworms in southern pines	Monitoring
Spruce budworms	Monitoring

The WPSB pheromone program progressed with unusual dispatch. It began in 1977 with efforts to identify the chemical structure of the pheromone, proceeded through several phases of field testing of mating disruption, and culminated in 1982 with EPA registration of two commercial formulations for implementing mating disruption treatments. Available commercial treatments now include two formulations for aerial application, and one that is suitable for hand application by ground crews. Both aerial and ground techniques have consistently yielded damage reductions of about 75%. Effective dosages have ranged from 5 to 20 g of pheromone per hectare. Effective treatments were maintained even with use of a less-expensive synthetic blend of pheromone that did not precisely match the insect's natural blend. We have found it is imperative that treatment areas either be isolated from other infested pines, or large enough to minimize the effects of mated females immigrating from untreated areas. Need for retreatment may be as infrequent as every second or third year.

Many factors contributed to the success of WPSB pheromone development. The low-population density and other aspects of this insect's biology and behavior make it particularly vulnerable to a pheromone-based system of control. More important, perhaps, to other programs was a committed partnership of researchers, developers, and users formed to carry out the research and development effort. The "partnership" included both funding and "hands-on" participation by developers and user-groups at an early stage in the research and development process. Likewise, the researchers remained active in the program through the large-scale field testing and EPA registration of development products. Such cooperative efforts contributed greatly to the success of the program, including transfer of technology to operational use. Similar cooperation contributed to the success of developing pheromone-baited traps as an operational system for monitoring Douglas-fir tussock moth populations, and it likely represents a critical principle for the success of most research and development programs.

Development of WPSB pheromone was further facilitated by use of public funds to pay a substantial share of the cost of toxicity testing required for EPA registration. Product registration with the EPA is a requirement for any material classified as a pesticide, and this includes semiochemicals, such as pheromones, when they are used in broadcast treatments. Unique to pheromones, however, is their target specificity for a few or even one pest species per pheromone compound or blend of compounds. This is a significant advantage in terms of environmental safety and acceptance. Economically, however, it greatly limits the profit incentive for a company to support the development and registration costs for a particular pheromone, because the resulting product can be marketed for only one or a few pests. Thus, to expedite development, and particularly for those classified as pheromones of "minor" pests, it may be necessary for public agencies to underwrite the costs of toxicity tests required for registration. The test results would, of course, become a matter of public record and be available to any individual or company for purposes of registering a related product.

Increased use of public agency funds to provide toxicity testing of pheromones, together with a streamlining of EPA's registration procedures (and experimental-use permit policies and processing), would stimulate further pheromone development. The end result would be to accelerate development of more semiochemical tools useful in pest management.

ACKNOWLEDGMENTS

I wish to thank C. W. Berisford, J. H. Borden, S. Burke, G. L. DeBarr, J. E. Dewey, B. S. Lindgren, M. D. McGregor, C. G. Niwa, D. L. Overhulser, C. Sartwell, T. D. Schowalter, P. J. Shea, L. L. Sower, and J. M. Wenz for their advice, information, or review of this manuscript.

REFERENCES

1. Sartwell, C., Daterman, G. E., Koerber, T. W., Stevens, R. E., and Sower, L. L. Distribution and hosts of *Eucosma sonomana* in the western United States as determined by trapping with synthetic sex attractants. *Ann. Entomol. Soc. Am. 73*:254–256 (1980).
2. Stoszek, K. J. Damage to ponderosa pine plantations by the western pine-shoot borer. *J. For. 71*:701–705 (1973).
3. Stevens, R. E. and Jennings, D. T. Western pine-shoot borer: A threat to intensive management of ponderosa pine in the Rocky Mountain Area and Southwest. USDA For. Serv. Gen. Tech. Rep. RM-45, 1977, 8 p.
4. Sower, L. L. and Shorb, M. D. Effect of western pine shoot borer (Lepidoptera: Olethreutidae) on vertical growth of ponderosa pine. *J. Econ. Entomol. 77*:932–935 (1984).
5. Daterman, G. E. Synthetic sex pheromone for detection survey of European pine shoot moth. USDA For. Serv. Res. Pap. PNW-180, 1974, 12 p.
6. Sower, L. L., Daterman, G. E., Sartwell, C., and Cory, H. T. Attractants for the western pine shoot borer, *Eucosma sonomana*,

and *Rhyacionia zozana* determined by field screening. *Environ. Entomol. 8*:265–267 (1979).
7. Overhulser, D. L., Daterman, G. E., Sower, L. L., Sartwell, C., and Koerber, T. W. Mating disruption with synthetic sex attractants controls damage by *Eucosma sonomana* (Lepidoptera: Tortricidae, Olethreutinae) in *Pinus ponderosa* plantations. II. Aerially applied hollow fiber formulation. *Can. Entomol. 112*: 163–165 (1980).
8. Sower, L. L., Overhulser, D. L., Daterman, G. E., Sartwell, C., Laws, D. E., and Koerber, T. W. Control of *Eucosma sonomana* by mating-disruption with synthetic sex attractant. *J. Econ. Entomol. 75*:315–318 (1982).
9. Sartwell, C., Daterman, G. E., Sower, L. L., Overhulser, D. L., and Koerber, T. W. Mating disruption with synthetic sex attractants controls damage by *Eucosma sonomana* (Lepidoptera: Tortricidae, Olethreutinae) in *Pinus ponderosa* plantations. I. Manually applied polyvinyl chloride formulation. *Can. Entomol. 112*:159–162 (1980).
10. Daterman, G. E., Sartwell, C., and Sower, L. L. Prospects for controlling forest Lepidoptera with controlled release pheromone formulations. In *Controlled Release of Bioactive Materials*. (Baker, R., ed.). Academic Press, New York, 1980, pp. 213–226.
11. Daterman, G. E. Control of western pine shoot borer damage by mating disruption–A reality. In *Insect Suppression with Controlled Release Pheromone Systems*. (Kydonieus, A. and Beroza, M., eds.). CRC Press, Boca Ration, 1982, pp. 155–163.
12. Sartwell, C., Daterman, G. E., Overhulser, D. L., and Sower, L. L. Mating disruption of western pine shoot borer (Lepidoptera: Tortricidae) with widely spaced releasers of synthetic pheromone. *J. Econ. Entomol. 76*:1148–1151 (1983).
13. Sower, L. L. and Overhulser, D. L. Recovery of *Eucosma sonomana* (Lepidoptera: Tortricidae) populations after mating-disruption treatments. *J. Econ. Entomol. 79*:1645–1647 (1986).
14. Daterman, G. E., Sower, L. L., and Sartwell, C. Challenges in the use of pheromones for managing western forest Lepidoptera. In *Insect Pheromone Technology: Chemistry and Applications*. (Leonhardt, B. A. and Beroza, M., eds.). ACS Symposium Series No. 190, American Chemical Society, Washington, D.C., 1982, pp. 243–254.
15. Linn, C. E., Jr., Campbell, M. G., and Roelofs, W. L. Male moth sensitivity to multicomponent pheromones: Critical role of female-released blend in determining the functional role of components and active space of the pheromone. *J. Chem. Ecol. 12*:659–668 (1986).

16. Linn, C. E., Campbell, M. G., and Roelofs, W. L. Pheromone components and active spaces: What do moths smell and where do they smell it? *Science* *237*:650–652 (1987).
17. Niwa, C. G., Daterman, G. E., Sartwell, C., and Sower, L. L. Control of *Rhyacionia zozana* (Lepidoptera: Tortricidae) by mating-disruption with synthetic sex pheromone. *Environ. Entomol.* (in press) 1988.
18. Sower, L. L. and Mitchell, R. G. Host-tree selection by western pine shoot borer (Lepidoptera: Olethreutidae) in ponderosa pine plantations. *Environ. Entomol.* *16*:1145–1147 (1987).
19. Furniss, M. M., Daterman, G. E., Kline, L. N., McGregor, M. D., Trostle, G. C., Pettinger, L. F., and Rudinsky, J. A. Effectiveness of the Douglas-fir beetle antiaggregative pheromone methylcyclohexenone at three concentrations and spacings around felled host trees. *Can. Entomol.* *106*:381–392 (1974).
20. McGregor, M. D., Furniss, M. M., Oaks, R. D., Gibson, K. E., and Meyer, H. E. MCH pheromone for preventing Douglas-fir beetle infestation in windthrown trees. *J. For.* *82*:613–616 (1984).
21. Rudinsky, J. A., Furniss, M. M., Kline, L. N., and Schmitz, R. F. Attraction and repression of *Dendroctonus pseudotsugae* (Coleoptera: Scolytidae) by three synthetic pheromones in traps in Oregon and Idaho. *Can. Entomol.* *104*:815–822 (1972).
22. Wickman, B. E. Tree mortality and top kill related to defoliation by the Douglas-fir tussock moth in the Blue Mountains outbreak. USDA For. Serv. Res. Pap. PNW-233, 1978, 47 p.
23. Daterman, G. E., Livingston, R. L., Wenz, J. M., and Sower, L. L. How to use pheromone traps to determine outbreak potential. USDA Forest Service, Douglas-fir Tussock Moth Handbook, Agric. Handbk. No. 546, 1979, 11 p.
24. Daterman, G. E. Pheromone responses of forest Lepidoptera: Implications for dispersal and pest management. In *Proceedings of II International Union of Forest Research Org. (IUFRO) Symposium on Dispersal of Forest Insects: Evaluation, Theory, and Management Implications.* (Berryman, A. and Safranyik, L., eds.). Canadian Forestry Service, USDA Forest Service, CANUSA-West, and Washington State Univ., 1980, pp. 251–265.
25. Shepherd, R. F., Gray, T. G., Chorney, R. J., and Daterman, G. E. Pest management of Douglas-fir tussock moth, *Orgyia pseudotsugata* (Lepidoptera: Lymantriidae): Monitoring endemic populations with pheromone traps to detect incipient outbreaks. *Can. Entomol.* *117*:838–848 (1985).
26. DeBarr, G. L., Barber, L. R., Berisford, C. W., and Weatherby, J. C. Pheromone traps detect webbing coneworms in loblolly pine seed orchards. *South. J. Appl. For.* *6*:122–127 (1982).

27. Weatherby, J. C., DeBarr, G. L., and Barber, L. R. Monitoring coneworms with pheromone traps: A valuable pest detection procedure for use in southern pine seed orchards. In *Proceedings of the 18th Southern Forest Tree Improvement Conf.* (Schmidtling, R. C. and Griggs, M. M., eds.). South. For. Tree Improvement Comm., Long Beach, Miss., 1985, pp. 206–220.
28. Smith, R. G., Daterman, G. E., Daves, G. D., McMurtrey, K. D., and Roelofs, W. L. Sex pheromone of the European pine shoot moth: Chemical identification and field tests. *J. Insect Physiol.* *20*:661–668 (1974).
29. Gray, T. G., Slessor, K. N., Shepherd, R. F., Grant, G. G., and Manville, J. F. European pine shoot moth, *Rhyacionia buoliana* (Lepidoptera: Tortricidae): Identification of additional pheromone components resulting in an improved lure. *Can. Entomol.* *116*:1525–1532 (1984).
30. Hill, A. S., Berisford, C. W., Brady, U. E., and Roelofs, W. L. Nantucket pine tip moth, *Rhyacionia frustrana*: Identification of two sex pheromone components. *J. Chem. Ecol.* 7:517–528 (1981).
31. Gargiullo, P. M., Berisford, C. W., and Godbee, J. F. Prediction of optimal timing for chemical control of the Nantucket pine tip moth, *Rhyacionia frustrana* (Comstock) (Lepidoptera: Tortricidae), in the southeastern coastal plain. *J. Econ. Entomol.* *78*:148–154 (1985).
32. Gargiullo, P. M., Berisford, C. W., Canalos, C. G., Richmond, J. A., and Cade, S. C. Mathematical descriptions of *Rhyacionia frustrana* (Lepidoptera: Tortricidae) cumulative catches in pheromone traps, cumulative eggs hatching, and their use in timing of chemical control. *Envir. Entomol.* *13*:1681–1685 (1984).
33. Kamm, J. A., Morgan, P. D., Overhulser, D. L., McDonough, L. M., Triebwasser, M., and Kline, L. N. Management practices for cranberry girdler (Lepidoptera: Pyralidae) in Douglas-fir nursery stock. *J. Econ. Entomol.* *76*:923–926 (1983).

21

Practical Use of Insect Pheromones to Manage Coniferous Tree Pests in Eastern Canada

CHRIS J. SANDERS / Forestry Canada, Ontario Region, Sault Ste. Marie, Ontario, Canada

I. INTRODUCTION

The objective of this book is to focus attention on the potential or operational use of pheromones in insect pest management. Currently, among the pests of coniferous trees in eastern Canada, the only operational use of pheromones is in monitoring of the spruce budworm (*Choristoneura fumiferana*) populations with pheromone traps, and this will be the main subject of this paper.

However, it would be remiss not to mention other work that is progressing toward the practical stage, and that, we hope, will be operational within a few years. This is summarized in Table 1. Monitoring systems are under development for most major forest pests in eastern Canada, but some are closer to reality than others. Among systems for the coniferophagous pests, two of major concern, those for the jack-pine budworm (*C. pinus*) and the hemlock looper (*Lambdina fiscellaria*), are held up because of problems in the identification of the pheromones. Attractants are available for species of *Zeiraphera* and *Dioryctria*, and work is underway to calibrate trap-catch with population density, but further work on pheromone identification is required for some species in each complex. Among seed and cone pests, in addition to *Dioryctria*, research is focusing on the spruce seed moth, *Cydia strobilella*.

Pheromones are also being developed for regulating or controlling forest pests in eastern Canada. The white pine weevil (*Pissodes strobi*) is a potential candidate because it has many of the attributes

TABLE 1 Research Underway in Eastern Canada on the Sex Pheromones of Coniferous Tree Pests

Common name	Name	Stage of research	Researchers
Spruce budworm	*Choristoneura fumiferana*	Monitoring	ForCan/USDA and cooperators
		Disruption	UNB/ForCan
Jack pine budworm	*Choristoneura pinus*	Pheromone ID	ForCan/RPC
Hemlock looper	*Lambdina fiscellaria*	Pheromone ID	
Budmoths	*Zeiraphera* spp,	Monitoring Disruption	ForCan RPC/ForCan
Coneworms	*Dioryctria* spp.	Monitoring	ForCan
Seed Moth	*Cydia strobilella*	Detection/timing	ForCan
White pine weevil	*Pissodes strobi*	Pheromone ID	
Root collar weevil	*Hylobius congener*	Pheromone ID	RPC/ForCan

that make it appropriate for control by pheromones—it is a pest at low densities on a high-value crop, and the insect is cryptic, making control by insecticides difficult. Progress is again delayed by problems in identification of the pheromone. The *Zeiraphera* complex in white spruce plantations has many of the same attributes as the white pine weevil. The pheromone of one of the four species has been characterized, and disruption trials in small plots appear promising. Because the spruce budworm is such a major pest, all possible methods of control, including mating disruption by pheromones, are being considered. Trials began in 1975 and have continued to date (Table 2). Up until 1981, although the trials had all produced good disruption to traps, attempts to demonstrate population reduction were inconclusive. A confounding factor is the possibility of invasion by mated female moths from outside the treated areas. To overcome this, Seabrook and his associates have used large cages to exclude invading moths (6). These trials, beginning in 1983, have demonstrated up to 80% reduction in mating success.

TABLE 2 Field Trials of the Aerial Application of Sex Pheromones to Disrupt Spruce Budworm-Mating Behavior, 1975–1981

Year	Location	Formulation	AI/ha (g)	Area treated	Refs.
1975	Ontario	NCR microcapsules	7	12 ha	4
1977	Ontario	Conrel fibers	1.5	10 ha	2, 3
		Conrel fibers	9	250 ha	
		Conrel fibers	14	10 ha	
1978	New Brunswick/ Nova Scotia	Conrel fibers	0.1–25	8 plots × 10 ha	No report, but see 4
1980	Maine	Hercon flakes (+ Pherotec)	25	145 and 30 ha	No report, but see 5
		Hercon flakes (+ Pherotec)	250	9 ha	
1981	Ontario	Hercon flakes (+ Pherotec)	70	30 ha	4
1983	New Brunswick	Hercon flakes (+ Pherotec)	250	Cages in 2-, 3-ha plots	6
1984	New Brunswick	Hercon flakes (+ AR 1990)	250	Cages in 2-, 5-ha plots	
1985	New Brunswick	Hercon flakes (+ AR 1990)	250	5 ha	Seabrook and Kipp, Unpublished Report, ESA 1987 meeting
1986	New Brunswick	Hercon flakes (+ AR 1990)	250	5 ha	Seabrook and Kipp, Unpublished Report, ESA 1987 meeting
1987	New Brunswick	ICI microcapsules	100	16 ha	In progress

II. MONITORING SPRUCE BUDWORM POPULATIONS

A. History

Research on the use of sex pheromone traps to monitor spruce budworm populations began soon after the identification of the major pheromone component in 1971 (7).

At first, sticky traps were used (8), but problems with saturation (9) led researchers to explore the use of high-capacity, nonsaturating traps. This approach was begun by Ramaswamy and Cardé (10) but soon blossomed into a major area of research. Much of this work was funded by the Canada–United States joint spruce budworms research program (CANUSA), and was coordinated by D. C. Allen (SUNY, Syracuse). As a result of this work, plans were formulated for a trapping system to cover the range of the budworm throughout eastern North America. To ensure comparability among the different regions, a standardized method of trap deployment was chosen (11). The Multi-pher trap was selected (12), baited with a polyvinyl chloride lure (13). The release rate chosen was low, about 20 ng/ha (close to that of a virgin female moth), to keep the catches low. Traps are deployed in clusters of three in a triangle with 40 m between traps to minimize trap interference. Although catches are highest in the tree canopy, traps are hung for convenience at head height. They are deployed 1 or 2 weeks before moth flight and are left untouched until the flight is over, some 6 weeks later.

The traps are deployed by numerous different agencies (Table 3). The results are being used for local interpretation of the budworm situation, but they are also being collated to give national and international coverage with D. Souto and C. J. Sanders as coordinators.

The first year of the operational program was 1985. Traps were placed out in about 500 locations, but the results were inconsistent; in many areas catches were lower than anticipated from larval population densities. The various components of the system were examined carefully, and it was concluded that the probable explanation lay in the fact that the dog flea-collars used as convenient sources of insecticide in the traps in 1985 contained the nonvolatile tetrachlorvinphos and not dichlorvos, as intended.

This was rectified in 1986 with the use of Vaportape II as used in gypsy moth traps, and the 1986 and 1987 results were more as expected. We now have 2-years of data, but because the 1987 data are still being processed at the time of writing, discussion will focus on the 1986 data.

TABLE 3 Agencies in Canada and the United States Cooperating in the Operational Monitoring of Spruce Budworm Populations with Sex Pheromone Traps

Canada
J. D. Irving Co.
New Brunswick, Department of Natural Resources and Energy
Nova Scotia, Department of Lands and Forests
Quebec, Ministry of Energy and Resources
Forestry Canada
Newfoundland Forestry Centre
Maritimes Forestry Centre
Laurentian Forestry Centre
Great Lakes Forestry Centre
United States
International Paper Co.
Maine, Forest Service
New Hampshire, Division of Forests and Lands
Vermont, Department of Forests, Parks, and Recreation
New York, Department of Environmental Conservation
Minnesota, Department of Natural Resources
U.S. Forest Service, State and Private Forestry
U.S. Forest Service, Research
U.S. Forest Service:
White Mountain National Forest
Green Mountain National Forest
Ottawa National Forest
Superior National Forest

B. The 1986 Results

The numbers of traps deployed in 1986 by jurisdiction are shown in Table 4. As can be seen, coverage is not uniform. This is due to differences in the economic importance of the budworm in the different jurisdictions and to differences in budworm population densities. For example, in Ontario, the jurisdiction with forested land-mass second only to Quebec, the budworm is of less concern than further east, and budworm populations are still high in some locations, leading to visible defoliation that makes traps redundant. In New Brunswick, concern is high, but so are population densities, and again traps are

TABLE 4 Numbers of Spruce Budworm Sex Pheromone Monitoring Plots by Jurisdiction

Newfoundland	44
Prince Edward Island	1
Nova Scotia	11
New Brunswick	28
Quebec	273
Ontario	54
Maine	214
New Hampshire	7
Vermont	12
New York	23
Michigan	6
Minnesota	37
	710

largely redundant. In Quebec and Maine, concern over the budworm is high, populations have collapsed over large areas, and the critical question in these two jurisdictions is: When will budworm outbreaks return? Here, pheromone traps have been accepted enthusiastically as tools for monitoring low-density populations between outbreaks to determine when populations start to increase again. It is probable that when the merit of pheromone traps as a method of monitoring low-density population is realized, there will be further expansion of their use.

The processing of all data is computerized and plot locations are recorded by Universal Transverse Mercator (UTM) grid coordinates. This will enable the data to be displayed on computer-generated Geographic Information System maps now being developed by M. Power that incorporate the UTM grid. Once this technique is available, it will be possible to overlay the trap catches on site characteristics and on prevous years' results. An example of the output, prepared by hand, is shown in Figure 1.

C. Objectives of the Program

1. Outbreak Prediction

The primary objective of the current program is to provide early-warning of when populations may return to outbreak levels. Because

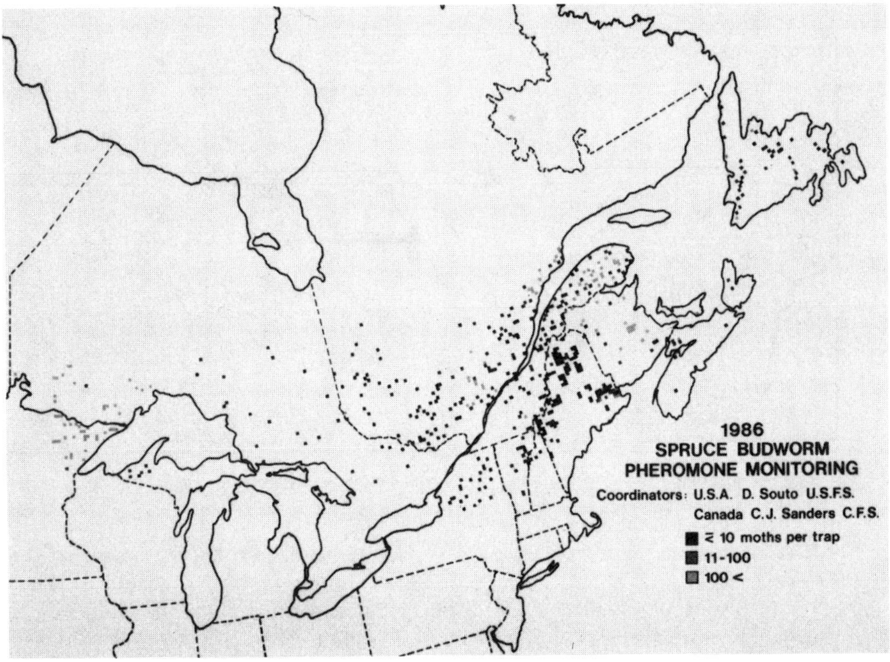

FIGURE 1 Eastern North America showing location of spruce budworm sex pheromone-monitoring plots, 1986.

the program has been operational for only 2 years, it is not yet possible to demonstrate the effectiveness on a broad scale. However, its potential can be seen in the data from a single plot in northwestern Ontario, near Black Sturgeon Lake, for which the population has been monitored continuously for over 20 years (Fig. 2). During the initial years, the synthetic pheromone was not available and, therefore, female moths were used as lures. Also, over the years, the type of trap in use has changed. Nevertheless, the results demonstrate clearly that the traps have monitored the changes in population density accurately over the years.

Regression of $\log_{10}$ moth catch versus $\log_{10}$ larval density in the following generation gives a coefficient of determination (r^2) of 81%. During the early 1960s, larval populations were too low to be measured accurately. Frequently, several thousand branches were sampled but yielded only a few budworms, yet the traps consistently caught a few moths. As population densities rose in the 1970s, culminating in an outbreak that first caused visible defoliation in 1983,

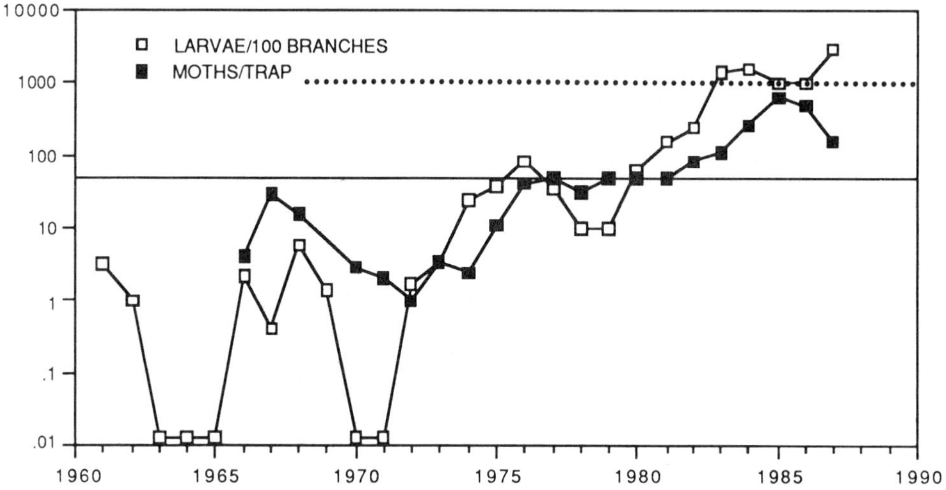

FIGURE 2 Spruce budworm larval densities and catches of moths in pheromone traps at Black Sturgeon Lake, northwestern Ontario, 1960–1987. Upper horizontal line denotes threshold larval density for defoliation, lower line threshold of 50 moths per trap.

so the pheromone catches also rose. With the wisdom of hindsight, it is now possible to say that the pheromone trap-catches could have been used to predict the onset of this outbreak. By using the criterion of 3 successive years of increasing moth catch, or a threshold of 50 moths per trap, it would have been possible to predict the outbreak in 1977, 6 years in advance. Allowing 4 years from the start of the outbreak until the beginning of tree mortality, this would have provided forest management with 10-years advanced warning of tree mortality.

The importance of this in forestry may not be appreciated by those more familiar with agriculture. In contrast to most agricultural pests, which are chronic, persistent problems, forest pests tend to be cyclical, existing for long periods at low density and causing no economic damage. The spruce budworm is a prime example. Outbreaks have appeared in the past at intervals of 30 to 100 years, depending on locations. Until this last outbreak, many foresters had never experienced a budworm outbreak. Populations are now decreasing over much of the budworm's range. As the populations decline so does the concern, and many of the current foresters will have retired before the next outbreak. There is a danger that, in spite of our experience now, the next outbreak will come as a surprise, taking the forest community unprepared.

Forestry cannot afford such miscalculations. A modern pulp and paper mill requires a capital outlay of up to 400 million dollars and provides many jobs in areas in which other employment is scarce. A continuous predictable wood supply is essential. Roads have to be built (at costs of up to 200,000 dollars/km for a primary access road) and camps established for the logging crews. Therefore, it is critical that the stands scheduled for harvesting are available when needed, and that a budworm outbreak does not remove them first. Forest management works on 5- to 10-year operational plans. Given a 10-year warning of budworm outbreaks, plans can be adjusted to allow for them.

2. *Thresholds*

A 10-year warning is therefore of great potential value for midterm planning, but the closer the outbreak gets, the more precise the predictions have to be to allow for the rescheduling of harvesting to remove the most vulnerable stands and to plan for control operations if they are warranted. For these purposes, more intensive assessment is necessary, involving the collection of branches to count larvae. An additional role for the pheromone-trapping system will be a threshold trap-catch to alert the forest manager to the need for a switch to a more intensive assessment.

The accepted technique for estimating larval densities is to collect branches containing the overwintering second-instar larvae (L_2) and to remove the larvae for counting by washing the branches in a mild caustic solution. Unfortunately, at low densities these samples are very variable, correlations between moth catch and L_2 density are not good, and the establishment of thresholds is difficult (e.g., 14).

The relationship between moth catch and L_2 density for three jurisdictions in 1986 are shown in Figures 3–5. The reliability of three arbitrary threshold levels of moth catch as indicators that L_2 populations remain below a given density, is shown in Table 5. If we chose a threshold catch of 10 moths per trap as an indication that densities are below one L_2 per branch, we would have been wrong in only 1 out of 8 locations in New Brunswick, 5 out of 77 in Quebec, and 2 out of 8 in Ontario. Reliability is equally good using 30 moths as an indicator of densities remaining below 10 L_2 per branch, or 100 moths for 30 L_2 per branch. Before the reliability and usefulness of such thresholds can be established, however, data will have to be examined for several more years.

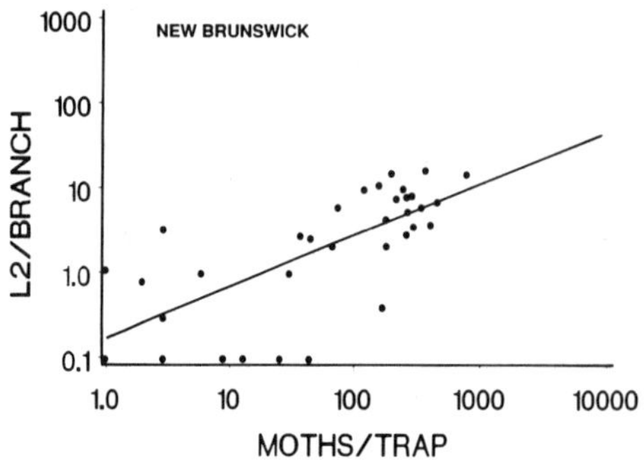

FIGURE 3 Relationship between moth catches and density of second-instar larvae (L_2), New Brunswick, 1986.

3. *Density Estimation*

The ultimate question asked of a sex pheromone-trapping program is whether or not the trap-catches can be used to predict population densities, thus replacing other more costly sampling techniques. As with the establishment of thresholds, it is too early to say. Previous data have given correlations between moth catch and larval densities in the following generation with coefficients of determination (r^2) of 60% and higher (14,15), but the relationships varied from year to year and with location. The 20 years of trapping data at Black Sturgeon

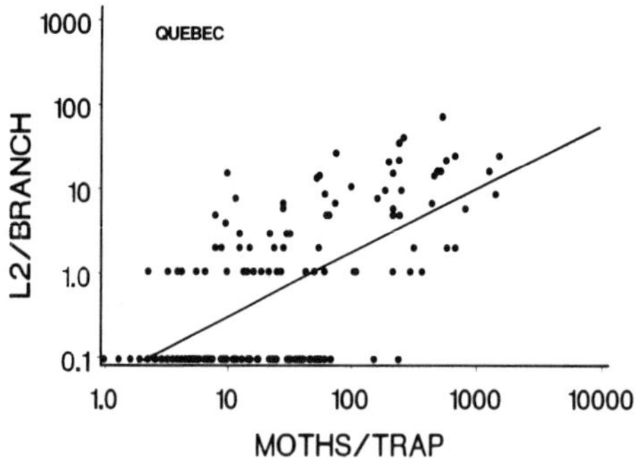

FIGURE 4 Relationship between moth catch and density of second-instar larvae (L_2), Quebec, 1986.

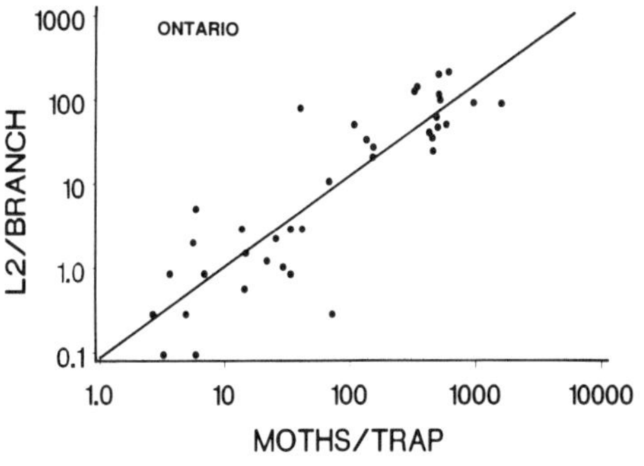

FIGURE 5 Relationship between moth catch and density of second-instar larvae (L_2), Ontario, 1986.

Lake (see Fig. 2) gave an r^2 of 81%. The 1986 data gave an r^2 of 47% for New Brunswick (see Fig. 3), 47% for Quebec (see Fig. 4) and 76% for Ontario (see Fig. 5). Both the New Brunswick and Quebec coefficients are greatly affected by the large number of zero counts for L_2 estimates (transformed in the graphs to $\log_{10}$ (n + 0.1).

D. Problems

The monitoring program for spruce budworm can now be considered as operational and ongoing, but there are a number of areas in which improvements could be made.

1. Lures

The choice of polyvinyl chloride (PVC) lures for the program was largely one of expediency; they had been used for a long time, had given satisfactory results, and are simple to formulate; thus, a manufacturer could always be found. However, they do have disadvantages. They have a first-order decay rate of sufficient magnitude to show up in reduced catches as the lures age (16). Accordingly, other formulations are being tested. Of these, Biolures (Consep Membranes Inc.), which have an almost constant release rate, show great promise. The first Biolures tested had release rates that were in the order of 2000 ng/hr (48 μg/day)—far higher than the target of 20 ng/hr. However, in 1987, Biolures were provided with release rates in the same order as the PVC. Field trials showed them to be quite comparable in potency to the PVC (Table 6). If subsequent

TABLE 5 Reliability of Three Arbitrary Levels of Moth Catch, 10, 30, and 100, as Indicators that Population Densities Remain Below 1, 10, and 30 Larvae per Branch, Respectively[a]

Threshold catch	Total number of plots	Larval threshold densities 1	10	30
New Brunswick				
10	8	1		
30	10		1	
100	16			1
Quebec				
10	77	5		
30	119		5	
100	155			5
Ontario				
10	8	2		
30	14		2	
100	20			3

[a]Figures indicate the number of plots in which larval densities exceed the expected.

tests over a wider range of population densities confirm this, then a change to Biolures would be appropriate.

2. *Number of Traps Per Location*

Previous work has shown that correlations between average catches in clusters of three traps are virtually the same as in clusters of five (17). The users have now raised the question of whether or not this can be reduced to two or even one. Accordingly, in 1987, traps were placed singly, and in clusters of two and three with 40 m between traps and at least 200 m between sites in a number of plots in Ontario, covering a wide range of population densities. Average catches from the clusters of three were then allocated to categories, 0 to 1, 2 to 3, 4 to 10, 11 to 30, 31 to 100, 101 to 300, and over 301, and the catches in the single trap and pairs of traps were compared to determine if they coincided with the same category. In 11 out of 13 locations, the paired traps gave the same result as the cluster of three. In 4 of 13, the single trap gave a different estimate, but in

TABLE 6 Catches During 1 Week in Multi-pher Traps Baited with Biolures or PVC Pellets[a]

	Biolure			
	A (25)	B (25)	C (25)	PVC (5)
Catch	9.4 ± 5.8	11.0 ± 3.8	13.2 ± 6.0	14.2 ± 5.93
Release rate (μg/day)	1	7–9	15–18	0.5

[a]Release rates of pheromone provided by manufacturers.

only one was the estimate out by more than one category (Table 7). Therefore, if the availability of traps were a limiting factor, the use of single traps is a possibility to give more extensive coverage with only a slight loss of accuracy. This should, however, be a last resort. Each year there is some attrition of traps because of vandalism or bears; deployment of single traps runs the risk of forfeiting any estimate, and multiple traps should be used if possible.

3. Number and Spacing of Trap Location

Because the primary objective of the program is the monitoring of long-term trends, it is very important that the locations where the traps are deployed should be carefully selected. They should be in sites representative of the budworm-prone forests in that area, easily accessible, clearly marked, and if possible, in stands not likely to be harvested in the near future. The number of locations and their proximity to one another is a difficult question because factors, such as site, topography, variability of cover type, the level of economic concern, and cost, all play a role. Data previously obtained from northwestern Ontario (Figs. 6 and 7) suggest that for widespread general trends, plots could be as few as one for every 100 × 100-km block of the UTM grid (i.e., 1/10,000 km^2). For the accessible, productive spruce budworm-prone forests of eastern Canada, this would necessitate only 85 locations. For long-term general trends, this might be enough, although where topography and climate are more varied, such as in the Maritimes, even long-term trends may vary on a finer scale. For determining differences in growth rates of populations to locate where defoliation will occur first, a much finer resolution will be required. Unfortunately, this

TABLE 7 Comparisons Between Catches in Single Traps, Pairs, and Clusters of Three in 13 Locations in Ontario

Catch per trap		
Three traps	Two traps	One trap
4.3	6.5	4
6.3	1.0	9
8.3	5.5	8
11.3	12.5	9
25.0	23.0	39
45.0	28.5	22
77.7	75.0	81
87.0	77.0	306
110.3	131.5	144
153.0	203.5	108
273.0	192.5	237
585.0	485.5	541

[a]Forty meters between the paired traps and between traps in the cluster of three, 200 m between treatments.

can only come from experience. Meanwhile forest pest specialists should be encouraged to establish as many locations as is practicable, using such aids as Geographic Information System maps and overlays to ensure representative coverage of susceptible forest-cover types.

III. CONCLUSION

Sex pheromones are under development in eastern Canada for use in monitoring and controlling several forest pests. However, at present, the only fully operational program is the use of traps for monitoring fluctuations in spruce budworm population densities. A standardized program using low–release-rate polyvinyl chloride lures and Multi-pher traps is now in operation covering six eastern U. S. states and six provinces. Early results show significant relationships

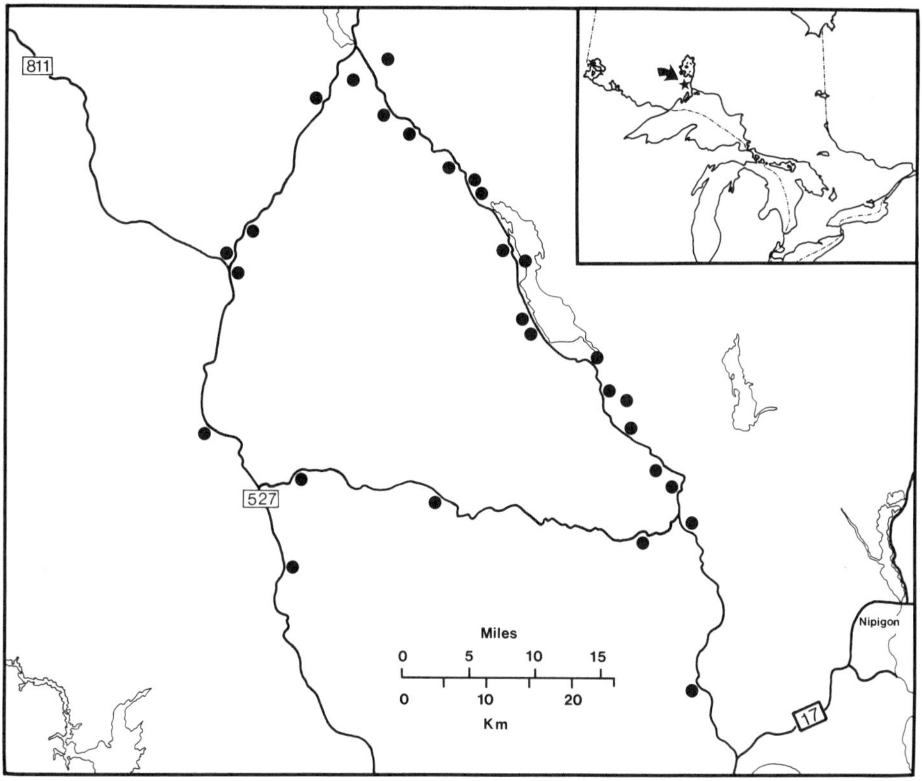

FIGURE 6 Location of sampling points in northwestern Ontario (see Fig. 7).

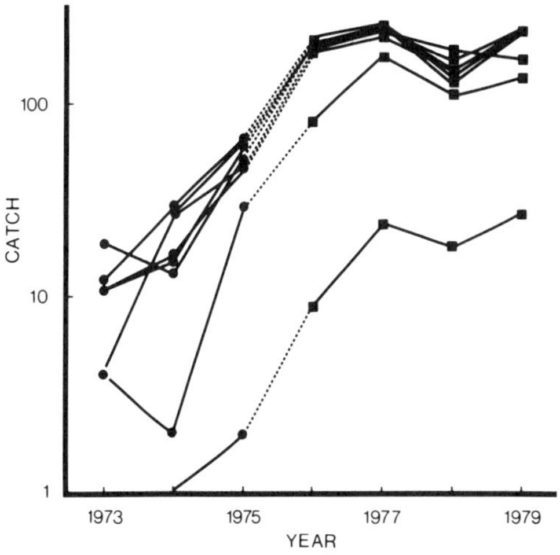

FIGURE 7 Trap catches over a 7-year period in every third plot shown in Figure 6.

between trap catch and subsequent population density and indicate that the program has great potential as an "early-warning" system, which will provide up to 10-years warning of tree mortality. With data from a few more years, it should be possible to establish levels for threshold catches which can be used to indicate the need for more intensive sampling to delineate problem areas.

There is a need for more consideration of the location and number of trapping sites, but this requires the cooperation of the users (mostly government agencies), which will come about only as they gain confidence in the system.

REFERENCES

1. Sanders, C. J. Disruption of sex attraction in the eastern spruce budworm. *Environ. Entomol.* 5:868–872 (1976).
2. Sanders, C. J. The role of the sex pheromone in the management of spruce budworm. In *Chemical Ecology: Odour Communication in Animals.* (Ritter, F. J., ed.). Elsevier/North-Holland Biomedical Press, Amsterdam, 1979, pp. 281–289.
3. Sanders, C. J. Mate location and mating in eastern spruce budworm. *Bi-Mon. Res. Notes Can. For. Serv.* 35(1):2–3 (1979).
4. Sanders, C. J. and Silk, P. J. Disruption of spruce budworm mating by means of Hercon plastic laminated flakes, Ontario 1981. Information Report No. 335. Great Lakes Forest Research Centre, Sault Ste. Marie, Ontario, 1982, 22 pp.
5. Wiesner, C. J. and Silk, P. J. Monitoring the performance of eastern spruce budworm pheromone formulations. In *Insect Pheromone Technology: Chemistry and Applications.* ACS Symposium Series 190. American Chemical Society, Washington, D.C., 1982, pp. 209–218.
6. Seabrook, W. D. and Kipp, L. R. The use of a two component blend of the spruce budworm sex pheromone for mating suppression. In Proc. 13th Int. Symp. Controlled Release of Bioactive Materials. (Chandry, I. A. and Thies, C., eds.). Controlled Release Society, Lincolnshire, Ill., 1986, pp. 128--129.
7. Weatherston, J., Roelofs, W., Comeau, A., and Sanders, C. J. Studies of physiologically active arthropod secretions. X. Sex pheromone of the eastern spruce budworm, *Choristoneura fumiferana* (Lepidoptera: Tortricidae). *Can. Entomol.* *103*:1741–1747 (1971).
8. Sanders, C. J. Evaluation of sex attractant traps for monitoring spruce budworm populations (Lepidoptera: Tortricidae). *Can. Entomol.* *110*:43–50 (1978).

9. Houseweart, M. W., Jennings, D. T., and Sanders, C. J. Variables associated with pheromone traps for monitoring spruce budworm populations (Lepidoptera: Tortricidae). *Can. Entomol.* *113*:527–537 (1981).
10. Ramaswamy, S. B. and Cardé, R. T. Non-saturating traps and long-life attractant lures for monitoring spruce budworm males. *J. Econ. Entomol.* *75*:126–129 (1982).
11. Allen, D. C., Abrahamson, L. P., Jobin, L., Souto, D. J., and Sanders, C. J. Use of sex pheromone traps to monitor spruce budworm populations. Can. For. Serv., Govt. of Can., Spruce Budworm Handbook, 1986, 16 pp.
12. Sanders, C. J. Evaluation of high capacity, non-saturating sex pheromone traps for monitoring population densities of spruce budworm (Lepidoptera: Tortricidae). *Can. Entomol.* *118*:611–619 (1986).
13. Sanders, C. J. Release rates and attraction of PVC lures containing synthetic attractant of the spruce budworm, *Choristoneura fumiferana*. *Can. Entomol.* *113*:103–111 (1981).
14. Allen, D. C., Abrahamson, L. P., Eggen, D. A., Lanier, G. N., Swier, S. R., Kelley, R. S., and Auger, M. Monitoring spruce budworm (Lepidoptera: Tortricidae) populations with pheromone-baited traps. *Environ. Entomol.* *15*:152–165 (1986).
15. Sanders, C. J. Monitoring spruce budworm population density with sex pheromone traps. *Can. Entomol.* *120*:175–183.
16. Sanders, C. J. and Meighen, E. A. Controlled-release sex pheromone lures for monitoring spruce budworm populations. *Can. Entomol.* *119*:305–313 (1987).
17. Sanders, C. J. Sex pheromone traps and lures for monitoring spruce budworm populations—The Ontario experience. USFS Publ. Gen. Tech. Rep. 88, 1984.

22

Use of Disparlure in the Management of the Gypsy Moth

DOUGLAS M. KOLODNY-HIRSCH / Forest Pest Management, Maryland Department of Agriculture, Annapolis, Maryland

CHARLES P. SCHWALBE / Animal and Plant Health Inspection Service, U.S. Department of Agriculture, Otis Air National Guard Base, Massachusetts

I. INTRODUCTION

The gypsy moth, *Lymantria dispar* L., is a devastating pest affecting over 500 species of hardwood forest, shade, and ornamental trees in Europe and the United States. Outbreaks of the gypsy moth are cyclic, frequently extensive, and generally last 2 to 3 years. High-density populations cause significant ecological and economic impact in forest and urban areas. In 1981, a record 13 million acres were defoliated in the United States. Total losses from pest damage during that year were estimated at 250 to 350 million dollars. The current approach to gypsy moth management in the United States relies heavily on the aerial application of insecticides to protect foliage and reduce pest populations in high-value sites (1). In recent years the use of insecticides has become increasingly controversial with the heightened awareness of potential environmental and health hazards resulting from their use. These concerns have emphasized the need for more selective forms of pest control. Accordingly, development of alternative improved management technology for the gypsy moth has been the goal of a broad research effort since the early 1970s (2). The use of the female sex pheromone for survey and control has been an important element of that research program. Following the isolation, identification and synthesis of *cis*-7,8-epoxy-2-methyl-octadecand (disparlure) by Bierl et al. (3), major advances in the technology for survey and management of this pest have evolved.

It is our purpose in this chapter to summarize recent advances in the development of disparlure for mating disruption, mass trapping, survey, and monitoring and to discuss the remaining challenges related to field applications of pheromones for managing gypsy moth.

II. MATING DISRUPTION

A. Field Trials

The optically active (+) enantiomer of disparlure was synthesized by Iwaki et al. (4) and was later determined to be clearly superior to the racemic mixture as an attractant (5). However, small-scale field tests have demonstrated that racemic disparlure is as effective as (+)-disparlure for inhibiting gypsy moth mating in the field (6). Because of its lower costs, virtually all published studies have used the racemic form in mating-disruption trials. Mating disruption is based on the premise that permeating the insect's habitat with its synthetic sex pheromone will modify the normal behavior of males and, thus, disrupt proper orientation to the female (3). Ultimately, the incidence of mating is reduced, and it is implicitly assumed that the reproductive potential and growth of the pest population is diminished. Although mating disruption of the gypsy moth has been attempted since 1971 (7), little attention has been awarded the type of behavioral modification that mediates communication disruption. Proposed mechanisms of mating disruption within the Lepidoptera and the gypsy moth have been reviewed by Bartell (8), Bednyi et al. (9) and Cardé (Chap. 4). Schwalbe and Mastro (10) recently presented evidence suggesting camouflage of natural source points as a plausible mechanism and rule out the possibility of false-trail following.

Although the gypsy moth is not an ideal candidate for control by mating disruption, aspects of its biology make it amenable to the technique. Because of its high fecundity (300–1200 eggs per mass) and the fact that males are highly polygamous, mating must be reduced to a greater degree than with other less-fecund Lepidoptera to effect population change. However, flightless females can not disperse far from their emergence sites, and mark-recapture studies (11,12) indicate that few male moths disperse farther than a few hundred meters; therefore, large numbers of adults would not be expected to reinvade treated sites. Because the insect is univoltine, and the adult males appear in a distinct, predictable flight over a 6-week period, disparlure applications can be accurately timed to coincide with male activity. Recent studies suggest that in contrast with current theory (13) sexual communication between the sexes is completely mediated by olfactory cues (Cardé, personal communication). However, the probability of random encounters between individuals increases proportionately with higher-population densities. Thus, ecological and theoretical (14,15) considerations suggest that control of the gypsy moth by mating disruption is best suited to low-population densities at which random contacts between the sexes is minimized.

The use of disparlure as a mating disruptant has received considerable research effort, which has been reviewed by Beroza (16), Granett (17), Cameron (18), Webb et al. (19), Roelofs (20), and most recently by Plimmer (21).

Research with the mating disruption technique initially involved testing pheromones in small woodlots with simulated gypsy moth populations. Field trials conducted by Stevens and Beroza (7) on Dauphin Island, Alabama, in 1971 reported that aerial application of a silicone-treated paper formulation of disparlure at 50 mg active ingredient (AI) per hectare effectively reduced male trap catch for several weeks after application. Further tests in Massachusetts confirmed these results. This study confirmed the feasibility of mating disruption with the gypsy moth. However, the disruptant treatments did not remain active for more than 3 weeks, and an effort was mounted to develop a controlled-release formulation of greater persistence. Also, since the potential area to be treated for gypsy moth was large, formulations that were suitable to aerial application were explored. Subsequent laboratory and field studies in the United States, Europe, and Russia (22–30) demonstrated the usefulness of several controlled-release formulations, developed reproducible bioassay techniques for quantifying efficacy, and identified several factors influencing the effectiveness of mating disruption.

Out-of-season field studies by Beroza et al. (22) in Alabama and Massachusetts evaluated several candidate formulations including granulated cork, molecular sieves, and microcapsules applied by aircraft at application rates of 1.8 to 11.1 g active ingredient (AI) disparlure per hectare. This study demonstrated biological activity from all candidate formulations. The highest degree of disruption achieved was from the cork formulation (8.2 g (AI)/ha), which reduced male captures by 99% for 7 weeks. Pre- and postseason tests conducted later that year on Cape Cod revealed that one molecular-sieve and three microcapsule formulations suppressed catch of released males by 90% up to 6 to 8 weeks posttreatment. Cameron (23), in out-of-season field tests in Pennsylvania, demonstrated that mating of gypsy moths was significantly suppressed after the broadcast application of a granular cork formulation at 7.5 and 25 g (AI)/ha (28% and 57%, respectively). More encouraging was the complete suppression of mating of laboratory females in plots aerially treated with a microencapsulated formulation at a rate of 5 g disparlure (AI)/per hectare. Results were sufficiently promising to persuade other researchers to conduct large field trials with the microencapsulated formulations in sparse populations.

Beroza et al. (24) sprayed a 60-km^2 woodlot in Massachusetts with a microencapsulated formulation of disparlure (microcapsules contained 2.2% disparlure in xylene) at a rate of 5 g (AI)/ha. He reported that application of the pheromone greatly reduced communication between the sexes (97%), as indicated by reduction in moth trap capture. Mating of females was significantly suppressed up to 5 weeks after application, except for a 2-week interval during peak male activity. Equally encouraging was that intensive sampling (n = 100 0.1-ha fixed radius plots) revealed that egg mass populations in the treated plots remained at pretreatment levels, whereas populations in control woodlots significantly increased in the following year. Furthermore, 15% of the postseason egg masses in the treated area were infertile, compared with only 2% in the control plots. Concurrent evaluations by Cameron et al. (25) and Schwalbe et al. (26) in simulated gypsy moth populations in Pennsylvania reported that a microencapsulated formulation of disparlure aerially broadcast at rates of 5.0 and 15 g (AI)/ha to 16-ha plots suppressed mating as long as 6 weeks after application. These tests further confirmed that low doses of microencapsulated disparlure could remain biologically active over the course of male flight, and that mating incidence generally declined as dosage was increased. Successes of these early disruption trials with disparlure indicated the coordinated use of insecticides and the mating disruption technique as a promising strategy for controlling gypsy moth populations over a range of conditions (14,24). Beroza et al. (27) followed these studies with a test of the effect of variable dose rates and applications using the microencapsulated attractant similar to that applied earlier (24) (microcapsules contained 2% disparlure in 1:3 amyl acetate:xylene). Results obtained with 5 g (AI) lure per hectare were considered inadequate. However, application of 20 g disparlure per hectare was more successful; mating success of tethered females was reduced 94% to 97%, and low-level populations (10 to 15 egg masses per hectare) were suppressed by 67%, thereby confirming earlier reports of a dose–response relationship.

After the application of microencapsulated disruptant with a knapsack blower at 18 g (AI)/ha, Granett and Doane (28) reported complete disruption of mating among wild-type females for an entire season, even in dense populations. Pre- and postseason egg mass counts indicated suppression of pest populations in treated plots. However, no statistical tests were performed on these data. Nevertheless, their results suggested that disparlure had promise in reducing moderately dense populations.

Field observations demonstrated that although microencapsulated formulations provided a relatively uniform distribution of pheromone, the active ingredient was inefficiently released (19,21). Indeed, field bioassays (21) revealed that some microcapsule formulations lost half their disparlure content in 10 to 34 days. Accordingly, more recent studies in the United States and in Europe (30–33) concentrated on the evaluation and refinement of alternative formulations, including hollow plastic fibers, polymeric 3-layer laminated plastic tape, and plastic laminated flakes. In 1976, after initial laboratory screening, seven disparlure formulations, including microcapsules, were field tested in Massachusetts and Maryland (30). Objectives of this study were to test the relative efficacy (as measured by disorientation of males and inhibition of female mating) of these formulations at different densities and population histories. Although all candidate formulations reduced the incidence of mating, significant differences in effectiveness of individual formulations were observed. Results showed that the 1976-NCR 2% microcapsules were more effective in disrupting successful mating than the other treatments tested. Differences in effectiveness among the seven formulations were ascribed to differential disruptant release and environmental stability. Curves relating the incidence of mating in test plots over the course of the season illustrate that most mating in pheromone-treated plots occurred when mating in the control areas was greatest. Also, peak mating was observed during peak male flight, when mating reduction dropped to levels considered inadequate to suppress populations. The degree of mating reduction was also greater in Maryland, where substantially lower pretreatment gypsy moth populations were found. The level of mating disruption in Massachusetts was 83%, whereas in Maryland, 97% reduction was achieved.

Later work by Schwalbe et al. (31) was conducted in sparsely infested areas of Massachusetts to assess increased application rates of microencapsulated disparlure and to test two new formulations of plastic laminate flakes and hollow fibers. Increasing the application rate of the 1978-NCR formulation from 5 to 50 g (AI)/ha resulted in a significant increase in biological activity, confirming results from earlier studies (25–27). As in their previous study (30), most mating in treated plots occurred when male activity was highest in the control plots.

Using a granulated laminated plastic-flake formulation applied by air, Webb et al. (33) recently demonstrated a dose–response for racemic disparlure, both for disruption of mating communication and for female mating. Their results showed the incidence of mating disruption to be positively correlated with dose rate and inversely related to population density (Fig. 1). Furthermore, although the results were not statistically tested, they recorded a 35.5-fold increase in pest populations in untreated woodlots but only a 0.7-fold increase in woodlots receiving a 75 g (AI)/ha application rate. Similarly, Schwalbe and Mastro (10) recently reported female-mating success to be reduced

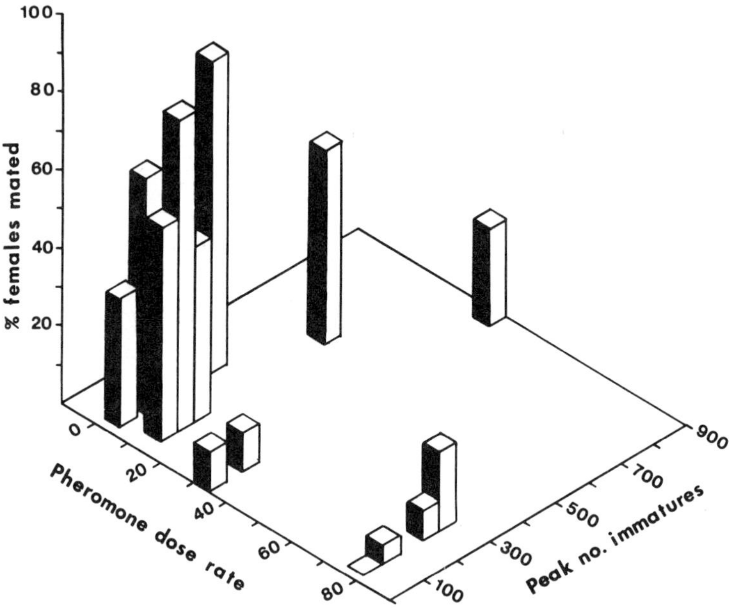

FIGURE 1 Influence of disruptant dose rate and population density on levels of mating disruption among field-collected females (from Ref. 33).

6.5%, 34.5%, and 84% in plots treated with a plastic laminated tape at 5, 50, and 500 g ha, respectively (i.e., release rate of 15, 150, and 1500 mg/ha per day, respectively).

Most recently, a study conducted in sparsely infested woodlots in Maryland showed that applications of Luretape in 1984 and 1986 at 50 g (AI)/ha effectively disrupted male catch (>93%) in (+)-disparlure-baited traps and mating of untethered laboratory-reared females (>90%) for 2 consecutive years between applications (Kolodny-Hirsch and Webb, unpublished data). Further, as illustrated in Figure 2, levels of mating disruption were shown to be curvilinear over the course of a season. Higher values were observed during periods preceding and following peak male flight. These observations support a hypothesis proposed by Schwalbe et al. (30), and upheld by others (31,33), that the degree of mating reduction is inversely related to male population density. Although not statistically significant, the Maryland data showed the growth rate of the gypsy moth populations in treated plots, to be half that seen in control plots over the 4-year period.

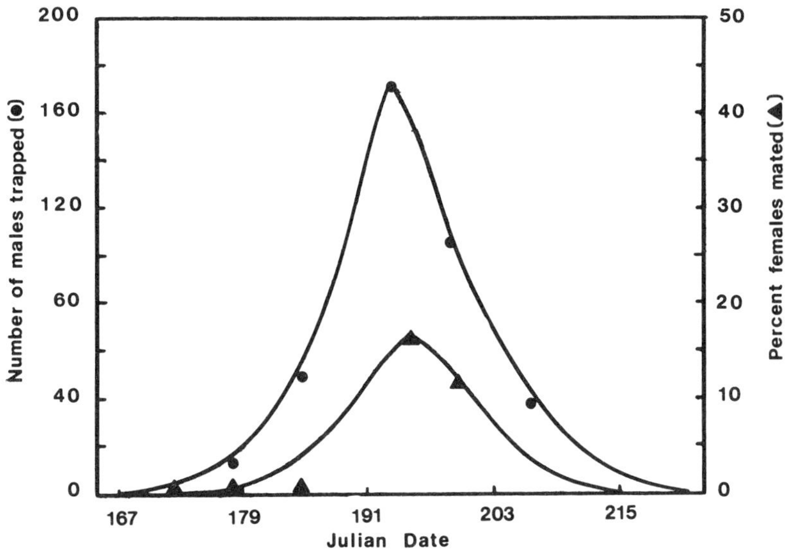

FIGURE 2 Relation between mating disruption and male trap catch from mating disruption trials conducted in Maryland during 1987 (Kolodny-Hirsch and Webb, unpublished data).

In 11 of 15 mating disruption trials with the gypsy moth, published between 1972 and 1988, evidence for mating disruption has been based entirely on the reduction of male trap-catch and on the mating success of laboratory-reared or field-collected females placed in the test plots. These indirect methods have proved useful in demonstrating the feasibility of the technique and in assessing the relative performance of candidate disruptant formulations; however, evidence for suppression of gypsy moth (i.e., efficacy as defined here) is circumstantial because measurements of population trends were not made. Workers who have attempted to measure the effect of treatment on actual population change by counting egg masses or larvae under burlap have been frustrated by the limited usefulness of these methods. Egg-mass sampling is imprecise at low densities unless large sample numbers are taken, and larval counts under burlap cannot be accurately related to absolute population estimates. Because of these sampling difficulties, there are few published accounts in which attempts were made to estimate pre- and posttreatment egg-mass population levels. In these cases (24,27,28,32), data suggest that, under certain conditions, population trends may be lowered. However, in only one report (24) were changes in population trends statistically tested and shown to be significant. Theory suggests that levels of mating disruption exceeding 93% should result in population reduction (i.e., population trend <1; 14). Further, because each mated female produces one egg mass, even moderate levels of mating inhibition should dampen population increase. Indeed, levels of mating disruption exceeding 90% have been reported with several formulations; unfortunately, estimates of egg-mass populations were not conducted. Thus, after 17 years of research with mating disruption of the gypsy moth, we still lack sufficient data documenting suppression or dampening of populations.

In conclusion, a substantial body of data unequivocally demonstrates the utility of several controlled-release formulations of disparlure as a means for effectively disrupting pheromonal communication and reducing the incidence of mating between sexes. However, variable and often conflicting reports on similar formulations preclude the selection of optimal materials with confidence. Furthermore, although many of these formulations are biologically active at very low dose rates, more practical mating disruption is realized at higher-dose rates (>50 g (AI)/ha) for which application costs are currently prohibitively expensive. Theoretical and empirical results indicate that ultimately, disruption of mating will be most efficacious at low-population densities (<20 egg masses per hectare) and during intervals when male pressure is lowest (i.e., before and after peak male flight or in female-biased populations). However, limited data (10) suggest that acceptable levels of disruption could be achieved at higher-population levels with release rates of 1500 mg disruptant per hectare per day.

B. Operational Use

Currently, the three-layer plastic laminated tape (Hercon Luretape) and granulated flake (Hercon Disrupt) are registered with the EPA for gypsy moth-mating disruption (34). Table 1 provides a summary of demonstrational and operational use of disparlure for mating disruption since 1979. Although there is a lack of overwhelming confidence in mating disruption, this table illustrates the long track record of its operational use and indicates that many agencies are becoming increasingly interested in incorporating this technique within their suppression programs. Unfortunately, lack of adequate documentation precludes an objective assessment of mating disruption efficacy. Presumably because of its relative ease of application, the plastic laminated tape has been the primary formulation chosen by agencies for mating disruption. Moreover, cost of this material is comparable with other control agents. Data from demonstrations in Maryland show that with two seasons of effectiveness from a single

TABLE 1 Summary of Operational Use of Mating Disruption for the Gypsy Moth in the United States

Year	State	Formulation	Area treated (ha)
1979	Oconomowoc, WI	Granulated flakes	170
1979	Reedsville, PA	Microencapsulated	4050
1981	Oconomowoc, WI	Granulated flakes	41
1982	Great Falls, VA	Laminated tape	81
1983	Fairfax, VA	Laminated tape	37
1983	S.W. Virginia	Granulated flakes	1620
1984	Fairfax, VA	Laminated tape	65
1985	Fairfax, VA	Laminated tape	234
1985	Virginia Beach, VA	Laminated tape	?
1985	W. Mooreland, VA	Laminated tape	73
1985	Great Falls, VA	Laminated tape	29
1985	Severna Park, MD	Laminated tape	73
1986	Fairfax, VA	Laminated tape	151
1986	Rockcreek, VA	Laminated tape	9
1986	Bethesda, MD	Laminated tape	12
1987	Fairfax, VA	Laminated tape	203
1987	Great Falls, VA	Laminated tape	58
1987	Rockville, MD	Laminated tape	49
1987	Severna Park, MD	Laminated tape	70
		total =	7025

application and dose rates of 50 g (AI)/ha, cost of this material is 50 dollars/ha or 1 dollar/g (Kolodny-Hirsch, unpublished data). Although these expenses could be further reduced by maximizing the distance between manually deployed releasers, necessary studies relating spacing to efficacy are currently incomplete.

Nevertheless, despite the economic feasibility and the ecological advantages of the mating-disruption technique (i.e., low toxicity, specificity, compatibility with other tactics; 14), Table 1 shows that disparlure has been used on only a limited scale by organizations charged with managing or regulating the gypsy moth. There are several constraints hindering the widespread adoption of the disruption technique in operational programs. First, the concept of mating disruption runs counter to the current approach to gypsy moth management, which centers around the suppression of high-density populations as opposed to low-density populations (1). Consequently, mating disruption does not fit into an acceptable management niche. Second, lack of data showing population suppression tends to erode confidence in the technique. Although greater efficacy could certainly be achieved with increased release rates of disruptant, this is not economically feasible using currently registered formulations. Third, as discussed by Lewis (35), semiochemicals do not lend themselves as readily to development, adoption, and implementation by industry as do conventional insecticides. Fourth, the principal formulation of pheromone purchased by agencies is the three-layer plastic laminated tape (see Table 1), which is not practical to apply to large acreages. Although granulated flake formulations are amenable to aerial application, they require specialized application equipment. Finally, after 17 years of study, researchers have become increasingly aware of the problems encountered in mating disruption trials and the complexity of the system. Accordingly, their optimism has waned considerably. This is reflected by the decline in publications on this subject noted by Rothschild from 1976 to 1980 (36). The foregoing discussion indicates several areas of research that must be addressed before mating disruption can be considered a viable operational tactic. These research requirements are presented in the following section.

C. Suggested Areas of Research

1. Mechanism of Communication Disruption

First and foremost we must understand how communication disruption works. This information would have enormous implications in identifying optimal formulations and delivery systems. In addition, a more complete understanding of the changes in the behavior of males in the presence of racemic disparlure would indicate optimal placement of controlled-release materials and traps for evaluating treatment effectiveness, and perhaps the minimum size of plots for pheromone treatment.

2. *Atmospheric Dispersion*

Work needs to be continued in the field to elucidate the atmospheric dispersion of disparlure from various formulations (37). Although it has been assumed that more even distribution of pheromone might be achieved by the use of many small dispersed sources, such as microcapsules and chopped hollow fibers, localized turbulence close to the ground may render uniform permeation of the air with pheromone very difficult to achieve. One perceived limitation of manual application of pheromone dispensers at ground level is their inability to provide sufficient concentrations in the canopy of trees to effectively disrupt orientation and mating of females located there. Limited studies with microencapsulated formulations have shown that concentrations of disparlure decrease with increasing height above the ground (37,38). However, the biological significance of these observations is unclear because levels of mating disruption were not monitored. Similarly, Sartwell et al. (39) found that hand-treatment with three-layer laminated tape against the western pine shoot borer, *Eucosma sonomana*, was less effective in plots with numerous tall trees.

3. *Optimal Release Rate*

It is necessary to continue studies to determine the optimal release rate of racemic disparlure, similar to those reported by Schwalbe and Mastro (10). Also, we need to find out if changes in release rate modify the behavior that brings forth communication disruption. Behavioral observations by Palaniswamy et al. (40) showed that at lower release rates per releaser, communication disruption of the forest tent caterpillar, evidenced by reduced orientation to baited traps, was partly due to confusion and partly due to a reduction in male searching; however, at higher release rates, disruption was mainly due to a reduction in searching. Similar behavior may exist for the gypsy moth.

4. *Formulations*

Improvements in pheromone formulations must be made if we hope to expand the role disparlure will play in pest management (41). Relatively few formulations have been field tested from the standpoint of emission rates, comparative longevity, compatible stickers, and total quantities of pheromone emitted. These data are necessary before a rational choice of formulations can be made. Earlier field studies (37) showed that encapsulation of disparlure in gelatin permitted vaporization of only a small percentage of the lure present in the capsules. Laboratory and field observations also suggest the need for the more rapid release of disruptant from the laminated tape to adequately suppress female mating during peak male flight (Kolodny-Hirsch and Webb, unpublished data; Leonhardt, personal communication). Field experiments in Massachusetts showed (10) that effective mating disruption could be achieved by treatment with plastic laminate dispensers releasing a total of 1500 mg disparlure per hectare per day. Current efforts are aimed at developing alternative formulations that allow practical application (Leonhardt, personal communication).

5. *Efficacy*

As we have already discussed, data demonstrating efficacy, as indicated by a decrease in population trend, are limited. Because of the density-dependent nature of mating disruption, most studies with the gypsy moth have concerned themselves with low-density populations, for which even substantial changes in population density between years are difficult to detect without an inordinate amount of sampling effort. Accordingly, most researchers opted not to collect detailed life stage data. Furthermore, although gypsy moth populations may remain at innocuous levels in control plots for several consecutive years, previous studies were limited to 1 year of evaluation. These concerns could be adequately addressed by intensifying sampling effort and conducting multiyear studies with repeated applications. Finally, it is essential in future tests that decisions for dose rates of disruptant treatment be based on the need to demonstrate efficacy and not on rates perceived as being economically feasible.

6. *Theoretical Models*

Because the assumptions underlying Beroza and Knipling's (14) model are highly speculative and deterministic and the factors governing gypsy moth population dynamics are multifaceted and highly stochastic, their predictions were necessarily quite general and need to be interpreted with caution. Indeed, with the current advances in modeling the gypsy moth life system, a more comprehensive description of the processes effecting mating disruption could be accomplished.

III. MASS TRAPPING

The idea of using pheromone-baited traps directly as control tools for gypsy moth has been given most attention since the development of (+)-disparlure as a trap bait. The superior attractiveness of this compound (compared with previous materials used for trapping) led to optimism that mass trapping could be used to control certain gypsy moth populations. It is important to emphasize that the goal of most of this research was to develop mass-trapping techniques for use in infestations that are geographically isolated from the area generally infested by the gypsy moth. Most of these infestations result from inadvertent transport of egg masses by man, often with outdoor household articles moved from generally infested areas harboring high-density populations. These isolated infestations are discovered through a network of detection traps and are typically very well defined through extensive survey, small (on the order of 5 to 50 ha); and usually very sparse (generally less than 10 egg masses per hectare). On theoretical grounds, such populations are ideally suited for control by mass trapping.

Most field tests that have assessed factors that influence the effectiveness of mass trapping have been done with released insects (to obtain maximum control of test population density). Those studies indicated that at relatively low trap densities (about 2.5 traps per hectare), high proportions of the released males were captured, but mating success of females was also high. At higher trap densities of 25 traps per hectare, trap catch was only slightly improved over lower-density plots, but mating incidence was very low (Table 2). Therefore, it appears that at sufficiently high trap densities, moths are caught faster than they are able to locate and mate with females. Indeed, those experiments showed that males are captured significantly sooner after release in high trap density plots than in those containing fewer traps. This accelerated capture rate at high trap densities contributes to the reduced mating success of females.

Peak mating activity of females normally occurs around 1300 to 1400 hr. Hourly examination of traps in trapped plots revealed that

TABLE 2 Percentage Male Recovery (SEM), Female Mating Success, and Male Precapture Period (PCP) of 10 Pairs of Gypsy Moths Released into 4-ha Plots, Cape Cod, Massachusetts, 1983

Traps/ha	Intertrap distance	% Males recovered	% Females mated	PCP (hr)
2.5	63.2	67.5 (11.2)	23.2 (6.2)	3.50
7.5	36.5	75.6 (18.3)	8.9 (9.3)	2.45
25.0	20.0	86.0 (5.5)	4.0 (8.4)	2.25
Control			67.0 (35.7)	

male captures tended to occur earlier in the day than did mating in untrapped plots, and this effect was more pronounced when the mornings were warm. Thus, mass-trapping efficacy is further enhanced by the capture of males (and their removal from the population) earlier in the day than at the time when mating activity normally occurs.

Behavioral observations of individual males in released populations showed that females (which mate only once) were most often mated by multiple mating males, and that a large proportion of males did not mate at all. Conversely, in mass-trapped plots, multiple mating by males was rare. Most males that were observed mating females in trapped plots were captured in traps shortly after mating (Schwalbe, unpublished data). Thus, traps minimize the time available to males for multiple mating, and this has a substantial impact on female-mating success. This also demonstrates the variability in mating behavior among males in a population, which is seldom recognized in attempts to modify behavior for control.

An isolated infestation of gypsy moths was discovered in Monona, Wisconsin in 1981 and, based on trap survey results, the male population was estimated to be 2360 individuals. In 1982, 1983, and 1984, 7.5 traps per hectare were deployed over the area infested. During those years a steady reduction in the number of males caught, the number of larvae and pupae found in the area and, based on locations where males were captured, the number of hectares infested, were observed. Since 1985, no moths have been captured, indicating that the population has been eliminated. Similar projects in other isolated infestations have yielded equally positive results and, currently, mass trapping is one of the preferred tactics used by states in eradication of sparse, localized gypsy moth populations. The confidence of users in the technology has heightened to the point where chemical spraying has been supplanted by the mass-trapping approach. In certain situations very localized treatment with microbial insecticides is made so that adult populations are sufficiently low for traps to be effective in preventing mating.

IV. SURVEY AND MONITORING

A. Population Estimation

Decisions about the need for and the selection of intervention tactics for the gypsy moth are primarily based on estimates of egg mass density (42). While several methods for sampling egg masses are available to managers (43–45), they are limited by imprecision of estimates at low-population densities and the high cost of sampling on a regional basis. Disparlure-baited traps, however, are cost-effective and sensitive at low population levels and have been shown to have potential in monitoring yearly changes in the density of gypsy moth populations (46). Accordingly, a concerted effort is being made to develop procedures for quantifying the relationship between pheromone trap-catch of adult males and densities of other life stages. Efforts, to date, have produced mixed results. Regression-based models have been successful in correlating the number of male moths caught in pheromone traps with other preadult life stages (47–50). However, possibly because of differences in male behavior in high-density populations, these relationships have been more obscure in areas with established or epidemic populations (49,51; J. Elkinton, personal communication). Figure 3 shows the relation between male trap-catch from traps deployed on a systematic 1-km grid of pheromone traps in Maryland and egg mass estimates obtained within the 1-km^2 cells formed by the grid (Kolodny-Hirsch, unpublished data). Although

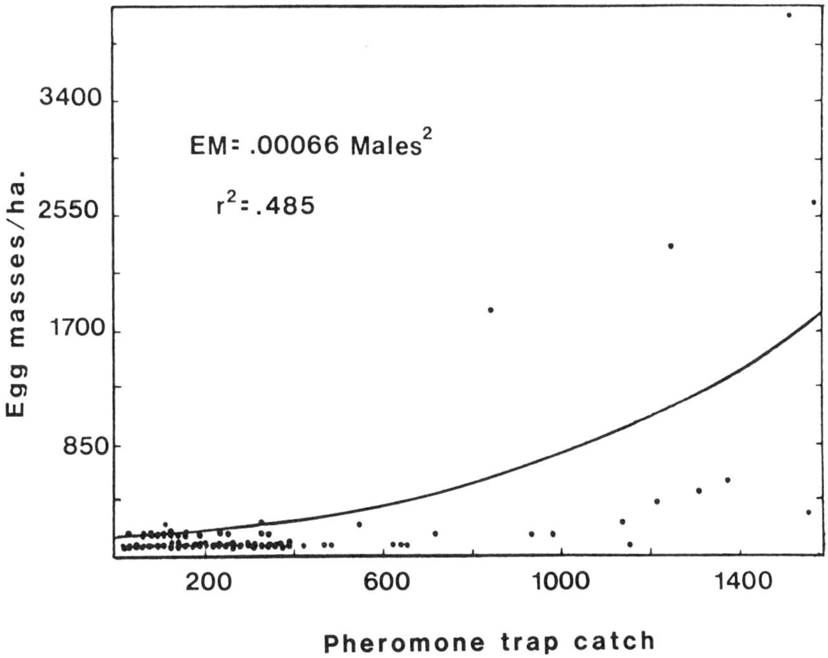

FIGURE 3 Relation between gypsy moth pheromone trap catch from traps placed on a 1-km grid and estimates of egg-mass density within 1-km^2 cells (data from Maryland Gypsy Moth Integrated Pest Management Project 1985).

statistically significant, the correlation coefficient indicates that the number of males trapped is only a partial predictor of subsequent population levels. Estimates of adult sex ratio derived from observations of pupal exuviae under burlap bands could be used to improve this correlation (F. W. Ravlin, personal communication). Recently, Bellinger et al. (52) reported that wing length of male moths caught in pheromone traps in Virginia provided a good correlation between eggs per mass and egg masses per unit area. Another source of variability in the regression model shown in Figure 3 is that the efficiency of pheromone traps can be affected by many external factors (i.e., trap placement, tree species, tree size, trap density; 42), so that trap catch can (to some extent) be independent of population density. This source of variation could be reduced by either accounting for these variables in the regression or by standardizing trap placement. Efficiency of trap catch is also radically influenced by trap design and population density. Elkinton (53) has recently shown that the large capacity USDA milk-carton traps baited with (+)-disparlure, which are used by most agencies for monitoring the distribution and abundance of gypsy moth populations, become less efficient at higher-population densities. Similarly, Miao et al. (54) showed an inverse relationship between the efficiency of pheromone traps baited with natural pheromone and synthetic disparlure and moth density in China. Figure 3 provides graphic evidence for this finding and shows that the relationship between egg mass densities and trap-catch above 300 males per trap becomes increasingly dubious. Until quantitative models of greater utility are developed, male moth surveys conducted within the generally infested area will be limited to generating profiles of gypsy moth populations and monitoring changes in the insect population between years. Nevertheless, this information can be useful as an early-warning monitoring tool and in delineating areas warranting egg mass surveys (1).

B. Detection and Delimitation

The mobile nature of the gypsy moth as it hitchhikes on man's commodities is almost legendary. Articles that are outdoors during periods of high populations in the Northeast are almost certain to be infested with eggs. When these items are transported elsewhere and exposed to favorable environmental conditions, these eggs can hatch and, if the habitat is suitable, new infestations can develop. A trapping system for detecting low-level, isolated gypsy moth infestations enables southern, western, and midwestern states, in cooperation with federal agencies, to conduct effective programs designed to prevent the establishment of gypsy moth populations in those areas.

Before 1970, surveys had been made with various types of extracts of female abdominal tips, and the performance of such baits was unreliable, at best. The discovery of disparlure, in 1970, led to the ready availability of pure attractant, which ushered in an era in which gypsy moth survey expanded. In the years 1972 to 1977, over 500,000 gypsy moth traps were placed, and numerous infestations were discovered. Much was learned about designing inexpensive traps suitable for survey purposes, and it was during this period that the delta trap, which has become the standard since its introduction in 1975, was developed. More basically, states developed the capability to organize and conduct large-scale survey. However, in 1977, synthesis of the most attractive optically active (+) enantiomer of disparlure revolutionized our survey programs. The use of this material added considerable sensitivity to detection surveys. In 1977, survey with (+)-disparlure revealed gypsy moth infestations where they previously had not been suspected. Over the next few years, many states faced an uphill battle in detecting, delimiting, and eradicating isolated infestations that had existed, unnoticed, for years. However, diligent surveys utilizing 200,000 to 250,000 traps annually have been conducted since then, and we are now at a point where new infestations are found within a couple of years after they have been introduced. These traps are a fundamental complement to the regulatory program and enable the conduct of programs designed to keep presently uninfested states free of gypsy moth.

It is important to design and use a survey system around the behavior of the target insect. Although the gypsy moth is a powerful flyer, it tends not to fly over great distances. The probability of detecting an isolated infestation of gypsy moths decreases as traps are spaced more widely (11,12). We have adopted a trap density of one trap per square mile as being logistically reasonable and adequately sensitive to detect isolated infestations. Following this guideline, infestations are generally detected before they are much larger than 100 acres. Moths captured in such detection surveys *only* indicate that a gypsy moth infestation may exist somewhere in the vicinity. It is highly unlikely that a moth in a trap originated from an infestation more than 1 mile away and, therefore, small infestations more than 1 mile away from a trap site are unlikely to be detected. Recognizing that it is often not feasible to continually trap areas at the rate of one trap per square mile every year, some states have opted to rotate such surveys; trapping, for example, each third of the state every third year. Thus, every 3 years, the entire area is trapped according to guidelines. The important point is that the probability of detecting moths increases exponentially as the trap is placed closer to the infestation.

As mentioned earlier, the discovery of male moths in a detection trap merely indicates that a population may exist somewhere in the vicinity. To more accurately define the exact location of that infestation requires additional survey effort. This phase of the survey is called delimitation because its purpose is to delimit the precise boundaries of the infestation. This tactic also takes advantage of the male moth's behavior, which generally is to be captured in the nearest trap. At least 95% of the males captured are expected to be found in traps within 200 m of the point of male moth origin (11,12,54). Population distribution is deduced from the pattern of male moth catches in grids of traps. Testing and experience have led to the adoption of 32 traps per square mile over the area suspected of being infested, as the desired trap density for delimitation, but delimitation densities have ranged from 16 to 81 traps per square mile, depending upon available resources and the sensitivity of the area. Traps surrounding areas of highest population density can be expected to catch the most moths, and trap catches will tend to decrease in areas closer to the edge of the infestation. The line of demarcation between areas with trap catches and those where moths are no longer caught can be taken as the outer limits of the infestation. The use of this technique for circumscribing the area of infestation facilitates accurate planning of eradication treatments, making it possible to eliminate control treatments in areas where gypsy moth populations do not exist.

All supplies used in gypsy moth survey programs are commercially available. The delta trap is marketed by a number of companies. In the late 1970s (+)-disparlure was very rare and very expensive. However, large-scale synthesis procedures have been devised, and now the attractant is readily available.

The technology for conducting effective and sensitive detection and delimitation surveys is as well developed for the gypsy moth as it is for any pest program. Both the power and the limitations of the trapping system are well understood, and it is possible to interpret trapping results in terms of population distribution and density. The systematic deployment of traps according to prescribed guidelines is a powerful probe for monitoring gypsy moth introductions and is the fundamental basis for conducting a program designed to exclude the gypsy moth from areas of the country.

V. CONCLUSION

In conclusion, as a survey device, male traps baited with (+)-disparlure placed in a systematic and standardized fashion have been used routinely and with great success for the detection and delimitation of incipient gypsy moth populations. Furthermore, a systematic array of large-capacity pheromone traps can be a valuable tool for monitoring trends of pest populations in IPM programs (1,46). As a control tactic, mass trapping of gypsy moth is used in many states for the eradication of low-density isolated populations. However, the use of disparlure-baited traps for absolute estimates of gypsy moth populations and the application of racemic disparlure for disrupting mating and effecting population reduction have been less successful and are still in an early stage of development. Possibly with the advent of a pheromone trap that is equally efficient at low- and high-population levels and the inclusion of other ancillary predictor variables, male moth counts could be a practical means for classifying broad categories of infestation densities. Ultimately, in an IPM program this information would aid in pinpointing areas for which more intensive surveys would be warranted. The future of mating disruption of the gypsy moth, however, is less certain. Some of the data are very encouraging, but there is much to be learned before this technique can be used with confidence. Research is warranted to improve and better understand this potentially useful control tactic. To this end, we should resolve to demonstrate its technological feasibility rather than to restrict our studies around economic constraints.

REFERENCES

1. Reardon, R., McManus, M., Kolodny-Hirsch, D., Tichenor, R., Raupp, M., Schwalbe, C., Webb, R., and Meckley, P. Development and implementation of a gypsy moth integrated pest management program. *J. Arboric.* *13*:209–216 (1987).
2. Doane, C. C. and McManus, M. L. (eds.). *The Gypsy Moth: Research Toward Integrated Pest Management.* USDA Tech. Bull. 1584, 1981, 757 pp.
3. Bierl, B. A., Beroza, M., and Collier, C. W. Potent sex attractant of the gypsy moth: Its isolation, identification and synthesis. *Science* *170*:87–89 (1970).
4. Iwaki, S., Marumo, S., Saito, T., Yamada, M., and Katagiri, K. Synthesis and activity of optically active disparlure. *J. Am. Chem. Soc.* *96*:7842–7844 (1974).
5. Plimmer, J. R., Schwalbe, C. P., Paszek, E. C., Bierl, B. A., Webb, R. E., Marumo, S., and Iwaki, S. Contrasting effectiveness of (+) and (-) enantiomers of disparlure for trapping native populations of gypsy moth in Massachusetts. *Environ. Entomol.* *6*:518–522 (1977).
6. Plimmer, J. R., Leonhardt, B. A., and Webb, R. E. Management of the gypsy moth with its sex attractant pheromone. In *Insect Pheromone Technology: Chemistry and Application.* (Leonhardt, B. A. and Beroza, M., eds.). ACS Symp. Series. No. 190, American Chemical Society, Washington, D.C., 1982, pp. 231–242.

7. Stevens, L. J. and Beroza, M. Mating-inhibition field tests using disparlure, the synthetic gypsy moth sex pheromone. *J. Econ. Entomol.* *65*:1090–1095 (1972).
8. Bartell, R. J. Mechanisms of communication disruption by pheromone in the control of Lepidoptera: A review. *Physiol. Entomol.* *7*:353–364 (1982).
9. Bednyi, V. D., Chernichuk, L. L., Chekanov, M. I., and Chekrizova, V. L. The influence of preliminary maintenance of males of the gypsy moth in an atmosphere saturated with disparlure on their mating ability. *Khimoretseptsiya Nasekomykh.* *5*: 123–125 (1980).
10. Schwalbe, C. P. and Mastro, V. C. Gypsy moth mating disruption: Dosage effects. *J. Chem. Ecol.* *14*:581–588 (1988).
11. Schwalbe, C. P. Disparlure-baited traps for survey and detection. In *The Gypsy Moth: Research Toward Integrated Pest Management.* (Doane, C. C. and McManus, M. L., eds.). USDA Tech. Bull. 1584, 1981, pp. 542–548.
12. Elkinton, J. S. and Cardé, R. T. Distribution, dispersal and apparent survival of the male gypsy moths as determined by capture in pheromone-baited traps. *Environ. Entomol.* *9*:729–737 (1980).
13. Richerson, J. V., Brown, E. A., and Cameron, E. A. Premating sexual activity of gypsy moth males in small plot field tests [*Lymantria (=Porthetria) dispar* (L.): Lymantriidae]. *Can. Entomol.* *108*:439–448 (1976).
14. Beroza, M. and Knipling, E. F. Gypsy moth control with the sex attractant pheromone. *Science* *177*:19–27 (1972).
15. Knipling, E. F. The basic principles of insect population suppression and management. *USDA, Agric. Handbk.* 512, 1979.
16. Beroza, M. Control of the gypsy moth and other insects with behavior-controlling chemicals. In *Pest Management with Insect Sex Attractants and Other Behavior-Controlling Chemicals.* (Beroza, M., ed.). ACS Symposium Series No. 23. American Chemical Society, Washington, D.C., 1976, pp. 99–118.
17. Granett, J. A pheromone for managing gypsy moth populations. In *Perspectives in Forest Entomology.* (Anderson, J. F. and Kaya, H. K., eds.). Academic Press, New York, 1976, pp. 137–148.
18. Cameron, E. A. The use of disparlure to disrupt mating. In *The Gypsy Moth: Research Towards Integrated Pest Management.* (Doane, C. C. and McManus, M. L., eds.). USDA Tech. Bull. 1584, 1981, pp. 554–560.
19. Webb, R. E., McComb, C. W., Plimmer, J. R., Bierl-Leonhardt, B. A., Schwalbe, C. P., and Altman, R. M. Disruption along the "leading edge" of the infestation. In *The Gypsy Moth:*

Research Toward Integrated Pest Management. (Doane, C. C. and McManus, M. L., eds.). USDA Tech. Bull. 1584, 1981, pp. 560--570.

20. Roelofs, W. L. Establishing efficacy of sex attractants and disruptants for insect control. Entomological Society of America, 1979, 97 pp.
21. Plimmer, J. R. Disruption of mating in gypsy moth. In *Insect Suppression with Controlled Release Pheromone Systems*, Vol. II. (Kydonieus, A. F. and Beroza, M., eds.). CRC Press, Boca Raton, 1982, pp. 135–154.
22. Beroza, M., Stevens, L. J., Bierl, B. A., Philips, F. M., and Tardif, J. G. R. Pre- and postseason field tests with disparlure, the sex pheromone of the gypsy moth to prevent mating. *Environ. Entomol.* *2*:1051–1057 (1973).
23. Cameron, E. A. Disparlure: A potential tool for gypsy moth population manipulation. *Bull. Entomol. Soc. Am.* *19*:15–19 (1973).
24. Beroza, M., Hood, C. S., Trefrey, D., Leonard, D. E., Knipling, E. F., Klassen, W., and Stevens, L. J. Large field trial with microencapsulated sex pheromone to prevent mating of the gypsy moth. *J. Econ. Entomol.* *67*:569--664 (1974).
25. Cameron, E. A., Schwalbe, C. P., Beroza, M., and Knipling, E. F. Disruption of gypsy moth mating with microencapsulated disparlure. *Science* *183*:972–973 (1974).
26. Schwalbe, C. P., Cameron, E. A., Hall, D. J., Richerson, J. V., Beroza, M., and Stevens, L. J. Field tests of microencapsulated disparlure for suppression of mating among wild and laboratory-reared gypsy moths. *Environ. Entomol.* *3*:589--592 (1974).
27. Beroza, M., Hood, C. S., Trefrey, D., Leonard, D. E., Knipling, E. F., and Klassen, W. Field trials with disparlure in Massachusetts to suppress mating of the gypsy moth. *Environ. Entomol.* *4*:705–711 (1975).
28. Granett, J. and Doane, C. C. Reduction of gypsy moth male mating potential in dense populations by mistblower sprays of micro-encapsulated disparlure. *J. Econ. Entomol.* *68*:435--437 (1975).
29. Bednyi, V. D. and Kovalev, B. G. The disorientation of males of the gypsy moth with disparlure. In *Proceedings of the II All-Union Symposium on Chemoreception*. Vilnius, 25--27 June, 1975, pp. 177–179.
30. Schwalbe, C. P., Paszek, E. C., Webb, R. E., Bierl-Leonhardt, B. A., Plimmer, J. R., McComb, C. W., and Dull, C. W. Field evaluation of controlled release formulations of disparlure for gypsy moth, *Lymantria dispar*, mating disruption. *J. Econ. Entomol.* *72*:322--326 (1979).

31. Schwalbe, C. P., Paszek, E. C., Bierl-Leonhardt, B. A., and Plimmer, J. R. Disruption of gypsy moth (Lepidoptera: Lymantriidae) mating with disparlure. *J. Econ. Entomol. 76*:841–844 (1983).
32. Maksimovic, M. The use of pheromone for controlling the gypsy moth by disrupting mating. *Zast. Bilja 31*:303–307 (1980).
33. Webb, R. E., Tatman, K. M., Leonhardt, B. A., Plimmer, J. R., Boyd, V. K., Bystrak, P. G., Schwalbe, C. P., and Douglass, L. W. Effect of aerial application of racemic disparlure on male trap catch and female mating success of gypsy moth (Lepidoptera: Lymantriidae). *J. Econ. Entomol. 81*:268–273 (1988).
34. Quisumbing, A. R. and Kydonieus, A. F. Laminated structure dispensers. In *Insect Suppression with Controlled Release Pheromone Systems.* (Kydonieus, A. F. and Beroza, M., eds.). CRC Press, Boca Raton, 1982, pp. 213–236.
35. Lewis, W. J. Semiochemicals: Their role with changing approaches to pest control. In *Semiochemicals: Their Role in Pest Control.* (Nordlund, D. A., Jones, R. L., and Lewis, W. J., eds.). John Wiley & Sons, New York, 1981, pp. 3–12.
36. Rothschild, G. H. L. Mating disruption of lepidopterous pests: Current status and future prospects. In *Management of Insect Pests with Semiochemicals.* (Mitchell, E. R., ed.). Plenum Press, New York, 1981, pp. 207–228.
37. Caro, J. H., Bierl, B. A., Freeman, H. P., Glotfelty, D. E., and Turner, B. C. Disparlure: Volatilization rates of two microencapsulated formulations from a grass field. *Environ. Entomol. 6*:877–881 (1977).
38. Plimmer, J. R., Caro, J. H., and Freeman, H. P. Distribution and dissipation of aerially applied disparlure under a woodland canopy. *J. Econ. Entomol. 71*:155–157 (1978).
39. Sartwell, C., Daterman, G. E., Overhulser, D. L., and Sower, L. L. Mating disruption of western pine shoot borer (Lepidoptera: Tortricidae) with widely spaced releasers of synthetic pheromone. *J. Econ. Entomol. 76*:1148–1151 (1983).
40. Palaniswamy, P., Chisholm, M. D., Underhill, E. W., Reed, D. W., and Peesker, S. J. Disruption of forest tent caterpillar (Lepidoptera: Lasiocampidae) orientation to baited traps in aspen groves by air permeation with (5*Z*,7*E*)-5,7-dodecadienal. *J. Econ. Entomol. 76*:1159–1163 (1983).
41. Plimmer, J. R., Bierl-Leonhardt, B. A., and Inscoe, M. N. Evolution of formulations. In *The Gypsy Moth: Research Toward Integrated Pest Management.* (Doane, C. C. and McManus, M. L., eds.). USDA Tech. Bull. 1584, 1981, pp. 536–542.
42. Ravlin, F. W., Bellinger, R. G., and Roberts, E. A. Gypsy moth management programs in the United States: Status,

evaluation, and recommendations. *Bull. Entomol. Soc. Am. 33*: 90–98 (1987).

43. Wilson, R. W., Jr. and Fontaine, G. A. Gypsy moth egg-mass sampling with fixed- and variable-radius plots. USDA *Agric. Handbk.* 523, 1978, 46 pp.
44. Eggen, D. A. and Abrahamson, L. P. Estimating gypsy moth egg mass densities. State University of New York Coll. of Environ. Science & Forestry. Misc. Publ. 1 (ESF 83-002), 1983.
45. Kolodny-Hirsch, D. M. Evaluation of methods for sampling gypsy moth (Lepidoptera: Lymantriidae) egg mass populations and development of sequential sampling plans. *Environ. Entomol. 15*:122–127 (1986).
46. Elkinton, J. S. and Cardé, R. T. The use of pheromone traps to monitor distribution and population trends of the gypsy moth. In *Management of Insect Pests with Semiochemicals.* (Mitchell, E. R., ed.). Plenum Press, New York, 1981, pp. 41–55.
47. Maksimovic, M. Contribution to the investigation of the numerousness of the gypsy moth by means of the trap method. [In Serbo-Coroatian, English summary.] *Zast. Bilja 49/50*:41 (1958).
48. Maksimovic, M. Sex attractant traps with female odor of the gypsy moth used for forecasting the inrease of populations of gypsy moth. *Proc. 12th Int. Congr. Entomol.*, 1965, p. 398.
49. Bednyi, V. D. and Kovalev, B. G. Basis for the use of disparlure for determination and prediction of numbers of the gypsy moth. *Khimoretseptsiya Nasekomykh. 3*:147–151 (1978).
50. Granett, J. Estimation of male mating potential of gypsy moths with disparlure baited traps. *Environ. Entomol. 3*:383–385 (1974).
51. Luciano, P. and Prota, R. Two methods of evaluating population density in *Lymantria dispar* L. *Acad. Naz. Ital. Entomol.* 599–606 (1985).
52. Bellinger, R. G., Ravlin, F. W., and McManus, M. L. Seasonal wing length variation used to predict gypsy moth [*Lymantria dispar* (L.)] population fecundity. Paper presented at the Annual Meeting of the Entomological Society of America. Boston, Mass., November 29–December 3, 1987.
53. Elkinton, J. S. Changes in efficiency of the pheromone-baited milk-carton trap as it fills up with male gypsy moths (Lepidoptera: Lymantriidae). *J. Econ. Entomol. 80*:754–757 (1987).
54. Miao, J. C., Wang, H. L., and Zheng, Z. T. Study on the attraction effect and the biological activation of disparlure of *Porthetria dispar* L. *J. Northeast. For. Inst. 2*:49–57 (1982).

Part IV
Pests of Field Crops

23

Application of the Sex Pheromone of the Rice Stem Borer Moth, *Chilo suppressalis*

SADAHIRO TATSUKI* / University of Tsukuba,
Tsukuba, Ibaraki, Japan

I. INTRODUCTION

The female sex pheromone of the rice stem borer moth, *Chilo suppressalis* (Walker), the most serious rice borer in temperate and subtropical Asia, was first identified as a mixture of (Z)-11-hexadecenal (Z11-16:Ald) and (Z)-13-octadecenal (Z13-18:Ald) (1,2). Since then, several studies have been made on the application of the synthetic pheromone and analogues to direct control of this pest by means of mating disruption (3–8). We have determined that (Z)-5-hexadecene, a deoxygenated analogue of Z11-16:Ald, could be used as a novel disruptant against *C. suppressalis* (5,7).

On the other hand, it was shown that the synthetic two-component blend was less attractive to wild males than were live virgin females, although the two compounds were shown to be essential for male attraction (9). These results suggested the presence of other synergistic pheromone component(s), as found in other lepidopterous species (10), leading us to reexamine the pheromone system of this species. (Z)-9-Hexadecenal (Z9-16:Ald), a positional isomer of Z11-16:Ald, was identified as a third pheromone component which drastically enhanced attractiveness to male moths in the field when combined with the two known components (11). The potent attractiveness of the three-component blend led us to expect it to be more effective than the two-component synthetic blend, not only for mating disruption, but also in other applications such as population monitoring.

**Current affiliation*: The University of Tokyo, Tokyo, Japan

In the present paper, I describe the identification of the third pheromone component, studies on the effectiveness of the sex pheromone in population monitoring, and recent progress of mating disruption trials using a single pheromone component (Z11-16:Ald) and the three-component blend.

II. THIRD PHEROMONE COMPONENT

A. Identification of Pheromone-Related Chemicals

As it was anticipated that the additional components, if any, would have chemical structures related to the known pheromone components (10), we tried at first to detect such chemicals in pheromone gland extracts by instrumental analyses and then to evaluate their biological activities.

From insects reared successively on rice seedlings (12), ovipositor tips of virgin females were carefully removed so that no scales or hairs were included. The removed tips were then soaked in redistilled *n*-hexane for 0.5–1 hr at room temperature. This procedure permitted analysis of the crude extract by capillary gas chromatography (GC) without any purification. The extract (approximately 2000 FE) was analyzed by GC using a glass capillary column coated with CHDMS. With nine major peaks obtained, retention times (Rt) of P1, P3, P5, P6, and P7 were quite similar to those of authentic hexadecanal (16:Ald), Z11-16:Ald, (*Z*)-11-hexadecen-1-ol (Z11-16:OH), octadecanal (18:Ald), and Z13-18:Ald, respectively (Fig. 1). P2 was expected to be one of the isomers of hexadecenal, as it was located in the chromatogram between the peaks of 16:Ald and (*E*)-11-hexadecenal (see Fig. 1). An exact comparison of Rt of P2 and several isomers of hexadecenal including (*Z*)-7-, (*Z*)-9-, (*E*)-10-, and (*E*)-11- isomers was made by means of cochromatography and indicated that the Rt of P2 coincided completely with that of Z9-16:Ald.

Further characterization of suspected compounds was conducted by GC-mass spectrometry (GC-MS) with a fused-silica capillary column coated with PEG-20M. Both mass spectra obtained with the two peaks corresponding to P2 and P3 showed the same pattern as that obtained with authentic Z11-16:Ald. Therefore, P2 and P3 were identified as Z9-16:Ald and Z11-16:Ald, respectively. The other peaks were also identified by GC-MS to be 16:Ald (P1), Z11-16:OH (P5), 18:Ald (P6), and Z13-18:Ald (P7). Thus, four pheromone-related compounds were detected, in additionn to the two known pheromone compounds. The ratio of 16:Ald/Z9-16:Ald/Z11-16:Ald/Z11-16:OH/18:Ald/Z13-18:Ald in the extract was estimated from the GC peak area to be approximately 26:5:48:5:1:6.

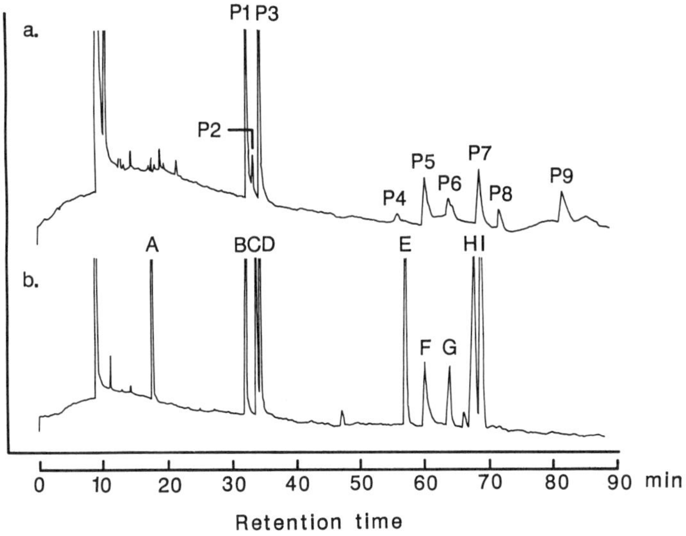

FIGURE 1 Capillary gas chromatographic traces obtained with (a) pheromone gland extract and (b) with mixture of authentic compounds. P1-P9, see text. A, BHT; B, hexadecanal; C, (*E*)-11-hexadecenal; D, (*Z*)-11-hexadecenal; E, (*Z*)-11-hexadecenyl acetate; F, (*Z*)-11-hexadecen-1-ol; G, octadecenal; H, (*E*)-13-octadecenal; I, (*Z*)-13-octadecenal. (From Ref. 11.).

B. Field Tests for Pheromonal Activity

A preliminary field test was conducted during the end of the second flight of *C. suppressalis* in Okayama Prefecture, Japan, in 1982 to compare the attractiveness of a mixture of the six components so far detected with that of the two known components and of virgin females (11). Table 1 clearly shows that the six-component blend was more attractive to male moths than either the two-component blend or the virgin females. However, further field tests could not be conducted because of a decrease in the number of moths.

The next field test was conducted in Bogor, Indonesia, in 1983 in collaboration with the Indonesian Atomic Energy Agency to determine which one(s) among the four pheromone-related compounds had pheromonal activity. The result indicated that a three-component blend containing Z9-16:Ald, in addition to the two known pheromone components, was as attractive as the six-component blend and caught significantly more males than the two-component blend (see Table 1).

TABLE 1 Attractiveness of Various Combinations of Compounds Identified from Pheromone Gland Extract and of Virgin Females

Source (μg/septum)	Total no. males caught at		
	Okayama, Japan[a]	Bogor, Indonesia[b]	Los Baños, Philippines[c]
Z11-16:Ald(250) + Z13-18:Ald(30) +			
none	2b[d]	4b	40b
Z9-16:Ald(25)		37a	959a
16:Ald(150) + 18:Ald(5)		2b	16b
Z11-16:OH(25)		0b	25b
Z9-16:Ald(25) + 16:Ald(150) + 18:Ald(5) + Z11-16:OH(25)	43a	31a	1237a
Virgin females	11b	16ab	156b
Control	2b	0b	0c

Source: [a]Ref. 11; [b]M. Hoedaya, personal communication, [c]Ref. 13.
[d]Values followed by the same letter within each column are not significantly different (DMRT, $P < 0.05$).

Similar results, although with many more male catches, were obtained later in a field test conducted at the International Rice Research Institute, Los Baños, the Philippines (13), which is also shown in Table 1. These results clearly show that Z9-16:Ald is the only component that enhances the attractiveness of the two pheromone components when they are mixed together. It is also notable that in all the field tests the three- and/or six-component blends always caught more males than did live virgin females.

C. Laboratory Wind Tunnel Experiment

With use of a laboratory wind tunnel (7), we observed male behavioral responses to the same combinations of synthetic compounds as used in the field tests and to the crude pheromone extract (11). Either a rubber septum with the synthetic compounds or a filter paper tip with the extract was placed on a piece of white cardboard near the upwind end of the tunnel. For one or two test series, about 50 male moths were introduced into the tunnel. One-minute observations were made of their behavioral responses to each source, presented in random order, followed by an interval of more than 10 min in which no stimulus was offered. The male behavior elicited by the crude extract in the wind tunnel was composed of the following components: (1) upwind orientation flight to within 20 to 50 cm of the pheromone source, (2) reduction of flight speed, followed by hovering and/or backing up, (3) zigzag flight while slowly approaching the source, (4) landing near the source and wing fanning while walking (mating dance), and (5) contact with the source. Table 2 shows that the three-component blend containing Z9-16:Ald added to Z11-16:Ald and Z13-18:Ald elicited the complete mating behavior pattern, as did the crude extract and the six-component blend. On the other hand, the two-component blend usually elicited only the first two behavioral responses. Neither the mixture of saturated aldehydes, 16:Ald and 18:Ald, nor the alcohol, Z11-16:OH, had any appreciable effect on male behavior when mixed with the two-component blend in ratios similar to those in the crude extract. These results also showed that of the four newly identified components only Z9-16:Ald was behaviorally active and that it played an important role in eliciting close-range orientation behavior such as landing and wing fanning in the male *C. suppressalis*. This may explain the synergism of the two-component blend by Z9-16:Ald in the field tests. Conversely, the low male catches by the two-component blend could be the result of the lack of or decrease in close-range orientation behavior of the males.

III. POPULATION MONITORING

A. Attractiveness of Three-Component Pheromone

1. Effects of Component Ratios and Septum Dosage

After the potent attractiveness of the three-component blend was demonstrated, we conducted field tests in Okayama Prefecture to find a better formulation for population monitoring. As pheromone dispensers, rubber septa were used throughout because they had already been shown to be more suitable dispensers than others such as polyethylene capsules (9); BHT (2,6-di-*tert*-butyl-*p*-cresol (10%) was mixed with the synthetic pheromone mixture as an antioxidant.

Proportion of Z9-16:Ald: The effects of the ratio of Z9-16:Ald to the two-component blend were examined by adding varying amounts of Z9-16:Ald to a fixed amount of Z11-16:Ald and Z13-18:Ald in the natural ratio (8:1). Table 3 shows that no significant effects were detected even when the amount of Z9-16:Ald was one-third or three

TABLE 2 Behavioral Responses of Male Moths to Various Combinations of Compounds Identified from Pheromone Gland Extract and to Pheromone Gland Extract in Laboratory Wind Tunnel

Source	Response index[a,b]				
(μg/septum or F.E./filter paper)	OF	SR/H	ZF	L/MD	C
Z11-16:Ald(250) + Z13-18:Ald(30) +					
none	1.6 a	1.6 b	0.2 b	0.0 b	0.0 b
Z9-16:Ald(25)	3.0 a	3.0 a	2.8 a	2.6 a	2.6 a
16:Ald(150) + 18:Ald(5)	2.0 a	1.8 ab	0.0 b	0.0 b	0.0 b
Z11-16:OH(25)	1.6 a	1.2 bc	0.0 b	0.0 b	0.0 b
Z9-16:Ald(25) + 16:Ald(150) + 18:Ald(5) + Z11-16:OH(25)	3.0 a	3.0 a	2.8 a	2.4 a	1.2 ab
Pheromone gland extract(10)	3.0 a	3.0 a	2.8 a	2.8 a	1.6 ab
Control	0.0 b	0.0 c	0.0 b	0.0 b	0.0 b

[a]Two or more males responded simultaneously: 3; two or more males responded, although separately: 2; one male responded: 1; no response: 0. OF, upwind orientation flight; SR/H, speed reduction and hovering; ZF, zigzag flight; L/MD, landing and mating dance; C, contact with the source.
[b]Values followed by the same letter within each column are not significantly different (DMRT, $P < 0.05$).
Source: Ref. 11.

TABLE 3 Effect of the Ratio of Z9-16:Ald Added to the Two-Component Blend on Male Catches

Amount of Z9-16:Ald (μg/septum)	Amount of 2-component blend[a] (μg/septum)	Total no. males caught[b]
	270	1 d
8.3	270	84 ab
25[c]	270	90 a
75	270	67 abc
250	270	32 bcd
Virgin females		16 cd
Control		0 d

[a]Z11-16:Ald and Z13-18:Ald (8.1).
[b]Values followed by the same letter are not significantly different (DMRT, $P < 0.05$).
[c]Natural ratio.
Source: Tanaka et al., unpublished data.

times that in the natural ratio. However, the attractiveness significantly decreased when the proportion of Z9-16:Ald was increased to 10 times the natural ratio. We discovered no other ratios with higher activity than the natural ratio. Again, the attractiveness of the synthetic pheromone was higher than that of virgin females in this experiment.

Septum dosage: In our previous field test using the two-component blend, the optimum dosage range for male attraction was rather narrow (9). Consequently, we tested the effect of septum dosage of the three-component blend on male catches. In the first trial, however, there were no significant differences in the attractiveness, with dosages between 0.1 and 1 mg/septum, although 3 mg/septum showed a little lower attractiveness (Table 4, Test A). Therefore, we conducted the next field trial using a wider range of dosages, and found a lower threshold level of less than 0.01 mg and a trend toward decreased attractiveness at 10 mg, although this was with a very low population density (see Table 4, Test B).

From these results, we have tentatively been using a rubber septum containing 0.3 mg of the three-component blend (natural ratio) and 0.03 mg BHT as a standard formulation for population monitoring.

TABLE 4 Effect of Septum Dosage of the Three-Component Blend on Male Catches

Test	Amount of 3-component blend[a] (mg/septum)	Total no. males caught[b]
A	0.1	87 a
	0.3	90 a
	1.0	83 a
	3.0	52 b
	Control	0 c
B	0.001	0 b
	0.01	5 ab
	0.1	20 a
	1.0	9 ab
	10.0	1 b
	Control	0 b

[a]Z11-16:Ald, Z13-18:Ald, and Z9-16:Ald (48:6:5).
[b]Values followed by the same letter within each test are not significantly different (DMRT, $P < 0.05$ for test A and Freedman's test followed by multiple-range test, $0.05 < P < 0.10$ for test B.
Source: Tanaka et al., unpublished data.

This standard formulation retains the same level of attractiveness for about 50 days under the field conditions (14).

2. *Comparison with Light Trap*

The attractiveness of a pheromone trap baited with the standard formulation was compared with a standard light trap, which conventionally has been used to monitor the occurrence of the rice stem borer in Japan (15–17). Figure 2 shows some typical results. In most cases, a pheromone trap always caught several times more moths than the light trap throughout the moth occurrence period; in other words, the fluctuation patterns of both moth catch curves were generally parallel, with more moth catches by the former (see Fig. 2a,b). However, although in very few cases, specifically in the second flight period, moth catches by the light trap sometimes exceeded those by a pheromone trap (see Fig. 2c). This may have been due to some unknown ecological factors such as the bias of local population density or to microclimatic factors inside the test fields. Although we

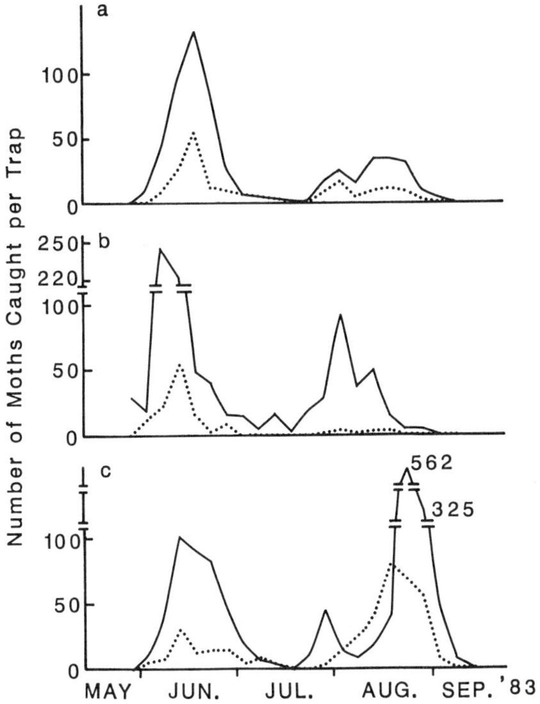

FIGURE 2 Comparison of trap catches with sex pheromone trap (male moths, solid lines) and with light trap (male and female moths, dotted lines) at three locations of Japan. (a) Yasuzuka Town, Niigata Pref.; (b) Suzaka City, Nagano Pref.; (c) Ogata Town, Niigata Pref. [(a) and (b) from Ref. 16; (c) Kanno et al., unpublished data.]

must clarify those problems, it has been apparent that a pheromone trap can be used as a much more sensitive monitoring tool than the light trap. There should be no major obstacles in the use of a pheromone trap in place of a light trap for the estimation of timing of infestation by the subsequent generation larvae.

IV. MATING DISRUPTION

Previously, by screening various pheromone-related chemicals, we found (*Z*)-5-hexadecene to be an effective disruptant with greater stability under field conditions than the pheromone components, because the aldehyde formulations were very unstable in the field (3,

5,7). Since the 1981 tests, however, the use of the aldehyde pheromone component(s) has become possible with the development of a controlled-release technique and the development of effective stabilizing agents by our collaborative companies.

A. Screening of Disruptant

Because the major pheromone component, Z11-16:Ald, had previously been shown to have higher disruptive activity than Z13-18:Ald (3,5,7), and also to be as effective as a mixture of these two components (Kanno et al., unpublished data), we had used Z11-16:Ald alone as a disruptant for larger-scale experiments in 1981 and 1982 as described later. However, the discovery of the third pheromone component, Z9-16:Ald, led us to further investigations in search of better disruptants containing Z9-16:Ald.

To select a better disruptant we used a small-scale orientation disruption test in paddy fields in Niigata Prefecture. In each test plot, a synthetic pheromone trap was surrounded by 16 evenly spaced polyethylene capillary dispensers containing a test compound or mixture as shown in Figure 3. The disruptive activity was evaluated by the percentage inhibition of male attraction to the pheromone trap, compared with the control trap. The results obtained so far have shown

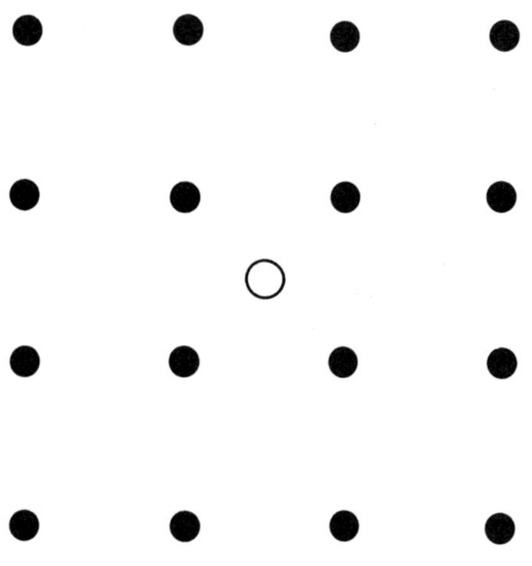

FIGURE 3 Arrangement of virgin female/pheromone trap (open circle) and 16 evaporation points (closed circles) of compounds in orientation disruption experiments.

that the three-component blend in the natural ratio has higher disruptive activity than either single component, Z11-16:Ald (Table 5) or Z9-16:Ald (Table 6). Also, these results suggested that Z9-16:Ald was less active than Z11-16:Ald, although no direct comparison was made.

B. Release Rate and Distance Between Dispensers

In larger-scale mating disruption trials, in general, there have been two distinct methods of spacing dispensers of a disruptant. One is the "spray method" of which the use of microcapsules is representative; these should be placed close together to obtain almost uniform permeation of the air with the disruptant vapor. Another is the "point-source" method in which dispensers with relatively high disruptant release rates may be sparsely spaced. We have adopted the point-source method for *C. suppressalis*, since the spray method may be less efficient in paddy fields because of the presence of irrigated water. As it is desirable in the point-source method that the evaporation points are spaced as far apart as possible from the viewpoint of labor and working in the paddies, we examined the effects of the

TABLE 5 Comparison of Orientation Disruption Effect of Z11-16:Ald and the Three-Component Blend

Test	Distance between dispensers[a] (m)	Release rate in unit area (mg/ha/day)	% Orientation disruption[b]	
			Z11-16:Ald	3-component
A	1	1999	100 a	100 a
	4	125	99.6 a	100 a
	16	8	67.5 b	92.8 a
B[c]	1	1499	100 a	100 a
	4	94	94.0 a	100 a
	16	6	31.1 b	91.4 a

[a]Release rate: ca. 0.2 mg/day/dispenser.
[b]Values followed by the same letter within each test are not significantly different (DMRT, $P < 0.01$).
[c]Arrangement of a pheromone trap and dispensers was the same as in Figure 3 except for removal of the inner four dispensers.

TABLE 6 Comparison of Orientation Disruption Effect of Z9-16:Ald and the Three-Component Blend[a]

Release rate		% Orientation disruption[b]	
From each dispenser (mg/d)	In unit area (mg/ha/d)	Z9-16:Ald	3-component
0.1	250	77.8 b	99.9 a
0.03	75	74.7 b	98.9 a
0.01	25	40.6 c	98.4 a

[a]Distance between dispensers was 2 m.
[b]Values followed by the same letter are not significantly different (DMRT, $P < 0.01$).
Source: Kanno et al., unpublished data.

release rate of Z11-16:Ald and distance between evaporation points on male orientation disruption in the field in Niigata Prefecture (18).

Arrangement of a virgin female trap and evaporation points in each test plot was the same as shown in Figure 3. Up to a distance of 16 m between evaporation points, a greater than 90% orientation disruption was obtained when the release rate was higher than 50 mg/ha per day regardless of the distances (Table 7, Test A). At distances farther than 32 m, the percentage of disruption tended to decrease even when the overall release rate was the same (see Table 7, Test B). However, at 32 m, the percentage rose to nearly 90% when the release rate was increased to 234 mg/ha per day. Furthermore, a significant disruptive effect was obtained even at 64 m. These results suggest the possibility that distances between evaporation points can be further increased while maintaining effective orientation disruption if the release rate is sufficiently high.

C. Mating Disruption over Paddies

Larger-scale disruption experiments using the major pheromone component, Z11-16:Ald, were conducted in 1981 and 1982 in paddy fields at Okayama Prefecture, Japan, where rather high populations of the borer had been observed (19). Since 1983, the three-component blend has been used as a disruptant in the same areas. Throughout these

TABLE 7 Effects of Release Rate of Z11-16:Ald and Distance Between Evaporation Points on Male Orientation Disruption

Test	Distance between evaporation points (m)	Release rate: From each point (mg/d)	Release rate: In unit area (mg/ha/d)	% Orientation disruption[a]
A[b]	1	0.08	799	100 a
	2	0.08	199	99 ab
	4	0.08	50	96 ab
	8	0.08	12	84 b
	16	0.08	3	59 c
	4	0.32	199	97 ab
	8	0.32	50	91 ab
	16	0.32	12	50 c
	8	1.28	199	97 ab
	16	1.28	50	91 ab
B[c]	16	1.5	59	72 ab
	32	6.0	59	42 b
	32	24.0	234	89 a
	64	96.0	234	73 ab

[a]Values followed by the same letter within each test are not significantly different (DMRT, $P < 0.05$).
[b]1980 test; polyethylene capillary dispensers were used.
[c]1981 test; plastic laminate dispensers were used.
Source: Ref. 18.

trials, we have used the point-source method, with either sealed polyethylene capillary tubes or plastic laminate films as disruptant dispensers. The dispensers were evenly spaced in the grid with distances between dispensers of 5 or 3 m. Release rates from the dispensers were estimated by measuring either the decrease in fluid length of disruptant in the polyethylene capillary tube or the weight loss of the laminated dispensers. The disruptive effect in the test paddies during treatment was monitored by comparing the results with monitor traps (virgin female or synthetic pheromone) and tethered females in the treated paddy with those in the control paddy unless otherwise stated. The degree of control obtained by the disruption treatment was evaluated by examining the extent of infestation of rice stems by the next-generation larvae. The population density of the insect in the test paddies before treatment was estimated from the extent of infestation caused by the previous larvae of the adults concerned. The dispensers used in the second flight in 1981 and 1982 were renewed once after approximately 15 days from the onset of the treatment because of a gradual decrease of release rate. However, the polyethylene capillary dispensers with thicker walls that have been used since 1983 keep a very stable release rate for longer than 1 month even under field conditions in the summer season, and have not needed renewal.

1. Mating Disruption Against First-Flight Moths

Disruption treatment with Z11-16:Ald for the first-flight moths (i.e., second or overwintering generation) was made using the paddies of Okayama Prefectural Agricultural Experiment Station, San-Yo Town, from June 23 to July 20, 1981. The mean release rate of Z11-16:Ald from polyethylene capillary dispensers during the first 2 weeks, in which the most of moths would mate, was 0.8 g/ha per day. Table 8 shows that a very satisfactory disruptive effect was suggested by the monitoring tests. However, all the feral female moths collected from both the treated and the control paddies during the treatment were mated individuals, and no apparent control was seen in the examination of rice stems after the treatment. This may be explained by the movement of overwintering insects; full-grown larvae usually bore into rice straws to hibernate in them. Rice straws are often transferred to other places (e.g., barns, greenhouses, orchards, and such) which, consequently, would be emergence sites of the overwintering generation adults. As female moths usually mate near emergence sites, most of the females in paddies may be immigrants that have already mated. Thus, disruption treatment over paddies during the first-flight period of *C. suppressalis* may be of no use for control. Nevertheless, this experiment appeared to show that the pheromone component itself could be used as a disruptant for a relatively long period when formulated into suitable dispensers, such as polyethylene capillaries.

2. Mating Disruption Against Second-Flight Moths

Some control by the disruption treatment over paddies against the second-flight moths (i.e., first generation) was expected because all of the adults should emerge and mate throughout the paddies. Experiments have been conducted since 1981 using the paddies of either the Okayama Prefectural Agricultural Experiment Station, San-yo Town or private paddies at Soja City where the highest population density in Okayama Prefecture had been observed several years before or the

TABLE 8 Mating disruption with Z11-16:Ald Against First-flight Moths

Treatment (ha)	Monitoring tests during treatment: VF trap (no. males/trap/night)	Tethered female (% mated)	Feral female (% mated)	Control effect after treatment (% infested stems)
Treated (1.2)	0.3	1.1 (n = 159)	100 (n = 18)	0.72
Control (1.0)	6.1	45.1 (n = 163)	100 (n = 15)	0.67

Source: Ref. 19.

paddies of University of Okayama, Kurashiki City. Treatment with the disruptant was continued for about 1 month from the end of July or the first of August, during which time most of the moths emerged.

In some experiments, we obtained rather good results, not only in the monitoring tests, but also in the control effect, whereas in other cases we failed to achieve positive control. To analyze these results, the various factors concerning experimental conditions, the results of the monitoring tests during the treatment, and the control effects are summarized in Table 9. Several important relationships are suggested. The control effect should be closely related to both the pretreatment population density of the larvae, which would be reflected in the moth population density during treatment, and the release rate of the disruptant in the unit area. Namely, rather good control effects were obtained at a release rate of around 0.5 g/ha per day, with a low population density (<1% of infested stems; 1983–1985), whereas no appreciable effects were noted with 0.5–1.0 g/ha per day with the middle level of population density (1–3% of infested stems: 1981A, 1982A). However, control was obtained with the higher level of release rate, 1.2–1.6 g/ha per day, even with a high population density (>3% of infested stems; 1981B).

In contrast to the results obtained with first-flight moths, the results of the monitoring tests correlated well with the degree of control. Briefly, to obtain good control, it may be necessary to achieve orientation disruption greater than 95% or mating disruption in tethered females greater than 90% in the monitoring tests. In addition, these relationships were observed even though only relatively small areas were used, suggesting that mated females of this generation did not disperse as readily as did the overwintering females.

TABLE 9 Summary of Mating Disruption Trials Against Second-Flight Moths from 1981 to 1985 in Okayama Prefecture

Year	Location[a]	Treated area (ha)	Disruptant[b]/ formulation[c]	Release rate (g/ha/d)	Population before treatment (% infestation)	Monitoring tests (% disruption)		Control effectiveness (% suppression of infestation)
						Orientation	Mating	
1981A	OAS	1.0	S/C	0.7–0.9	1–2	80	93	5.9
1981B	SOJ	0.19	S/L	1.2–1.6	4–7	100	85	77.1
1982A	SOJ	0.18	S/L	0.6–0.9	2–3		62	-77.1
1982B	SOJ	0.3	S/L	0.6–0.9	2–5		88	1.9
1983	OAS	1.0	B/C	0.3–0.7	<1	100	100	75.0
1984	SOJ	0.7	B/C	0.4–0.6	<1	88	94	71.1
1985	UNO	0.24	B/C	1.0–1.2	<1	100	100	78.6

[a]OAS, Okayama Prefectural Agricultural Experiment Station; SOJ, Soja City; UNO, University of Okayama.
[b]S, Z11-16:Ald; B, three-component blend.
[c]C, polyethylene capillary tubes; L, plastic laminate films.
Source: Ref. 19.

In these experiments, there seemed to be no appreciable difference in control obtained with Z11-16:Ald alone or with the three-component blend, although a direct comparison could not be made.

D. Supplemental Small-Scale Disruption Experiment

To examine the effects of release rate of a disruptant in a given area on the disruptive effect, a small-scale disruption experiment was conducted at Soja City in 1986 with a rather high moth population density. In each test plot, a trap baited with the synthetic pheromone and polyethylene capillary tubes containing the three-component blend were spaced as in Figure 3 with 3-m intervals between evaporation points. Tethered females were also placed near the trap. Three different levels of release rate, 0.5, 1.0, and 2.0 g/ha per day, were used considering the data shown in Table 9. Although there were no significant differences in either the number of males caught or the percentages of the females mated with the three levels of release rate, the percentages tended to decrease as the release rate increased, especially the percentage of mating (Table 10). If greater than 90% mating disruption in tethered females is necessary to achieve good

TABLE 10 Effects of Release Rate of Three-Component Blend in a Unit Area on Disruptive Effects in a Small-Scale Experiment

	Pheromone trap		Tethered female	
Release rate (g/ha/d)	Total no. males caught[a]	% Orientation disruption	% Mated[a]	% Mating disruption
0.5	6 b	93.0	19.7 ab (n = 71)	64.4
1.0	3 b	96.5	14.0 b (n = 60)	74.7
2.0	2 b	97.7	3.0 b (n = 63)	94.6
Control	86 a		55.3 a (n = 68)	

[a]Values followed by the same letter within each column are not significantly different (DMRT, $P < 0.05$).

Source: Tatsuki et al., unpublished data.

control, as already suggested, these results may indicate that a release rate of nearly 2 g/ha per day is necessary with a high moth population density. This release rate roughly coincided with that discussed before.

V. CONCLUDING REMARKS

Since the finding of the third pheromone component, the potent attractiveness of the three-component blend has raised the possibility of using the pheromone in population monitoring. It is expected that, in the near future, pheromone traps as monitoring tools will be used in a government forecasting program for the rice stem borer throughout Japan. It may also be worth reexamining the use of mass trapping under some conditions, because of the greater attractiveness of the three-component blend than that of virgin females.

In mating disruption, both the technical development of formulating stable dispensers containing the aldehyde pheromone components and the use of the three-component blend seem to have increased the feasibility of using this method to controlling the rice stem borer. Some information on the relationship between the moth population density and the release rate required for appreciable control has been obtained. However, further understanding of the mechanism of disruption, as well as more information on the movement or dispersal of the adults, will be important for the practical application of mating disruption.

REFERENCES

1. Nesbitt, B. F., Beevor, P. S., Hall, D. R., Lester, R., and Dyck, V. A. Identification of the female sex pheromones of the moth, *Chilo suppressalis*. *J. Insect Physiol. 21*:1883–1886 (1975).
2. Ohta, K., Tatsuki, S., Uchiumi, K., Kurihara, M., and Fukami, J. Structures of sex pheromones of rice stem borer. *Agric. Biol. Chem. 40*:1897–1899 (1976).
3. Kanno, H., Tatsuki, S., Uchiumi, K., Kurihara, M., Fukami, J., Fujimoto, Y., and Tatsuno, T. Disruption of courtship behavior of the rice stem borer moth, *Chilo suppressalis* Walker, with components of the sex pheromone and their related chemicals. *Appl. Entomol. Zool. 13*:321–323 (1978).
4. Beevor, P. S. and Campion, D. C. The field use of 'inhibitory' components of lepidopterous sex pheromones and pheromone mimics. In: *Chemical Ecology: Odour Communication in Animals*. Ritter, F. J. (ed.). Elsevier/North-Holland, Amsterdam (1979), pp. 313–325.

5. Kanno, H., Hattori, M., Sato, A., Tatsuki, S., Uchiumi, K., Kurihara, M., Fukami, J., Fujimoto, Y., and Tatsuno, T. Disruption of sex pheromone communication in the rice stem borer moth, *Chilo suppressalis* Walker (Lepidoptera: Pyralidae), with sex pheromone components and their analogues. *Appl. Entomol. Zool. 15*:465–473 (1980).
6. Beevor, P. S., Dyck, V. A., and Arida, G. S. Formate pheromone mimics as mating disruptants of the striped rice borer moth, *Chilo suppressalis* (Walker). In: *Management of Insect Pests with Semiochemicals.* Mitchell, E. R. (ed.). Plenum Press, New York (1981), pp. 305–312.
7. Tatsuki, S. and Kanno, H. Disruption of pheromone communication in *Chilo suppressalis* with pheromone and analogs. In: *Management of Insect Pests with Semiochemicals.* Mitchell, E. R. (ed.). Plenum Press, New York (1981), pp. 313–325.
8. Lee, J. O., Park, J. S., Goh, H. G., Kim, J. H., and Jun, J. G. Field study of mating confusion of synthetic sex pheromone in the striped rice borer, *Chilo suppressalis* (Lepidoptera: Pyralidae). *Korean J. Plant Prot. 20*:25–30 (1981).
9. Tatsuki, S., Kurihara, M., Uchiumi, K., Fukami, J., Fujimoto, Y., Tatsuno, T., and Kishino, K. Factors improving field trapping of male rice stem borer moth, *Chilo suppressalis* Walker (Lepidoptera: Pyralidae) by using synthetic attractant. *Appl. Entomol. Zool. 14*:95–100 (1979).
10. Tamaki, Y. Complexity, diversity, and specificity of behavior-modifying chemicals in Lepidoptera and Diptera. In *Chemical Control of Insect Behavior – Theory and Application.* Shorey, H. H. and McKelvey, J. J., Jr. (eds.). John Wiley & Sons, New York (1977), pp. 253–285.
11. Tatsuki, S., Kurihara, M., Usui, K., Ohguchi, Y., Uchiumi, K., Fukami, J., Arai, K., Yabuki, S., and Tanaka, F. Sex pheromone of the rice stem borer, *Chilo suppressalis* (Walker) (Lepidoptera: Pyralidae): The third component, *Z*-9-hexadecenal. *Appl. Entomol. Zool. 18*:443–446 (1983).
12. Uchiumi, K. Mass rearing method of the rice stem borer, *Chilo suppressalis* W. (Lepidoptera: Pyralidae). *Rikagaku Kenkyusho Hokoku 50*:70–74 (1974).
13. Mochida, O., Arida, G. S., Tatsuki, S., and Fukami, J. A field test on a third component of the female sex pheromone of the striped stem borer, *Chilo suppressalis*, in the Philippines. *Entomol. Exp. Appl. 36*:295–296 (1984).
14. Kanno, H., Abe, N., Tatsuki, S., and Fukami, J. The active term of the synthetic sex pheromone of the rice stem borer moth, *Chilo suppressalis* Walker (Lepidoptera: Pyralidae). *Proc. Assoc. Plant Prot. Hokuriku No. 32*:42–43 (1984).

15. Kanno, H., Onozuka, K., Mizusawa, M., Saeki, Y., Koike, K., Tatsuki, S., and Fukami, J. Comparison of trap efficiency between the synthetic sex pheromone and the light trap in the rice stem borer moth, *Chilo suppressalis* Walker (Lepidoptera: Pyralidae). *Proc. Assoc. Plant Prot. Hokuriku No. 32*:44–46 (1984).
16. Kanno, H., Abe, N., Mizusawa, M., Saeki, Y., Koike, K., Kobayashi, S., Tatsuki, S., and Usui, K. Comparison of the trap efficiency and of the fluctuation pattern of moth catches between the synthetic sex pheromone and the light-trap in the rice stem borer moth, *Chilo suppressalis* Walker (Lepidoptera: Pyralidae). *Jpn. J. Appl. Entomol. Zool. 29*:137–139 (1985).
17. Nakano, K., Ando, T., Koyama, S., and Emura, K. Practical use of the synthetic sex pheromone for population monitoring of the rice stem borer moth, *Chilo suppressalis* Walker. *Proc. Assoc. Plant Prot. Hokuriku No. 34*:12–15 (1986).
18. Kanno, H., Hattori, M., Sato, A., Tatsuki, S., Uchiumi, K., Kurihara, M., Ohguchi, Y., and Fukami, J. Release rate and distance effects of evaporators with *Z*-11-hexadecenal and *Z*-5-hexadecene on disruption of male orientation in the rice stem borer moth, *Chilo suppressalis* Walker (Lepidoptera: Pyralidae). *Appl. Entomol. Zool. 17*:432–438 (1982).
19. Tanaka, F., Yabuki, S., Tatsuki, S., Tsumuki, H., Kanno, H., Hattori, M., Usui, K., Kurihara, M., Uchiumi, K., and Fukami, J. Control effect of communication disruption with synthetic pheromones in paddy fields in the rice stem borer, *Chilo suppressalis* (Walker) (Lepidoptera: Pyralidae). *Jpn. J. Appl. Entomol. Zool. 31*:125–133 (1987).

24

The Use of Pheromones for the Control of Cotton Bollworms and *Spodoptera* spp. in Africa and Asia

LAWRENCE J. McVEIGH, DEREK G. CAMPION, and BRIAN R. CRITCHLEY / Overseas Development Natural Resources Institute, Chatham Maritime, Chatham, Kent, England

I. INTRODUCTION

Although pheromones have been used successfully for the control of insect pests in the developed world (1,2), their use as a control technique in the developing world has largely been neglected. This may be because their methods of use are considered too sophisticated for use by Third World peasant farmers or because of perceived problems with their registration and marketing.

This paper describes work by staff of the Overseas Development Natural Resources Institute (ODNRI) to demonstrate the effectiveness and suitability of pheromones in the developing world. The use of pheromones has been tested within differing pest control systems in Egypt, Pakistan, and east Africa. The importance of cooperation with local plant protection agencies and enlightened agrochemical companies is also described.

II. CONTROL OF *PECTINOPHORA GOSSYPIELLA* IN EGYPT

Three years of trials in Egypt, from 1981 until 1983, examined the possibilities for control of the pink bollworm of cotton, *Pectinophora gossypiella* (Saunders) using laminate flake, microencapsulated, and hollow fiber formulations of the pheromone supplied, respectively, by BASF, ICI, and Sandoz. Each formulation was applied aerially or by hand three to five times during the cotton season at rates of between 20 and 50 g/ha per season. Levels of control comparable with, and sometimes better than, a conventional pesticide application program were achieved (3). It has also been demonstrated that populations of beneficial insects are much higher in pheromone treated areas than in those treated with broad-spectrum insecticides (3,4).

As a result of these trials, *P. gossypiella* pheromones were first used commercially in Egypt in 1984. By 1986, the area under treatment with pheromones had grown to 25,000 ha. This was the successful culmination of a collaborative project between ODNRI and the Egyptian Academy of Science and Technology, together with support from three major agrochemical companies, BASF (West Germany), ICI (UK), and Sandoz (Switzerland). The use of pheromones on such a scale was unique in the developing world and has been undertaken on a comparable scale only in the United States.

More recently the "twist-tie" pheromone formulation marketed by the Mitsubishi Corporation (Japan), which requires only one application for season-long control of *P. gossypiella*, has been successfully evaluated in large-scale trials in Egypt (G. Moawad, personal communication). Only hand application is possible with this formulation, but this may be of particular relevance in developing countries in which modern technologies are not always available and labor is cheap.

In Egypt, full responsibility for controlling the cotton pest complex rests with the Ministry of Agriculture. Although an environmentally acceptable control strategy is favored, possible risks of failure with a relatively new technique, coupled with a strong lobby from the private sector promoting conventional insecticides, has resulted in a policy of caution and the need for further demonstrations of efficiency.

In the absence of conventional pesticides in pheromone-treated areas in Egypt, it has been argued that there could be an upsurge of the Egyptian cotton leafworm *Spodoptera littoralis* (Boisduval) in midseason and of the spiny bollworm *Earias insulana* (Boisduval) late in the season.

Spodoptera littoralis infestations early in the season are usually controlled throughout the country by the hand collection of egg masses using teams of children (5). This policy, in addition to strict sanitation at the end of the season, has enabled standard applications of insecticide for the control of *S. littoralis*, *P. gossypiella*, and *E. insulana* to be restricted to between four and five for the entire cotton-growing season.

An integration of mechanical and pheromonal selective control measures, followed by the application of broad-spectrum insecticides later in the season would enable further reductions to be made in insecticide usage, but would maintain the "safety net" of insecticide use. This strategy was first shown to be effective in Egypt in 1983 in a series of 40-ha blocks of cotton (6).

During 1987, a similar strategy was adopted by the Egyptian Ministry of Agriculture, under ODNRI supervision, in a total area of 3200 ha. After hand collection of egg masses for *S. littoralis* control, two or three *P. gossypiella* pheromone applications were made using microcapsules and hollow fibers, followed by one or two sprays of conventional insecticide.

The results (7) showed that mean infestations by bollworms were not significantly different in the pheromone-treated areas when compared with areas subjected to the standard insecticide regimen. Estimated yields of seed cotton from the pheromone-treated areas were similar to, if not greater than, those from insecticide-treated areas. In one area, the estimated yield of seed cotton in the pheromone-treated plots was nearly 33% higher than in the insecticide-treated check plots. The extra yield from the pheromone-treated areas was due to the cotton plants having produced more and bigger bolls. A survey of boll weights showed that, in one area, bolls were nearly 17% heavier than those from nearby insecticide-treated areas. It is possible that the increase in boll weight in the pheromone-treated area was due to increased cross-pollination of the cotton plants as the result of the survival of the bees in the absence of broad-spectrum insecticides. Cross-pollination as the result of pollinating bees has been reported elsewhere to increase hybrid vigor, resulting in both a greater percentage of boll setting and increased boll weight (8).

An additional benefit, which is important to the small farmers, was that, as a result of the preservation of the bee population, honey production in hives sited in the locality of the pheromone trial sites increased from zero in 1986, when the area was treated with insecticide only, to more than 10,000 kg in 1987.

Commercial applications of *P. gossypiella* pheromone will resume in Egypt during 1988 on a total of 12,000 ha, based on a program of two applications of the microencapsulated or hollow fiber formulations of the pheromone or one application of the twist-ties, followed by one or two applications of conventional insecticide.

III. CONTROL OF *PECTINOPHORA GOSSYPIELLA* AND *EARIAS* SPECIES IN PAKISTAN

The hollow fiber, microencapsulated, and twist-tie formulations of *P. gossypiella* pheromone have been tested in Pakistan in cooperation with the Pakistan Central Cotton Committee. Because aerial application is not available, the microcapsules are applied by motorized knapsack sprayers and the fiber and twist-tie formulations by hand. The

P. gossypiella pheromone formulations used in conjunction with early-season applications of systemic insecticides and late-season applications of pyrethroids or organophosphates gave control of the pest complex equal to, or better than, that of the conventional insecticide program (9). Sole use of pyrethroids for cotton pest control was observed, on occasion, to result in outbreaks of two-spotted mites that necessitated acaricidal treatments. No such outbreaks occurred in the pheromone-treated blocks. After trials in 1987 that have confirmed these results (10), the three formulations are expected to gain registration for sale on the open market in 1988.

Both the spiny bollworm *E. insulana* and the spotted bollworm *E. vittella* occur in Pakistan and are now controlled with insecticides. Multicomponent pheromones for both species have been identified and synthesized at ODNRI (11,12). They have a common main component, (*E*,*E*)-10,12-hexadecadienal, and preliminary trials in Pakistan indicated that this single component can be used as a common disruptant (13). Season-long pheromonal control of the bollworm complex is therefore possible.

Trials, in 1987, in Pakistan that used twist-tie and black hollow fiber formulations of the main component of *Earias* pheromone have demonstrated their effectiveness in controlling both of the *Earias* spp. in a large area (10). The twist-tie formulation was applied once only and provided season-long control of both *Earias* spp. The hollow fibers were applied three or four times at 10-day intervals. At the end of the season, numbers of bolls per plant and the estimated yield of seed cotton in plots that were treated with both the *P. gossypiella* pheromone and the *Earias* pheromone were greater than in plots treated with *P. gossypiella* pheromone only or plots treated with insecticides. *Earias* damage cannot be related to boll infestations alone. Both *E. insulana* and *E. vittella* attack the pin-squares and flowers, causing premature shedding. The early-season protection of the cotton by the pheromone against attack by *Earias* spp. may well be the reason for the higher numbers of bolls and higher estimated yields in the treated plots.

Night observations using regular "walk-rounds" showed a complete absence of moth flight activity and mating pairs in the pheromone-treated areas, unlike the situation in insecticide-treated areas where *Earias* activity was readily seen (10).

Insecticide applications were required against sucking pests in all the plots. Nevertheless, it was possible to show that insecticide applications were reduced by two or three in the pheromone-treated plots, when compared with plots at which no pheromone was applied. This reduction in insecticide usage is important in this kind of agricultural system for which the pest complex can be controlled only by the application of a combination of pesticides and for which the number of applications of insecticide per season is increasing rapidly.

IV. CONTROL OF *SPODOPTERA* SPP. IN AFRICA

Much attention has been paid to the pheromones of *Spodoptera* spp. The main components of the pheromones of two of the *Spodoptera* spp. were identified and synthesized at ODNRI. These are the pheromones of the Egyptian cotton leafworm *S. littoralis* (14) and the African armyworm *S. exempta* (Walker) (15). Both of these insects are of major economic importance in the African continent. Direct control of these insects with pheromones has proved difficult, partly because their pheromones are multicomponent and are relatively unstable and partly because the insects themselves are polyphagous. In addition, *S. exempta* is a vigorous flyer and has the potential to migrate many hundreds of miles (16,17).

Spodoptera littoralis has been shown to have a limited migratory potential (18,19); therefore, the prospects for its control by use of pheromones are greater. Work in Egypt with microencapsulated formulations of its pheromone supplied by ICI showed that mating disruption of *S. littoralis* is possible, but only at rates of active ingredient of 40 to 80 g/ha^{-1} per application (20,21). At today's prices for the pheromone, these rates are too high to be economically viable.

Field observations, using night vision equipment, showed that adult male *S. littoralis* moths will land on, and remain in contact with, point sources of pheromone at densities of 500–1000/ha^{-1} for an average of 3 sec (B. W. Bettany, unpublished data). At higher numbers of pheromone sources, trail masking or confusion occurs such that the frequency of moth arrival at the sources is reduced. At the same time, laboratory studies showed that up to 100% of moths exposed to insecticide-spiked pheromone sources for 10 to 20 seconds will die within 24 hr, and the mating capability of survivors is seriously impaired (K. de Souza, personal communication). It has already been shown that sublethal doses of insecticide mixed with a pheromone formulation adversely affect *P. gossypiella* moths (22), and, therefore, it was thought that a "lure-and-kill" technique could be developed for *S. littoralis*. The levels of pheromone required for this technique are not likely to exceed 2.5 g/ha^{-1} per application and, therefore, would be economically acceptable. The levels of insecticide required would also be very low. The effectiveness of such a lure-and-kill technique for *S. littoralis* is being tested in the field in Egypt.

The trials have so far demonstrated that it may be possible to control *S. littoralis* by this technique, at a density of 500 pheromone and insecticide sources per hectare (L. J. McVeigh, unpublished data). Trap catch in treated plots is reduced, and mating of virgin female moths tethered in treated plots can be reduced by 100%. However, because of the unstable nature of the pheromone, persistence of the joint pheromone and insecticide formulation in the field is a problem that still remains to be overcome. The polyphagous habits of the cotton leafworm also poses practical problems to the application of this technique, and these will require further examination.

Nevertheless, these trials have provided a model for a lure-and-kill technique of insect control that is practically and technically feasible. It may also be possible to incorporate this technique into other systems for other pests, such as the diamond-back moth *Plutella xylostella* (L.) for which a reduction in the use of insecticides is required, but control with pheromones alone may not be possible.

Spodoptera exempta is a migrant pest in central, eastern, and southern Africa and can devastate large areas of cereals and grasslands. Because of its migratory nature and the difficulties in predicting likely sites of outbreaks, it is not a suitable candidate for direct control with pheromones. However, its pheromone is used extensively in a trap network (23) set up to monitor its movements, as part of a forecasting system being developed jointly by ODNRI and the Desert Locust Control Organization in east Africa. There is interest in extending this network, which at present utilizes more than 3500 lures per annum, to include west Africa, where *S. exempta* has recently become important.

For maximum efficiency, it is essential that the trap network uses a standardized and effective trap, incorporating the most sensitive pheromone blend available. A recent reexamination of the pheromone of *S. exempta* has disclosed the presence of four hitherto undetected components (24). One of these has been shown to enhance catch significantly when compared with the originally identified binary mixture that is still used as the standard lure in the trap network. Significant differences in catch have also been shown to arise from the different trap types used in the network. Field trials have also shown that a plastic funnel trap is the most effective type (25). In addition, funnel traps are robust, require a minimum of maintenance, and are cheap. Replacement of the existing mixture of traps in the network with funnel traps has already begun. If comparisons between the two- and three-component pheromone blends in different localities in 1988 confirm the enhanced attraction of the three-component mixture, it will replace the standard binary mixture that is now used.

V. CONCLUSIONS

The work of ODNRI has clearly shown that pheromones have considerable potential in the control of insect pests under a variety of conditions in developing countries. Governmental pest control organizations and farmers, along with local representatives of the agrochemical companies, have demonstrated their willingness to test pheromones as an alternative to conventional pesticides. Their enthusiasm stems from an appreciation of the lack of human or environmental damage caused by pheromones and, in some cases, the additional benefits resulting from their use. Nor are they unaware of the spiraling costs of conventional pest control and the cost benefits that the use of pheromones will bring. As a consequence, registration of the pheromones for use as selective pesticides has not proved a problem.

Although this paper has concentrated on pheromones for cotton pest control in developing countries, pheromone research in ODNRI is not restricted to cotton. Other projects are concerned with the use of pheromones on a wide variety of crops including rice, maize, cocoa, and sugar; orchard crops such as macadamia; and stored products, and these are showing promising results.

REFERENCES

1. Campion, D. G. Survey of pheromone uses in pest control. In *Techniques in Pheromone Research.* Hummel, H. E. and Miller, T. A. (eds.). Springer Verlag, New York, Berlin, Heidelberg, Tokyo, 1985, pp. 405–469.
2. Kydonieus, A. F. and Beroza, M. (eds.). *Insect Suppression with Controlled Release Pheromone Systems*. CRC Press, Boca Raton, Fl., 1982, pp. 274; 312.
3. Critchley, B. R., Campion, D. G., McVeigh, L. J., McVeigh, E. M., Cavanagh, G. G., Hosny, M. M., Nasr, El Sayed A., Khidr, A. A., and Naguib, M. Control of the pink bollworm, *Pectinophora gossypiella* (Saunders) (Lepidoptera: Gelechiidae) in Egypt by mating disruption using hollow fibre, laminate flake and microencapsulated formulations of synthetic pheromone. *Bull. Entomol. Res.* *75*:329–345 (1985).
4. El Adl, M. A., Hosny, M. M., and Campion, D. G. Mating disruption for control of the pink bollworm *Pectinophora gossypiella* (Saunders) in the Delta growing region of Egypt. *Trop. Pest Manage.* *34*:210–214 (1988).
5. Isa, A. L. Cotton pest problems in Egypt. In *Proceedings of Symposia IX International Congress of Plant Protection*, Vol. 2. Kommedahl, T. (ed.). Washington, D.C., 1979, pp. 545–547.
6. Critchley, B. R., Campion, D. G., McVeigh, E. M., McVeigh, L. J., Jutsum, A. R., Gordon, R. F. S., Marrs, G. J., Nasr, El Sayed, A., and Hosny, M. M. Microencapsulated pheromones in cotton pest management. In *Proceedings of British Crop Protection Conference*, Brighton, UK, 1984, pp. 241–245.
7. Critchley, B. R., Campion, D. G., and McVeigh, L. J. (Unpublished report, 1988).

8. McGregor, S. E. *Insect Pollination of Cultivated Crop Plants.* USDA Agricultural Handbook No. 496, Washington, D.C., 1976, 411 pp.
9. Campion, D. G., Critchley, B.R., and McVeigh, L. J. In *Chemical Modification of Insect Behaviour in Plant Protection.* Jutsum, A. R. and Gordon, R. F. S. (eds.). John Wiley & Sons, Chichester, England, 1989, pp. 89–119.
10. Chamberlain, D. J., Critchley, B. R., Cavanagh, G. G., and Campion, D. G. (Unpublished report, 1988).
11. Hall, D. R., Beevor, P. S., Lester, R., and Nesbitt, B. F. (*E*,*E*)-10,12-Hexadecadienal: A component of the sex pheromone of the spiny bollworm *Earias insulana* (Boisd.) (Lepidoptera: Noctuidae). *Experientia 36*:152–153 (1980).
12. Cork, A., Beevor, P. S., Hall, D. R. Nesbitt, B. F., and Campion, D. G. A sex attractant for the spotted bollworm, *Earias vittella. Trop. Pest Manage. 31*:158 (1985).
13. Critchley, B. R., Campion, D. G., Cavanagh, G. G., Chamberlain, D. J. and Attique, M. R. Control of three species of bollworm pests of cotton in Pakistan by a single application of their combined sex pheromones. *Trop. Pest. Manag.* 33:374 (1987).
14. Nesbitt, B. F., Beevor, P. D., Hall, D. R., and Lester, R. Sex pheromones of two noctuid moths. *Nature 244*:208–209 (1973).
15. Beevor, P. S., Hall, D. R., Lester, R., Poppi, R. G., Read, J. S., and Nesbitt, B. F. Sex pheromones of the armyworm moth, *Spodoptera exempta* (Wlk.). *Experientia 31*:22–23 (1975).
16. Brown, E. S., Betts, E., and Rainey, R. C. Seasonal changes in the distribution of the African armyworm, *Spodoptera exempta* (Wlk.) (Lep.:Noctuidae), with special reference to eastern Africa. *Bull. Entomol. Res.* 58:661–728 (1969).
17. Tucker, M. R., Mwandoto, S., and Pedgley, D. E. Further evidence for windborne movement of armyworm moths, *Spodoptera exempta* in east Africa. *Ecol. Entomol.* 7:463–473 (1982).
18. Campion, D. G., Bettany, B. W., McGinnigle, J. B., and Taylor, L. R. The distribution and migration of *Spodoptera littoralis* (Boisduval) (Lepidoptera: Noctuidae), in relation to meteorology on Cyprus, interpreted from maps of pheromone trap samples. *Bull. Entomol. Res.* 67:501--522 (1977).
19. Nasr, El Sayed A., Tucker, M. R., and Campion, D. G. Distribution of moths of the Egyptian cotton leafworm, *Spodoptera littoralis* (Boisduval) (Lepidoptera: Noctuidae), in the Nile Delta interpreted from trap catches in a pheromone trap network in relation to meteorological factors. *Bull. Entomol. Res.* 74:487--494 (1984).
20. Campion, D. G., McVeigh, L. J., Hunter-Jones, P., Hall, D. R., Lester, R., Nesbitt, B. F., Marrs, G. J., and Alder, M. R. In

Management of Insect Pests with Semiochemicals. Mitchell, E. R. (ed.). Plenum Publishing, New York, 1981, pp. 253–265.
21. Campion, D. G. and Murlis, J. Sex pheromones for the control of insect pests in developing countries. *Med. Fac. Landbouww. Rijksuniv. Gent.* *50*:203–209 (1985).
22. Haynes, R. F., Li, W. G., and Baker, T. C. Control of pink bollworm moth (Lepidoptera: Gelechiidae) with insecticide and pheromones (attracticide): Lethal and sublethal effects. *J. Econ. Entomol.* *79*:1466–1471 (1986).
23. Rose, D. J. W. Warning system for armyworm in Africa: Physical and biological considerations. In *Proceedings of IX International Congress of Plant Protection*, Vol. 1. Kommedahl, T. (ed.). Washington, D.C., August 1979, pp. 168–170.
24. Cork, A., Murlis, J., and Megenasa, T. Identification and field testing of additional components of the female sex pheromone of the African armyworm, *Spodoptera exempta* (Lepidoptera: Noctuidae) *J. Chem. Ecol.* *15*:1349–1364 (1989).
25. Murlis, J., Gibbons, D. B., Chamberlain, D. J., Bettany, B. W., Cavanagh, G. G., de Souza, K. R., Casci, F., and Cork, A. (Unpublished report, 1986.)

25

Use of Pink Bollworm Pheromone in the Southwestern United States

THOMAS C. BAKER / University of California, Riverside, California

ROBERT T. STATEN / Animal and Plant Health Inspection Service, U.S. Department of Agriculture, Phoenix, Arizona

HOLLIS M. FLINT / Agricultural Research Service, U.S. Department of Agriculture, Phoenix, Arizona

I. INTRODUCTION

Commercial use of pink bollworm pheromone (gossyplure) in the Southwestern United States for disruption of mating, unprecedented and unsurpassed for any moth species in terms of its success, grower acceptance, and acreage treated, had an early history that was marked by a persistent skepticism that somebody was hiding something, that the technique did not really work, or at least that its efficacy was never adequately demonstrated. Part of this perception came from a criticism about lack of untreated control fields in the landmark efficacy papers by Shorey's group at the University of California, Riverside, in 1977 (1,2). This criticism seems unreasonable, considering the cost to growers or to the funding agency of absorbing the costs incurred from damage in the untreated check fields. Additionally, in the opinion of many who have worked in pheromones in this region for a long time and who know the politics of cotton research here, there were other reasons for the unbending skepticism. It arose from early political differences between those who were and who were not working with pheromones, and was often based on a reluctance of some to scientifically scrutinize the existing, albeit somewhat variable, data. The perception that pheromone disruption was or was not effective was also erroneously and simplistically tied to its economic performance in the markeplace, where many other factors unfamiliar to biologists determine success or failure in displacing a highly organized infrastructure, such as insecticide sales.

In the late 1970s, the combined pioneering efforts of Shorey's pheromone group and the fledgling company Conrel, led by Bill Brooks resulted in a hollow-fiber formulation that was effective at controlling pink bollworm and reducing the number of pesticide sprays in the process (3,4). Their efforts were historically important to the pheromone field because at that time these researchers were working more intensively than others toward commercialization of pheromones in a period of general disappointment; during this period pheromones were perceived as having failed to fulfill their promise as powerful new tools for insect control. Shorey's and Brooks' efforts resulted in the pink bollworm fiber pheromone formulation becoming the first disruptant formulation in the United States to receive Environmental Protection Agency (EPA) registration in 1978 and gave pheromone research needed encouragement at a critical time.

II. HISTORY OF USE

A. Use by Growers

Growers accepted the new insect control technology, and its use grew mainly by word of mouth. The acreage receiving at least one commercial application of the fiber formulation Nomate PBW, for instance, doubled each year from about 8000 ha in 1978 to over 40,000 ha in Arizona and Southern California in 1981 (C. C. Doane, personal communication). Grower acceptance was great enough that, in 1982, after a disastrous 1981 season of pyrethroid-spawned whitefly outbreaks, a general movement of growers in the Imperial Valley (Fig. 1) of California resulted in a vote in favor of implementing a Pest Abatement District through the County Agricultural Commissioner's Office headed by Commissioner Claude Finnell. Every grower of the 18,225 ha that were planted to cotton that year was required to put on at least four applications of mating disruptant against the pink bollworm. The objective was to suppress early-season populations by preventing mating and, thus, to refrain from early season use of insecticides (see Sec. II.C for results of this project).

Recently, skepticism once again surfaced after the sale of Albany International's Controlled Release Division in 1984. The sale of this Division, formerly Conrel, renewed the claim among long-term doubters that gossyplure for pink bollworm control does not work and, again, mistakenly linked economic performance with biological efficacy. Existing data show, however, that use of gossyplure is still high

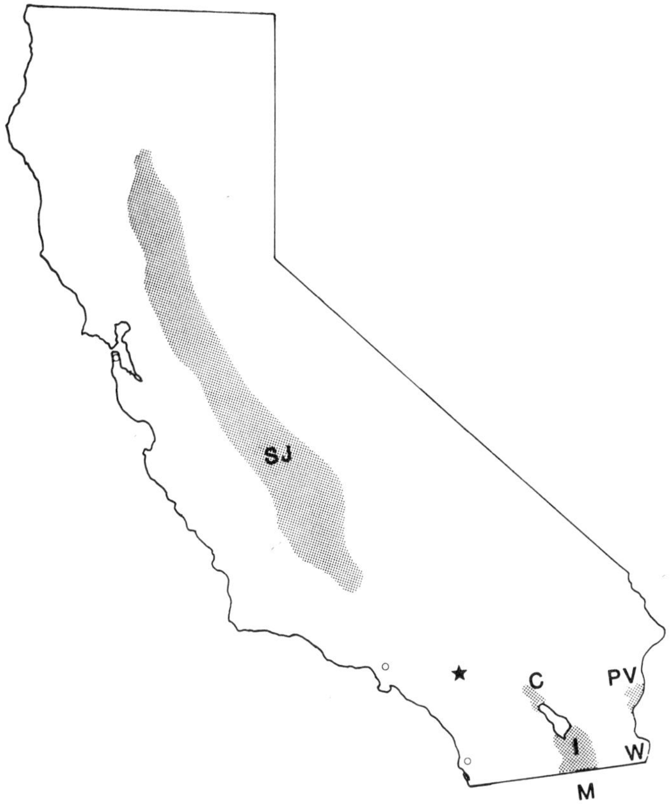

FIGURE 1 Map of California showing the major cotton-growing irrigation districts (shaded areas). Riverside is marked with a star, with Los Angeles and San Diego as small circles to the west and southwest of Riverside, respectively. SJ, San Joaquin Valley, PV, Palos Verdes Valley; C, Coachella Valley; and I, Imperial Valley. Mexicali Valley (M) in Mexico is just a continuation of the Imperial Valley across the border. W, Winterhaven, a small cotton-growing district on the Colorado River.

(Table 1). Since 1982, the acreage planted to cotton in Southern California and Arizona has steadily declined, and in 1986 was almost one-half that in 1982. The causes of this reduction were the implementation of the Payment In Kind program (PIK) in 1983, which encouraged growers not to plant cotton, declining commodity prices, and increasing costs of insecticide control. However, despite the shrinkage in acreage, the percentage of acreage receiving at least one application

TABLE 1 Cotton Acreage and Pink Bollworm Pheromone Use, by Year, in Southern California and Arizona

Year	Total ha cotton	Total kilos Nomate	Average g/ha Nomate	Total ha-appl. Nomate[a]	Average appls/ season	Treated hectares Nomate	Total ha pheromone (%)	
							Nomate	Total (est.)
1982	243,013	4000	29.6	134,874	4.5	29,972	12	15
1983	162,009	2500	27.2	92,344	4.0	23,086	14	18
1984	170,109	1700	29.6	68,043	3.5	19,441	11	17
1985	157,959	1150	22.2	51,033	3.0	17,011	11	19
1986	145,808	850	19.8	42,932	2.0	21,466	15	29

[a]Values in this column are referred to as "hectare-applications."
Source: Courtesy of C. C. Doane, Scentry Corporation, Buckeye, Arizona.

of Nomate PBW (Scentry Inc.), a commercial pheromone disruptant, increased 3% (see Table 1). When estimated use of other disruptants is added to these values, the percentage of cotton acreage treated at least once with gossyplure was nearly 30% in 1986.

What *has* decreased over this period has been the number of gossyplure treatments applied each year, so that hectare-applications have also declined (see Table 1). From the very beginning of the commercial use of gossyplure, growers and pest-control advisors (PCAs) tried to reduce the number of applications and the amount of gossyplure in each application to gain a competitive edge and increase profits. They also disregarded label instructions from the very start by adding small amounts of insecticide to the sticker and changing the timing of applications. The experimenting evolved to the point at which now many growers use two instead of four applications, but begin much later in the season than the label recommendations [recommended start is at "pin-square" (the first appearance of flower buds, or "pinhead squares")] and use the product in conjunction with concurrent broadcast insecticide sprays (dual applications). The reduction in hectare-applications is thus part of the normal evolution of the use of a product, although this may have now resulted in a use pattern that is not optimally efficacious, either alone or as an IPM component.

B. Use by the California Department of Food and Agriculture

For years a very important, but relatively unknown, user of gossyplure has been the California Department of Food and Agriculture (DFA). The pink bollworm has not yet become established in the San Joaquin (Central) Valley, although every summer adults are blown into the valley on storm systems originating in Mexico and the Southern California desert cotton-growing regions (see Fig. 1). The threat of these blown-in pink bollworms initiating established populations in the more than 284,000 ha of San Joaquin Valley cotton is great, because currently the growers enjoy high profits and low insecticide usage. Growers apply only a few sprays each year, compared with the many required for pink bollworm control in the infested southern desert regions. The CDFA wants to avoid the secondary effects of broadcast insecticides to eliminate pink bollworm populations, and so in conjunction with the USDA-APHIS sterile moth-rearing facility in Phoenix they have, since 1968, conducted a successful program of sterile moth releases, recently backed up by pheromone disruption when necessary, that has kept populations from becoming established around blown-in moths.

The CDFA monitors an extensive grid of pheromone traps extending at various densities throughout the Valley, and when a male moth is caught in a trap, an airdrop of sterile moths is implemented. A minimum effective ratio for control has been established at 60:1 steriles/natives, which is checked by male captures in the grid of pheromone traps. All sterile males are dyed pink by the diet they ingest at the rearing facility, and sterile males are identified in the traps by their pink color. When the minimum ratio of sterile/native moths cannot be attained, the CDFA applies gossyplure disruptant-attracticide, either in fibers or flakes. Up to 2025 ha are treated with pheromone each year to eliminate moth populations. The CDFA has used the disruptant as a backup to sterile release every year since 1982 and is satisfied with its effectiveness. The CDFA personnel believe it is an optimal use for a technique that works best at lower population densities.

C. The California Pest Abatement District

Once the pink bollworm became an economic pest in the Imperial Valley of California in 1966, as elsewhere in the Southwest, yields declined drastically (Table 2). This occurred despite increasingly heavy spray regimens of from 10 to 15 sprays per season, up from about two per season before 1965. Initially, the sprays were organophosphate-oriented, but they became pyrethroid-oriented in recent years. The use of pyrethroid insecticides is correlated with outbreaks of whitefly, *Bemesia tabacci* (Genn.) (5). A severe whitefly outbreak in the Imperial Valley in 1981 resulted in lower yields of cotton fiber; the fiber was also of lower quality because of contamination by whitefly honeydew ("sticky cotton") (5). Neighboring melon, lettuce, and other vegetable crops were destroyed by whitefly-vectored virus disease. The heavy reliance on pyrethroid insecticides has continued; a spray record from the Agricultural Commissioner's Office for a field in the Imperial Valley in 1984 showing 19 insecticide applications, 12 of them pyrethroid, is not at all unusual for the Southern California Desert valleys. In addition to the monocrotophos (Azodrin), dicafol (Kelthane), acephate (Orethene), and azinphosmethyl (Guthion) applied April 22, May 5, June 19 and 30, and July 6, respectively, fenvalerate (Pydrin) was applied July 13, 18, 21, 23, 29, August 4, 11, 16, 22, 28, September 2 and 10, mixed variously with propargite (Comite), cote, and chlordimeform (Galecron). Pyrethroid-resistance has been correlated with pyrethroid usage and was documented in the Imperial Valley using pheromone traps containing insecticide-laced sticker (6). Interestingly, resistance was not present in the neighboring Mexicali Valley of Mexico, where pyrethroid insecticides are rarely used because of their cost (6).

After the whitefly outbreak in 1981, the Imperial Valley growers mandated a program for 1982 requiring at least four pheromone disruption sprays, after which they could either continue with pheromone or

TABLE 2 Reduction in Cotton Yield in the Imperial Valley of California After Arrival of Pink Bollworm as a Pest in 1966

Years	Avg. no. 227-kg bales/ha	Dollar value at $1.32/kg
1961–1965	3.43	1029
1966–1970	2.25	675
1971–1975	2.13	639
1976–1980	2.23	669
1981–1984	2.49	747

Source: Imperial County Agricultural Commissioner's Office.

switch over to conventional insecticides. The goal was for the growers to use minimal insecticides until August. Also, growers were permitted to apply chlordimeform until they sprayed a Class 1 insecticide for the first time. The entire 18,226 ha in the Valley received at least four applications of Nomate PBW (fibers; Albany International) or Disrupt PBW (flakes; Herculite Products Inc.), nearly all of it in the usual attracticide formulation. Growers could choose from three regimens: 18.5 g/ha hollow fibers applied a minimum of six times beginning at first pin-square (the first "hostable" material on the plant), 37.0 g/ha fibers applied a minimum of four times beginning at first bloom (about 2–3 weeks after first pin-square); and 148 g/ha laminated flakes applied a minimum of four times beginning at first bloom. The amount of pheromone in active ingredient (AI) per hectare in all three treatments was similar (Table 3). The USDA-APHIS monitored fields in the Valley throughout the season, and a report was filed at the Imperial County Agricultural Commissioner's Office concerning the efficacy of the abatement effort (7).

The report showed that overall, the program accomplished its goals (8). In the abatement year, 1982, early-season use of insecticides was curtailed, and the whitefly and associated sticky cotton and disease were not a problem. Moreover, average yield in the Valley increased to 6.7 bales/ha from 5.4 bales in 1981 (8). Beneficial insects were preserved well into August, which many thought was impossible after high midsummer temperatures. Although there were no control plots because of the area-wide treatment, the fields in the Winterhaven district only 75 km away (see Fig. 1), all of which were treated by conventional insecticide programs, were a disaster; larval infestation of

TABLE 3 Boll Infestation Levels in the 1982 Imperial Valley Cotton Pest Abatement District

Amt./ha treatment	Larvae/100 bolls last week of treatment	Larvae/100 bolls week ending 6 Aug	% Fields under pheromone treatment after 6 Aug	On 6 Aug., no. fields with over 5% Damage	10% Damage
Albany, 18.5 g at "pin-square"[a]	1.5	0.8	83	0/6	0/6
Albany, 37.0 g at bloom[b]	6.5	6.7	38	10/27	4/27
Hercon, 148.0 g at bloom[c]	5.3	10.2	10	5/10	4/10

[a]Nomate PBW hollow fiber formulation 1.4 g AI/ha per application, minimum six required applications, average first application 11 May; six fields totaling 247 ha were sampled.
[b]Nomate PBW hollow fiber formulation, 3.7 g AI/ha per application, minimum four required applications, average first application 6 June; 27 fields totaling 912 ha were sampled.
[c]Disrupt PBW laminated flake formulation, 3.7 g AI/ha per application, minimum four required applications, first application 10 May; 10 fields totaling 252 ha were sampled.
Source: Ref. 7.

bolls was 32.6% (six fields sampled), compared to 5.8% in the Imperial Valley (20 fields sampled) (7), and yields were less than 4.9 bales/ha (8). Additionally, the Palos Verde Valley, 100 km away, yields and damage similar to those in the abatement district were obtained by growers using pheromone disruptant during that same year. Those using a conventional insecticide program (11 fields sampled) incurred 7.5% infestation, whereas those using a Nomate PBW program (12 fields sampled) had 3.8% infested bolls. Thus, the abatement program with its use of pheromones had brought damage down to a level comparable with that experienced normally in neighboring valleys such as the Palos Verdes (7).

Within the abatement district, it was clear that those growers who chose the pin-square regimen had the least damage (see Table 3). If these squares are left unprotected, larvae originating from them, although not contributing directly to later damage counts, can result in economically important populations earlier in the season. Fields treated at pin-square, although beginning gossyplure treatments earlier than the others, were also maintained longer on the gossyplure treatments (see Table 3). This is another indication that in these fields, the recommendation to change from gossyplure treatments to insecticides could be made later in the season (generally PCAs use 1--5% damaged bolls as an economic threshold). Finally, the average costs for control, insecticides included, were comparable for the sampled fields in the abatement program (7). Total costs for the 18.5 or 37.0 g/ha of Nomate PBW product or 148 g/ha of Disrupt PBW product plus insecticides were 523, 533, or 529 dollars/ha, respectively. Although 500 dollars/ha appears high, this was nearly 250 dollars/ha less than was spent the previous year for 1.2 bales/acre less yield. Many growers felt that the 500 dollars/ha cost was equivalent to their average cost of a conventional insecticide program during a typical year.

By any measure reported to the County Agricultural Commissioner's Office, including the lower cost per hectare for control, the reduced insecticide usage, increased preservation of beneficial insects, and improved quantity and quality of yields, the California Pest Abatement District's mandatory areawide use of gossyplure must be considered a success. Still, uncertainty managed to follow this program. Skeptics pointed to the lack of an abatement district in 1983 as proof that the 1982 program had somehow been inadequate. They failed to recognize that most growers in the valley again used pheromones in 1983. Growers did not want to be restrained another year by the lack of choice built into such a program, and so did not vote to formalize their gossyplure use into an official abatement district in 1983. In addition, some of the nonpheromone acreage that year went to a newly touted program of delayed planting.

III. NEW FORMULATIONS, AND UNDERSTANDING OLD ONES

A. The Shin-Etsu/Mitsubishi PBW-ROPE

In 1985, after a highly promising initial test in 1984 (9), Imperial Valley cotton growers supported an extensive test of a promising new formulation, PBW-ROPE, consisting of a wire-based, sealed polyethylene tube filled with gossyplure manufactured by Shin-Etsu Chemical Industry Co., Ltd. (10). The use of 78 AI/ha, a relatively large dose of gossyplure, was not a new approach for this species (1). However, recent reductions in the cost of synthesizing gossyplure allowed a high-dose formulation to become commercially feasible. The test was performed in both the Mexicali Valley of Mexico (see Fig. 1) and the Imperial Valley, which traditionally have moderate to heavy infestations. Many consider the Imperial Valley to be the most severe test area in the United States for any pink bollworm control technique. The PBW-ROPE dispensers were twist-tied around the cotton plants at first pin-square, evenly spaced at a density of 1000 dispensers per hectare (78 g AI/ha). With this formulation, the dispensers remained about 5--10 cm above the ground throughout the season, although the plants grow to 1.5 m in height. Growers or their pest control advisors decided when to spray with insecticides (10).

In the PBW-ROPE-treated fields in both valleys, damage to bolls, as well as the number of insecticide sprays, was reduced, compared with that in conventional insecticide-treated fields (Table 4; 10). The use of insecticides before August was virtually eliminated by the PBW-ROPE application; most insecticide applications began in August, after the pheromone in the dispensers had completely evaporated. Boll damage was 0.9–1.7% in the conventional insecticide-treated fields, whereas the PBW-ROPE-treated fields had only 0.3–0.7% damage. The reduction in insecticide use, already impressive when coupled with the significantly reduced damage, was even more remarkable when considering the pressure on PCAs to take action with insecticides rather than risk damage. These results were so impressive that a more ambitious project was carried out in 1986 in the Coachella Valley (see Fig. 1) (11).

The goal of the 1986 experiment was to have areawide coverage of the Coachella Valley with a coordinated pin-square treatment of the PBW-ROPE. All 446 ha in 31 fields were treated with the PBW-ROPE, at 1000 dispensers per hectare (78 g AI/ha) (11). Eighteen of the fields were planted early in the season and had first pin-square in early May. Such full-season fields usually have more pressure from pink bollworm than do late-planted fields. Because of regulatory and labor problems, one of these 18 fields did not receive a PBW-ROPE application until 1 June at bloom, 3 weeks after hostable squares were present. Most of the remaining 13 fields, all late-planted, received their PBW-ROPE treatments before first pin-square, whereas three of these received treatment between pin-square and bloom. All fields were monitored with gossyplure-baited Delta traps and weekly boll samples taken by APHIS personnel and by the growers' PCAs. The PCAs could then decide when insecticide spraying was needed.

Through 12 August, only 23 larvae were found in boll samples over the entire valley, for a valley-wide infestation rate of less than 0.3% (Table 5) (11). More impressively, 19 of these larvae were found in the one full-season field that received its gossyplure treatment late. Only two other fields in the entire valley had a measurable larval population. Pheromone trap captures of both native males and some

TABLE 4 Effects of PBW-ROPE Treatment on Boll Damage and Insecticide Usage in the Imperial and Mexicali Valleys in 1985

Treatment	Larvae/100 bolls		No. insecticide treatments
	Aug	Sept	
	Imperial Valley[a]		
Conventional insecticide	0.85	0.88	11.4 a
Conventional pheromone	0.90	2.1	10.4 a
PBW-ROPE pheromone	0.32	0.39	6.6 b
	Mexicali Valley[b]		
Conventional insecticide	1.72 a	1.55 b	4.9 a
PBW-ROPE pheromone	0.7 b	0.72 b	2.9 b

[a]There were seven PBW-ROPE-treated fields, eight conventional insecticide-treated fields, and eight conventional pheromone-treated fields (Nomate PBW or Disrupt). Means in same column having no letters in common are significantly different according to ANOVA followed by Duncan's multiple range test ($P = 0.01$).
[b]There were 16 PBW-ROPE-treated fields and 14 conventional insecticide-treated fields. Means in same column having no letters in common are significantly different according to Student's t test ($P = 0.01$).
Source: Ref. 10.

sterile males released to serve as an extra indicator of effective gossyplure emission, revealed that the PBW-ROPE treatments lost their efficacy by the first week in August. This was over 60 days after application for the early treatments. Several weeks later there was a big increase in the larval densities in sampled bolls (11) (see Table 5).

Particularly instructive in this Coachella Valley experiment was the drastically reduced use of insecticide by the growers. Through the end of July, only one field, the full-season field that received late treatment, was sprayed (11). No other field received insecticide until 23 August, long after the dispensers had lost their efficacy. In the previous year insecticides had been first applied on 1 June and then applied regularly through July and August. During the 1985 season insecticide usage on the two ranches making up most of these cotton fields had averaged 6 and 11 treatments per field, respectively,

TABLE 5 Effects of PBW-ROPE Application on Boll Damage in Coachella Valley Areawide Pheromone Treatment Study, 1986

Date	No. bolls sampled	No. fields sampled	No. fields with larvae	Total larvae	% Boll infestation
6/19	50	1	1	3	6.0
6/25	500	5	1	3	0.6
7/2	960	12	1	2	0.2
7/9	1180	15	2	9	0.8
7/16	1520	19	1	1	0.1
7/21	1520	19	2	3	0.2
7/28[a]	1600	20	0	0	0.0
8/5[b]	1520	19	1	1	0.1
8/12	1600	20	1	1	0.1
8/18	1760	22	10	43	2.4
8/27	2200	28	14	138	6.3

[a]Through 28 July, 19 of the 21 larvae found valley-wide came from a single field that received a PBW-ROPE application ca. 3 weeks too late.
[b]At 5 Aug., gossyplure was no longer released from the formulations at effective levels, and disruption broke down, as registered in increased pheromone trap catches.
Source: Ref. 11.

which is typical of most years for the Coachella Valley. In 1986, however, 14 fields received just one insecticide application through August, and through 17 September, no field received more than four insecticide applications. Fourteen of the 31 fields (45%) received no insecticides whatsoever, including four of the 18 long-season fields, for an overall valley-wide average of a little over 1.5 applications per field (11).

In 1987, the PBW-ROPE formulation was again tested areawide on the 405 ha of cotton in the Coachella Valley as part of a program to integrate sterile moth-releases into the pink bollworm control strategy (11). Sterile moths were released (about 1480 moths/ha per day) beginning before pin-square in the 24 fields of 28 that were at the pin square stage before 1 June. The strategy was to apply gossyplure at pin-square only on those fields in which the 60:1 sterile/native ratio could not be maintained, which occurred on 202 ha. Therefore, by definition, the pheromone was applied to the fields with the highest density of native moths. The PCAs used their judgment in recommending application of conventional insecticides using boll samples and pheromone trap counts as criteria.

Through August, 21 of 27 fields received no insecticide treatments, and season-long, through September, 20 of 27 fields were completely insecticide-free (11). Boll infestation valley-wide was quite low throughout the season (Table 6). This is even more impressive considering that the preponderance of larvae found through 18 August

TABLE 6 Effect of PBW-ROPE Treatment, Integrated with Sterile Moth Releases, on Boll Infestation in Area-Wide Coachella Valley Study, 1987[a]

Date	No. bolls sampled	No. fields sampled	No. fields with larvae	Total larvae	% Boll infestation
6/30	500	5	0	0	0
7/6-9	1700	17	0	0	0
7/13-17	2400	24	3	7[c]	0.1
7/20-23	2400	24	6	33[d]	0.8
7/27-30[b]	2500	25	6	34	1.4
8/3-6	2700	27	4	11	0.4
8/10-13	2700	27	5	6	0.2
8/18	2700	27	5	5	0.2
8/24-25	1350	27	4	25	1.8
8/31	1350	27	7	41	3.0

[a]202 of the 405 hectares in the valley received PBW-ROPE dispensers at pin-square after sterile releases could not maintain the proper 60:1 sterile/native ratio needed for effective control.
[b]At the end of August, only 6 of 27 fields had received any insecticide application.
[c]Five larvae in one boll.
[d]All 20 infested bolls containing 33 larvae were found in three contiguous fields.
Source: Ref. 11.

occurred in a contiguous block of three fields that had lost the desired 60:1 ratio in mid-May. In July, these three fields received insecticides, based on rapidly escalating pheromone trap captures, larval counts in bolls, and captures in light traps. In the fourth week of sampling, 15 of the 20 infested bolls valley-wide were found in these three fields, accounting for 21 of the 33 larvae found that week. These fields had had a measurable percentage of infested blooms in preboll counts. Another block of two adjacent fields had measurable infested blooms and accounted for the remainder of the larvae found during the last 2 weeks of boll counts in August (see Table 6).

Some conclusions can be drawn from the results of these experiments with the PBW-ROPE dispensers during 1985–1987. First, applications of PBW-ROPE are most effective when put on at the first pin-square stage. Second, with only one application of PBW-ROPE, control of pink bollworm equivalent to or better than conventional insecticides is generally achieved well into August. Third, the success of the PBW-ROPE application is due to communication disruption alone, because no insecticide is included in the formulation. Fourth, insecticide usage was drastically reduced, or even eliminated, when the dispenser was used. Fifth, secondary outbreaks of whitefly were observed only in fields heavily treated with insecticides for pink bollworm, demonstrating that beneficial insects are preserved in fields treated only with pheromone. Sixth, and importantly, that the formulation is hand-applied does not seem to be an obstacle to growers. The cost of a single PBW-ROPE treatment is competitive with a conventional insecticide program when fields normally require heavy insecticide use. In 1987, the growers' cooperative in Brawley was able to sell the rope application at 135 dollars/ha, including application costs. Of course, growers are reluctant to spend money early in a season, because they always must worry about being prepared for later possible outbreaks of pest X in August. The saving of more than four insecticide sprays by a PBW-ROPE application begins to overcome this reluctance to spend money "up-front." In general, the psychology of up-front expenditures for pheromone disruptants on pests of any crop is a problem that must be overcome to facilitate widespread commercial use of pheromones.

B. Efficacy of Attracticide Formulations

From the time a commercial gossyplure formulation first became available, growers added small amounts of pyrethroid insecticides to the sticker used with the commercial product to reduce the amount of pheromone applied and lessen the cost of each application. The efficacy of the "attracticide" or "attract-and-kill" approach was explained as being due to males being attracted to the fibers or flakes, touching the pyrethroid contained in the sticker, and then becoming incapacitated. Some questioned whether the attracticide technique really incapacitates sufficient numbers of males to result in control above and beyond the communication disruption effect of the pheromone alone. Others felt that the insecticide-laced formulation might reduce the populations of beneficial insects to the point where a key advantage offered by gossyplure treatment alone—preservation of beneficials—would be negated.

Both concerns have proved to be groundless. Extensive laboratory data have shown that not only can significant mortality be inflicted on males that are attracted to and touch a fiber coated with insecticide-treated sticker, but significant sublethal effects also may add to the efficacy of the attracticide formulation. In the admittedly ideal environment of a wind tunnel, using freshly formulated sticker plus insecticide plus gossyplure dispenser, male pink bollworm moths flew to and touched a gossyplure fiber coated with pyrethroid just as readily as they did to a fiber coated with untreated sticker. High percentages of these males, sometimes approaching 80%, depending on the insecticide, died within 24 hr (12). Importantly, males that touched the insecticide-laced source, but survived, often exhibited a 90% reduction in flight up to and touching of the pheromone source compared with those that had flow up and touched the fiber with untreated sticker (biotac) the previous day (12). These survivors demonstrated an apparent reduced sensitivity to pheromone in the lower probability of wing fanning and taking flight they exhibited when placed in the pheromone plume. Otherwise these survivors of contact with the attracticide appeared normal in every way, including their ability to perform coordinated flight (12).

Thus, if false-trail following (See Chap. 4) were a significant mechanism in disruption of mating of pink bollworm, as it appears to be (R. T. Staten, unpublished observations), then surviving males poisoned by the attracticide would be less likely to locate a female the following night than they would if the disruptant alone had been applied. Even if males recovered enough by the second night post-poisoning (12) to fly upwind to a pheromone source, they would be just as likely to do so to an attracticide source again.

Concern has been expressed about the potential negative impact of attracticide on beneficial insects. Available evidence refutes the idea that beneficial insects are sufficiently reduced in numbers to interfere with biological control. Butler and Las (13) found that when treated fields were compared with untreated fields, there was no measurable effect of permethrin-laced Nomate on the populations of more than five species of beneficial insects. Broadcast commercial insecticide applications, on the other hand, reduced these populations by 60—100% immediately following treatment. Similarly, Beasley and Henneberry (14) found only 26—29% reduction of the populations of beneficial species they sampled when pyrethroid insecticides were added to standard insecticide-free disruptant formulations. In contrast, conventional insecticide applications reduced populations of these same insects by up to 100%. Thus, attracticide formulations are unlikely to contribute to secondary pest outbreaks in the same way that conventional insecticide applications can (15).

C. Use of a Pure *Z,Z* Isomer Formulation

Apart from the many new controlled-release dispensers being developed and tested, a fundamentally different approach to creating a better pink bollworm disruptant, that of changing the pheromone blend in the formulation, has been pursued by Flint and colleagues (16–18). The two components of the pink bollworm pheromone were identified as (*Z,Z*)-7,11-hexadecenyl acetate and (*Z,E*)-7-11-hexadecenyl acetate (19). These isomers are normally emitted by females in a ratio of about 60:40 *Z,Z*/*Z,E* (20,21). Commercial formulations emit gossyplure, a 50:50 mixture, which effectively prevents males from locating females, presumably by a combination of false-trail following (R. T. Staten, unpublished observations) and habituation of the central nervous system (see Chap. 4). The latter mechanism, thus far, has not been experimentally demonstrated for the commercial disruptant, although it is heavily implicated. The gossyplure, although slightly off-ratio compared with the natural blend, evokes upwind flight and source contact in males at levels similar to the natural ratio (22,23).

Flint's work has focused on the field application of the pure *Z,Z* isomer as the disruptant, either in flakes or rope dispensers. The *Z,Z* isomer alone elicits no trap catch (22) or even upwind flight (23) of males. Thus, among the advantages of using this isomer alone to disrupt mating communication would be the possibility that males might become habituated without being activated and, thus, not be stimulated "to activity wherein they might find females" as could occur with the standard 50:50 blend (24). Also, use of the *Z,Z* isomer would reduce the possibility that males would be attracted into the disruptant-treated field where they might locate females.

Interestingly, in *Z,Z*-treated fields, the trap catch of males by the normally optimal 50:50 (female-emitted) blend is reduced to near zero, and the capture of males shifts over to traps emitting a 9:1 ratio of *Z,Z*/*Z,E* (17,18). This is explained by habituation of the male's sensory system to the *Z,Z* isomer; an excess of *Z,Z* in the emitted blend is then required to create sensory stimulation similar to that of the optimal 50:50 blend in an unhabituated system (17,18). These results suggest that, under field conditions, habituation, in addition to false-trail following, very likely is a significant mechanism in pink bollworm mating disruption by commercial gossyplure formulations. An additional advantage offered by such an effect in *Z,Z*-treated fields is that monitoring traps emitting a 9:1 blend would be able to capture males, making it possible to follow moth population trends, whereas females would become unable to communicate with males at all. In effect, there would be a communication black-out between male and female moths but not between the growers and the males. Currently, in gossyplure-treated fields, the shutdown of monitoring traps prevents the PCA from sampling moth populations by trap catch.

It remains to be seen whether such advantages would be outweighed by potential disadvantages. Thus far, with moths the natural blend of pheromone components is the most effective in causing disruption at the lowest dosages (25). This could mean that as formulations age and emission rates diminish, a gossyplure formulation might have several days' longer efficacy than a Z,Z formulation, because of the lower threshold for the natural blend. Likewise, at the outset more AI might be required when using Z,Z alone, compared with the natural blend, because only one mechanism, habituation, is available to produce disruption by the Z,Z isomer. The cost of producing pure Z,Z, compared with gossyplure, is a final factor in determining the potential utility of this approach to disruption. Initial estimates indicate that the synthesis would not be significantly more expensive than current syntheses of gossyplure (A. Yamamoto, personal communication to H. M. F.).

IV. CONCLUSIONS

The use of the pink bollworm sex pheromone for commercial control of pink bollworm in the Southwestern United States is widespread, despite the reduction in cotton acreage that has occurred since 1982. Growers and PCAs accepted the new technology, but experimented with it to reduce the amount of AI applied and, hence, their costs. They tried many things, including delaying applications and adding insecticide to the sticker. Sometimes these efforts resulted in erratic control, but in these instances they blamed their application technique, not the pheromone control itself. Attracticide applications preserve beneficial predatory insects nearly as well as pheromone disruptant without insecticide added. Moreover, the attracticide formulations have a demonstrated ability to not only attract and kill males, but also to impart significant sublethal effects, reducing pheromone responsiveness of survivors.

The 1982 Imperial Valley Pest Abatement District program enacted by the growers was a success, reducing insecticide applications and overall season-long pest control costs, while raising the cotton yield by 1.2 bales/ha and eliminating whitefly-caused sticky-cotton and disease. Although the abatement program was officially discontinued the following year because growers wanted greater freedom of choice, most cotton acreage in the valley was again under early-season pheromone control on an individual grower-choice basis.

New formulations continue to be developed, and a very promising one is the hand-applied PBW-ROPE dispenser from Shin-Etsu Chemical Industry Co. Ltd. which works without insecticide on the pure disruption principle. The reductions in damage and in the use of insecticides in the Imperial, Mexicali, and Coachella Velleys, when the dispensers were applied at pin-square, have been dramatic and unprecedented. The cost per hectare for the single application, including application cost, is becoming competitive with an insecticide regimen. Integration of the PBW-ROPE dispenser with sterile moth releases is also promising, as is the possibility of disrupting mating by application of the single *Z,Z* isomer of the pheromone. Another new use of gossyplure is in monitoring traps containing insecticide-treated sticker to quickly and easily sample native populations for insecticide resistance (6). We anticipate the increasing use of pheromones, including gossyplure, in integrated pest management programs on cotton and other crops.

ACKNOWLEDGMENTS

We thank Charles C. Doane for permission to use his unpublished data, and Rick Vetter for drawing Figure 1.

REFERENCES

1. Shorey, H. H., Gaston, L. K., and Kaae, R. S. Air-permeation with gossyplure for control of the pink bollworm. In *Pest Management with Insect Sex Attractants.* Beroza, M. (ed.). ACS Symposium Series 23, American Chemical Society, Washington, D.C. (1976), po. 67–74.
2. Gaston, L. K., Kaae, R. S., Shorey, H. H., and Sellers, D. Controlling the pink bollworm by disrupting sex pheromone communication between adult moths. *Science 196*:904–905 (1977).
3. Brooks, T. W. and Kitterman, R. L. Gossyplure H.F. – pink bollworm suppression with male sex attractant pheromone released from hollow fibers -- 1976 experiments. In *Proceedings of the Beltwide Cotton Production-Mechanization Conference, National Cotton Council of America*, Memphis, Tenn. (1977), pp. 79–82.
4. Brooks, T. W., Doane, C. C., and Staten, R. T. Experience with the first commercial pheromone communication disruptive for suppression of an agricultural pest. In *Chemical Ecology: Odour*

Communication in Animals. Ritter, F. M. (ed.). Elsevier/North-Holland Biomedical, Amsterdam (1979), pp. 375–388.

5. Johnson, M. W., Toscano, N. C., Reynolds, H. T., Sylvester, E. S., Kido, K., and Natwick, E. T. Whiteflies cause problems for southern California growers. *Calif. Agric.* Sept.-Oct.: 24–26 (1982).
6. Haynes, K. F., Miller, T. A., Staten, R. T., Li, W.-G., and Baker, T. C. Pheromone trap for monitoring insecticide resistance in the pink bollworm moth (Lepidoptera: Gelechiidae): New tool for resistance management. *Environ. Entomol.* *16*:84–89 (1987).
7. Staten, R. T., Finnel, C., and Jensen, L. Monitoring of the Imperial Valley pheromone program. Report to Imperial County Agricultural Commissioner's Office, June 23, 1983, 14 pp.
8. Heinrichs, T. Mandatory pheromone program dropped. *Calif. Ariz. Farm Press* Feb: pp. 20 (1983).
9. Flint, H. M., Merkle, J. R., and Yamamoto, A. Pink bollworm (Lepidoptera: Gelechiidae): Field testing a new polyethylene tube dispenser for gossyplure. *J. Econ. Entomol.* *78*:1431–1436 (1985).
10. Staten, R. T., Flint, H. M., Weddle, R. C., Quintero, E., Zarate, R. E., Finnell, C. M., Hernandes, M., and Yamamoto, A. Pink bollworm (Lepidoptera: Gelechiidae): Large-scale field trials with a high-rate gossyplure formulation. *J. Econ. Entomol.* *80*:1267–1271 (1987).
11. Staten, R. T., Miller, E., Grunnet, M., and Andress, E. The use of pheromones for pink bollworm management in western cotton. *Proceedings, Beltwide Cotton Production Research Conferences* (1988) pp. 206–209.
12. Haynes, K. F., Li, W.-G., and Baker, T. C. Control of pink bollworm moth (Lepidoptera: Gelechiidae) with insecticides and pheromones (attracticide): Lethal and sublethal effects. *J. Econ. Entomol.* *79*:1466–1471 (1986).
13. Butler, G. D. and Las, A. S. Predaceous insects: Effect of adding permethrin to the sticker used in gossyplure applications. *J. Econ. Entomol.* *76*:1448–1451 (1983).
14. Beasley, C. A. and Henneberry, T. J. Evaluations of gossyplure formulation for pink bollworm control under commercial conditions in California's Palo Verde Valley. In *Proceedings, Beltwide Cotton Production Research Conferences*, San Antonio, Texas (1983), pp. 171–172.
15. Van Steenwyk, R. A., Toscano, N. C., Ballmer, G. R., Kido, K., and Reynolds, H. T. Increase of *Heliothis* spp. in cotton under various insecticide treatment regimes. *Environ. Entomol.* *4*:993–996 (1975).

16. Flint, H. M. and Merkle, J. R. Pink bollworm (Lepidoptera: Gelechiidae): Communication disruption by pheromone composition imbalance. *J. Econ. Entomol.* *76*:40–46 (1983).
17. Flint, H. M. and Merkle, J. R. The pink bollworm (Lepidoptera: Gelechiidae): Alteration of male response to gossyplure by release of its component Z,Z-isomer. *J. Econ. Entomol.* *77*:1099–1104 (1984).
18. Flint, H. M., Curtice, N. J., and Yamamoto, A. Pink bollworm (Lepidoptera: Gelechiidae): Further tests with (*Z*,*Z*)-isomer of gossyplure. *J. Econ. Entomol.* *81*:679–683 (1988).
19. Hummel, H. E., Gaston, L. K., Shorey, H. H., Kaae, R. S., Byrne, K. J., and Silverstein, R. M. Clarification of the chemical status of the pink bollworm sex pheromone. *Science 181*: 873–875 (1973).
20. Haynes, K. F., Gaston, L. K., Pope, M. M., and Baker, T. C. Potential for evolution of resistance to pheromones: Interindividual and interpopulational variation in chemical communication system of pink bollworm moth. *J. Chem. Ecol.* *10*:1551–1565 (1984).
21. Haynes, K. F. and Baker, T. C. Potential for evolution of resistance to pheromone: World-wide and local variation in the chemical communication system of the pink bollworm moth, *Pectinophora gossypiella*. *J. Chem. Ecol.* *14*:1547–1560.
22. Flint, H. M., Balasubramanian, M., Campero, J., Strickland, G. R., Ahmad, Z., Barral, J., Barbosa, S., and Khail, A. F. Pink bollworm: Response of native males to ratios of *Z*,*Z*- and *Z*,*E*-isomers of gossyplure in several cotton growing areas of the world. *J. Econ. Entomol.* *72*:758–762 (1979).
23. Linn, C. E. Jr. and Roelofs, W. L. Response specificity of male pink bollworm moths to different blends and dosages of sex pheromone. *J. Chem. Ecol.* *11*:1583–1590 (1985).
24. Shorey, H. H. Application of pheromone for manipulating insect pests of agricultural crops. In *Proceedings of Symposium on Insect Pheromones and Their Applications*. Kono, T. and Ishii, S. (eds.) (1976), pp. 97–108.
25. Roelofs, W. L. and Novak, M. A. Small-plot disorientation tests for screening potential mating disruptants. In *Management of Insect Pests With Semiochemicals*. Mitchell, E. R. (ed.). Plenum Press, New York (1981), pp. 229–242.

26

Role of the Boll Weevil Pheromone in Pest Management

RICHARD L. RIDGWAY and MAY N. INSCOE / Agricultural Research Service, U.S. Department of Agriculture, Beltsville, Maryland

WILLARD A. DICKERSON* / Agricultural Research Service, U.S. Department of Agriculture, Raleigh, North Carolina

I. INTRODUCTION

The boll weevil, *Anthonomus grandis* Boheman, is indigenous to Mexico and Central America; it was first reported in southern Texas in 1894. However, a careful survey of the infested area indicated that the boll weevil had been causing damage there since 1892 (1). The boll weevil caused serious losses as it spread north and eastward. In 1907, it crossed the Mississippi River, and by 1917 it had infested cotton in the remaining southeastern states (2). The boll weevil began to expand its range into far western Texas in 1961 (3) and became a major concern near the High Plains of Texas in 1963 (4). The distribution of the boll weevil in cultivated cotton in the western United States began a substantial expansion in 1978, and by 1982 it occurred in southern California and much of southern Arizona (5). However, areawide management or eradication programs initiated in the Southeast in 1977 and in the Southwest in 1985 have reduced the range of the boll weevil. It has essentially been eliminated from Virginia, North Carolina, most of eastern South Carolina, southern California, and southwestern Arizona. However, the boll weevil continues to be one of the major insect pests in the United States. The National Cotton Council has estimated that the total aggregate losses caused by the boll weevil are about 12 billion dollars (6). Losses were estimated to be between 200 and 300 million dollars per year in the mid-1970s (7) and between 150 and 300 million

**Current affiliation*: North Carolina Department of Agriculture, Raleigh, North Carolina dollars per year (combined control costs and yield loss) in the late 1980s (8,6). In addition, broad-spectrum insecticides applied for boll weevil control also destroy natural enemies that could otherwise limit damaging infestations of the bollworm, *Heliolithis zea* (Boddie), and the tobacco budworm, *Heliolithis virescens* (F.) (9,10). Losses owing to destruction of beneficial insects and development of pesticide resistance as a result of insecticides applied to cotton have been estimated to be 150 million dollars (11). A comprehensive review of cotton insect management with special reference to the boll weevil is available (12).

The development of resistance to the chlorinated hydrocarbon insecticides in the mid-1950s had perhaps the greatest impact on events related to the boll weevil. This development was a major factor in the establishment of the U.S. Department of Agriculture's (USDA's) Boll Weevil Research Laboratory at Starkville, Mississippi (13). A major research thrust initiated there in the early 1960s led to the identification of the pheromone of the boll weevil and the preparation of the synthetic version, grandlure, by Tumlinson et al. (14). This discovery has had a major impact on the design and implementation of boll weevil management programs. Several reviews are available (15–20). However, our knowledge of the nature of the boll weevil pheromone, and particularly of the use of grandlure, continues to expand. Therefore, an updated review of the nature of the pheromone, design of traps and dispensers, applications, and the role of grandlure in operational programs in the United States is presented here.

II. DISCOVERY AND NATURE OF THE BOLL WEEVIL PHEROMONE

Cross and Mitchell (21) made observations, in a simple but elegant experiment conducted in 1963, that suggested the presence of a male-produced, wind-borne boll weevil pheromone. Keller et al. (22) subsequently conducted laboratory studies in which an active airborne substance emitted by males was collected by drawing air continuously over caged males and through a column of activated charcoal for 11 weeks. The residue obtained, after extraction of the charcoal and evaporation of the solvent, stimulated and attracted female weevils. This finding generated much interest in the isolation of the attractive material because of the potential value of an attractant for the boll weevil in management programs.

A. Isolation, Identification, and Synthesis of Pheromone Components

In 1966, efforts were initiated to isolate and determine the structures of the compounds responsible for the pheromone stimulus and to confirm the postulated structures by synthesis (23,24). The development of a reliable laboratory method for bioassay (25) provided an essential tool for following the isolation and purification of the attractive material and made possible the eventual identification of its components.

Various procedures for obtaining the attractant were investigated (24); some of the most active materials were obtained by steam distillation of extracts of weevils. Steam distillation of extracts of fecal material, readily available in large quantities from the boll weevil rearing facility, also yielded an attractive distillate. Extracts of 67,000 males, 4.5 million weevils of mixed sexes, and 54.7 kg of fecal material were used to obtain active material in sufficient quantities to achieve the identification and confirmation of the identity of the boll weevil pheromone components. Active material obtained from steam distillates was fractionated by column chromatography. Although none of the resulting fractions was active, the combination of two of the fractions was very active, indicating that more than one pheromone component was involved. Through further purification by column chromatography, followed by gas-liquid chromatography, four active components were isolated (14,15,26).

Identification of the four components was accomplished through examination of mass, nuclear magnetic resonance (NMR), and infrared (IR) spectra and by chemical reactions. The compounds were shown to be

I. (1*R*-*cis*)-1-Methyl-2-(1-methylethenyl)cyclobutaneethanol
[*cis*-2-isopropenyl-1-methylcyclobutaneethanol]
II. (*Z*)-2-(3,3-Dimethylcyclohexylidene)ethanol
[(*Z*)-3-3-dimethyl-$\Delta^{1,\beta}$-cyclohexaneethanol]
III. (*Z*)-(3,3-Dimethylcyclohexylidene)acetaldehyde
[(*Z*)-3,3-dimethyl-$\Delta^{1,\alpha}$-cyclohexaneacetaldehyde]
IV. (*E*)-(3,3-Dimethylcyclohexylidene)acetaldehyde
[(*E*)-3,3-dimethyl-$\Delta^{1,\alpha}$-cyclohexaneacetaldehyde]

These structures were confirmed by synthesis. The mass, NMR, and IR spectra of the synthetic compounds were identical with those of the natural materials, as was the biological activity of the synthetic mixtures in laboratory bioassays. For activity, both compounds I and II and either III or IV were essential; all four pheromone components were reported to be necessary for optimum activity (14,15). The synthetic four-component mixture was called grandlure (27–29), and the individual components were termed grandlure I (or grandisol), grandlure II, grandlure III, and grandlure IV, respectively. The activity of the four-component synthetic mixture was confirmed in traps in the field (28).

Synthetic grandlure preparations from different sources have been found to vary in attractancy (30). Investigation showed that several synthetic intermediates or byproducts were inhibitory (30), demonstrating the need for careful purification.

B. Component Ratios

The four components of the boll weevil pheromone were present in fecal matter from mixed sexes in concentrations of 0.76, 0.57, 0.06, and 0.06 ppm (13:9:9:1), respectively (14); concentrations in whole boll weevils were about 0.1 of these values. In male fecal matter, the ratio of the components was 6:6:2:1 (31).

A laboratory bioassay was used to evaluate the response of females to various combinations of the identified compounds (14). Mixtures containing compounds I and II and either III or IV were approximately equivalent to male weevils in attractancy. All single components and two-component mixtures, as well as three-component mixtures that did not contain both alcohols, showed very little activity. The mixture containing the four synthetic components, grandlure, exceeded male weevils in attractancy; the optimum concentrations of the four components in laboratory tests were 0.09, 0.07, 0.12, and 0.12 μg, respectively. This ratio of components (approximately 9:7:12:12) was initially considered to be the standard (29). In trapping tests, Coppedge et al. (29) showed that increases in the percentages of the two alcohol components improved performance. Hardee and associates (32) confirmed this and demonstrated that the component ratio could be varied considerably. Ratios of 2:6:1:1, 3:5:1:1, and 8:6:3:3 were considered most promising, on the basis of cost and effectiveness.

Subsequently, grandlure utilized in field research and operational programs consisted of about 40%, 30%, 15%, and 15% of compounds I, II, III, and IV, respectively, with compounds III and IV being synthesized as a mixture (G. H. McKibben, personal communication). In some recent electrophysiological studies, Dickens and Prestwich (33) reported that weevils were much less sensitive to grandlure III than to grandlure IV and that omission of grandlure III (the (*Z*) aldehyde) from the mixture had no siginificant effect on attractancy in the field.

C. Optical Activity of Grandisol (Grandlure I)

Although the natural (*cis*)-1-methyl-2-(1-methylethenyl)cyclobutaneethanol isolated by Tumlinson et al. (14) was optically active, the synthetic material they prepared, as well as material prepared in other syntheses, reviewed by Katzenellenbogen (34) and Henrick (35), was an optically inactive mixture of the two optical isomers (i.e., a racemate). This material was biologically active as a pheromone component in laboratory bioassays and field tests, and a synthetic grandlure formulation containing the racemate was competitive with caged male weevils in a field-trapping test (18). Because of the difficulty of synthesizing the optically active isomer, racemic grandisol has been used in all field applications of grandlure.

Synthesis of the enantiomers (optical isomers) of grandisol was attempted by several scientists (36–39), but yields were generally very low and optical purity greater than 80% was not achieved. In 1978, Mori et al. (40) reported that (-)-grandisol (topical purity 80%) was equal in biological activity to the natural (+) enantiomer. This finding was most unusual; in no other instance had an insect producing an optically active pheromone or pheromone component responded equally to the opposite isomer. Generally, the opposite isomer has been either biologically inactive or inhibitory.

Recently, two improved syntheses of the two grandisol enantiomers have been reported (41–43). Dickens and Mori (44), using the optically pure enantiomers synthesized by Mori and Miyake (43), demonstrated through electrophysiological recordings and trapping tests that (-)-grandisol has little, if any, biological activity. They suggest that the activity observed previously with (-)-grandisol could be explained by the fact that the material was not optically pure and by observations that mixtures of the other three components of grandlure (II, III, and IV) are somewhat attractive alone (14,44). Trap captures of released weevils using optically pure (+)-grandisol in combination with the other three grandlure components were somewhat reduced when (-)-grandisol was also present, but the difference was not statistically significant (44).

The difference between the optical activity of the naturally occurring component I and of synthetic grandlure I (or grandisol) has generally been ignored. Grandisol is frequently, but erroneously, referred to as the (+) isomer in publications and in *Chemical Abstracts*. Up to the present, grandlure used in the field has contained the optically inactive mixture of the two grandisol isomers.

D. Seasonal Responses

In the original field studies of the boll weevil pheromone (performed in June and July, 1963 and 1964), female boll weevils were shown to respond to material released by the male (21). Later it was demonstrated that both males and females responded later in the season (45). These seasonal variations in the responses of boll weevils to the natural male-produced pheromone were confirmed by other workers (46–49); in spring, traps baited with males attracted both sexes of overwintered adults, and again in the fall, weevils of both sexes responded to males, but mostly females responded in midseason during the fruiting period. Thus, the natural pheromone is both a sex attractant pheromone and an aggregating pheromone. The synthetic pheromone, grandlure, also evokes both a sex attractant and an aggregation response (50,51).

E. Involvement of Host Chemicals

The question of whether boll weevils locate cotton by random flight or by specific attraction has been a subject of debate for many years. In 1912, Hunter and Pierce (52) concluded that weevils could not use sensory perception to locate cotton fields, but in 1918, Viehoever et al. reported the isolation of a material attractive to the boll weevil from cotton plants (53). Nevertheless, with the demonstration that boll weevils were preferentially attracted to male weevils, rather than to weevil-free fruiting cotton plants (54), the view became prevalent that a major factor in cotton location by weevils was the pheromone produced by males that chanced on the cotton in random flight. However, evidence of attraction by the cotton plant or by plant constituents continued to mount, and it became apparent that the boll weevil's ability to locate the cotton plant involved olfactory perception of host plant volatiles as well as of the pheromone (55).

Numerous studies have been published on the fractionation and identification of cotton plant constituents and on investigations of their attractancy. In the 1960s (56,57) it was shown that a water extract of cotton seedlings contained material that was attractive to laboratory-reared boll weevils of both sexes and that the antennal club was the site of chemoreception for the attractant (58). Oil obtained from cotton plants was shown to contain some volatile materials that were attractive in the laboratory and in the field (59,60). Numerous volatile constituents were isolated and identified (61,62), and several of these were found to be somewhat attractive. In other studies, volatiles from the air around a cotton plant were investigated (55–63). From these various investigations, it became very clear that numerous components of the cotton plant had some attractancy for the boll weevil, although their effects were often short-ranged. It was suggested that multicomponent mixtures, rather than single compounds, were important in the attraction of boll weevils to cotton (59,60).

Attractancy of the pheromone is increased in the presence of host chemicals; this was clearly demonstrated in the field by Coppedge and Ridgway (64). Different systems for attracting or destroying overwintered boll weevils were compared: untreated plants, plants treated with insecticide aldicarb, treated plants and male-baited wing traps, and male-baited wing traps located at some distance from cotton. From four to six times as many weevils were attracted by cotton plants and baited traps as were attracted to cotton plants without traps, and the total number attracted to cotton and baited traps was nearly twice that captured by wing traps alone.

Increased attractancy has also been demonstrated with the synthetic pheromone and plant extracts or mixtures of plant components. Hardee and co-workers (27), who use firebrick as a carrier, reported in 1971 that aqueous extracts of cotton squares (buds) enhanced the attractancy of grandlure to boll weevils in a laboratory bioassay. However, at least some of the increased attractancy can be attributed to the added water, for a later report from the same laboratory indicated that the attractancy of grandlure-impregnated firebrick was markedly increased by the addition of water (28). Recently, it was found (65,66) that significantly more weevils were captured by traps containing both grandlure and a mixture of cotton monoterpenes than by traps baited with grandlure alone.

The role of host chemicals in pheromone production has also been a subject of much investigation. Boll weevils feeding on cotton squares have been shown to be more attractive than unfed weevils or those feeding on laboratory diet (23,46,54,67) or on terminals, cotyledons, and leaves (68). It has been speculated (15,68) that this might occur because the weevils obtain substances necessary for biosynthesis of pheromone from the cotton squares. On the other hand, Mitlin and Hedin (69) demonstrated that the insect can synthesize the pheromone *de novo* from acetate, mevalonate, and glucose. Despite a number of studies (31, 70–72), a clear connection between cotton components and biosynthesis has not been established.

The demonstration by Keller et al. (73) that aqueous extracts of cotton plants contained material with arrestant or feeding-stimulant action gave rise to investigations of the materials involved, summarized by Hedin et al. (61,62). Aqueous extracts of cotton squares and sucrose were used to alter feeding behavior and increase the efficacy of systemic insecticides (74). Cottonseed oil and ground cottonseed were also used in developing a bait for the boll weevil that was used to increase efficacy of microbial agents and for ecological studies (75–77). More recently, McKibben et al. (78) and Lusby et al. (79) have identified a number of cotton constituents having feeding-stimulant activity.

III. TRAPS AND DISPENSERS

A. Traps

In the first detailed study of trap designs for the boll weevil, Cross et al. (67) evaluated 12 types of traps baited with caged males or with extracts of males. The two most efficient traps were an adhesive-coated plywood wing trap, painted dark green to approximate the color of a cotton plant, and a clear plastic trap with four oblique screen funnels. However, a trap designed more specifically to take the behavior of the weevil into account (80) was later found to be more effective and selective. This trap, which came to be called the Leggett trap, had a conical body surmounted by a screen funnel and collecting container in which the lure was placed. The body of the trap was painted fluorescent yellow because of 1969 findings that this color was more attractive to the weevils (81). Boll weevils responded to the airborne pheromone and the color of the trap. On reaching the cone, they moved upward through the funnel into the collection container and were then unable to leave because they could not find the funnel opening. Subsequent trap designs have generally been based on the principles of the Leggett trap.

In studies of the effect of color, boll weevils responded best to highly reflective fluorescent yellow pigments (81–86). Taft and associates (87) found that boll weevil response to reflected light is greatest in the 500 to 525-nm region of the spectrum, corresponding to the region at which the intensity of reflected light from the fluorescent yellow pigments is greatest (84). This is also the region of peak reflectance of the cotton plant (87).

A trap (50), similar to the Leggett trap, but smaller so it could be placed in fields with minimum interference with farming practices, became known as the "infield trap" and was widely used. Modifications of this trap (88) resulted in a trap that captured 42% more weevils than the standard infield trap. Further modifications have resulted in a six-component molded plastic boll weevil trap (89,90), more suitable for use in operational programs and at least equal in efficiency to previous traps. A wire-mesh funnel is used to provide better ventilation and prevent heat build-up, thus increasing pheromone longevity (91). The trap components can be easily manufactured and can be stored and transported in much less space than is required for assembled traps. Other advantages are the ease of assembly in the field and the replaceability of damaged parts.

B. Dispensers

Initial results obtained with grandlure in the field were disappointing; traps baited with grandlure impregnated on dry firebrick caught no more weevils than did control traps (28). Addition of water to the firebrick made the traps competitive with traps baited with live males, but activity lasted no more than 2 days. A controlled-release formulation was clearly needed before the synthetic pheromone could be effectively used in traps because the four grandlure components are quite volatile, and the two aldehyde components are readily oxidized (92).

Various formulations (e.g., tablets or capsules of paraffin, microcrystalline wax, polyurethan foam, polyethylene glycol, polyvinyl pyrrolidone, or cellulose acetate) were tested in the laboratory or in the field (28,93); an aqueous formulation containing polyethylene glycol in a cotton dental wick or a cellulose acetate cigarette filter was nearly as attractive as caged males. Gas chromatographic analysis of residual grandlure components, however, showed that the half-life of the attractant in the formulation was fewer than 3 days (94).

Trapping studies with a variety of formulations and modifications (29,95–97) continued, resulting in three formulations that, over 3 weeks, were more effective than caged males; these were a cigarette filter impregnated with a solution of grandlure and contained in a shielded glass vial, a commercial gel formulation in plastic cups, and a semisolid gel formulation. A cardboard cartridge, and later a polyester wrapping, were found to be preferable to the glass vial for controlling the rate of release from the cigarette filter (98,99). Other formulations, including a plastic laminate and a polyvinyl chloride dispenser, were later found to be competitive (100–103). Thus, various types of lure dispenser formulations are available for use in boll weevil traps. Evaluation of these dispensers has been greatly improved by correlation of data obtained in the field with quantitative chemical data determined in the laboratory (29,95,104); factors requiring consideration in comparing dispensers include relative biological activity, rates of release of the various components, geographic location, weather conditions, and condition of the crop. Through such evaluations, performance criteria and specifications to aid in the procurement of boll weevil pheromone dispensers have been established (see Chap. 7).

Recent efforts have resulted in the development of an improved plastic laminate dispenser and a polyvinyl chloride dispenser acceptable for large-scale use (103,104). Large quantities of these dispensers have been procured for use in organized, ongoing boll weevil control programs (T. Barlow, personal communication; F. Myers, personal communication). Several other dispensers, also on the market, are used by individual growers, primarily for monitoring on a field-by-field basis.

IV. APPLICATIONS

The demonstration of the existence of a pheromone emitted by male boll weevils quickly stimulated research on potential applications using live insects. The identification and synthesis of grandlure stimulated research, particularly in trapping methods. Additionally, other uses of the pheromone, such as mating disruption and applications in combination with a toxicant (trap crops and attracticides), have been examined. Numerous factors affect the response of the boll weevil to traps. Variables already mentioned include trap design, pheromone formulations, seasonal variations, and effects of diet. Some additional variables affecting the weevil's response to traps include the concentrations of pheromone produced by native weevils, the nature and condition of the host crop, geographic location, weather, farming practices and, especially, the differences between laboratory-reared and native insects. The inability to control these many variables under field conditions has led to some conflicting results. Nevertheless, substantial progress has been made in developing possible applications for the synthetic pheromone, grandlure, in traps, for mating disruption, in trap crops, and in attracticides. Also, pheromone traps baited either with male weevils or with grandlure have been important tools in studies of biology, behavior, and population dynamics of the boll weevil (4,18,105).

A. Trapping

The potential value of pheromone traps in detection, monitoring, and control by mass trapping was an early topic of extensive investigation. Traps currently used for survey or mass trapping (90,106) are adaptations of the Mitchell infield trap (50). The lure is generally a controlled-release dispenser containing about 10 mg of grandlure.

The requirements for a trap used for survey (i.e., the assessment of boll weevil populations) are determined by the purpose of the survey. In detection, when the intent is to determine the presence or absence of the boll weevil in a given area, sensitivity of the trap is of major interest. In monitoring, to determine the extent and magnitude of a population, a quantitative relationship between trap captures and population is desired.

To understand some of the factors involved in effective use of traps, it is necessary to take the life cycle of the insect into account (12,107, 108). Diapausing (hibernating) adult weevils spend the winter in leaf litter and other protected sites close to cotton fields. In the spring, emerging weevils move into cotton fields in search of food, because cotton is their primary host plant. The timing of emergence from overwintering varies with geographic location and takes place over several weeks. The weevils mate in the fields and the females lay their eggs in the cotton flower buds (squares), where the larvae develop and pupate. Emerging adults eat their way out of the squares and soon are ready to mate. A generation occurs about once each month under optimum conditions. Toward the end of the cotton-growing season, the weevils leave the fields to find shelter for overwintering, in most areas of the United States. Therefore, in the spring, the most effective placement of traps is along the borders of the cotton fields to intercept the emerging weevils as they search for cotton. As the cotton grows, traps placed in the fields become more efficient (50). In the fall, as weevils move to hibernating sites, border traps again become more efficient.

1. Detection

The boll weevil trap was probably first used for detection related to boll weevil management near the High Plains of Texas in 1970 (4). The trap continues to be used around fields in this area for detection and to aid in program management. A number of other studies were conducted in which the trap was used very effectively for detection in dispersal studies (105,109,110). After the development of the in-field trap (50) a number of studies were conducted demonstrating that the trap was much more efficient in detecting weevils than manual surveys (50,111). Later, experiments were conducted to compare traps placed around the border of fields with traps within fields. Results indicate that at low-population densities results are similar, but that at higher densities, particularly during midseason when boll weevil reproduction is occurring, the in-field trap is more efficient (20, 112,113). Given results on probability of detection from field experiments (19,114) and a cotton insect—cotton crop model, the probability of detecting a boll weevil population if one existed in a specific eradication trial was estimated to be 0.9983 (115). The trap has also been used for detection in other management programs (6).

Thus, the boll weevil pheromone has been a valuable tool for detecting new boll weevil infestations where the boll weevil does not normally occur and for detecting reintroductions into areas from which the boll weevil has been eliminated.

2. Monitoring

Whereas detection usually addresses the presence or absence of boll weevils, monitoring attempts to use numbers captured as a guide to decision making. Most often, the decision is associated with whether or not to apply an insecticide. Specific studies have been primarily related to investigation of the relationship between trap captures, insecticide applications for overwintered boll weevil control, and numbers of boll weevil-damaged squares (116—119). From data obtained over several years on cumulative trap captures and subsequent square damage, a trap index system was derived, based on the number of weevils captured over a 6-week period, for predicting the results of treatment for weevil control (Fig. 1; 118).

Although comparative scientific data are limited, monitoring with pheromone traps as a guide to application of insecticides for diapause boll weevil control and even for midseason applications is practiced in some areas. Some specific guidelines have been reported (120).

Thus, boll weevil pheromone traps are useful for monitoring populations. However, additional research to make possible the estimation of actual boll weevil population levels at different times of the year would greatly increase the value of the trap as a monitoring tool.

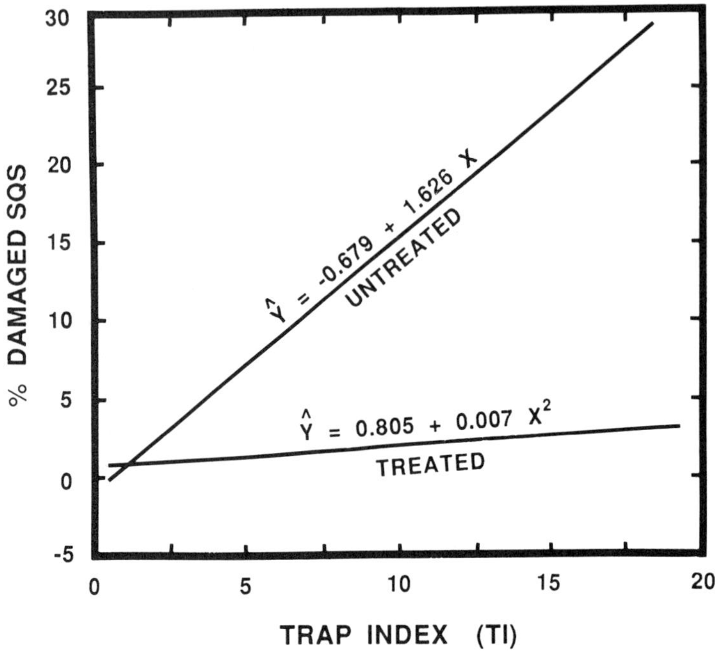

FIGURE 1 Trap index system for predicting treatment results of weevil control (from Ref. 118).

3. *Mass Trapping*

Efforts at mass trapping for population suppression with the boll weevil pheromone utilizing live insects and the wing trap were initiated in 1968. Up to 2.5 traps per hectare placed around fields in Texas apparently resulted in some delay in the development of boll weevil populations in trapped fields (121). A number of subsequent mass trapping experiments were conducted with male-baited traps placed around fields (27,122,123). Results of these studies indicated that low-density populations of boll weevils were substantially reduced, but the percentage of the boll weevil population captured declined substantially as the population density increased. However, in reviewing the results of these experiments in 1976, Bottrell and Rummel (124) concluded that the actual value of pheromone traps as a weevil suppression tool was poorly defined and that the methods used provided unrealistically high-suppression estimates. Later, Gilliland and Rummel (18) concluded that the general experimental results and the lack of

TABLE 1 Suppression of a Simulated First-Generation Population of Boll Weevils by Mass Trapping in 1.6-ha Cotton Fields

No. of traps/ 0.4 ha	Avg. no. of female weevils captured[a]	% Reduction of weevils
1.9	2.0	27
3.6	3.0	41
4.2	5.5	75
5.7	6.7	92

[a]Assuming 50% of the emerging boll weevils were females and an estimated average of 7.3 female weevils emerged in each field.
Source: Adapted from Ref. 20.

definitive guidelines for population management precluded widespread acceptance of grandlure as a management tool. The use of the synthetic pheromone in additional experiments and the development of the in-field trap did not greatly clarify the situation (50,113,125). Perhaps the most definitive experiment conducted on mass trapping involved the placing of cotton squares infested with boll weevil larvae into cotton fields otherwise uninfested with boll weevils and recording trap captures obtained when various numbers of traps were located in the fields. The percentage of emerging boll weevil adults captured ranged as high as 92% with 5.7 traps per 0.4 ha (Table 1; 19). This experiment provided solid evidence that high trap densities could be effective in suppressing boll weevil populations, although the parameters involved are not likely to accommodate practical use in high boll weevil densities because of the many traps required. Another, more recent series of experiments directed toward low-density populations indicated that one trap per acre eliminated boll weevils from 80% of the fields where the populations were estimated to be fewer than 2.5/ha (126).

Existing experimental data indicate that high trap densities are often needed to obtain high levels of population suppression. Therefore, mass trapping is probably not practical against many boll weevil populations. However, considerable evidence exists that mass trapping is useful at low population densities. Therefore, its application is restricted to programs for which elimination of reproduction or eradication is the program goal. Trap densities (one per 0.4 ha) that are likely to result in population suppression are currently in use, primarily in the Southeastern Boll Weevil Eradication Program (127).

B. Mating Disruption

Laboratory tests have demonstrated that high concentrations of grandlure can prevent females from locating males (128). Field tests showed that grandlure treatment of isolated cotton fields significantly reduced (93% fewer) trap captures of released, laboratory-reared virgin females (129). Subsequently, treatment of fields with 200 point sources of 48 mg of grandlure each per 0.4 ha significantly reduced oviposition by released laboratory-reared weevils (130). However, further research and development on mating disruption has not occurred, presumably because of the perceived high cost and because this approach interferes with the effectiveness of the pheromone traps used for survey.

C. Trap Cropping and Attracticides

Trap cropping, the use of a small portion of the principal crop or of a preferred crop to attract potential pests that can then be destroyed, has long been suggested as a means for cultural control of boll weevils (131,132). The placement of pheromone sources within insecticide-treated areas of a field has been found to increase the effectiveness of such trap crop areas in suppressing low-density populations of overwintered boll weevils (122,133,134) or in delaying the development of damaging infestations (135). These findings have led to additional studies of ways to improve the effectiveness of this technique: combinations of insecticide-treated, early planted or early maturing cotton trap crops and grandlure bait stations have been found to concentrate and suppress boll weevil populations (136); combinations of insecticide-treated, grandlure-baited trap crops and peripheral grandlure-baited traps have provided season-long boll weevil suppression (132,137,138); and creation of artificial in-field trap crops by use of grandlure bait stations in restricted areas of a field have precluded the need for an early-planted trap crop (50,100,132). Although this approach is promising and conceptionally desirable, it is not in general use because of intrinsic operational difficulties. However, continued research could lead to adoption in some areas (139).

Attracticides, which combine some type(s) of chemical attractant or feeding stimulant with an insecticide, represent a suppression strategy with obvious parallels to trap cropping. Recently discovered host-plant-derived attractants, and feeding stimulants (66,78, 79), used in combination with synthetic grandlure, provide an opportunity for developing a highly effective, multicomponent attracticide for the boll weevil. An invention report has been filed with the

Agricultural Research Service requesting that a patent application be filed for the formulation of an impregnated plastic dispenser that attracts and kills boll weevils; the dispenser contains grandlure, a feeding stimulant, a plant-derived attractant, and a toxicant (J. W. Smith, personal communication).

V. OPERATIONAL PROGRAMS

Practical applications of the boll weevil pheromone have primarily involved its use in traps. The specific role of the pheromone trap has varied considerably depending on the importance of the boll weevil in a particular cotton-growing area and the approach to boll weevil management that is used. This variation in use is quite consistent with the concept of targeted insect management programs that evolved in 1980–1981 from the evaluation of alternative cotton insect management programs (140,141). The diverse roles of pheromone traps in various management programs in the United States are illustrated by the following examples.

A. Overwintered Control by Individual Farmers

The concept of overwintered boll weevil control, which involves the application of insecticides to kill adult weevils before cotton begins fruiting, was first proposed by Ewing and Parencia (142). Until recent years, it had been applied primarily in a prophylactic manner because of the difficulty of estimating and predicting boll weevil populations. The development of pheromone trap indexes (116–118) for predicting boll weevil populations and the general availability of pheromone traps for monitoring overwintered boll weevils have substantially improved decision making. Pheromone traps are now used as an aid to decision making in a number of states. Parametric analysis with boll weevil and crop simulation models has confirmed the value of well-managed overwintered boll weevil control (143). Although the specifics vary, traps placed around fields early in the growing season are recommended. Treatment thresholds are in the range of two to five boll weevils per trap before fruiting (Table 2; 120).

B. Texas High Plains

In 1964, a fall diapause program was initiated near the High Plains of Texas. The objective of the program was to kill potential overwintering boll weevils with insecticides toward the end of the growing season (144). This tactic has been credited with preventing the spread of the boll weevil onto the High Plains of Texas (145). The boll weevil pheromone trap has played an important role in the design, operation, and

TABLE 2 Recommended Treatment Thresholds for Control of Overwintered Boll Weevils

State	Sampling period	Trap density	Treatment threshold[a]
Georgia	From planting to pinhead squares	1/4 ha	>5/trap/wk
Mississippi	4 wk in May	1/8 ha	>3/trap
Texas	6 wk before pinhead squares	6-8/field	(1, no treatment) 2.5/trap
Arkansas	1 wk before pinhead squares	1 every 28–46 m around field	>2/trap
Oklahoma	1 wk before pinhead squares		<2/trap

[a]Thresholds may be adjusted based on site-specific conditions.
Source: Adapted from Ref. 120.

evaluation of this program since the first trap was available. Currently, about 80,000 of the 400,000 ha of cotton in the Rolling Plains of Texas are treated with insecticides in the fall of each year to reduce the probability of boll weevils infesting the cotton on the High Plains (6). Approximately 400 pheromone traps are operated each year as a line of traps between the High and Rolling Plains. Trap captures are used to guide visual inspections of cotton for reproducing boll weevil populations. An additional 75 traps are operated throughout the Rolling Plains to provide a general assessment of boll weevil populations. The information obtained from the pheromone traps substantially reduces costs of manual surveys and improves the operational efficiency of the program (R. Edwards, personal communication).

C. South Texas

A boll weevil suppression program based primarily on use of fast-maturing varieties and early cotton stalk destruction has evolved in south Texas since 1982 (146,147). Data obtained with pheromone traps have contributed significantly to the evaluation and design of this program. For instance, trap captures of 81.0, 17.7, and 1.2 boll weevils per trap in 1982, 1983, and 1984, respectively, confirm the areawide reduction in the boll weevil population (146). The program is being implemented through a pest management district in the Lower Rio Grande Valley and in several other similar districts in south and west Texas. A total of 28 counties are involved (K. R. Summy, personal communication).

D. Southeast

The most extensive use of boll weevil pheromone traps has been in the Southeastern Boll Weevil Eradication Program that began in 1978. The program has essentially proceeded in three phases: (1) eradication trial in North Carolina and Virginia, (2) expanded program into southern North Carolina and South Carolina, and (3) expanded program into Georgia, Alabama, and Florida (Fig. 2). The boll weevil eradication trial was conducted in North Carolina and Virginia in 1978–1980. The technology applied included pheromone traps for

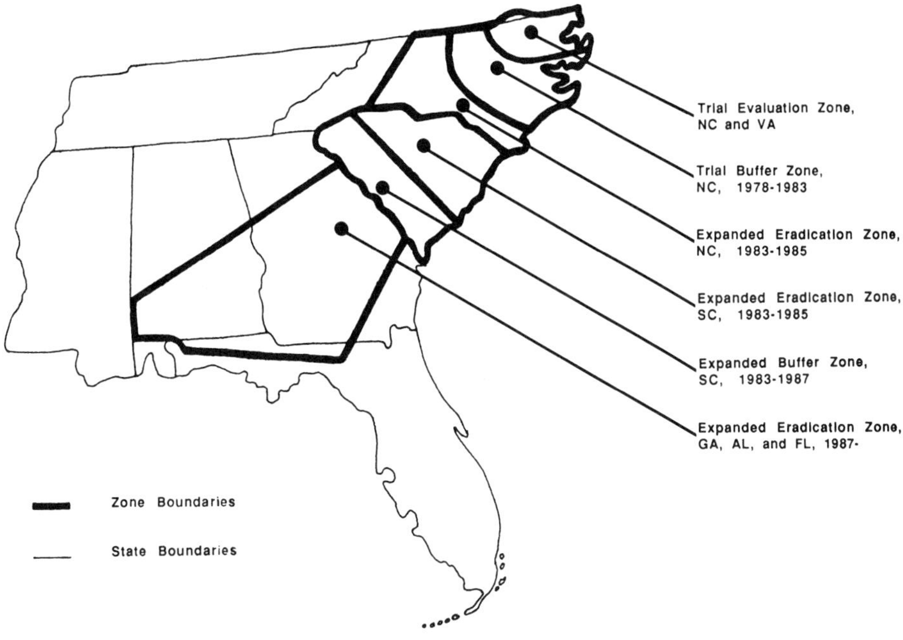

FIGURE 2 Diagram of the state boundaries (light line) and zone boundaries (heavy line) within the Southeastern Boll Weevil Eradication Program (adapted from Refs. 6 and 120).

surveillance and suppression, release of sterile insects, and diapause control applied on a mandatory basis (110). The cotton acreage involved in this trial area ranged from about 4,800 ha in 1978 to 28,000 in 1987 (127). The trial was generally considered to be an economic (148) and biological (110) success, although there was some disagreement about the interpretation of biological results (110,149).

In the eradication trial, traps were initially deployed at the rate of one per 3–4 ha around fields to aid in scheduling fall diapause boll weevil applications. In the spring of 1979 and 1980, one trap per 0.4 ha was placed around fields to monitor populations. After cotton was growing, two traps per 0.4 ha were placed in the fields to maximize detection and for possible suppression through mass trapping. In the fall, traps were again placed around fields at one trap per 0.4 ha. When reproduction of the boll weevil had been eliminated in an area, the trap density was reduced to one trap per 2–4 ha, primarily to detect possible reintroductions. Also, insecticide applications for boll weevil diapause control were applied as needed in the adjoining buffer zone to reduce weevil populations and to lower the possibility of reintroductions into the trial area. The extremely low numbers of boll weevils that were detected in the original trial evaluation area reflected the efficacy of the program in eliminating reproducing boll weevil populations (Table 3; 120). The increased trap captures in the buffer area in 1982 and 1983 and overall increases in boll weevil populations outside the program area in southeastern North Carolina during those years probably reflect the reduced boll weevil suppression efforts in the buffer area. The value of the pheromone traps in detecting reintroduced boll weevils so that measures can be taken to prevent reestablishment of boll weevil populations has been demonstrated repeatedly in the trial evaluation area. Specific cases for 1984 and 1985 are discussed by Dickerson et al. (150).

The program was expanded in 1983 to include the remainder of North Carolina and all of South Carolina. The technology used in the expanded program was similar to that used in the original trial area, except that sterile boll weevils were not used and the number of traps was reduced. Also, there was no routine use of infield traps. Traps were placed in the fields only as a result of border captures. During the second year, only one trap per 0.4 ha was placed around the fields. In addition, more emphasis was placed on stalk destruction because of the longer growing season (151). The total cotton acreage monitored in this phase ranged from 35,200 ha in 1983 to 60,800 in 1987 (127). The pheromone trap was again of critical importance in managing and evaluating the expanded program. Trap captures were used as a guide for insecticide applications (Table 4). During the spring of year 2, insecticides were applied if trap captures exceeded an average of 0.1 weevil per trap (120). Mass trapping was considered adequate for population elimination when boll weevils were

TABLE 3 Boll Weevil Capture in Fields in the Southeastern Boll Weevil Eradication Trial Evaluation Zone, 1977–1984

Year	No. of ha	No. of fields	% Fields from which indicated no. of weevils were captured		
			0	1–5	>5
1979	6,080	1,020	99.12	0.88	0.00
1980	10,680	1,775	99.98	0.06	0.06
1981	14,280	2,600	99.00	0.96	0.04
1982	15,120	3,000	91.47	6.66	1.87
1983	14,360	2,500	95.60	4.00	0.40
1984	25,200	4,300	99.95	0.00	0.05
1985	25,840	4,500	>99.9	0.0	<0.1
1986	20,200	4,100	>99.9	<0.1	0.0
1987	28,780	4,400	>99.9	<0.1	0.0
1988	37,170	5,680	>99.9	<0.1	0.0

Source: Modified from Refs. 90 and 120.

at low densities (fewer than 0.1 weevil per trap). Although precise quantitative information is difficult to obtain at such low population densities, data obtained by Leggett et al. (126; J. Leggett, personal communication) tend to support the efficacy of one trap per 0.4 ha for suppression at these low densities. Also, the results of the expanded program in southern North Carolina and South Carolina provided additional evidence that the combined components of the program were effective in eliminating reproducing boll weevil populations (Table 5; 90).

The southeastern program was expanded again in 1987 into most of Georgia and parts of Alabama and Florida. This expansion initially involved 158,800 ha of cotton (127). Technologies similar to those used in the expanded program in North Carolina and South Carolina are being used. The pheromone trap continues to play a prominent role throughout the southeastern program. About 590,000 traps and 8.25 million dispensers were used in 1988 (T. Barlow, personal communication). The current program, for a variety of reasons, will likely require more time to eliminate reproducing boll weevil populations than was needed on Virginia, North Carolina, and South Carolina.

TABLE 4 Treatment Thresholds for the Expanded Boll Weevil Eradication Program in North Carolina and South Carolina

Sampling period	Trap density/ha	Treatment threshold[a]	Comments
Fall of year 1	1/4.0		Apply insecticides at regular intervals until frost, or until stalks are destroyed
3-4 wk before pinhead squares, year 2	1/0.4	<0.1/trap	No treatment
		0.1–1.0/trap	Apply series of insect growth regulators and/or organophosphate insecticides
		>1.0/trap	Apply series of organophosphate insecticides
Fall of year 2	1/0.4	>5/field/wk	Apply insecticides at regular intervals until frost, or until stalks are destroyed

[a]These are general thresholds that may vary considerably, depending on conditions.
Source: Modified from Ref. 151.

E. Southwest

An organized areawide boll weevil program was initiated in southern California, southwestern Arizona, and part of Mexico in 1985. Extensive trapping, insecticide applications, and cultural controls led to elimination of reproducing populations in these areas by 1987 (6). In 1988, the program was expanded to cover the remainder of Arizona and adjoining areas of Mexico. In this southwestern program, the boll weevil pheromone trap is used primarily for detection and to aid in decision making related to insecticide application (152). About 50,000 traps, deployed at one or two traps per 4 ha, and 1.2 million dispensers were used in 1988 (F. Myers, personal communication).

TABLE 5 Boll Weevil Captures in Fields in the Expanded Eradication Zones in North Carolina,[a] 1983–1986

Location by year/zone (expanded eradication)	No. of ha	No. of fields	% Fields from which indicated no. of weevils were captured		
			0	1–5	>5
1983					
N.C.	6,600	800	0	1	99
S.C.	21,240	2,000	2	5	93
1984					
N.C.	9,200	1,000	21	35	44
S.C.	32,000	2,300	22	25	53
1985					
N.C.	9,400	1,000	90	7	3
S.C.	37,000	4,300	81	12	7
1986					
N.C.	8,600	1,000	>99.9 <0.1[b]	0.0	
S.C.	34,400	4,200	>97.3	2.3	<0.4[c]

[a]Excluding the expanded buffer zone.
[b]A total of 3 boll weevils captured in 3 fields.
[c]A total of 365 boll weevils captured in 15 fields.
Source: From Refs. 90 and 150.

F. Benefits

Information is not available to calculate specific monetary benefits associated with the use of the boll weevil pheromone in management programs. However, studies have been conducted on a number of programs in which the boll weevil pheromone trap is a component. For instance, in the Southeastern Boll Weevil Eradication Program, in which the pheromone trap is a major component, estimates of reduced insecticide applications, increased cotton acreage, and increased cotton yield indicate that the program in North Carolina, Virginia, and South Carolina is producing increased annual returns of over 75 dollars/acre (Table 6; 153). From an environmental viewpoint, the 50–70% reduction in insecticides associated with the program in the three states also represents a substantial and continuing benefit. Other studies have also shown that substantial benefits have resulted from programs

TABLE 6 Benefits Associated with Reduced Insecticides, Increased Cotton Acreage, and Yield Increases for the Original Trial Eradication Zone in Virginia and North Carolina and the Expanded Eradication Zones in North Carolina and South Carolina

	Dollars per acre	
Benefit	Original eradication zone (Va, N.C.)	Expanded eradication zones (N.C., S.C.)
Net reduced pesticide	28.87	30.01
Acreage expansion	13.28	13.80
Yield effect	34.50	34.50
Total benefits	76.65	78.31

Source: Modified from Ref. 153.

in which pheromone traps have had some role (154–156). Furthermore, the use of the boll weevil pheromone trap for monitoring throughout the Cotton Belt has resulted in more efficient use of insecticides and has produced direct economic and environmental benefits.

VI. CONCLUSIONS

The availability of a synthetic pheromone and of effective formulations for dispensing it has contributed to all phases of boll weevil management, including surveillance, suppression, and program evaluation, as well as to basic research on the biology, behavior, and population dynamics of the boll weevil. Operational programs rely heavily upon pheromone-baited traps to detect the presence of an individual boll weevil and to monitor the relative density of boll weevils in a general population. Traps are also used as a survey tool in evaluating the efficacy of control efforts. Use of the pheromone directly in suppression strategies, such as mass trapping, mating disruption, trap cropping and attracticides, is less widespread, but these strategies continue to be potentially important in managing boll weevil populations.

The impact of the current use of the boll weevil pheromone is difficult to quantify because it is used in such a variety of ways as a component in management programs. However, the very fact that it is so widely used is a strong indication of its importance. Certainly, a number of boll weevil management programs are greatly enhanced by or, in some cases, would not be possible without the boll weevil pheromone.

ACKNOWLEDGMENTS

The assistance of John C. Davis in the preparation of the manuscript and of Thomas Barlow, Frank Myers, Ronald Edwards, and Kenneth R. Summy in providing information on operational programs is gratefully acknowledged.

REFERENCES

1. Townsend, C. H. T. Report on the Mexican cotton boll weevil in Texas (*Anthonomus grandis* Boh.). *Insect Life 7*:295–309 (1895).
2. Hunter, W. D., and Coad, B. R. The boll weevil problem. *USDA Farmer's Bull.* pp. 1329–1330 (1923).
3. Robertson, O. T., Noble, L. W., and Orr, G. E. Spread of the boll weevil and its control in far west Texas. *J. Econ. Entomol. 59*:754–756 (1966).
4. Bottrell, D. G., White, J. R., Moody, D. S., and Hardee, D. D. Overwintering habitats of the boll weevil in the Rolling Plains of Texas. *Environ. Entomol. 1*:633–638 (1972).
5. Bergman, D., Henneberry, T. J., and Bariola, L. A. Distribution of the boll weevil in southwestern Arizona cultivated cotton from 1978–1981. In *Proceedings, Beltwide Cotton Production Research Conferences, 1982.* National Cotton Council of America, Memphis, 1982, pp. 204–207.
6. National Cotton Council of America. *Boll Weevil Eradication: A Cooperative Federal, State, and Producer Program,* 1989, 12 pp.
7. Coker, R. R. Economic impact of the boll weevil. In *Boll Weevil Suppression, Management, and Elmination Technology* (Davich, T. B., ed.). Agriculture Research Service, Southern Region, USDA, New Orleans, 1976, pp. 3–4.
8. King, E. G., Phillips, J. R., and Head, R. B. Forty-first annual conference report on cotton insect research and control. In *Proceedings, Beltwide Cotton Production Presearch Conferences,* New Orleans, Jan. 3–8, 1988. National Cotton Council of America, Memphis, 1988, pp. 188–202.
9. Knipling, E. F. Technically feasible approaches to boll weevil eradication. In *Proceedings, Beltwide Cotton Production-Mechanization Conferences,* Hot Springs, Ark., Jan. 11–12, 1968. National Cotton Council of America, Memphis, 1968, pp. 14–18.

10. Ridgway, R. L. Control of the bollworm and tobacco budworm through conservation and augmentation of predacious insects. In *Proceedings, Tall Timbers Conference Ecology Animal Control Habitat Management.* 1969, pp. 127–144.
11. Pimentel, D., Andow, D., Gallahan, D., Schreiner, I., Thompson, T. E. Dyson-Hudson, R., Jacobson, S. N. Irish, M. A., Kroop, S. F., Moss, A. M., Shepard, M. D., and Vinzant, B. G. Pesticides: Environmental and social costs. In *Pest Control: Cultural and Environmental Aspects* (Pimentel, D., and Perkins, J. H., eds.). Westview Press, Boulder, Colo. 1980, pp. 99–158.
12. Ridgway, R. L., Lloyd, E. P., and Cross, W. H. (eds.). *Cotton Insect Management with Special Reference to the Boll Weevil.* U.S. Government Printing Office, Washington, D.C. 1983.
13. Davich, T. B., Foreword. In *Boll Weevil Suppression, Management, and Elimination Technology.* Memphis, Feb. 13–15, 1974 (Davich, T. B., ed.). Agricultural Research Service, Southern Region, USDA, New Orleans, 1976, pp. i–ii.
14. Tumlinson, J. H., Hardee, D. D., Gueldner, R. C., Thompson, A. C., Hedin, P. A., and Minyard, J. P. Sex pheromones produced by male boll weevil: Isolation, identification, and synthesis. *Science 166*:1010–1012 (1969).
15. Tumlinson, J. H., Gueldner, R. C., Hardee, D. D., Thompson, A. C., Hedin, P. A., and Minyard, J. P. The boll weevil sex attractant. In *Chemicals Controlling Insect Behavior* (Beroza, M., ed.). Academic Press, New York, 1970, pp. 41–59.
16. Hedin, P. A., Gueldner, R. C., and Thompson, A. C. Utilization of the boll weevil pheromone for insect control. In *Pest Management with Insect Sex Attractants.* Chicago, Aug. 26, 1975 (Beroza, M., ed.). American Chemical Society, Washington, D.C., 1976, pp. 30–52.
17. Texas Agricultural Experiment Station. *Detection and Management of the Boll Weevil with Pheromone.* Texas A&M University System, College Station, Tex., 1976, 68 pp.
18. Gilliland, F. R., Jr., and Rummel, D. R. The role of pheromone for boll weevil detection and suppression. In *The Boll Weevil: Management Strategies* (Warren, L. O., Chairman, Compiling Committee). Arkansas Agricultural Experiment Station, Fayetteville, Ark. June 1978, pp. 84–95.
19. Lloyd, E. P., McKibben, G. H., Witz, J. A., Hartstack, A. W., Lockwood, D. F., Knipling, E. F., and Leggett, J. E. Mass trapping for detection, suppression and integration with other suppression measures against the boll weevil. In *Management of Insect Pests with Semiochemicals: Concepts and Practice* (Mitchell, E. R., ed.). Plenum Press, New York, 1981, pp. 191–203.

20. Lloyd, E. P., McKibben, G. H., Leggett, J. E., and Hartstack, A. W. Pheromones for survey, detection, and control. In *Cotton Insect Management with Special Reference to the Boll Weevil* (Ridgway, R. L., Lloyd, E. P., and Cross, W. H., eds.). U.S. Government Printing Office, Washington, D.C., 1983, pp. 179–205.
21. Cross, W. H., and Mitchell, H. C. Mating behavior of the female boll weevil. *J. Econ. Entomol.* *59*:1503–1507 (1966).
22. Keller, J. C., Mitchell, E. B., McKibben, G. H., and Davich, T. B. A sex attractant for female boll weevils from males. *J. Econ. Entomol.* *57*:609–610 (1984).
23. Hardee, D. D., Mitchell, E. B., and Huddleston, P. M. Field studies of sex attraction in the boll weevil. *J. Econ. Entomol.* *60*:1221–1224 (1967).
24. Tumlinson, J. H., Hardee, D. D., Minyard, J. P., Thompson, A. C., Gast, R. T., and Hedin, P. A. Boll weevil sex attractant: Isolation studies. *J. Econ. Entomol.* *61*:470–474 (1968).
25. Hardee, D. D., Mitchell, E. B., and Huddleston, P. M. Procedure for bioassaying the sex attractant of the boll weevil. *J. Econ. Entomol.* *60*:169–171 (1967).
26. Tumlinson, J. H., Gueldner, R. C., Hardee, D. D., Thompson, A. C., Hedin, P. A., and Minyard, J. P. Identification and synthesis of the four compounds comprising the boll weevil sex attractant. *J. Org. Chem.* *36*:2616–2621 (1971).
27. Hardee, D. D., Wilson, N. M., Mitchell, E. B., and Huddleston, P. M. Factors affecting activity of grandlure, the pheromone of the boll weevil, in laboratory bioassays. *J. Econ. Entomol.* *64*:1454–1456 (1971).
28. Hardee, D. D., McKibben, G. H., Gueldner, R. C., Mitchell, E. B., Tumlinson, J. H., and Cross, W. H. Boll weevils in nature respond to grandlure, a synthetic pheromone. *J. Econ. Entomol.* *65*:97–100 (1972).
29. Coppedge, J. R., Bull, D. L., House, V. S., Ridgway, R. L., Bottrell, D. G., and Cowan, D. B., Jr. Formulations for controlling the release of synthetic pheromone (grandlure) of the boll weevil. 2. Biological studies. *Environ. Entomol.* *2*:837–843 (1973).
30. Hedin, P. A., Burks, M. L., and Thompson, A. C. Synthetic intermediates and byproducts as inhibitors of boll weevil pheromone attractancy. *J. Agric. Food Chem.* *33*:1011–1017 (1985).
31. Hedin, P. A., Hardee, D. D., Thompson, A. C., and Gueldner, R. C. An assessment of the life time biosynthesis potential of the male boll weevil. *J. Insect Physiol.* *20*:1707–1712 (1974).
32. Hardee, D. D., McKibben, G. H., Rummel, D. R., Huddleston, P. M., and Coppedge, J. R. Response of boll weevils to component ratios and doses of the pheromone, grandlure. *Environ. Entomol.* *3*:135–138 (1974).

33. Dickens, J. C., and Prestwich, G. D. Different recognition of geometric isomers by the boll weevil, *Anthonomus grandis* Boh. (Coleoptera: Curculionidae): Evidence for only three essential components in aggregation pheromone. *J. Chem. Ecol. 15*:529–540 (1989).
34. Katzenellenbogen, J. A. Insect pheromone synthesis: New methodology. *Science 194*:139–148 (1976).
35. Henrick, C. A. The synthesis of insect sex pheromones. *Tetrahedron 33*:1845–1889 (1977).
36. Hicks, D. R. Approaches to the synthesis of (+)- and (-)-grandisol. *Diss. Abstr. Int. 36B*:2225 (1975).
37. Hobbs, P. D., and Magnus, P. D. Studies on terpenes. 4. Synthesis of optically active grandisol, the boll weevil pheromone. *J. Am. Chem. Soc. 98*:4594–4600 (1976).
38. Mori, K. Synthesis of the both enantiomers of grandisol, the boll weevil pheromone. *Tetrahedron 34*:915–920 (1978).
39. Jones, J. B., Finch, M. A. W., and Jakovac, I. J. Enzymes in organic synthesis. 26. Synthesis of enantiomerically pure grandisol from an enzyme-generated chiral synthon. *Can. J. Chem. 60*:2007–2011 (1982).
40. Mori, K., Tamada, S., and Hedin, P. A. (-)-Grandisol, the antipode of the boll weevil pheromone, is biologically active. *Naturwissenschaften 65* (1978).
41. Webster, F. X., and Silverstein, R. M. Grandisol and Lineatin enantiomers. In *Insect Pheromone Technology: Chemistry and Applications*, New York (Leonhardt, B. A., and Beroza, M., eds.). American Chemical Society, Washington, D.C., 1982, pp. 87–106.
42. Webster, F. X., and Silverstein, R. M. Synthesis of optically pure enantiomers of grandisol. *J. Org. Chem. 51*:5226–5231 (1986).
43. Mori, K., and Miyake, M. Pheromone synthesis. 100. A new synthesis of both the enantiomers of grandisol, the boll-weevil pheromone. *Tetrahedron 43*:2229–2239 (1987).
44. Dickens, J. C., and Mori, K. Receptor chirality and behavioral specificity of the boll weevil, *Anthonomus grandis* Boh. *J. Chem. Ecol. 15*:517–528 (1989).
45. Bradley, J. R., Jr., Clower, D. F., and Graves, J. B. Field studies of sex attraction in the boll weevil. *J. Econ. Entomol. 61*:1457–1458 (1968).
46. Hardee, D. D., Cross, W. H., Mitchell, E. B., Huddleston, P. M., Mitchell, H. C., Merkl, M. E., and Davich, T. B. Biological factors influencing responses of the female boll weevil to the male sex pheromone in field and large cage tests. *J. Econ. Entomol. 62*:161–165 (1969).

47. Hardee, D. D., Cleveland, T. D., Davis, J. W., and Cross, W. H. Attraction of boll weevils to cotton plants and to males fed on three diets. *J. Econ. Entomol.* *63*:990–991 (1970).
48. Mitchell, E. B., Hardee, D. D., Cross, W. H., Huddleston, P. M., and Mitchell, H. C. Influence of rainfall, sex ratio, and physiological condition of boll weevils on their response to pheromone traps. *Environ. Entomol.* *1*:438–440 (1972).
49. Mitchell, E. B., Lloyd, E. P., Hardee, D. D., Cross, W. H., and Davich, T. B. In-field traps and insecticides for suppression and elimination of populations of boll weevils. *J. Econ. Entomol.* *69*:83–88 (1976).
50. Mitchell, E. B., and Hardee, D. D. In-field traps: A new concept in survey and suppression of low populations of boll weevils. *J. Econ. Entomol.* *67*:506–508 (1974).
51. Hardee, D. D. Boll weevil population management, detection, or elimination with in-field traps. In *Proceedings, Beltwide Cotton Production Research Conferences*, New Orleans, Jan. 6–8, 1975. National Cotton Council of America, Memphis, 1975, pp. 132–135.
52. Hunter, W. D., and Pierce, W. D. *The Mexican Cotton-Boll Weevil: A Summary of the Investigations on This Insect up to December 31, 1911.* 1912.
53. Viehoever, A., Chernoff, L. H., and Johns, C. O. Chemistry of the cotton plant with special reference to upland cotton. *J. Agric. Res.* *13*:345–352 (1918).
54. Hardee, D. D., Cross, W. H., amd Mitchell, E. B. Male boll weevils are more attractive than cotton plants to boll weevils. *J. Econ. Entomol.* *62*:165–169 (1969).
55. Chang, J. F., Benedict, J. H., and Payne, T. L. Trapping of cotton terpene volatiles and boll weevil pheromone. In *Proceedings, Beltwide Cotton Production Research Conferences*. National Cotton Council of America, Memphis, 1985, pp. 397–398.
56. Keller, J. C., Maxwell, F. G., Jenkins, J. N., and Davich, T. B. A boll weevil attractant from cotton. *J. Econ. Entomol.* *56*: 110–111 (1963).
57. Hardee, D. D., Mitchell, E. B., Huddleston, P. M., and Davich, T. B. A laboratory technique for bioassay of plant attractants for the boll weevil. *J. Econ. Entomol.* *59*:240–241 (1966).
58. Hardee, D. D., MItchell, E. B., and Huddleston, P. M. Chemoreception of attractants from the cotton plant by boll weevils, *Anthonomus grandis* (Coleoptera: Curculionidae). *J. Econ. Entomol.* *59*:867–868 (1966).
59. Minyard, J. P., Hardee, D. D., Gueldner, R. C., Thompson, A. C., Wiygul, G., and Hedin, P. A. Constituents of the cotton bud components attractive to the boll weevil. *J. Agric. Food Chem.* *17*:1093–1097 (1969).

60. McKibben, G. H., Mitchell, E. B., Scott, W. P., and Hedin, P. A. Boll weevils are attracted to volatile oils from cotton plants. *Environ. Entomol. 6*:804–806 (1977).
61. Hedin, P. A., Thompson, A. C., and Gueldner, R. C. The boll weevil–cotton plant complex. *Toxicol. Environ. Chem. Rev. 1*:291–351 (1973).
62. Hedin, P. A., Thompson, A. C., and Gueldner, R. C. Cotton plant and insect constituents that control boll weevil behavior and development. *Recent Adv. Phytochem. 10*:271–350 (1976).
63. Keller, J. C., Davich, T. B., Maxwell, F. G., Jenkins, J. N., Mitchell, E. B., and Huddleston, P. Extraction of boll weevil attractant from the atmosphere surrounding growing cotton. *J. Econ. Entomol. 58*:588–589 (1965).
64. Coppedge, J. R., and Ridgway, R. L. *The Integration of Selected Boll Weevil Suppression Techniques in an Eradication Experiment. Prod. Res. Rept. No. 152 (USDA)*: 26 pp. (1973).
65. Chang, J. F., Benedict, J. H., Payne, T. L., and Camp, B. J. Attractiveness of cotton volatiles and grandlure to boll weevils in the field. In *Proceedings, Beltwide Cotton Production Research Conferences*, Dallas, Jan. 4–8, 1987. National Cotton Council of America, Memphis, 1987, pp. 102–104.
66. Dickens, J. C. Green leaf volatiles enhance aggregation pheromone of boll weevil, *Anthonomus grandis. Entomol. Exp. Appl. 52*: 191-203 (1989).
67. Cross, W. H., Hardee, D. D., Nichols, F., Mitchell, H. C., Mitchell, E. B., Huddleston, P. M., and Tumlinson, J. H. Attraction of female boll weevils to traps baited with males or extracts of males. *J. Econ. Entomol. 62*:154–161 (1969).
68. Hardee, D. D. Pheromone production by male boll weevils as affected by food and host factors. *Contrib. Boyce Thompson Inst. 24*:315–322 (1970).
69. Mitlin, N., and Hedin, P. A. Biosynthesis of grandlure, the pheromone of the boll weevil, *Anthonomus grandis*, from acetate, mevalonate, and glucose. *J. Insect Physiol. 20*:1825–1831 (1974).
70. Thompson, A. C., and Mitlin, N. Biosynthesis of the sex pheromone of the male boll weevil from monoterpene precursors. *Insect Biochem. 9*:293–294 (1979).
71. Hedin, P. A. A study of factors that control biosynthesis of the compounds which comprise the boll weevil pheromone. *J. Chem. Ecol. 3*:279–289 (1977).
72. Hedin, P. A., Lindig, O. H., and Wiygul, G. Enhancement of boll weevil *Ahthonomus grandis* Boh. (Coleoptera: Curculionidae) pheromone biosynthesis with JH III. *Experientia 38*:375–376 (1982).
73. Keller, J. C., Maxwell, F. G., and Jenkins, J. N. Cotton extracts as arrestants and feeding stimulants for the boll weevil. *J. Econ. Entomol. 55*:800–801 (1962).

74. Ridgway, R. L., Jones, S. L., and Gorzycki, L. J. Tests for Boll weevil control with a systemic insecticide and a boll weevil feeding stimulant. *J. Econ. Entomol.* *59*:149–153 (1966).
75. Daum, R. J., McLaughlin, R. E., and Hardee, D. D. Development of the bait principle for boll weevil control: Cottonseed oil, a source of attractants and feeding stimulants for the boll weevil. *J. Econ. Entomol.* *60*:321–325 (1967).
76. Lloyd, E. P., Daum, R. J., McLaughlin, R. E., Tingle, F. C., McKibben, G. H., McCoy, J. R., Bell, M. R., and Cleveland, T. C. A red dye to evaluate bait formulations and to mass mark field populations of boll weevils. *J. Econ. Entomol.* *61*:1440–1444 (1968).
77. Daum, R. J., Gast, R. T., and Davich, T. B. Marking adult boll weevils with dyes fed in a cottonseed oil bait. *J. Econ. Entomol.* *62*:943–944 (1969).
78. McKibben, G. H., Thompson, M. J., Parrott, W. L., Thompson, A. C., and Lusby, W. R. Identification of feeding stimulants for boll weevils from cotton buds and anthers. *J. Chem. Ecol.* *11*:1229–1238 (1985).
79. Lusby, W. R., Oliver, J. E., McKibben, G. H., and Thompson, M. J. Free and esterified sterols of cotton buds and anthers. *Lipids* *22*:80–83 (1987).
80. Leggett, J. E., and Cross, W. H. A new trap for capturing boll weevils. *Coop. Econ. Insect Rep.* *21*:773–774 (1971).
81. Hardee, D. D., Cross, W. H., Mitchell, E. B., Huddleston, P. M., and Mitchell, H. C. Capture of boll weevils in traps baited with males: Effect of size, color, location, and height above ground level. *Environ. Entomol.* *1*:162–166 (1972).
82. Roach, S. H., Agee, H. R., and Ray, L. Influence of position and color of male-baited traps on captures of boll weevils. *Environ. Entomol.* *1*:530–532 (1972).
83. Leggett, J. E., Cross, W. H., Mitchell, H. C., Johnson, W. L., and McGovern, W. L. Improved traps for capturing boll weevils. *J. Ga. Entomol. Soc.* *10*:52–61 (1975).
84. Cross, W. H., Mitchell, H. C., and Hardee, D. D. Boll weevils: Response to light sources and colors on traps. *Environ. Entomol.* 5:565–571 (1976).
85. Leggett, J. E., and Cross, W. H. Boll weevils: The relative importance of color and pheromone in orientation and attraction to traps. *Environ. Entomol.* *7*:4–6 (1978).
86. Leggett, J. E. Boll weevil: Competitive and non-competitive evaluation of factors affecting pheromone trap efficiency. *Environ. Entomol.* *9*:416–419 (1980).
87. Taft, H. M., Hopkins, A. R., and Agee, H. R. Response of overwintered boll weevils to reflected light, odor, and electromagnetic radiation. *J. Econ. Entomol.* *62*:419–424 (1969).

88. Dickerson, W. A., McKibben, G. H., Lloyd, E. P., Kearney, J. F., Lam, J. J., Jr., and Cross, W. H. Field evaluation of a modified in-field boll weevil trap. *J. Econ. Entomol.* *74*:280–282 (1981).
89. Dickerson, W. A. Boll weevil trap. U.S. Patent 4,611,425. Issued Sept. 16, 1986.
90. Dickerson, W. A., Ridgway, R. L., and Planer, F. R. Southeastern Boll Weevil Eradication Program: Improved pheromone traps and program status. In *Proceedings*, *Beltwide Cotton Production Research Conferences*, Dallas, Jan. 4–8, 1987. National Cotton Council of America, Memphis, 1987, pp. 335–337.
91. Rummel, D. R., Carroll, S. C., and Shaver, T. N. Influence of boll weevil trap design on internal trap temperature and grandlure volatilization. *Southwest. Entomol.* *12*:127–138 (1987).
92. Henson, R. D., Bull, D. L., Ridgway, R. L., and Ivie, G. W. Identification of the oxidative decomposition products of the boll weevil pheromone, grandlure, and the determination of the fate of grandlure in soil and water. *J. Agric. Food Chem.* *24*:2288–2301 (1976).
93. McKibben, G. H., Hardee, D. D., Davich, T. B., Gueldner, R. C., and Hedin, P. A. Slow-release formulations of grandlure, the synthetic pheromone of the boll weevil. *J. Econ. Entomol.* *64*:317–319 (1971).
94. McKibben, G. H., Gueldner, R. C., Hedin, P. A., Hardee, D. D., and Davich, T. B. Release characteristics of polymeric attractant and repellent compositions. *J. Econ. Entomol.* *65*:1512–1514 (1972).
95. Bull, D. L., Coppedge, J. R., Hardee, D. D., Rummel, D. R., McKibben, G. H., and House, V. S. Formulations for controlling the release of synthetic pheromone (grandlure) of the boll weevil. 3. Laboratory and field evaluations of three slow-release preparations. *Environ. Entomol.* *2*:905–909 (1973).
96. Bull, D. L., Coppedge, J. R., Ridgway, R. L., Hardee, D. D., and Graves, T. M. Formulations for controlling the release of synthetic pheromone (grandlure) of the boll weevil. 1. Analytical studies. *Environ. Entomol.* *2*:829–835 (1973).
97. Hardee, D. D., Graves, T. M., McKibben, G. H., Gueldner, R. C., and Olsen, C. M. A slow-release formulation of grandlure, the synthetic pheromone of the boll weevil. *J. Econ. Entomol.* *67*:44–46 (1974).
98. Bull, D. L. Formulations of grandlure. In *Detection and Management of the Boll Weevil with Pheromone.* Texas A&M University System, College Station, Tex., 1976, pp. 5–8.
99. McKibben, G. H., Johnson, W. L., Edwards, R., Kotter, E., Kearny, J. F., Davich, T. B., Lloyd, E. P., and Ganyard,

M. C. A polyester-wrapped cigarette filter for dispensing grandlure. *J. Econ. Entomol.* *73*:250–251 (1980).
100. Hardee, D. D., McKibben, G. H., and Huddleston, P. M. Grandlure for boll weevils: Controlled release with a laminated plastic dispenser. *J. Econ. Entomol.* *68*:477–479 (1975).
101. Johnson, W. L., McKibben, G. H., Rodriguez, V. J., and Davich, T. B. Boll weevil: Increased longevity of grandlure using different formulations and dispensers. *J. Econ. Entomol.* *69*:263–265 (1976).
102. Merkl, M. E. Grandlure for boll weevils: Test of new formulations in Mississippi. *J. Ga. Entomol. Soc.* *12*:273–276 (1977).
103. Dickerson, W. A., Leonhardt, B. A., and Ridgway, R. L. Field bioassay and laboratory chemical analysis of controlled-release dispensers for the pheromone of the boll weevil, *Anthonomus grandis*. In *Proceedings*, *International Symposium on Controlled Release of Bioactive Materials*. Toronto, Ontario, Canada, Aug. 2–5, 1987. Controlled Release Society, Lincolnshire, Ill., 1987, pp. 249-250.
104. Leonhardt, B. A., Dickerson, W. A., Ridgway, R. L., and DeVilbiss, E. D. Laboratory and field evaluation of controlled release dispensers containing grandlure, the pheromone of the boll weevil (Coleoptera: Curculionidae). *J. Econ. Entomol.* *81*: 937–943 (1988).
105. Davich, T. B., Hardee, D. D., and Alcala, J. M. Long-range dispersal of boll weevils determined with wing traps baited with males. *J. Econ. Entomol.* *63*:1706–1708 (1970).
106. Forey, D. E., and Quisumbing, A. R. Newly designed boll weevil SCOUT trap. In *Proceedings*, *Beltwide Cotton Production Research Conferences*, Jan. 4–8, 1987, Dallas, National Cotton Council of America, Memphis, pp. 139–141.
107. Cross, W. H. Biology, control, and eradication of the boll weevil. *Annu. Rev. Entomol.* *18*:17–46 (1973).
108. Davich, T. B. *Boll Weevil Suppression, Management, and Elimination Technology*. Memphis, Feb. 13–15, 1974. Agricultural Research Service, Southern Region, USDA, 1976.
109. Ridgway, R. L., Bariola, L. A., and Hardee, D. D. Seasonal movement of boll weevils near the High Plains of Texas. *J. Econ. Entomol.* *64*:14–19 (1971).
110. U.S. Department of Agriculture (Biological Evaluation Team of the USDA Interagency Working Group on the Boll Weevil. Cross, W. H., Team Leader), *Biological Evaluation of Alternative Beltwide Boll Weevil/Cotton Insect Management Programs. Overall Evaluation*. Appendix A. Apr. 22, 1981.
111. Merkl, M. E., Cross, W. H., and Johnson, W. L. Boll weevil: Detection and monitoring of small populations with in-field traps. *J. Econ. Entomol.* *71*:29–30 (1978).

112. Leggett, J. E. Detection probability and efficiency of infield and border traps for capturing overwintered boll weevils (Coleoptera: Curculionidae) at low population levels. *Environ. Entomol. 13*:324–328 (1984).
113. Lloyd, E. P., Leggett, J. E., Ridgway, R. L., and Dickerson, W. A. Boll weevil pheromone traps for detection and suppression of low level boll weevil populations. In *Proceedings, Beltwide Cotton Production Research Conferences*, New Orleans, Jan. 6–11, 1985. National Cotton Council of America, Memphis, 1985, pp. 141–142.
114. Leggett, J. E., Lloyd, E. P., and Witz, J. A. Efficiency of infield traps in detection or suppressing low population levels of boll weevils. *Environ. Entomol. 10*:125–130 (1981).
115. McKibben, G. H., and Cross, W. H. Use of pheromone traps to estimate probability of zero populations of boll weevils. *Southwest. Entomol. 9*:371–374 (1984).
116. Rummel, D. R., White, J. R., Carroll, S. C., and Pruitt, G. R. Pheromone trap index system for predicting need for overwintered boll weevil control. *J. Econ. Entomol. 73*:806–810 (1980).
117. Benedict, J. H. Segers, J. C., Anderson, D. A., Parker, R. D., Walmsley, M. R., and Hopkins, S. W. *Use of Pheromone Traps in the Management of Overwintered Boll Weevil on the Lower Gulf Coast of Texas*. Texas Agricultural Experiment Station Miscellaneous Publication MP-1576. Texas A&M University System, College Station, Tex., 1985.
118. Benedict, J. H., Urban, T. C., George, D. M., Segers, J. C., Anderson, D. J., McWhorter, G. M., and Zummo, G. R. Pheromone trap thresholds for management of overwintered boll weevils (Coleoptera: Curculionidae). *J. Econ. Entomol. 78*:169–171 (1985).
119. Henneberry, T. J., Meng, T., and Bariola, L. A. Boll weevil: Grandlure trapping and early-season insecticide applications in relation to cotton infestations in Arizona, USA. *Southwest. Entomol. 13*:251–260 (1988).
120. Ridgway, R. L., Dickerson, W. A., Brazzel, J. R., Leggett, J. F., Lloyd, E. P., and Planer, F. R. Boll weevil pheromone trap captures for treatment thresholds and population assessments. In *Proceedings, Beltwide Cotton Production Research Conferences*, New Orleans, Jan. 6–11, 1985. National Cotton Council of America, Memphis, 1985, pp. 138–141.
121. Hardee, D. D., Cross, W. H., Huddleston, P. M., and Davich, T. B. Survey and control of the boll weevil in west Texas with traps baited with males. *J. Econ. Entomol. 63*:1041–1048 (1970).
122. Lloyd, E. P., Scott, W. P., Shaunak, K. K., Tingle, F. C., and Davich, T. B. A modified trapping system for suppressing

low-density populations of overwintered boll weevils. *J. Econ. Entomol. 65*:1144–1147 (1972).
123. Boyd, F. J., Jr., Brazzel, J. R., Helms, W. F., Moritz, R. J., and Edwards, R. R. Spring destruction of overwintered boll weevils in west Texas with wing traps. *J. Econ. Entomol. 66*: 507–510 (1973).
124. Bottrell, D. G., and Rummel, D. R. Suppression of boll weevil populations with pheromone traps. In *Detection and Management of the Boll Weevil with Pheromone.* Texas A&M University System, College Station, Tex., 1976, pp. 37–44.
125. Taft, H. M., and Hopkins, A. R. Boll weevils: Effects of various numbers of Leggett traps on small and large populations. *J. Econ. Entomol. 71*:598–600 (1978).
126. Leggett, J. E., Dickerson, W. A., and Lloyd, E. P. Suppressing low level boll weevil populations with traps: Influence of trap placement, grandlure concentration and population level. *Southwest. Entomol. 13*:205–216 (1988).
127. Planer, F. R. Southeast Boll Weevil Eradication Program. In *Proceedings, Beltwide Cotton Production Research Conferences*, New Orleans, Jan. 3–8, 1988. National Cotton Council of America, Memphis, 1988, pp. 239–240.
128. Huddleston, P. M., Mitchell, E. B., and Wilson, N. M. Disruption of boll weevil communication. *J. Econ. Entomol. 70*: 83–85 (1977).
129. Villavaso, E. J., and McGovern, W. L. Boll weevil: Disruption of pheromonal communication in the laboratory and small field plots. *J. Ga. Entomol. Soc. 16*:306–311 (1981).
130. Villavaso, E. J. Boll weevil: Isolated field plot studies of disruption of pheromonal communication. *J. Ga. Entomol. Soc. 17*:347–350 (1982).
131. Mally, F. W. The Mexican cotton-boll weevil. *USDA Farmers' Bull. 130*:11–12 (1901).
132. Gilliland, F. R., Jr., Lambert, W. R., Weeks, J. R., and Davis, R. L. Trap crops for boll weevil control. In *Boll Weevil Suppression, Management, and Elimination Technology* (Davich, T. B., ed.). Agricultural Research Service, Southern Region, USDA, New Orleans, 1976, pp. 41–44.
133. Bottrell, D. G. New strategies for management of the boll weevil. In *Proceedings, Fifth Annual Texas Conference on Insects, Plant Disease, Weed and Brush Control.* 1972. Texas A&M University, College Station, Tex., 1972, pp. 67–72.
134. Boyd, F. J. Progress report on the pilot boll weevil eradication experiment. In *Proceedings, Beltwide Cotton Production-Mechanization Conferences*, Phoenix, Jan. 9–10, 1973. National Cotton Council of America, Memphis, 1973, pp. 20–22.

135. Scott, W. P., Lloyd, E. P., Bryson, J. O., and Davich, T. B. Trap plots for suppression of low density overwintered populations of boll weevils. *J. Econ. Entomol.* *67*:281–183 (1974).
136. Gilliland, F. R., Jr., Lambert, W. R., Weeks, J. R., and Davis, R. L. *A Pest Management System for Cotton Insect Pest Suppression*. Auburn Univ. Agr. Expt. Sta. Prog. Rept. No. 105, 1973, 6 pp.
137. Gilliland, F. R., Jr. Traps and trap crops for boll weevil suppression. In *Proceedings, 27th Cotton Insect Research Conference*, 1974, pp. 128–130.
138. Rummel, D. R., McIntyre, R. C., and Neeb, C. W. Suppression of boll weevils with grandlure-baited trap crops. In *Detection and Management of the Boll Weevil with Pheromone*. Texas A&M University System, College Station, Tex., 1976, pp. 53–61.
139. Moore, L., and Watson, T. F. Trap crop effectiveness in community boll weevil control programs. In *Proceedings*, *Beltwide Cotton Production Research Conferences*, New Orleans, Jan. 3–8, 1988. National Cotton Council of America, Memphis, 1988, pp. 285–286.
140. Ridgway, R. L. Some proposed regional areawide cotton insect management programs. In *Proceedings*, *Beltwide Cotton Production Mechanization Conference*, Las Vegas, Jan. 3–7, 1982. National Cotton Council of America, Memphis, 1982, pp. 89–90.
141. U.S. Department of Agriculture. *Cotton Insect Management Programs*. A report to the Secretary of Agriculture. USDA, Washington, D.C., May 1982.
142. Ewing, K. P., and Parencia, C. R., Jr. *Early-Season Applications of Insecticides on a Community-wide Basis for Cotton-Insect Control in 1950*. E-810. USDA, Bureau of Entomology and Plant Quarantine, 1950.
143. Curry, G. L., and Cate, J. R. Strategies for cotton-boll weevil management in Texas. In *Pest and Pathogen Control: Strategic, Tactical, and Policy Models* (Conway, G. R., ed.). John Wiley & Sons, Chichester, Engl., 1984, pp. 169–183.
144. Brazzel, J. R., Davich, T. B., and Harris, L. D. A new approach to boll weevil control. *J. Econ. Entomol.* *54*:723–730 (1961).
145. Rummel, D. R., White, J. R., and Wade, L. J. Late season immigration of boll weevils into an isolated cotton plot. *J. Econ. Entomol.* *68*:616–618 (1975).
146. Summy, K. R., Hart, W. G., Davis, M. R., Cate, J. R., Norman, J. W., Jr., Wofford, C. W., Jr., Heilman, M. D., and Namken, L. N. Regionwide management of boll weevil in southern Texas. In *Proceedings*, *Beltwide Cotton Production Research Conferences*, New Orleans, Jan. 3–8, 1988. National Cotton Council of America, Memphis, 1988, pp. 240–247.

147. Cate, J. R. Population management of boll weevil in sustainable cotton production systems. In *Proceedings, Beltwide Cotton Production Research Conferences*, New Orleans, Jan. 3–8, 1988. National Cotton Council of America, Memphis, 1988, pp. 249–254.
148. Carlson, C. A., and Suguiyama, L. *Economic Evaluation of Area-Wide Cotton Insect Management: Boll Weevils in the Southeastern United States.* North Carolina Agricultural Research Service, North Carolina State University, Raleigh, N.C., 1985, 24 pp.
149. National Research Council. *Cotton Boll Weevil: An Evaluation of USDA Programs.* National Academy Press, Washington, D.C., 1981, 130 pp.
150. Dickerson, W. A., Ridgway, R. L., Planer, F. R., Brazzel, J. R., and Bradway, T. J. Pheromone trap captures and insecticide use in the Southeastern Boll Weevil Eradication Program. In *Proceedings, Beltwide Cotton Production Research Conferences*, Las Vegas, Jan. 4–9. 1986. National Cotton Council of America, Memphis, 1986, pp. 231–232.
151. Planer, F. R. Status of boll weevil eradication program. In *Proceedings, Beltwide Cotton Production Research Conferences*, Atlanta, Jan. 8–12, 1984. National Cotton Council of America, Memphis, 1984, pp. 183.
152. *Agrichemical Age*. The great boll weevil conspiracy. pp. 6–7, 24 (June 1989).
153. Carlson, G. A., Sappie, G., and Hammig, M. *Economic Returns to Boll Weevil Eradication.* Agricultural Economic Report No. 621. Economic Research Service, U.S. Department of Agriculture, Washington, D.C., 1989.
154. Lacewell, R. D., Casey, J. E., and Frisbie, R. E. An evaluation of integrated cotton pest management in Texas: 1964–1974. *Tex. Agric. Exp. Sta. Rep.* 77-4 6-44 (1977).
155. Lacewell, R. D., Bottrell, D. G., Billingsley, R. V., Rummel, D. R., and Larson, J. L. Impact of the Texas High Plains Diapause Boll Weevil Control Program. *Tex. Agric. Exp. Sta.* MP-1165. 1974.
156. Frisbie, R. E., Walker, J. K., and Metzer, R. B. The case for IPM. *Cotton Grower* pp. 24–25 (July 1988).

27

Population Monitoring of *Heliothis* spp. Using Pheromones

JUAN D. LÓPEZ, JR. and TED N. SHAVER / Pest Management Research Unit, U.S. Department of Agriculture, College Station, Texas

WILLARD A. DICKERSON* / Agricultural Research Service, U.S. Department of Agriculture, Raleigh, North Carolina

I. INTRODUCTION

The genus *Heliothis* represents an important group of pests of field crops on a worldwide basis. Three species are considered particularly important: the cotton bollworm or corn earworm, *H. (Helicoverpa) zea* (Boddie), tobacco budworm, *H. virescens* (F.), and the Old World cotton bollworm, *H. (Helicoverpa) armigera* (Hübner). Geographic distribution of *H. zea* and *H. virescens* is restricted to the Americas, including several Caribbean islands. *Heliothis armigera* is distributed from the Cape Verde islands in the Atlantic through Africa, Asia, and Australasia to the South Pacific islands and from Germany in the north to New Zealand in the south (1).

These three species attack numerous field crops such as cotton, corn, sorghum, numerous vegetables, and legumes. Reliance on

---

**Current affiliation*: North Carolina Department of Agriculture, Raleigh, North Carolina insecticides to control the larvae has complicated their management because of the development of insecticide resistance, especially for high-value crops on which multiple generations occur. Identification of sex pheromones for the three species has provided exciting opportunities for integrating these chemicals into pest management programs.

Three approaches for the use of sex pheromones in *Heliothis* management are mass trapping, communication (mating) disruption, and monitoring. Essentially no effort has been devoted to mass trapping, but considerable research has been devoted to evaluating sex pheromones for mating disruption. Results have been encouraging, but the high mobility of adults and difficulty in developing pheromone formulations with sufficient longevity for practical field use are cited as major problems (2). Sex pheromones and traps are available and in use for monitoring populations of the three species of *Heliothis*. This application is the subject of this chapter. The primary areas to be covered include identification of pheromone components and their role in moth attraction to traps, development of pheromone dispensers, and design of practical traps. The second part of the chapter deals with application of sex pheromone traps as monitoring tools.

II. SEX PHEROMONE COMPONENTS

Identification of essential sex pheromone components for male sexual response has been complicated by involvement of multiple components and importance of their ratios. The effort has been further complicated by the different approaches that have been taken to determine which components are essential and their role in eliciting a "complete" sexual response from the male.

A. *Heliothis zea*

The relative amounts of the four aldehydes, (*Z*)-11-hexadecenal, (*Z*)-9-hexadecenal, (*Z*)-7-hexadecenal, and hexadecanal, reported as components washed from glands or collected as volatiles are shown in Table 1 (3,4). A relatively high level of the alcohol, (*Z*)-11-hexadecen-1-ol was also found in gland washes (5).

In laboratory evaluations, the binary mixture of (*Z*)-11-hexadecenal and (*Z*)-9-hexadecenal was sufficient for optimum male response (4,6). Similar results were obtained in field evaluations with traps of different combinations of the four components in amounts shown in Table 1 (3,7,8). In one evaluation, there was a trend for increased captures when additional components were added to the binary blend (8). The alcohol, (*Z*)-11-hexadecen-1-ol, reduced captures when mixed with the other four components (5).

TABLE 1 Composition of the *H. zea* Sex Pheromone

	Amounts of chemical components (%)			
Components	Klun (3) gland	Pope (4) volatiles	Teal (5) gland	Klun (3) field blend (μg)
(Z)-11-Hexadecenal	92.4	90.3	66.9	86.5 (115.5)
(Z)-9-Hexadecenal	1.7	1.4	1.6	3.4 (4.5)
(Z)-7-Hexadecenal	1.1	1.2	0.9	1.9 (2.6)
Hexadecanal	4.4	7.0	2.8	8.2 (11.0)
(Z)-11-Hexadecen-1-ol			27.9	

B. *Heliothis virescens*

Two components, (Z)-11-hexadecenal and (Z)-9-tetradecenal, at a 16:1 ratio were initially identified as shown in Table 2 (9,10). Subsequent studies led to identification of five and eight additional components, but relative amounts of common components were very similar (11,12). Volatiles emitted by calling females contained all six of the aldehyde components in varying proportions, but did not contain (Z)-11-hexadecen-1-ol (12,13).

In laboratory evaluations, the binary mixture of (Z)-11-hexadecenal and (Z)-9-tetradecenal was sufficient to elicit precopulatory flight behavior or behavioral activity (11,14). Addition of hexadecanal to the binary mixture enhanced behavioral activity, but there was no effect of (Z)-11-hexadecen-1-ol observed at relatively low concentrations in flight tunnel experiments (14). At concentrations greater than 3%, the alcohol inhibited male response. In another laboratory bioassay, all six aldehydes affected close-range reproductive behavior (12). On the basis of these studies, it was concluded that the alcohol was not a component of the sex pheromone of *H. virescens* (12,14).

A blend of seven components was developed for field evaluations, the components and proportions of which are shown in Table 2 (11). This blend contained 5.9% of the alcohol, (Z)-11-hexadecen-1-ol, and was used in the development of pheromone dispensers.

Field evaluations were made of different combinations of the seven components in the proportions shown (see Table 2) on cotton dental rolls or cigarette filters that were replaced nightly. The seven-component blend was the most effective and was equal to, or better than,

TABLE 2 Identification of Pheromone Components for *H. virescens*

	Amounts of chemical components (%)					
Component	Roelofs (9) Tumlinson (10) gland	Klun (11) gland	Teal (12) gland	Klun (11) field blend (μg)	Pope (13) volatiles	Teal (12) volatiles
(Z)-11-Hexadecenal	93.7	81.4	79.5	76.1 (115.5)	74.5	60.0
(Z)-9-Tetradecenal	6.3	3.2	3.3	4.6 (6.9)	5.0	18.1
(Z)-7-Hexadecanal		1.0	0.6	1.7 (2.6)	0.9	0.6
(Z)-9-Hexadecenal		1.3	0.8	3.0 (4.5)	12.7	7.3
Hexadecanal		9.5	8.6	7.3 (11.0)	6.6	13.0
(Z)-11-Hexadecen-1-ol		3.2	4.8	5.9 (9.0)		
Tetradecanal		1.6	2.5	1.5 (2.25)		
(Z)-9-Tetradecen-1-ol			0.3			
Hexadecanol			0.4			
Tetradecanol			0.3			

virgin females (8,11,15). Addition of the alcohol, (*Z*)-11-hexadecen-1-ol, to either the binary mixture or the four unsaturated aldehyde components generally led to significant increases in male captures, but captures were usually lower than captures with the seven-component blend.

C. *Heliothis armigera*

One component, (*Z*)-11-hexadecenal, was initially identified for *H. armigera* as shown in Table 3 (16). Examination of abdominal tip extracts of females that were originally from, or cultured from, insects collected in Malawi, India, Sudan, and Botswana led to identification of two components, (*Z*)-11-hexadecenal and (*Z*)-11-hexadecen-1-ol, with an additional component, (*Z*)-9-hexadecenal, identified only in females from Malawi (17). A reexamination of ovipositor washes from females from Malawi, India, and Sudan showed that (*Z*)-9-hexadecenal was consistently present, as were previously identified components (see Table 3; 18). Similar proportions of (*Z*)-11-hexadecenal and (*Z*)-9-hexadecenal were found in *H. armigera* from Israel, but hexadecanal and hexadecanol were also found (19).

Initial laboratory and field evaluations with different components and combinations of components provided variable results (16,20–23). Field evaluations in Israel showed that a 90–99% blend of (*Z*)-11-hexadecenal with 1–10% of (*Z*)-9-hexadecenal was very effective (24). Increased levels of (*Z*)-9-hexadecenal (26.2%) and addition of (*Z*)-11-hexadecen-1-ol (8.7%) to the blend reduced captures, whereas (*Z*)-7-hexadecenal (2.3%) had no effect on captures.

TABLE 3 Identification of Pheromone Components for *H. armigera*

Component	Amounts of chemical components (%)			
	Piccardi (16)	Nesbitt (17)	Nesbitt (18)	Dunkelblum (19)
(*Z*)-11-Hexadecenal	X[a]	X	88.6	87.0
Hexadecanal		X		4.0
(*Z*)-11-Hexadecen-1-ol		X	8.5	
(*Z*)-9-Hexadecenal		X (Malawi)	2.9	3.0
Hexadecanol		X		6.0

[a]X indicates only the presence of the compound.

III. PHEROMONE DISPENSERS

Dispensers used in sex pheromone traps for monitoring *Heliothis* spp. must provide a high and consistent level of attractiveness over a desired period. It is also important that performance criteria, including biological, chemical, and physical characteristics, be determined for dispensers. This approach will ensure consistent results from year to year and provide a standard for comparing other dispensers or improvements in the standard dispenser.

Several difficulties have been encountered in the development of dispensers for the sex pheromones of *H. zea*, *H. virescens*, and *H. armigera*. Because of the number of essential components, it has been difficult to determine and maintain release of optimum ratios from a substrate over the desired period. Instability of aldehyde components has been a major obstacle. Variability in the release characteristics of several potentially useful substrates also has complicated the situation. Ideally, the dispenser substrate should release components at a slow and consistent rate, such that effectiveness will persist for 1–3 weeks and provide protection from exposure to direct solar ultraviolet radiation and atmospheric oxygen to prevent rapid degradation of the aldehydes (25).

A. *Heliothis zea*

A plastic laminate and rubber septa were used as substrates in the initial studies on the development of dispensers for field use (8). The laminate (Hercon Laboratories Inc., South Plainfield, New Jersey) was prepared by mixing an adhesive with a blend of all four pheromone components in the ratio used in the field, at a dosage of 1.5 mg/cm^2 (see Table 1; 3), and laminating the adhesive between two layers of vinyl polymer plastic sheet (26). Both the laminate and the septa were effective for 14 days. Although differences in male captures between dosages (0.32–20.00 mg) were not significant, a dosage–response relationship was observed for both laminate and septa, with 1.25 and 2.50 mg capturing higher numbers of males. A subsequent study showed that a thin (0.086-cm) laminate containing 2.5 mg of the four-component blend in 6.5 cm^2 was more effective than septa with 1.25, 2.00, or 5.00 mg and that aging for 11 days in a greenhouse did not decrease effectiveness (27). Recent tests showed that the 1.25- and 2.50-mg doses ($\bar{x}$ = 28.1 and 27.3 males per trap per night, respectively), captured statistically equal numbers of males, whereas 0.625 mg ($\bar{x}$ = 23.0) captured significantly fewer males over an 18-day period. Several independent tests have verified the effectiveness of this laminate containing the four-component blend at a dosage of 1.25 mg (28,29).

An intensive evaluation of septa as dispensers indicated that a binary mixture of (*Z*)-11-hexadecenal/(*Z*)-9-hexadecenal at 9.7:0.3 mg/septum was effective (30). Addition of 0.03 mg of (*Z*)-7-hexadecenal increased captures, but hexadecanal at 0.5, 1.5, or 4.5 mg did not. Blend dosages of 1, 3, and 10 mg/septum were best.

There are a number of commercial dispensers available for *H. zea*. Information on effectiveness of these dispensers is critical for determining whether they meet criteria for use in a trapping system for monitoring *H. zea*. Comparison of experimental septa prepared in the Agricultural Research Service, USDA (ARS septa) containing a 3.0:0.09 mg blend of (*Z*)-11-hexadecenal/(*Z*)-9-hexadecenal (30) with Trece septum, Scentry hollow fiber, Biolure/Consep membrane, and Raylo rubber septum commercial dispensers showed considerable differences in effectiveness (31). There were no significant differences between the Trece septa and the Biolure/Consep membrane dispensers, but these two dispensers were somewhat less effective than the ARS septa. Both the Scentry hollow fibers and Raylo septa captured significantly fewer males. Longevity of the dispensers evaluated was determined to range from 2 to 7 weeks; however, there was variability in the results during the 2-year evaluation. Fresh ARS septa (3–4 days old) captured significantly higher numbers of males than did older septa (6–7 days old). Therefore, it was suggested that septa be exposed in the field before being used to bait pheromone traps.

In another comparison of Trece septum, Biolure/Consep membrane, laminate, and Scentry experimental polyvinyl chloride (PVC) capsule dispensers (2.5 mg), the most effective dispenser was the laminate (30). The Scentry PVC capsule was the second most effective, but it captured about one-fourth to one-third fewer males than the laminate. Trap captures with Trece septum and Biolure/Consep membrane dispensers were much lower than trap captures with other dispensers.

Because the laminate was a very effective dispenser for trapping *H. zea*, efforts were made to develop performance criteria. New and 8-day-old laminate dispensers captured equal numbers of males over an 8-day period while 16- and 24-day old laminates captured significantly fewer males. Thus, the effective life of the laminate was 2 weeks (32). Chemical analyses of new and aged dispensers provided the preliminary performance criteria for a 2-week laminate; (1) 12 × 25 mm in size; (2) 16-mil PVC outer layer; (3) 1.25 mg of a blend of (*Z*)-11-hexadecenal, (*Z*)-9-hexadecenal, (*Z*)-7-hexadecenal, and hexadecanal in a ratio of 87:3:2:8; (4) release rate in the range of about 0.05–0.40 μg/hour per dispenser as measured in the laboratory at 35°C; and (5) pheromone content in the range of 0.25–1.25 mg/dispenser during the period of use (33).

B. *Heliothis virescens*

Laminate, septa, and PVC substrates, either in capsules or in molded forms, were evaluated for use in dispensers for *H. virescens*. Both 10- and 20-mg doses of the binary blend [(*Z*)-11-hexadecenal and (*Z*)-9-tetradecenal] in the laminate were effective in initial studies (34). A 22-mg dose of a 10:1 blend of the binary mixture had greater longevity in septa than in laminate (35). The numbers of males captured with a 20-mg dose of binary and seven-component blends (see Table 2; 11) in the laminate were equal to that for the binary blend in septa, but significantly fewer males were captured with the seven-component blend in septa (8). In another study, the pheromone dose was shown to affect the efficacy of the binary and seven-component blends in the laminate over a 2-week period (36). No difference was observed between the binary and seven-component blends for a 10-mg dose, but the binary blend was significantly better than the seven-component blend when the dose was increased to 20 mg. The PVC capsules containing 10 mg of the binary blend were significantly better during the second week of evaluation than those containing the seven-component blend; however, there were no significant differences between the efficacy of the binary mixture in PVC capsules or in the laminate (36). Differences in longevity of different dosages of the binary and six- or seven-component blends in the laminate were attributed to inhibitory effects of (*Z*)-11-hexadecen-1-ol (29). The authors suggested that the relative alcohol concentration may increase emissions from the seven-component blend dispensers with aging because of differences in volatility. All of these studies indicated that the substrates evaluated were useful as dispensers for the pheromone of *H. virescens*; but the importance of the five additional pheromone components remained unclear.

Effects of different blends in molded PVC substrates of different colors and shapes were evaluated for *H. virescens* (25). The basic dispensers weighed 0.65 ± 0.03 g each. They contained 1% of different pheromone blends in liquid vinyl chloride monopolymer. Red actually reduced captures, whereas black molded dispensers in different forms were better than green, yellow, or orange. The binary blend gave the most consistent performance and was effective for 2 weeks. In addition, the binary blend impregnated in black surrogate moth or sheet block forms captured two to six times more males than the binary or seven-component blend in laminate. Thus, a very effective substrate for dispensers of *H. virescens* pheromone was identified.

Two recent studies with dispensers have determined that (*Z*)-11-hexadecen-1-ol enhances captures in pheromone traps; however, effects of this alcohol are greatly influenced by the relative amount of alcohol in the blend (37,38). Addition of 0.7% of the alcohol to a three-component blend of 65.1% (*Z*)-11-hexadecenal, 1.6% (*Z*)-9-tetradecenal, and 32.6% hexadecanal at a total dose of 1.5 mg in septa gave an optimal blend, but addition of 10.0% alcohol to the blend significantly reduced male captures (37). Previous reports indicated that 3.0% or higher concentrations of alcohol in septa inhibited male response (14). Therefore, the conclusion was that, in septa, concentrations of (*Z*)-11-hexadecen-1-ol higher than 1% in the blend would reduce efficacy. Similar results were reported for the black molded PVC substrate, for which adding 0.25 or 1.00% of the alcohol to binary or six-component blends significantly increased captures, but addition of 5.95% to either blend significantly reduced captures (38).

Given the foregoing findings, it seems likely that the amount of (*Z*)-11-hexadecen-1-ol used in field evaluations may have contributed to the confusion about the optimum blend for use in pheromone dispensers for *H. virescens*. The initial seven-component field blend evaluated (see Table 2; 11) contained 5.9% of (*Z*)-11-hexadecen-1-ol. Differences in the efficacy of the binary and seven-component blends in septa and PVC dispensers (8,36) can be explained on the basis of the inhibitory effects of high levels of alcohol. Lack of immediate inhibitory effects of alcohol in seven-component blends in laminates indicates that the alcohol may not be immediately released, but may become more predominant in emissions as the total concentration of aldehyde components in the dispensers decreases with aging, as suggested previously (29). If this is true, reevaluation of the effects of (*Z*)-11-hexadecen-1-ol in laminates may be warranted.

Assessment of dispensers available for *H. virescens* is necessary to determine their value for use in monitoring field populations. Ramaswamy septa (37) were very effective initially, when compared with black molded PVC sheet forms, but their efficacy decreased considerably between the fourth and eighth days relative to that of PVC dispensers (32). By the 9th to 12th day, the efficacy of the Ramaswamy septa was very poor, whereas the efficacy of the PVC dispensers was more consistent relative to virgin females. Tripling the pheromone load did not increase longevity of the Ramaswamy septa; hence, longevity is not a simple dosage response. In the same comparison, Flint septa (35) were not as effective as black molded PVC dispensers, but they were relatively consistent in their performance.

A commercial dispenser patterned after the black molded PVC dispenser developed for *H. virescens* (25) is being marketed by Scentry, Inc., Buckeye, Arizona. The commercially prepared PVC dispenser performed as well as or better than the original dispenser (Table 4).

Black molded PVC dispensers have been selected as a possible standard for use in monitoring *H. virescens*. Preliminary performance criteria for these dispensers are (1) 1% volume/volume or 0.76% weight/

TABLE 4 Comparison of Molded PVC Sheet Block Dispensers Prepared by D. E. Hendricks (ARS, Weslaco, Texas) or by Scentry, Inc., Buckeye, Arizona, for *H. virescens*, 1987[a]

	$\bar{x}$ Males/trap per night[b]	
Dates	Scentry PVC	Hendricks PVC
Jul. 28–Aug. 10	8.2 A	7.3 A
Aug. 11–26	44.7 A	34.0 B
Aug. 29–Sep. 10	34.2 A	33.3 A

[a]Lopez, J. D., Jr., unpublished data.
[b]Means in same row followed by different letters are significantly different according to Student's *t* test, $P = 0.05$.

weight pheromone concentration; (2) ratio of (*Z*)-9-tetradecenal/(*Z*)-11-hexadecenal in the volatilized pheromone in the range of 1:5 or 1:1.7, which is achieved when the ratio remains 1:33 to 1:10 in the dispenser during the aging period; (3) emission rate of total pheromone from 0.03 to 0.50 μg/hr when measured in the laboratory at 35°C, and which is achieved when the pheromone concentration per 600-mg dispenser is 0.8–6.6 mg; and (4) initial pheromone dose of 5–7 mg/600-mg dispenser weight (33). A 1000-mg dispenser with 7.6 mg of a 1:15 ratio of (*Z*)-9-tetradecenal/(*Z*)-11-hexadecenal was recommended to ensure a 2-week field life.

C. *Heliothis armigera*

Field evaluations to identify sex pheromone components for *H. armigera* were conducted with dispensers. Blends of 90–99% of (*Z*)-11-hexadecenal and 10–1% of (*Z*)-9-hexadecenal at a total dose of 2 mg/septum were effective for 31 days in Israel (24). Subsequently, a pheromone blend of 97% (*Z*)-11-hexadecenal and 3% (*Z*)-9-hexadecenal at a dose of 2 mg/septum was reported to be the standard dispenser in use in Israel (39). A blend of (*Z*)-11-hexadecenal and 9% (*Z*)-9-hexadecenal at a total dose of 0.55 mg in red rubber tubing has been used in Australia (40). A red rubber tubing dispenser containing a blend of 90:10 (*Z*)-11-hexadecenal/(*Z*)-9-hexadecenal at a dose of 1 mg was compared with laminate, septum, and hollow fiber commercial dispensers (41). Laminates and septa were superior to the rubber tubing and hollow fibers. Aging for up to 4 weeks did not decrease the efficacy of the laminates and rubber septa, but a decrease in efficacy was especially evident with the hollow fibers. A commercial laminate has been adopted as the standard for monitoring *H. armigera* in Australia.

IV. TRAP DESIGNS

Pheromone trap designs for monitoring *H. zea*, *H. virescens*, and *H. armigera* go hand in hand with the development of pheromone dispensers. Because there is an interaction between the attractant and the pheromone trap design relative to trap efficiency (7,15,42), development and evaluation of an effective dispenser--trap combination is needed.

Desirable attributes of sex pheromone traps for use in monitoring are: (1) accuracy in measuring insect density, rather than magnitude of trap catch per se; (2) sensitivity at low population densities; (3) durability under prevailing weather conditions; (4) exclusion of nontarget insects, predators, and debris; (5) ease of handling and transportation; and (6) reasonable cost (43).

A. *Heliothis zea* and *Heliothis virescens*

Both *H. zea* and *H. virescens* appear to respond similarly to different traps; therefore, designs for the two species will be discussed together. Sex pheromone trapping of both species was initiated with virgin females as an attractant. Earlier identification of an effective sex pheromone blend for *H. virescens* resulted in most of the initial trapping with synthetic pheromones to be conducted with this species.

Two basic designs of sticky traps have been used (44–47). In one design, openings were cut in plastic or cardboard canisters and Stickem was placed on the inside to catch males that entered in response to the attractant. Another design consisted of two flat pieces of material (wood, aluminum, or styrofoam plates) that were attached together in the center, with a gap between for the attractant. The inside of each plate was treated with Stickem. These traps tend to lose efficiency as the Stickem ages and becomes contaminated. Some of the specific designs have been reviewed (48,49).

Electrocutor grid traps (50) have been very valuable for research purposes. Initial evaluations indicated that the grid traps were nine times as effective as sticky traps (51).

A series of nonelectric traps incorporating a cone design was evaluated (52). A hardware wire cone trap with a glass container at the top of the cone to hold the captured males was the most effective. Another effective trap design was a wind vane or directional trap. Nocturnal observations of male behavior relative to the trap resulted in design modifications that significantly improved the efficiency of both traps. Removing half the base of the wind vane trap increased captures of males by 4.4 times, compared with a trap with a solid bottom (53). An inner wire cone rim added to the base of the cone trap increased captures 2 to 3.5 times depending on the size of the original cone (54).

Studies of the effect of nocturnal wind on the efficacy of wind vane and wire cone traps demonstrated the importance of environmental factors on trap efficiency as related to trap design (55). The modified wire cone trap captured significantly more males (1.7 times) than the wind vane trap in a 64-day evaluation period. Percentages of males captured in the modified wind vane trap increased as wind velocity increased, but captures in the modified wire cone trap were greatest when wind speeds during the night were low. Captures in wind vane traps exceeded captures in wire cone traps on only three nights, and then wind velocity was greater than 16 km/hr.

Nocturnal observations were made to determine trap efficiency, that is, the percent of males captured of those that responded to within a defined distance of the trap (Table 5). The most efficient trap generally was considered to be the electrocutor grid trap. The pie plate trap was shown to be relatively inefficient. Modification of the wind vane trap increased efficiency from 16 to 42%, whereas modification of the wire cone trap increased captures from about 5–6% to 25–30%.

A commercial prototype of the cone trap was developed by Albany International and is now marketed by Scentry, Inc., Buckeye, Arizona. It is made of white flexible plastic so that the trap is collapsible and, therefore, easier to handle. The capture chamber was an integral part of the trap in the original design, and the only access to the captured males was through a slit in the plastic that was held closed with Velcro. A more recent design has a removable capture container that facilitates counting of the moths captured. Comparisons of the efficacy of the plastic trap with that of the 75–50 wire cone trap (54) for capturing *H. zea* showed that the wire cone trap captured significantly more males than the plastic trap (32). Comparisons have shown that the plastic cone trap captures about 56 times more males than two different designs of water traps (31).

The recent trend has been to use the wire cone trap to monitor *H. zea* and *H. virescens*. Both electrocutor grid and wind vane traps are relatively expensive and considerably more difficult to transport and operate than the wire cone trap. Although the wire cone trap design has advantages over other trap designs, it is still relatively large and cumbersome to handle. Reducing the size of the wire cone by

TABLE 5 Capture Efficiency of Different Pheromone Trap Designs for *H. zea* and *H. virescens* Based on Nocturnal Observations

Trap types	Species	Capture efficiency (%)	Ref.
Electrocutor grid	*H. virescens*	34	(42)
	H. virescens	45	(15)
	H. zea	58	(7)
Pie plate	*H. virescens*	6	(42)
	H. zea	6	(7)
Wind vane	*H. virescens*	16	(53)
Modified wind vane	*H. virescens*	42	(53)
Wire cone	*H. virescens*	5-6	(54)
Modified wire cone	*H. virescens*	25-30	(54)

one-third made the trap more manageable, but resulted in a decrease in efficiency of about 40% (54).

Efforts are continuing to identify commercially available, inexpensive pheromone traps that are more practical for use in monitoring programs. Five different trap designs for capturing *H. zea* and *H. virescens* were compared in 1987 at Oxford, North Carolina (Table 6). The wire cone trap captured significantly more males than the other trap designs. The efficacy of the International Pheromone Systems' Universal Moth Trap (Unitrap) relative to the plastic and wire cone traps was notable because it is a much smaller trap, is commercially available, and is not particularly expensive. Therefore, the Universal Moth Trap is a possible alternative to the use of the wire cone trap.

B. *Heliothis armigera*

Several designs of sticky and water or liquid traps have been used for sex pheromone trapping of *H. armigera* (16,17,20,21,56). A dry funnel trap has become the standard for *H. armigera* in Israel and Australia (41,57). In India and surrounding countries, the ICRISAT trap is being used to evaluate movement (58).

Two evaluations of different trap designs were conducted for *H. armigera* in Australia (41,59). Comparison of sticky, water, dry funnel, modified dry funnel, and commercial plastic cone traps indicated

TABLE 6 Comparison of Five Different Sex Pheromone Trap Designs for Capturing Males of *H. zea* and *H. virescens*, Oxford, North Carolina, 1987[a]

	$\bar{x}$ Males/trap per interval[b]	
Trap type	*H. zea*	*H. virescens*
Wire cone (75–50)	43.5 A	37.3 A
Plastic cone	28.1 B	24.3 B
Universal Pheromone	12.5 C	11.5 C
French	5.4 D	4.4 D
ICRISAT, India	0.0 E	0.0 E

[a]Dickerson, unpublished data.
[b]Five replicates; randomized complete block design; means in same column followed by different letters significantly different, DMRT, $P = 0.05$.

that similar numbers of males were captured with both plastic cone and dry funnel traps; however, captures were very low during the evaluation period (41). Although effective, water traps were not practical because they required frequent refilling in hot weather. The use of sticky traps was quickly discontinued because of problems with the sticky surface. Dry funnel traps were selected as the standard because they were robust in construction, relatively inexpensive, and could be assembled easily from readily available parts. Plastic cone traps cost more and did not last as long under the prevailing hot and windy conditions. The commercial plastic cone trap, a modified plastic trap with a removable plastic capture container, a dry funnel trap, and a sticky codling moth trap of a wing design were also compared (59). The plastic cone trap captured ten times more males than the other traps and similar numbers were captured in the other three trap designs. Rainy weather interfered with the efficacy of the dry funnel trap and contamination of the sticky surface caused the efficacy of the codling moth trap to decline rapidly.

V. APPLICATIONS

The uses of sex pheromone traps in monitoring of *H. zea*, *H. virescens*, and *H. armigera* can be grouped in several categories: (1)

collection, (2) detection, (3) surveillance, and (4) prediction. Although these categories are not mutually exclusive, nevertheless, they provide a framework by which to evaluate application of pheromone-trapping technology for these species.

A. Collection

Pheromone traps provide an ideal method for collecting male moths from field populations and have been frequently used in mark, release, and recapture studies. Use of pheromone traps for *Heliothis* movement and density determination studies would be expected to continue. This will be especially true if more techniques are developed to identify the geographic origin of males captured in pheromone traps.

Three recent examples demonstrate use of pheromone traps for collection. A technique has been developed that uses *H. zea* and *H. virescens* males captured in pheromone traps to monitor the level of insecticie resistance in field populations (60). This approach makes it possible to obtain results quickly, without having to do laboratory rearing, and will facilitate more widespread monitoring since tests can be conducted locally. Pheromone traps were also used to recapture marked moths for density estimates of overwintering *H. virescens* in the Mississippi Delta (61). Male *H. zea* captured in pheromone traps in Arkansas were examined for pollen; on the basis of the plant species involved, it was determined that some of the males examined had migrated from south Texas to Arkansas in early spring (62).

B. Detection

Detection refers to determination of the presence or absence of *Heliothis* spp. in an area. The use of pheromone traps may be practical in very high value crops such as seed corn, sweet corn, and other vegetables for which essentially no damage can be tolerated and the presence of adults may be sufficient to initiate application of control measures. The ability of pheromone traps to detect low-population densities makes this possible. Currently, there is an ongoing study in Maryland and four surrounding states to evaluate *H. zea* pheromone traps for use, instead of blacklight traps, as an aid in decision-making in sweet corn (Gaylon Dively, personal communication). Pheromone traps were used in an areawide monitoring system for *H. zea* and *H. virescens* in the United States to determine first occurrence in the season (63).

C. Surveillance

Surveillance includes detection, but it also includes a measure of changes in abundance over time, and with more refinement, population density estimation. From a practical standpoint, this is the most important application, but it is also more complex because qualitative and quantitative relationships between captures in pheromone trap and field populations or their damage have to be determined. Complexity is increased, depending on whether the application is on a field-by-field or an areawide basis.

On a field-by-field basis, variable results have been obtained for the relationship between captures of the three species of *Heliothis* in traps and the populations of adults or immature stages in the field or their damage (40,57,64–66). Some of these studies have indicated usefulness of the traps in detecting the start of an infestation, in providing qualitative information on probable infestation levels, or in giving an indication of which species is present in the field when more than one species of *Heliothis* is involved. None of these studies has simultaneously evaluated the activity of adults in the field, efficiency of the trapping system, and populations of immatures in the field. From an experimental standpoint, dealing with specific fields simplifies the system because factors that influence captures in traps and the behavior of the adults can be evaluated more easily.

The ultimate potential use of pheromone traps on a field-by-field basis is to provide information about species composition and numbers present in the field that will be useful in making treatment decisions. Species determination is important because of different levels of insecticide susceptibility of the different species. Selection of an effective insecticide for the species present is critical in obtaining adequate control, as well as in insecticide-resistance management. To this end, ARS initiated a 3-year pilot test in 1987 to determine the technical feasibility of using captures of *H. zea* and *H. virescens* in pheromone traps as an aid in developing decision-making technology for control of these pests in cotton.

Research with pheromone traps on an areawide basis has indicated that pheromone traps do provide a measure of *H. zea* and *H. virescens* activity. One study, in particular, reported that captures of *H. virescens* in pheromone traps in Texas during the overwintering and F_1 generations realistically represented the actual population (67). In South Carolina, a direct relationship existed between numbers of *H. virescens* males captured in traps and egg counts in a cotton-growing area (68). Another study concluded that species ratios of males captured in pheromone traps were substantiated by species composition of larvae in area fields (69). Captures in traps could be used to give 1 or 2 day warning of incipient oviposition in Arkansas and North Carolina. Pheromone traps were used also to monitor occurrence of *H. zea* and *H. virescens* populations over a large part of the United States (63).

A technique was developed to estimate field populations over a large area from captures of *H. virescens* in pheromone traps (70).

With the use of this technique, a sex pheromone emission and response model (SPERM) was developed that will be helpful in relating captures in pheromone traps to field populations (71).

D. Prediction

Use of pheromone traps for prediction has been developed primarily on an areawide basis. Numbers of males caught are used as input for population models to predict the pattern of occurrence of future generations. Short- and long-term forecasts have been useful in scheduling scouting and crop management activities. Pheromone traps have been used for predicting timing of *H. zea* and *H. virescens* population trends; but, not for predicting actual pest densities (69,72).

VI. CONCLUSIONS

Overall assessment of available pheromone dispensers and traps indicates that there is considerable variation in their effectiveness for trapping *H. zea*, *H. virescens*, and *H. armigera*. However, technology has been developed to the point that the most effective and practical combination of dispenser and trap should be used to investigate relationships between trap captures and field populations. If useful relationships are established, we can then look for a trapping system that will be more practical for widespread field uses. Development of automated trapping systems (73,74) also may considerably simplify field operation of pheromone traps.

There is considerable potential for practical use of pheromone traps in monitoring populations of *H. zea*, *H. virescens*, and *H. armigera*, provided the key biotic and abiotic factors influencing captures in pheromone traps are better understood. Understanding these factors will also improve the understanding of adult field populations that is needed for improved management of *Heliothis* spp.

ACKNOWLEDGMENT

Preparation of this chapter and research reported was done in cooperation with Texas Agricultural Experiment Station, Texas A&M University, College Station, Texas.

DISCLAIMER

Mention of a commercial or a proprietary product does not constitute an endorsement by USDA.

REFERENCES

1. Reed, W. and Pawar, C. S. *Heliothis*: A global problem. In *Proceeding, International Workshop on* Heliothis *Management* (Reed, W., ed.), ICRISAT, Patancheru, India, 1982, pp. 9–14.
2. McLaughlin, J. R. and Mitchell, E. R. Practical development of pheromones in *Heliothis* management. In *Proceedings, International Workshop on* Heliothis *Management* (Reed, W., ed.), ICRISAT, Patancheru, India, 1982, pp. 309–318.
3. Klun, J. A., Plimmer, J. R., Bierl-Leonhardt, B. A., Sparks, A. N., Primiani, M., Chapman, O. L., Lee, G. H., and Lepone, G. Sex pheromone chemistry of female corn earworm moth, *Heliothis zea*. *J. Chem. Ecol.* *6*:165–175 (1980).
4. Pope, M. M., Gaston, L. K., and Baker, T. C. Composition, quantification, and periodicity of sex pheromone gland volatiles from individual *Heliothis zea* females. *J. Insect Physiol.* *30*: 943–945 (1984).
5. Teal, P. E. A., Tumlinson, J. H., McLaughlin, J. R., Heath, R. R., and Rush, R. A. (*Z*)-11-hexadecen-1-ol: A behavior modifying chemical present in the pheromone gland of female *Heliothis zea* (Lepidoptera: Noctuidae). *Can. Entomol.* *116*: 777–779 (1984).
6. Vetter, R. S. and Baker, T. D. Behavioral responses of male *Heliothis zea* moths in sustained-flight tunnel to combinations of 4 compounds identified from female sex pheromone gland. *J. Chem. Ecol.* *10*:193–202 (1984).
7. Sparks, A. N., Carpenter, J. E., Klun, J. A., and Mullinix, B. D. Field responses of male *Heliothis zea* (Boddie) to pheromonal stimuli and trap design. *J. Ga. Entomol. Soc.* *14*:318–325 (1979).
8. Hartstack, A. W., Jr., Lopez, J. D., Klun, J. A., Witz, J. A., Shaver, T. N., and Plimmer, J. R. New trap designs and pheromone bait formulations for *Heliothis*. In *Proceedings, Beltwide Cotton Production and Research Conference*, National Cotton Council, Memphis, Tenn., 1980, pp. 132–132.
9. Roelofs, W. L., Hill, A. S., Cardé, R. T., and Baker, T. C. Two sex pheromone components of the tobacco budworm moth, *Heliothis virescens*. *Life Sci.* *14*:1555–1562 (1974).
10. Tumlinson, J. H., Hendricks, D. E., Mitchell, E. R., Doolittle, R. E., and Brennan, M. M. Isolation, identification, and synthesis of the sex pheromone of the tobacco budworm. *J. Chem. Ecol.* *1*:203–214 (1975).

11. Klun, J. A., Bierl-Leonhardt, B. A., Plimmer, J. R., Sparks, A. N., Primiani, M., Chapman, O. L., Lepone, G., and Lee, G. H. Sex pheromone chemistry of the female tobacco budworm moth, *Heliothis virescens*. *J. Chem. Ecol.* *6*:177–183 (1980).
12. Teal, P. E. A., Tumlinson, J. H., and Heath, R. R. Chemical and behavioral analyses of volatile sex pheromone components released by calling *Heliothis virescens* (F.) females (Lepidoptera: Noctuidae). *J. Chem. Ecol.* *12*:107–126 (1986).
13. Pope, M. M., Gaston, L. K., and Baker, T. C. Composition, quantification, and periodicity of sex pheromone gland volatiles from individual *Heliothis virescens* females. *J. Chem. Ecol.* *8*: 1043–1055 (1982).
14. Vetter, R. S. and Baker, T. C. Behavioral responses of male *Heliothis virescens* in a sustained-flight tunnel to combinations of seven compounds identified from female sex pheromone glands. *J. Chem. Ecol.* *9*:747–759 (1983).
15. Sparks, A. N., Raulston, J. R., Lingren, P. D., Carpenter, J. E., Klun, J. A., and Mullinix, B. G. Field response of male *Heliothis virescens* to pheromonal stimuli and traps. *Bull. Entomol. Soc. Am.* *25*:268–274 (1979).
16. Piccardi, P., Capizzi, A., Cassani, G., Spinelli, P., Arsura, E., and Massardo, P. A sex pheromone component of the Old World bollworm, *Heliothis armigera*. *J. Insect Physiol.* *23*:1443–1445 (1977).
17. Nesbitt, B. F., Beevor, P. S., Hall, D. R., and Lester, R. Female sex pheromone components of the cotton bollworm, *Heliothis armigera*. *J. Insect Physiol.* *25*:535–541 (1979).
18. Nesbitt, B. F., Beevor, P. S., Hall, D. R., and Lester, R. (*Z*)-9-hexadecenal: A minor component of the female sex pheromone of *Heliothis armigera* (Hübner) (Lepidoptera: Noctuidae). *Entomol. Exp. Appl.* *27*:306–308 (1980).
19. Dunkelblum, E., Gothilf, S., and Kehat, M. Identification of the sex pheromone of the cotton bollworm, *Heliothis armigera*, in Israel. *Phytoparasitica* *8*:209–211 (1980).
20. Rothschild, G. H. L. Attractants for *Heliothis armigera* and *H. punctigera*. *J. Aust. Entomol. Soc.* *17*:389–390 (1978).
21. Gothilf, S., Kehat, M., Jacobson, M., and Galun, R. Screening pheromone analogues by EAG technique of biological activity on males of *Earias insulana*, *Heliothis armigera*, and *Spodoptera littoralis*. *Environ. Entomol.* *7*:31–35 (1978).
22. Gothilf, S., Kehat, M., Jacobson, M., and Galun, R. Sex attractants for male *Heliothis armigera* (Hbn.). *Experientia* *34*: 853–854 (1978).
23. Gothilf, S., Kehat, M., Dunkelblum, E., and Jacobson, M. Efficacy of (*Z*)-11-hexadecenal and (*Z*)-11-tetradecenal as sex attractants for *Heliothis armigera* on two different dispensers. *J. Econ. Entomol.* *72*:718–720 (1979).

24. Kehat, M., Gothilf, S., Dunkelblum, E., and Greenberg, S. Field evaluation of female sex pheromone components of the cotton bollworm, *Heliothis armigera*. *Entomol. Exp. Appl.* *27*: 188–193 (1980).
25. Hendricks, D. E., Shaver, T. N., and Goodenough, J. L. Development and bioassay of molded polyvinyl chloride substrates for dispensing tobacco budworm (Lepidoptera: Noctuidae) sex pheromone bait formulations. *Environ. Entomol.* *16*:605–613 (1987).
26. Quisumbing, A. R. and Kydonieus, A. F. Laminated structure dispensers. In *Insect Suppression with Controlled Release Pheromone Systems*, Vol. 1 (Kydonieus, A. F. and Beroza, M., eds.) CRC Press, Boca Raton, 1982, pp. 213–235.
27. Lopez, J. D., Jr., Shaver, T. N., and Hartstack, A. W., Jr. Evaluation of dispensers for the pheromone of *Heliothis zea*. *Southwest. Entomol.* *6*:117–122 (1981).
28. Carpenter, J. E., Pair, S. D., and Sparks, A. N. Trapping of different noctuid moth species by one trap baited with two lures. *J. Ga. Entomol. Soc.* *19*:120–124 (1984).
29. Zvirgzdins, A. and Henneberry, T. J. *Heliothis* spp.: Sex pheromone trap studies. In *Proceedings, Beltwide Cotton Production and Research Conference*, National Cotton Council, Memphis, Tenn., 1983, pp. 176–180.
30. Halfhill, J. E. and McDonough, L. M. *Heliothis zea* (Boddie): Formulation parameters for its sex pheromone in rubber septa. *Southwest. Entomol.* *10*:176–180 (1985).
31. Hoffman, M. P., Wilson, L. T., Zalom, F. G., and McDonough, L. Lures and traps for monitoring tomato fruitworm. *Calif. Agric.* *40*:17–18 (1986).
32. Lopez, J. D., Jr., Shaver, T. N., and Goodenough, J. L. Research on various aspects of *Heliothis* spp. pheromone trapping. In *Proceedings, Beltwide Cotton Production and Research Conference*, National Cotton Council, Memphis, Tenn., 1987, pp. 300–307.
33. Leonhardt, B. A., Hendricks, D. E., Shaver, T. N., Lopez, J. D., Mastro, V. C., Schwalbe, C. P., and Ridgway, R. L. Controlled-release dispensers for effective monitoring of insect populations with pheromones. In *Proceedings, 14th International Symposium on Controlled Release of Bioactive Materials* (Lee, P. I. and Leonhardt, B. A., eds.). Controlled Release Society, Lincolnshire, Il., 1987, pp. 249–250.
34. Hendricks, D. E., Hartstack, A. W., and Shaver, T. N. Effect of formulations and dispensers on attractiveness of virelure to the tobacco budworm. *J. Chem. Ecol.* *3*:497–506 (1977).
35. Flint, H. M., McDonough, L. M., Salter, S. S., and Walters, S. Rubber septa: A long lasting substrate for (*Z*)-11-hexadecenal

and (Z)-9-tetradecenal, the primary components of the sex pheromone of the tobacco budworm. *J. Econ. Entomol.* *72*:798–800 (1979).

36. Hendricks, D. E. Polyvinyl chloride capsules: A new substrate for dispensing tobacco budworm (Lepidoptera: Noctuidae) sex pheromone bait formulations. *Environ. Entomol.* *11*:1005–1010 (1982).
37. Ramaswamy, S. B., Randle, S. A., and Ma, W. K. Field evaluations of the sex pheromone components of *Heliothis virescens* (Lepidoptera: Noctuidae) in cone traps. *Environ. Entomol.* *14*: 293–296 (1985).
38. Shaver, T. N., Hendricks, D. E., and Lopez, J. D., Jr. Enhancement of field performance of *Heliothis virescens* pheromone components by (Z)-11-hexadecen-1-ol in PVC dispensers. In *Proceedings, Beltwide Cotton Production and Research Conference*, National Cotton Council, Memphis, Tenn., 1987, pp. 307–309.
39. Gothilf, S., Kehat, M., and Dunkelblum, E. Sex pheromones for monitoring populations of *Heliothis armigera*. *Phytoparasitica* *9*:78–79 (1981).
40. Rothschild, G. H. L., Wilson, A. G. L., and Malafant, K. W. Preliminary studies on the female sex pheromones of *Heliothis* species and their possible use in control programs in Australia. In *Proceedings, International Workshop on Heliothis Management* (Reed, W., ed.), ICRISAT, Patancheru, India, 1982, pp. 319–327.
41. Wilson, A. G. L. Evaluation of pheromone trap design and dispensers for monitoring *Heliothis punctigera* and *H. armigera*. In *Proceedings, 4th Australian Applied Entomological Research Conference* (Bailey, P. and Swinger, D., eds.). 1984, pp. 74–81.
42. Lingren, P. D., Sparks, A. N., Raulston, J. R., and Wolf, W. W. Applications for nocturnal studies of insects. *Bull. Entomol. Soc. Am.* *24*:206–212 (1978).
43. Ramaswamy, S. B. and Cardé, R. T. Nonsaturating traps and long-life attractant lures for monitoring spruce budworm males. *J. Econ. Entomol.* *75*:126–129 (1982).
44. Snow, J. W., Cantelo, W. W., Burton, R. L., and Hensley, S. D. Populations of fall armyworm, corn earworm, and sugar cane borer on St. Croix, U.S. Virgin Islands. *J. Econ. Entomol.* *61*:1757–1760 (1968).
45. Stadelbacher, E. A., Laster, M. L., and Pfrimmer, T. R. 1972. Seasonal occurrence of populations of bollworm and tobacco budworm months in the central Delta of Mississippi. *Environ. Entomol.* *1*:318–323 (1972).

46. Hendricks, D. E., Hollingsworth, J. P., and Hartstack, A. W., Jr. Catch of tobacco budworm months influenced by color of sex lure traps. *Environ. Entomol.* *1*:48–51 (1972).
47. Hendricks, D. E. and Leal, M. P. Catch of adult tobacco budworms influenced by height of sex lure traps. *J. Econ. Entomol.* *66*:1218–1219 (1973).
48. Hendricks, D. E. and Hartstack, A. W. Pheromone trapping as an index for initiating control of cotton insects, *Heliothis* spp: A compendium. In *Proceedings, Beltwide Cotton Production and Research Conference*, National Cotton Council, Memphis, Tenn., 1978, pp. 116–120.
49. Hartstack, A. W., Hendricks, D. E., Lopez, J. D., Stadelbacher, E. A., Phillips, J. R., and Witz, J. A. Adult sampling. In *Economic Thresholds and Sampling of* Heliothis *Species on Cotton, Corn, Soybeans, and Other Host Plants* (Sterling, W. L., ed.). South Coop. Ser. Bull. 231, Dept. of Agricultural Communications, Texas A&M University, College Station, Texas, 1979, pp. 105–131.
50. Wolf, W. W., Toba, H. H., Kishoba, A. N., and Green, N. Antioxidants to prolong the effectiveness of cabbage looper sex pheromone in the field. *J. Econ. Entomol.* *65*:1039–1041 (1972).
51. Goodenough, J. L. and Snow, J. W. Increased collection of tobacco budworms by electric grid traps as compared with blacklight and sticky traps. *J. Econ. Entomol.* *66*:450–453 (1973).
52. Hollingsworth, J. P., Hartstack, A. W., Buck, D. R., and Hendricks, D. E. Electric and nonelectric moth traps baited with the synthetic sex pheromone of the tobacco budworm. USDA, ARS, Southern Region, ARS-S-173, 1978, pp. 1–13.
53. Raulston, J. R., Sparks, A. N., and Lingren, P. D. Design and comparative efficiency of a wind-oriented trap for capturing live *Heliothis* spp. *J. Econ. Entomol.* *73*:586–589 (1980).
54. Hartstack, A. W., Witz, J. A., and Buck, D. R. Moth traps for the tobacco budworm. *J. Econ. Entomol.* *72*:519–522 (1979).
55. Hendricks, D. E., Perez, C. T., and Guerra, R. J. Effect of nocturnal wind on performance of two sex pheromone traps for noctuid moths. *Environ. Entomol.* *9*:483–485 (1980).
56. Bourdouxhe, L. Comparison de deux types de pièges pour le piègeage sexuel de *Heliothis armigera* au Senegal. *Plant Prot. Bull.* *30*:131–136 (1982).
57. Kehat, M., Gothilf, S., Dunkelblum, E., and Greenberg, S. Sex pheromone traps as a means of improving control programs for the cotton bollworm, *Heliothis armigera* (Lepidoptera: Noctuidae). *Environ. Entomol.* *11*:727–729 (1982).
58. Tahhan, O. and Hariri, G. Preliminary study of trapping *Heliothis armigera* (Hb.) with pheromones at ICARDA, Syria. *ICRISAT Chickpea Newslett.* *6*:19–20 (1982).

59. Sage, T. L. and Gregg, P. C. A comparison of four types of pheromone traps for *Heliothis armigera* (Hübner) (Lepidoptera: Noctuidae). *J. Aust. Entomol. Soc. 24*:99–100 (1985).
60. Plapp, F. W., Jr., McWhorter, G. M., and Vance, W. H. Monitoring for pyrethroid resistance in the tobacco budworm in Texas – 1986. In *Proceedings, Beltwide Cotton Production and Research Conference*, National Cotton Council, Memphis, Tenn., 1987, pp. 324–326.
61. Laster, M. L., Kitten, W. F., Knipling, E. F., Martin, D. F., Schneider, J. C., and Smith, J. W. Estimates of overwintered population density and adult survival rates for *Heliothis virescens* (Lepidoptera: Noctuidae) in the Mississippi Delta. *Environ. Entomol. 16*:1076–1081 (1987).
62. Hendrix, W. H. III, Mueller, T. F., Phillips, J. R., and Davis, O. K. Pollen as an indicator of long-distance movement of *Heliothis zea* (Lepidoptera: Noctuidae). *Environ. Entomol. 65*: 1148–1151 (1987).
63. Goodenough, J. L., Witz, J. A., Lopez, J. D., and Hartstack, A. W., Jr. A system for wide-area survey of *Heliothis*. In *Proceedings, Beltwide Cotton Production and Research Conference*, National Cotton Council, Memphis, Tenn., 1987, pp. 291–300.
64. Raulston, J. R., Lingren, P. D., Sparks, A. N., and Martin, D. F. Mating interaction between native tobacco budworms and released backcross adults. *Environ. Entomol. 8*:349–353 (1979).
65. Tingle, F. C. and Mitchell, E. R. Relationships between pheromone trap catches of male tobacco budworm, larval infestations, and damage levels in tobacco. *J. Econ. Entomol. 74*:437–440 (1981).
66. Roltsch, W. J. and Mayse, M. A. Population studies of *Heliothis* spp. (Lepidoptera: Noctuidae) on tomato and corn in southeast Arkansas. *Environ. Entomol. 13*:292–299 (1984).
67. Hartstack, A. W., Jr., Hollingsworth, J. P., Witz, J. A., Buck, D. R., Lopez, J. D., and Hendricks, D. E. Relation of tobacco budworm catches in pheromone baited traps to field populations. *Southwest. Entomol. 3*:43–51 (1978).
68. Johnson, D. R. Relationship between tobacco budworm (Lepidoptera: Noctuidae) catches when using pheromone traps and egg counts in cotton. *J. Econ. Entomol. 76*:182–183 (1983).
69. Witz, J. A., Hartstack, A. W., King, E. G., Dickerson, W. A., and Phillips, J. R. Monitoring and prediction of *Heliothis* spp. *Southwest. Entomol. Suppl. 8*:56–70 (1985).
70. Hartstack, A. W., Jr. and Witz, J. A. Estimating field populations of tobacco budworm moths from pheromone trap catches. *Environ. Entomol. 10*:908–914 (1981).
71. Hartstack, A. W., Jr., Witz, J. A., Hollingsworth, J. P., and Bull, D. L. SPERM – A sex pheromone emission and response model. *Trans. ASAE 19*:1170–1174, 1180 (1976).

72. Hartstack, A. W., Jr., King, E. G., and Phillips, J. R. Monitoring and predicting *Heliothis* populations in Southeast Arkansas. In *Proceedings, Beltwide Cotton Production and Research Conference*, National Cotton Council, Memphis, Tenn., 1983, pp. 187–190.
73. Hendricks, D. E. Low-frequency sodar device that counts flying insects attracted to sex pheromone dispensers. *Environ. Entomol.* *9*:452–457 (1980).
74. Hendricks, D. E. Portable electronic detector system used with inverted-cone sex pheromone traps to determine periodicity and moth captures. *Environ. Entomol.* *14*:199–204 (1985).

Part V

Stored-Product Insects and Insects Affecting Animals

28

Practical Use of Pheromones and Other Attractants for Stored-Product Insects

WENDELL E. BURKHOLDER / Agricultural Research Service, U.S. Department of Agriculture, Madison, Wisconsin, and University of Wisconsin, Madison, Wisconsin

I. INTRODUCTION

The use of traps with pheromones or other attractants as a supplement to conventional inspection methods has increased steadily in recent years. Some reasons for this are the development and marketing of new pheromones and traps, the need for early detection of insects, insecticide resistance problems, and an awareness of new restrictions on pesticides. The components of biological control can include a pest product used against itself (e.g., pheromones) for regulation of the pest populations (1). Pest management specialists usually have zero tolerances for insects in food storage environments and, therefore, utilize traps for early detection and localization of infestation. Pheromone traps, however, are not practical in the storage environment when there is an obvious and large infestation of insects.

There are considerable cost savings associated with the use of effective baited traps. Early detection permits corrective measures to be taken before substantial losses occur, and allows treatments to be made on an "as needed" basis. Time and money are saved because the traps work continuously and reduce the need for manual inspections. When there are isolated infestations, less-costly spot treatments may be used. The effectiveness of grain bin fumigation may be monitored easily with traps. Live insects in the traps after fumigation would indicate treatment failure. Prompt retreatment would prevent continued product loss. Insect pests of stored products are generally considered to cause 5% to 10% losses to farmers, commercial food storage, processing and merchandising industries, as well as to consumers within their homes.

This chapter will summarize the state of the art regarding the practical use of pheromones, other attractants, and traps, in the food storage environment.

II. THE PHEROMONES AND FOOD LURES

The functions of pheromones in stored-product insects follow two broad patterns. These are sex pheromones for the short-lived adult insects and aggregation pheromones for the long-lived adult insects (2). The short-lived adults generally do not feed, and the females usually produce sex pheromones. The long-lived adults feed, and the males usually produce aggregation pheromones, although females of such species may also produce sex pheromones. For practical trapping efforts, the aggregation pheromones are especially useful because both male and female insects respond, whereas sex pheromones usually attract only the opposite sex. Table 1 provides a summary of some of the stored-product insect pheromone components. The short-lived insects, such as the moths, dermestids, and anobiid beetles, rely on female-produced sex pheromones for communication. The long-lived adults, such as the grain weevils, grain borers, flour and grain beetles, rely on male-produced aggregation pheromones.

The primary function of the sex pheromones is to promote mating, whereas the function of the aggregation pheromone is more complex. Aggregations of three or more grain weevils are commonly observed on a single grain kernel. The pheromone is produced only by the feeding males, and freshly cracked wheat has been determined to synergize the pheromone response (56). Responding males and females are alerted to potential feeding sites and mates.

The use of food baits and lures for stored-product insects is an old practice. Recently more intensive studies of these attractants have taken place. Bait bags containing food materials have been used in Great Britain and some other countries for insect detection (57). The plastic mesh bags contain approximately 80 g of equal parts of wheat, broken groundnuts, and crushed carobs (locust beans). Successful use of the bags for monitoring stored-product insects was summarized by Pinniger et al. (57). Hodges et al. (58) used the bait bags as an aid to pest management in Indonesian milled rice stores. The bags indicated population changes that were roughly in proportion to those indicated by spear sampling.

A problem associated with the use of the bags is that insect reproduction may occur if they are lost or unattended, and new infestations may result. One procedure that avoids this problem is to use seed oils or extracts of food, rather than whole grains or seeds.

Several volatile components of wheat germ oil responsible for initiating aggregating activity of *Trogoderma glabrum* (Herbst) were identified by Nara et al. (59). Studies by Freedman et al. (60), Mikolajczak et al. (61--63), and other workers (64) demonstrated that oats generate a number of stimuli that induce responses in *Oryzaephilus surinamensis* (L.). Both wheat germ and oat oil have been combined with mineral oil for trapping stored-product Coleoptera (65).

III. THE TRAPS

The use of corrugated traps with wheat flour bait to trap *Tribolium confusum* Jacquelin duVal was reported by DeCoursey (66). Wilson (67) utilized sticky traps treated with alcoholic fish meal extracts for trapping carpet beetle (Dermestidae) larvae in houses. Crevice traps constructed of corrugated paper or sacking were described by Howe (68). A probe trap devised by Loschiavo and Atkinson (69) and a similar one developed by Burkholder (70) are adapted for use in grain. Brady et al. (40) used sticky traps with pheromone to catch *Plodia interpunctella* (Hübner) and *Cadra cautella* (Walker). Wide varieties of sticky traps are used to catch both flying moths and beetles.

The perforated plastic probe trap has been developed commercially for use in bulk grain (71). The trap was used successfully with the aggregation pheromone of *Rhyzopertha dominica* (F.) to catch the grain borers in bagged sorghum (72). Lippert and Hagstrum (73) used the probe traps to detect and estimate insect populations in bulk-stored wheat. Insect infestations of wheat stored on 20 Kansas farms were estimated with the probe traps and 0.265-kg samples of wheat. The traps in wheat bins for 2 days were 1.7-, 2.6-, 2.4-, and 2.6-fold more likely than wheat samples to detect infestations of *Cryptolestes ferrugineus* (Stephens), *Tribolium castaneum* (Herbst), *Oryzaephilus surinamensis*, and *Rhyzopertha dominica*, respectively (73). The plastic traps were developed for use in combination with pheromones, yet work exceptionally well without them. It appears the daily movement of the insects results in their interception by the traps. The insects then fall through the slanted holes into the bottom of the device. Cracked wheat (56), wheat germ oil (65), or other cracked grains are being used as food lures. The upper end of the trap contains a plastic tube or rod suitable for holding pheromone lures. The trap, therefore, is not only physically attractive and effective in catching insects but can also hold food and pheromone baits for additional attractance. For those insects that produce aggregation pheromones, the trapped insects may also serve to attract others.

The use of multilayered, corrugated paper traps with pheromones for trapping stored-product insects was reported by Barak and Burkholder (65,74), Burkholder (75), and Williams et al. (12). The

TABLE 1 Pheromone Components of Some Stored-Product Insects

Insect species and sex that produces pheromone		Pheromone components	Refs.
Anobiidae			
Stegobium paniceum	♀	2,3-Dihydro-2,3,5-trimethyl-6-(1-methyl-2-oxobutyl)-4*H*-pyran-4-one	3,4
Lasioderma serricorne	♀	4,6-Dimethyl-7-hydroxynonan-3-one	5–7
	♀	2,6-Diethyl-3,5-dimethyl-3,4-dihydro-2*H*-pyran	8,9
	♀	2,3-*cis*-2,3-Dihydro-3,5-dimethyl-2-ethyl-6-(1-methyl-2-oxobutyl)-4*H*-pyran-4-one	10
	♀	2,3-*cis*-2,3-Dihydro-3,5-dimethyl-2-ethyl-6-(1-methyl-2-hydroxybutyl)-4*H*-pyran-4-one	10
Bostrichidae			
Rhyzopertha dominica	♂	1-Methylbutyl (*E*)-2-methyl-2-pentenoate[a]	11,12
	♂	1-Methylbutyl (*E*)-2,4-dimethyl-2-pentenoate[a]	11,12
Prostephanus truncatus	♂	1-Methylethyl (E)-2-methyl-2-pentenoate[a]	13
Bruchidae			
Acanthoscelides obtectus	♂	(*E*)-(-)-Methyl-2,4,5-tetradecatrienoate	14,15
Callosobruchus chinensis	♀	Multicomponent (see Refs.)	16–18

Cucujidae			
Cryptolestes ferrugineus	♂	(*E*,*E*)-4,8-Dimethyl-4,8-decadien-10-olide[a]	19,20
	♂	(3*Z*,11*S*)-3-dodecen-11-olide[a]	
Cryptolestes pusillus	♂	(*Z*)-3-Dodecenolide[a]	21
	♂	(*Z*)-5-Tetradecen-13-olide[a]	21
Cryptolestes turcicus	♂	(*Z*,*Z*)-5,8-Tetradecadien-13-olide[a]	21
Oryzaephilus surinamensis	♂	(*Z*,*Z*)-3,6-Dodecadien-11-olide[a]	22
	♂	(*Z*,*Z*)-3,6-Dodecadienolide[a]	22
	♂	(*Z*,*Z*)-5,8-Tetradecadien-13-olide[a]	22
Oryzaephilus mercator	♂	(*Z*,*Z*)-3,6-Dodecadien-11-olide[a]	22
	♂	(*Z*)-3-Dodecen-11-olide[a]	22
Curculionidae			
Sitophilus oryzae	♂	(4*S*,5*R*)-5-Hydroxy-4-methyl-3-heptanone[a]	23,24
Sitophilus zeamais	♂	(4*S*,5*R*)-5-Hydroxy-4-methyl-3-heptanone[a]	23,24
Sitophilus granarius	♂	(*R**,*S**)-1-Ethylpropyl-2-methyl-3-hydroxy-pentanoate[a]	25
Dermestidae			
Attagenus unicolor (=*megatoma*)	♀	(*E*,*Z*)-3,5-Tetradecadienoic acid	26,27

(continued)

TABLE 1 (Cont.)

Insect species and sex that produces pheromone		Pheromone components	Refs.
[Dermestidae]			
Attagenus brunneus (=*elongatulus*)	♀	(*Z*,*Z*)-3,5-Tetradecadienoic acid	28–30
Anthrenus flavipes	♀	(*Z*)-3-Decenoic acid	31,32
Anthrenus verbasci	♀	(*Z*)-5- and (*E*)-5-Undecenoic acid	33
Trogoderma inclusum	♀	(*Z*)-14-Methyl-8-hexadecen-1-ol	26,35
	♀	(*Z*)-14-Methyl-8-hexadecenal	34
Trogoderma variabile	♀	(*Z*)-14-Methyl-8-hexadecenal	34
Trogoderma glabrum	♀	(*E*)-14-Methyl-8-hexadecen-1-ol	26,36
	♀	(*E*)-14-Methyl-8-hexadecenal	34
Trogoderma granarium	♀	92:8 (*Z*:*E*)-14-Methyl-8-hexadecenal	34,37
Gelechiidae			
Sitotroga cerealella	♀	(*Z*,*E*)-7,11-Hexadecadien-1-ol acetate	38,39
Pyralidae			
Cadra cautella	♀	(*Z*,*E*)-9,12-Tetradecadien-1-ol acetate	40,41
	♀	(*Z*,*E*)-9,12-Tetradecadien-1-ol	42–45
	♀	(*Z*)-9-Tetradecen-1-ol acetate	46,47

Cadra figulilella	♀	(*Z*,*E*)-9,12-Tetradecadien-1-ol acetate	48
Plodia interpunctella	♀	(*Z*,*E*)-9,12-Tetradecadien-1-ol acetate	40,41
	♀	(*Z*,*E*)-9,12-Tetradecadien-1-ol	45,49
Anagasta kuehniella	♀	(*Z*,*E*)-9,12-Tetradecadien-1-ol acetate	50,51
	♀	(*Z*,*E*)-9,12-Tetradecadien-1-ol	41,42
Ephestia elutella	♀	(*Z*,*E*)-9,12-Tetradecadien-1-ol acetate	47
	♀	(*Z*,*E*)-9,12-Tetradecadien-1-ol	42
Amyelois transitella	♀	(*Z*,*Z*)-11,13-Hexadecadienal	52
Tenebrionidae			
Tribolium castaneum	♂	4,8-Dimethyldecanal[a]	53–55
Tribolium confusum	♂	4,8-Dimethyldecanal[a]	53–55

[a]Aggregation pheromone; all others are sex pheromones.

corrugations attract the insects by serving as harborage sites similar to crevices in floors and walls. There are multiple entry points formed by the corrugations that provide tactile stimuli, and air chambers within the trap provide circulation for the lures. The early designs utilized insecticides for killing the insects. Barak and Burkholder (65) improved the trap by utilizing a vegetable oil-based food attractant instead of an insecticide. The oils serve as food lures and also suffocate the insects when they fall or crawl into the receptacle. Lures that incorporate a combination oil lure with mineral-, oat-, and wheat-germ oils with options for several pheromone lures are effective in attracting both larvae and adults. These traps have been commercialized by Zoecon Corp. and subsequently by Trece Inc.

Lindgren et al. (76) developed an insect trap from an inverted, flat-bordered, plastic weighing boat. It contains eight inward-projecting, conical depressions. The traps were used for trapping *Cryptolestes ferrugineus* and *Tribolium castaneum* in controlled environmental chambers. Lindgren (77) also developed a multiple-funnel trap for scolytid beetles that was used successfully with *R. dominica* pheromone for trapping the insects in warehouses (78).

Traps for flying moths are designed to be suspended from supports in warehouses. Though the traps were intended originally for trapping moths, they are useful in trapping beetles when baited with their sex or aggregation pheromones. The traps have a protective cover over a surface covered with sticky material. The pheromone lures are placed in the center of the trap. Funnel traps have also been used for catching moths and beetles. Shapas and Burkholder (79) used funnel traps with sex pheromone for trapping *Trogoderma glabrum* (Herbst). Cogan and Hartley (80) reported on a new plastic funnel trap that used an insecticide-impregnated strip to kill the insects. The trap contents are hidden from view, and present a more appealing appearance in sensitive food-processing areas than sticky traps. The funnel trap has been developed commercially in Great Britain.

IV. DETECTION, MONITORING, MANAGEMENT, AND CONTROL

The Anobiidae pests include the drugstore beetle, *Stegobium paniceum* (L.), and the cigarette beetle, *Lasioderma serricorne* (Fabricius). The cigarette beetle is the major pest of the tobacco industry whose facilities have been used for testing pheromone monitoring programs. Faustini (81) presented evidence that sex pheromone traps were superior monitors when compared with blacklight traps. He suggested that a major barrier to overcome in the establishment of a pheromone program in an industrial environment was the development of an educational and training program. He believes that when pheromone traps are used in conjunction with intensive sanitation and selective pest control programs (insecticide treatments), a substantial reduction in target pests results (81).

The Bostrichidae stored-product pests include the lesser grain borer (*R. dominica*) and the larger grain borer, *Prostephanus truncatus* (Horn). The borers produce aggregation pheromones that attract both sexes (11,13). The lesser grain borer pheromone has been used in the United States and Mexico in traps. It was used in Africa for monitoring both the lesser and larger grain borer before the time the larger grain borer pheromone became available (13). The larger grain borer pheromone is now being used in Africa to monitor this imported pest.

The Cucujidae include the *Cryptolestes* spp. (flat and rusty grain beetles) and *Oryzaephilus* spp. (sawtoothed and merchant grain beetles). A complex series of seven macrolide lactone pheromones have been described by Pierce et al. (22). The pheromones were used by Loschiavo et al. (82) for trapping *C. ferrugineus*.

The Curculionidae include the *Sitophilus* spp. grain weevils (rice, maize, and granary weevils). Males of the rice weevil, *S oryzae* (L.), and its congener, the maize weevil, *S. zeamais* Motschulsky, produce the same aggregation pheromone (23,24). This compound is attractive to both sexes of these two species, as well as to the granary weevil, *S. granarius* (23,83). Males of the granary weevil, *S. granarius*, produce a structurally similar aggregation pheromone (25,83). Corrugated cardboard traps (Storgard) were used to test attractive properties of combinations of synthetic pheromone and various food odors to maize weevils in laboratory test arenas (65). Cracked wheat was a stronger attractant than either whole wheat, wheat germ oil, or water (56). A strong synergistic effect between the pheromone and cracked wheat was demonstrated. Maize weevils were successfully lured out of a food source only when both cracked wheat and pheromone were present in the trap. Only a slight reduction in catch was observed as weevils aged. Preliminary field studies in North Carolina with the pheromone in plastic grain-probe traps were successful (R. Hillman, personal communication). Further studies with the pheromones in traps are in progress.

The Dermestidae include the warehouse, cabinet, and carpet beetles, which have female-produced sex pheromones (26,31,84). Dermestid species were monitored with pheromone traps in Milwaukee, Wisconsin. The corrugated paper traps were treated with malathion and either 0.25 mg (*E*,*Z*)-3,5-tetradecadienoic acid, a sex attractant of the black carpet beetle, *Attagenus unicolor* (27), or (*Z*)-14-methyl-8-hexadecen-1-ol (35), a component of the sex pheromone of several *Trogoderma* species. The traps treated with pheromones caught significantly more male insects than the control traps in a cargo terminal, a grain elevator, and at a milling company. These traps also caught *T. variabile* Ballion, a species not previously recorded in the warehouse. A seasonal emergence of *A. unicolor* over a 2-year period was recorded through the use of the trap program (74). The aldehyde component of the *Trogoderma* spp. pheromone is released by calling females (85) and is effective in the detection of the insects at low-population levels (86). The pheromone is used routinely by regulatory personnel for monitoring the Khapra beetle, *T. granarium* Everts.

The *Trogoderma* spp. protozoan pathogen, *Mattesia trogodermae* Canning, was effectively dispersed to females following male attraction by sex pheromones to a site containing spores (87). Only one, or a few spores are needed to start an epizootic. The adult males that came to the pheromone source returned to the emerging females and mated with them. The contaminated adults, when dead, were subsequently available as food for offspring (87).

The male-produced aggregation pheromone of *Tribolium castaneum* and *T. confusum* (red and confused flour beetles) was used in small test arenas with corrugated paper pitfall traps to evaluate attraction of adult *T. confusum*. Significantly more beetles were trapped in the pheromone-baited traps than in the controls (65). The pheromone has been commercialized and is being used in warehouses to find sites of infestation and to evaluate pesticide applications or other insect management procedures.

The value of the moth pheromone (*Z*,*E*)-9,12-tetradecadien-1-ol acetate (TDA) to the food industry has been well established. It is the primary sex attractant for males of several species of *Plodia*, *Anagasta*, *Cadra*, and *Ephestia*. In a recent report, Suss and Trematerra (88) indicated that in Italy the pheromone has allowed workers to program control measures against these insects in the food industry and in stored products. They also reported that in some cases its use brought about a substantial reduction in chemical treatments.

Studies conducted by Wohlgemuth et al. (89) in Germany demonstrated that pheromone traps caught male *Plodia interpunctella*, *Anagasta kuehniella* (Zeller), *Ephestia elutella* (Hübner), and *Cadra cautella* out of doors but within the city zones. On the outskirts of the cities and in their park-like areas as well as in the rural countryside *Cadra cautella* was missing from the traps, presumably because of a lack of the specialized food they require. In specially baited traps (pheromone and broken almonds) and in the vicinity of heavily infested objects, a small number of female moths were also caught (89). Cogburn et al. (90) also trapped *R. dominica* outside bins. Trap-catches around the bins were inversely proportional to the height of the traps. Cogburn (personal communication) reports that pheromone traps will always catch *R. dominica* in field and forested areas some distance from the grain storage areas. *Trogoderma glabrum* males as well as other dermestids are easily trapped outside (79). These results and others show that stored-product insects occur outside storage facilities.

Traps for monitoring insects in grain should be deployed just after harvest. Monitoring should be a continuous process until the grain is moved to the next destination. Particular care should be taken to monitor the traps after treatments, such as fumigation, to assess insect control. Traps should be placed both outside and inside the bins before harvest to determine residual populations of insects. Traps placed outside the bin and in the perimeter of the bin site may also reduce the number of migrating insects that are attracted to freshly harvested grain.

Warehouses and food-processing plants are also highly attractive to insects. Traps should be placed in a grid pattern within the buildings. Pheromone traps should be kept away from open doors and windows to prevent insect entry. All traps should be checked at least once a week. The insect catch, trap location, and other conditions should be recorded to establish sites of infestation and development trends. Suspect areas should be monitored more intensely. The tolerance for stored product insects is low. Often one insect is enough to start a special cleanup and intensified management and control effort. The use of traps in management programs is not practical unless accompanied by thorough sanitation procedures. Continued trapping will determine the effectiveness of the sanitation programs. During warm months when insect flights are expected traps should also be placed outside the buildings. Traps can be placed on fences, posts or other places approximately 25 m from the building.

Grain bins are often considered a closed system convenient for trapping. They present special problems, however, because of the difficulty of inserting traps to the bin bottom. Some insects, such as *Tribolium* spp., are often near the surface, whereas many species may live deep within the bin. The perforated plastic probe traps (70) may be pushed several meters into the bin, but this is not necessarily deep enough. Either solid or perforated tubes may be placed deep in the bin by hanging them in the empty bin before filling (Fig. 1; W. E. Burkholder, unpublished data). Our optional design of perforated tubes allows insect entry through the entire length of the tube. The insects collect at the bottom of the tube, and a removable trap insert is placed at the bottom of the tube to retrieve the trapped insects. The same perforated plastic trap described previously (70), with some minor modifications, is ideal for this. Another option is to suspend more than one trap at various intervals within the tube to aid in localizing infestations.

A new perforated plastic tube trap has been devised that is much less expensive and may be considered disposable (W. E. Burkholder, unpublished data). These traps are larger, can be made in any length, and are designed to be used for in-transit grain shipments for which the return of the trap is impractical. The traps are ideal for determining the success of in-transit treatments such as fumigations.

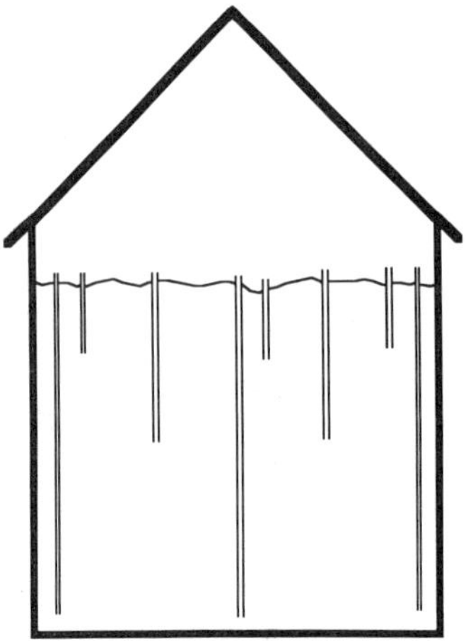

FIGURE 1 Schematic diagram of a grain bin showing plastic tubes that allow insect sampling at various depths and locations by means of trap inserts. Tubes for mid- to deep-bin sampling are suspended before filling. Tubes may be solid-wall or perforated to allow insects to enter the full length of the tube and also to allow better circulation of volatile lures.

V. SUMMARY

Pheromones and food attractants have dramatically changed the methods by which insect inspections are made. The baited traps are safe, easy to use, and are highly efficient in detecting infestations. The presence of insects is often determined before visual inspections reveal them. Although attractants are primarily a surveillance tool in a management program they have been, and likely will be, used more for direct control in the future. This will be possible only if sanitation measures and building facilities are excellent and certain innovative measures are employed. We have only begun to explore the possibilities.

ACKNOWLEDGMENTS

I thank H. C. Coppel, M. R. Strand, and J. K. Phillips for comments and suggestions on the manuscript. This research was supported by the College of Agriculture and Life Sciences, University of Wisconsin, Madison, and by a cooperative agreement between the University of Wisconsin and ARS-USDA.

REFERENCES

1. Research Briefings: Report of the research briefing panel on biological control in managed ecosystems. Vol. 6, COSEPUP, National Academy of Sciences, Engineering and Institute of Medicine, 1987, pp. 1–12.
2. Burkholder, W. E. Reproductive biology and communication among grain storage and warehouse beetles. *J. Ga. Entomol. Soc. 17*(suppl. 2):1–10 (1982).
3. Kuwahara, Y., Fukami, H., Howard, R., Ishii, S., Matsumura, F., and Burkholder, W. E. Chemical studies on the Anobiidae: Sex pheromone of the drugstore beetle, *Stegobium paniceum* (L.) (Coleoptera). *Tetrahedron 34*:1769–1774 (1978).
4. Kuwahara, Y., Fukami, H., Ishii, S., Matsumura, F., and Burkholder, W. E. Studies on the isolation and bioassay of the sex pheromone of the drugstore beetle *Stegobium paniceum* (Coleoptera: Anobiidae). *J. Chem. Ecol. 1*:413–422 (1975).
5. Chuman, T., Kato, K., and Noguchi, M. Synthesis of (±)-serricornin, 4,6-dimethyl-7-hydroxy-nonan-3-one, a sex pheromone of cigarette beetle (*Lasioderma serricorne* F.). *Agric. Biol. Chem. 43*:2005 (1979).
6. Chuman, T., Kohno, M., Kato, K., and Noguchi, M. 4,6-Dimethyl-7-hydroxy-nonan-3-one, a sex pheromone of the cigarette beetle (*Lasioderma serricorne* F.). *Tetrahedron Lett. 25*: 2361–2364 (1979).
7. Coffelt, J. A. and Burkholder, W. E. Reproductive biology of the cigarette beetle, *Lasioderma serricorne*. 1. Quantitative laboratory bioassay of the female sex pheromone from females of different ages. *Ann. Entomol. Soc. Am. 65*:447–450 (1972).
8. Levinson, H. Z., Levinson, A. R., Francke, W., MacKenroth, W., and Heemann, V. The pheromone activity of anhydroserricornin and serricornin for male cigarette beetles (*Lasioderma serricorne* (F.). *Naturwissenschaften 68*:148–149 (1981).
9. Levinson, H. Z., Levinson, A. R., Francke, W., MacKenroth, W., and Heemann, V. Suppressed pheromone responses of male tobacco beetles to anhydroserricornin in presence of serricornin. *Naturwissenschaften 69*:454–455 (1982).
10. Chuman, T., Mochizuki, K., Kato, K., Ono, M., and Okubo, A. Serricorone and serricorole, new sex pheromone components of cigarette beetle. *Agric. Biol. Chem. 47*:1413–1415 (1983).

11. Khorramshahi, A. and Burkholder, W. E. Behavior of the lesser grain borer *Rhyzopertha dominica* (Coleoptera: Bostrichidae): Male-produced aggregation pheromone attracts both sexes. *J. Chem. Ecol.* 7:33–38 (1981).
12. Williams, H., Silverstein, R. M., Burkholder, W. E., and Khorramshahi, A. Dominicalure 1 and 2: Components of aggregation pheromone from male lesser grain borer, *Rhyzopertha dominica* (F.) (Coleoptera: Bostrichidae). *J. Chem. Ecol.* 7: 759–780 (1981).
13. Hodges, R. J., Cork, A., and Hall, D. R. Aggregation pheromones for monitoring the greater grain borer *Prostephanus truncatus*. *Br. Crop Protect. Conf. Pests and Disease.* Brighton, 1984, pp. 255–260.
14. Hope, J. A., Horler, D. F., and Rowlands, D. G. A possible pheromone of the bruchid, *Acanthoscelides obtectus* (Say). *J. Stored Prod. Res.* 3:387–388 (1967).
15. Horler, D. F. (-)-Methyl-*n*-tetradeca-*trans*-2,4,5-trienoate, an allenic ester produced by the male dried bean beetle, *Acanthoscelides obtectus* (Say). *J. Chem. Soc.* C(6)859–862 (1970).
16. Honda, H. and Yamamoto, I. Evidence for and chemical nature of a sex pheromone present in Azuki bean weevil, *Callosobruchus chinensis* L. In *Proc. Symposium on Insect Pheromones and Their Application* (Nagaoka and Tokyo), 1976, p. 164.
17. Tanaka, K., Ohsawa, K., Honda, M., and Yamamoto, I. Copulation release pheromone, erectin, from the Azuki bean weevil (*Callosobruchus chinensis* L.). *J. Pestic. Sci.* 6:75–82 (1981).
18. Tanaka, K., Ohsawa, K., Honda, M., and Yamamoto, I. Synthesis of erectin, a copulation release pheromone of the azuki bean weevil, *Callosobruchus chinensis* L. *J. Pestic. Sci.* 7: 535–537 (1982).
19. Borden, J. H., Dolinski, M. G., Chong, L., Verigin, V., Pierce, H. D. Jr., and Oehlschlager, A. C. Aggregation pheromone in the rusty grain beetle, *Cryptolests ferrugineus* (Coleoptera: Cucujidae). *Can. Entomol.* 111:681–688 (1979).
20. Wong, J. W., Verigin, V., Oehlschlager, A. C., Borden, J. H., Pierce, H. D. Jr., Pierce, A. M., and Chong, L. Isolation and identification of two macrolid pheromones from the frass of *Cryptolestes ferrugineus* (Coleoptera: Cucujidae). *J. Chem. Ecol.* 9:451–474 (1983).
21. Millar, J. G., Oehlschlager, A. C., and Wong, J. W. Synthesis of two macrolide aggregation pheromones from the flat grain beetle, *Cryptolestes pusillus* (Schonherr). *J. Org. Chem.* 48: 4404–4407 (1983).
22. Pierce, A. M., Pierce, H. D. Jr., Millar, J. G., Borden, J. H., and Oehlschlager, A. C. Aggregation pheromones in the genus *Oryzaephilus* (Coleoptera: Cucujidae). In *Proceedings of the*

Third International Working Conference on Stored-Product Entomology. Kansas State University, Manhattan, Kansas, 1984, pp. 107--120.
23. Phillips, J. K., Walgenbach, C. A., Klein, J. A., Burkholder, W. E., Schmuff, N. R., and Fales, H. M. (*R**,*S**)-5-Hydroxy-4-methyl-3-heptanone: A male-produced aggregation pheromone of *Sitophilus oryzae* (L.) and *S. zeamais* Motsch. *J. Chem. Ecol.* *11*:1263--1274 (1985).
24. Schmuff, N., Phillips, J. K., Burkholder, W. E., Fales, H. M., Chen, C., Roller, P., and Ma, M. The chemical identification of the rice weevil and maize weevil aggregation pheromones. *Tetrahedron Lett.* *25*:1533–1534 (1984).
25. Phillips, J. K., Miller, S. P. F., Andersen, J. F., Fales, H. M., and Burkholder, W. E. The chemical identification of the granary weevil aggregation pheromone. *Tetrahedron Lett.* *28*:6145–6146 (1987).
26. Burkholder, W. E. and Dicke, R. J. Evidence of sex pheromones in females of several species of Dermestidae. *J. Econ. Entomol.* *59*:540--543 (1966).
27. Silverstein, R. M., Rodin, J. O., Burkholder, W. E., and Gorman, J. E. Sex attractant of the black carpet beetle. *Science* *157*:85–87 (1967).
28. Barak, A. V. and Burkholder, W. E. Behavior and pheromone studies with *Attagenus elongatulus* Casey (Coleoptera: Dermestidae). *J. Chem. Ecol.* *3*:219--237 (1977).
29. Barak, A. V. and Burkholder, W. E. Studies on the biology of *Attagenus elongatulus* Casey (Coleoptera: Dermestidae) and the effects of larval crowding on pupation and life cycle. *J. Stored Prod. Res.* *13*:169--175 (1977).
30. Fukui, H., Matsumura, F., Barak, A. V., and Burkholder, W. E. Isolation and identification of a major sex attractant component of *Attagenus elongatulus* Casey (Coleoptera: Dermestidae). *J. Chem. Ecol.* *3*:541--550 (1977).
31. Burkholder, W. E., Ma, M., Kuwahara, Y., and Matsumura, F. Sex pheromone of the furniture carpet beetle, *Anthrenus flavipes* (Coleoptera: Dermestidae). *Can. Entomol.* *106*:835--839 (1974).
32. Fukui, H., Matsumura, F., Ma, M. C., and Burkholder, W. E. Identification of the sex pheromone of the furniture carpet beetle, *Anthrenus flavipes* LeConte. *Tetrahedron Lett.* *40*:3563–3566 (1974).
33. Kuwahara, Y. and Nakamura, S. (*Z*)-5- and (*E*)-5-Undecenoic acid: Identification of the sex pheromone of the varied carpet beetle, *Anthrenus verbasci* L. (Coleoptera: Dermestidae). *Appl. Entomol. Zool.* *20*:354–356 (1985).
34. Cross, J. H., Byler, R. C., Cassidy, R. F. Jr., Silverstein, R. M., Greenblatt, R. E., Burkholder, W. E., Levinson, A. R.,

and Levinson, H. Z. Porapak-Q collection of pheromone components and isolation of (*Z*)- and (*E*)-14-methyl-8-hexadecenal, potent sex attracting components, from females of four species of *Trogoderma* (Coleoptera: Dermestidae). *J. Chem. Ecol. 2*: 457--468 (1976).

35. Rodin, J. O., Silverstein, R. M., Burkholder, W. E., and Gorman, J. E. Sex attractant of female dermestid beetle *Trogoderma inclusum* LeConte. *Science 165*:904–906 (1969).
36. Yarger, R. G., Silverstein, R. M., and Burkholder, W. E. Sex pheromone of the female dermestid beetle *Trogoderma glabrum* (Herbst). *J. Chem. Ecol. 1*:323--334 (1975).
37. Levinson, H. Z. and Bar Ilan, A. R. Function and properties of an assembling scent in the khapra beetle *Trogoderma granarium*. *Riv. Parassitol. 28*:27–42 (1967).
38. Keys, R. E. and Mills, R. B. Demonstration and extraction of a sex attractant from female Angoumois grain moths. *J. Econ. Entomol. 61*:46–49 (1968).
39. Vick, K. W., Su, H. C. F., Sower, L. L., Mahany, P. G., and Drummond, P. C. (*Z,E*)-7,11-Hexadecadien-1-ol acetate: The sex pheromone of the Angoumois grain moth, *Sitotroga cerealella*. *Experientia 30*:17–18 (1974).
40. Brady, U. E., Tumlinson, J. H., Brownlee, R. G., and Silverstein, R. M. Sex stimulant and attractant in the Indian meal moth and in the almond moth. *Science 171*:802–803 (1971).
41. Kuwahara, Y., Kitamura, C., Takahashi, S., Hara, H., Ishii, S., and Fukami, H. Sex pheromone of the almond moth and the Indian meal moth: *cis*-9,*trans*-12-Tetradecadienyl acetate. *Science 171*:801--802 (1971).
42. Kuwahara, Y. and Casida, J. E. Quantitative analysis of the sex pheromone of several phycitid moths by electron-capture gas chromatography. *Agric. Biol. Chem. 37*:681–684 (1973).
43. Read, J. S. and Beevor, P. S. Analytical studies on the sex pheromone complex of *Ephestia cautella* (Walker) (Lepidoptera: Phycitidae). *J. Stored Prod. Res. 12*:55--57 (1976).
44. Read, J. S. and Haines, C. P. The functions of the female sex pheromones of *Ephestia cautella* (Walker) (Lepidoptera, Phycitidae). *J. Stored Prod. Res. 12*:49–53 (1976).
45. Vick, K. W., Coffelt, J. A., Mankin, R. W., and Soderstrom, E. L. Recent developments in the use of pheromones to monitor *Plodia interpunctella* and *Ephestia cautella*. In *Management of Insect Pests with Semiochemicals*. (Mitchell, E. R., ed.). Plenum Press, New York, 1981, pp. 19–40.
46. Brady, U. E. Isolation, identification and stimulatory activity of a second component of the sex pheromone system (complex) of the female almond moth, *Cadra cautella* (Walker). *Life Sci. 13*: 227–235 (1973).

47. Brady, U. E. and Nordlund, D. A. *cis*-9,*trans*-12-Tetradecadien-1-yl acetate in the female tobacco moth *Ephestia elutella* (Hübner) and evidence for an additional component of the sex pheromone. *Life Sci.* *10*:797–801 (1971).
48. Brady, U. E. and Daley, R. C. Identification of a sex pheromone from the female raisin moth, *Cadra figulilella*. *Ann. Entomol. Soc. Am.* *65*:1356–1358 (1972).
49. Sower, L. L., Vick, K. W., and Tumlinson, J. H. (*Z*,*E*)-9-12-Tetradecadien-1-ol: A chemical released by female *Plodia interpunctella* that inhibits the sex pheromone response of male *Cadra cautella*. *Environ. Entomol.* *3*:120–122 (1974).
50. Brady, U. E., Nordlund, D. A., and Daley, R. C. The sex stimulant of the Mediterranean flour moth *Anagasta kuehniella*. *J. Ga. Entomol. Soc.* *6*:215–217 (1971).
51. Kuwahara, Y., Hara, H., Ishii, S., and Fukami, H. The sex pheromone of the Mediterranean flour moth. *Agric. Biol. Chem.* *35*:447–448 (1971).
52. Coffelt, J. A., Vick, K. W., Sonnet, P. E., and Doolittle, R. E. Isolation, identification, and synthesis of a female sex pheromone of the navel orangeworm, *Amyeolois transitella* (Lepidoptera: Pyralidae). *J. Chem. Ecol.* *5*:955–966 (1979).
53. Suzuki, T. 4,8-Dimethyldecanal: The aggregation pheromone of the flour beetles, *Tribolium castaneum* and *T. confusum* (Coleoptera: Tenebrionidae). *Agric. Biol. Chem.* *44*:2519–2520 (1980).
54. Suzuki, T. Identification of the aggregation pheromone of flour beetles *Tribolium castaneum* and *T. confusum* (Coleoptera: Tenebrionidae). *Agric. Biol. Chem.* *45*:1357–1363 (1981).
55. Suzuki, T. and Mori, K. (4*R*,8*R*)-(-)-4,8-Dimethyldecanal: The natural aggregation pheromone of the red four beetle, *Tribolium castaneum* (Coleoptera: Tenebrionidae). *Appl. Entomol. Zool.* *18*:134–136 (1983).
56. Walgenbach, C. A., Burkholder, W. E., Curtis, M. J., and Khan, Z. A. Laboratory trapping studies with *Sitophilus zeamais* (Coleoptera: Curculionidae). *J. Econ. Entomol.* *80*:763–767 (1987).
57. Pinniger, D. B., Stubbs, M. R., and Chambers, J. The evaluation of some food attractants for the detection of *Oryzaephilus surinamensis* (L.) and other storage pests. In *Proceedings of the Third International Working Conference on Stored-Product Entomology*. Kansas State University, Manhattan, Kansas, 1984, pp. 640–650.
58. Hodges, R. J., Halid, H., Rees, D. P., Meik, J., and Sarjono, J. Insect traps tested as an aid to pest management in milled rice stores. *J. Stored Prod. Res.* *21*:215–229 (1985).
59. Nara, J. M., Lindsay, R. C., and Burkholder, W. E. Analysis of volatile compounds in wheat germ oil responsible for an

aggregating response in *Trogoderma glabrum* larvae. *J. Agric. Food Chem.* *29*:68–72 (1981).

60. Freedman, B., Mikolajczak, K. L., Smith, C. R. Jr., Kwolek, W. F., and Burkholder, W. E. Olfactory response of *Oryzaephilus surinamensis* (L.) to extracts from oats. *J. Stored Prod. Res.* *18*:75–82 (1982).
61. Mikolajczak, K. L., Freedman, B., Zilkowski, B., Smith, C. R. Jr., and Burkholder, W. E. Effect of oat constituents on aggregation behavior of *Oryzaephilus surinamensis* (L.). *J. Agric. Food Chem.* *31*:30–33 (1983).
62. Mikolajczak, K. L., Zilkowski, B. W., Smith, C. R. Jr., and Burkholder, W. E. Volatile food attractants for *Oryzaephilus surinamensis* (L.) from oats. *J. Chem. Ecol.* *10*:301–309 (1984).
63. Mikolajczak, K. L., Zilkowski, B. W., Khan, M. Z. A., Smith, C. R. Jr., and Burkholder, W. E. Attractants for *Oryzaephilus surinamensis* (L.) (Coleoptera: Cucujidae): Dimethyl succinate, glutarate, and adipate. *J. Agric. Food Chem.* *33*:1029–1032 (1985).
64. Pierce, A. M., Borden, J. H., and Oehlschlager, A. C. Olfactory response to beetle-produced volatiles and host-food attractants by *Oryzaephilus surinamensis* and *O. mercator*. *Can. J. Zool.* *59*:1980–1990 (1981).
65. Barak, A. V. and Burkholder, W. E. A versatile and effective trap for detecting and monitoring stored-product Coleoptera. *Agric. Ecosystems Environ.* *12*:207–218 (1985).
66. DeCoursey, J. D. A method of trapping the confused flour beetle, *Tribolium confusum* DuVal. *J. Econ. Entomol.* *24*:1079–1081 (1931).
67. Wilson, H. F. Lures and traps to control clothes moths and carpet beetles. *J. Econ. Entomol.* *33*:651–653 (1940).
68. Howe, R. W. Studies on beetles of the family Ptinidae. III. A two-year study of the distribution and abundance of *Ptinus tectus* Boield. in a warehouse. *Bull. Entomol. Res.* *41*:371–394 (1950).
69. Loschiavo, S. R. and Atkinson, J. M. A trap for detection and recovery of insects in stored grain. *Can. Entomol.* *99*:1160–1163 (1967).
70. Burkholder, W. E. Stored-product insect behavior and pheromone studies: Keys to successful monitoring and trapping. In *Proceedings of the Third International Working Conference on Stored-Product Entomology*. Kansas State University, Manhattan, Kansas, 1984, pp. 20–33.
71. Burkholder, W. E. The use of pheromones and food attractants for monitoring and trapping stored-product insects. In *Insect Management for Food Storage and Processing*. (Baur, F. J.,

ed.). American Association of Cereal Chemists, St. Paul, Minn., 1984, pp. 69–85.
72. Leos-Martinez, J., Granovsky, T. A., Williams, H. J., Vinson, S. B., and Burkholder, W. E. Pheromonal trapping methods for the lesser grain borer, *Rhyzopertha dominica* (F.) (Coleoptera: Bostrichidae). *Environ. Entomol. 16*:747–751 (1987).
73. Lippert, G. E. and Hagstrum, D. W. Detection or estimation of insect populations in bulk-stored wheat with probe traps. *J. Econ. Entomol. 80*:601–604 (1987).
74. Barak, A. V. and Burkholder, W. E. Trapping studies with dermestid sex pheromones. *Environ. Entomol. 5*:111–114 (1976).
75. Burkholder, W. E. Application of pheromones for manipulating insect pests of stored products. In *Proceedings Symposium on Insect Pheromones and their Applications.* Nagaoka and Tokyo, December 8--11. (Kono, T. and Ishii, S., eds.). 1976, pp. 111–122.
76. Lingren, B. S., Borden, J. H., Pierce, A. M., Pierce, H. D. Jr., Oehlschlager, A. C., and Wong, J. W. A potential method for simultaneous, semiochemical-based monitoring of *Cryptolestes ferrugineus* and *Tribolium castaneum* (Coleoptera: Cucujidae and Tenebrionidae). *J. Stored Prod. Res. 21*:83–87 (1985).
77. Lindgren, B. S. A multiple funnel trap for scolytid beetles (Coleoptera). *Can. Entomol. 115*:299–302 (1983).
78. Leos-Martinez, J., Granovsky, T. A., Williams, H. J., Vinson, S. B., and Burkholder, W. E. Estimation of aerial density of the lesser grain borer (Coleoptera: Bostrichidae) in a warehouse using dominicalure traps. *J. Econ. Entomol. 79*:1134–1138 (1986).
79. Shapas, T. J. and Burkholder, W. E. Patterns of sex pheromone release from adult females, and effects of air velocity and pheromone release rates on theoretical communication distances in *Trogoderma glabrum. J. Chem. Ecol. 4*:395–408 (1978).
80. Cogan, P. M. and Hartley, D. The effective monitoring of stored product moths using a funnel pheromone trap. In *Proceedings of the Third International Working Conference on Stored-Product Entomology.* Kansas State University, Manhattan, Kansas, 1984, pp. 631–639.
81. Faustini, D. L. Practical aspects of a pheromone trapping program in an industrial environment. *J. Econ. Entomol.* 1989 (in press).
82. Loschiavo, S. R., Wong, J., White, N. D. G., Pierce, H. D. Jr., Borden, J. H., and Oehlschlager, A. C. Field evaluation of a pheromone to detect adult rusty grain beetles, *Cryptolestes ferrugineus* (Coleoptera: Cucujidae), in stored grain. *Can. Entomol. 118:*1–8 (1986).

83. Faustini, D. L., Giese, W. L., Phillips, J. K., and Burkholder, W. E. Aggregation pheromone from the male granary weevil, *Sitophilus granarius* (L.). *J. Chem. Ecol. 8*:679–687 (1982).
84. Burkholder, W. E. and Ma, M. Pheromones for monitoring and control of stored product insects. *Ann. Rev. Entomol. 30*:257–272 (1985).
85. Cross, J. H., Byler, R. E., Silverstein, R. M., Greenblatt, R. E., Gorman, J. E., and Burkholder, W. E. Sex pheromone components and calling behavior of the female dermestid beetle, *Trogoderma variable* Ballion (Coleoptera: Dermestidae). *J. Chem. Ecol. 3*:115–125 (1977).
86. Smith, L. W. Jr., Burkholder, W. E., and Phillips, J. K. Detection and control of stored food insects with traps and attractants: The effect of pheromone-baited traps and their placement on the number of *Trogoderma* species captured. Natick Technical Report TR 83/008: 1983, 13 pp.
87. Shapas, T. J., Burkholder, W. E., and Boush, G. M. Population suppression of *Trogoderma glabrum* by using pheromone luring for protozoan pathogen dissemination. *J. Econ. Entomol. 70*:469–474 (1977).
88. Suss, L. and Trematerra, P. Control of some Lepidoptera Phycitidae infesting stored-products with synthetic sex pheromone in Italy. In *Proceedings of International Conference on Stored Product Protection*. Vol. IV. (Donahaye, E. and Navarro, S., eds.). Tel Aviv, 1986, pp. 606–611.
89. Wohlgemuth, V. R., Reichmuth, C., Rothert, H., and Bode, E. Auftreten vorratsschädlicher Motten der Gattungen Ephestia und Plodia ausserhalb von Lagern und lebensmittelverarbeitenden Betrieben in Deutschland. *Anz. Schädlingskde. Pflanzenchutz Umweltshutz 60*:44–51 (1987).
90. Cogburn, R. R., Burkholder, W. E., and Williams, H. J. Field tests with the aggregation pheromone of the lesser grain borer (Coleoptera: Bostrichidae). *Environ. Entomol. 13*:162–166 (1984).

29

Use of Host Odor Attractants for Monitoring and Control of Tsetse Flies

DAVID R. HALL / Overseas Development Natural Resources Institute, Chatham Maritime, Chatham, Kent, England

I. INTRODUCTION

Tsetse flies, *Glossina* spp. (Diptera: Glossinidae) are hematophagous vectors of trypanosomes that cause sleeping sickness in man and a similar disease, "nagana," in cattle. Tsetse are found only in Africa, but it is estimated that over 11 million km^2 are rendered virtually uninhabitable because of the presence of tsetse, and that over 45 million people are at risk from human trypanosomiasis (1).

There are no vaccinations currently available for trypanosomiasis. Drug treatment is expensive and demands an extensive infrastructure for effective administration. Eradication of the vector provides a long-term solution to the problem, and this is feasible because of the characteristic lifecycle of the tsetse fly. The female tsetse needs to mate only once and then produces an egg approximately every 10 days. The egg and larva are retained in the uterus, and the fully grown larva is deposited in a suitable site. It burrows into the ground to pupate, and the adult fly emerges some 30 days later. Both sexes of the adult fly must feed on a host animal every few days.

In the past, tsetse have been cleared by destruction of vegetation which the flies require as resting sites, and by extermination of alternative host animals. These methods have been replaced by use of insecticides, either ground spraying of residual insecticide or sequential aerosol spraying of nonresidual insecticides from the air (2). Both techniques aim to kill flies soon after they emerge, and this must be maintained for the maximum pupal period to allow all the flies to emerge. However, because of the slow reproductive rate of tsetse, a population can be made to decline toward extinction by causing a low rate of mortality, as low as 1% to 2%/day (3), over a prolonged period, such as might be achieved by a trapping system.

Such trapping systems have been based on attraction of tsetse to the host animals on which they depend for blood meals or to devices designed to simulate host animals (4). In the early 1970s, Challier and Laveissière designed their eponymous biconical trap (5). Flies are attracted by the blue lower cone, enter through openings into the black interior, and are attracted upward by light shining through the upper cone and into a collection cage. These traps have been used extensively throughout Africa for monitoring the presence of a wide range of tsetse species, and they are particularly effective for species of the *G. palpalis* group. The latter are typically confined to linear, riverine habitats, essentially isolated from neighboring infestations, and biconical traps have also been used as effective components in control operations against these species. For control, the trap can be made more effective by spraying it with insecticide, typically a pyrethroid, to kill flies that alight on, but do not enter, the trap. More simply, a blue or blue and black cloth screen impregnated with insecticide can be used, providing a visual attractive stimulus to flies, which pick up a lethal dose of insecticide on contacting the cloth. Insecticide-impregnated traps and screens have been used to eradicate or control tsetse in Ivory Coast, Burkina Faso, and Cameroon (2). More recently, screens at linear spacings of 100 to 200 m. have been used to reduce populations of *G. palpalis palpalis* in Nigeria (6) and of *G. palpalis gambiensis* and *G. tachinoides* in Burkina Faso (7) to levels at which subsequent release of sterile males gave eradication. In the latter, biconical traps were then used to provide a barrier to reinvasion of the 3000 km^2 area cleared of tsetse (8).

Although useful for population monitoring, unbaited traps have proved inadequate for control of savannah species of tsetse fly, which range over a much wider habitat than the riverine species. The barrier of traps at 100-m spacing referred to earlier prevented reinvasion by the riverine species but was totally ineffective against the savannah species, *G. morsitans submorsitans* (8). Alternating traps and screens at a density of 33/km^2 provided an effective barrier to this species, but this density was uneconomical for a large-scale operation (9).

II. IDENTIFICATION OF HOST ODOR ATTRACTANTS FOR TSETSE

Historically, there have been numerous reports of attraction of tsetse to odors of host animals or their residues (10), but it was the work of

Vale, in Zimbabwe, that established the importance of odor in attracting savannah species of tsetse to their hosts and stationary baits (11), and provided a basis for subsequent work on chemical identification of the attractive components. By using electrified nets invisible to tsetse as trapping devices (12), he showed that numbers of tsetse attracted to a simple model were increased by up to 50 times by odors exhausted from an underground pit containing a single ox. By using similar devices, he showed that the Zimbabwe tsetse species, *G. pallidipes* and *G. morsitans morsitans*, are repelled by the odors and upright stance of men (13). Catches of tsetse increased with the weight of host animals used over the range 500 to 11,500 kg, with up to 7100 tsetse caught in 3 hr at the highest dose (14). Recently, even higher catches have been obtained with odor derived from up to 60,000 kg of livestock (Holloway, personal communication). Changes in attractiveness of host animals with diet were also studied, and a fattening diet after a week of starvation often produced an increase in attractiveness (15).

One obvious component of host odors, carbon dioxide, increased catches of tsetse by two to six times at doses between 2.5 and 15 L/min. Whereas synthetic carbon dioxide released at the rate produced by an ox attracted as many nontsetse flies as the total odor from the ox, it attracted only one-quarter as many tsetse flies (16). Acetone released at 0.3 to 30 g/hr increased catches by up to six times, and catches were further increased when both carbon dioxide and acetone were released together (16). A wide range of other chemicals was screened, but none was consistently attractive, although acetophenone and carboxylic acids, particularly hexanoic acid, were shown to be repellent (16). Lactic acid was found in human sweat and probably accounts for some of the repellency of humans (16,17).

Vale also showed, in a field experiment, that removal of the antennae from tsetse reduced the number attracted to acetone and carbon dioxide by 85%, and increased their availability to human baits by three times (18). These results indicated that the receptors for odor attractants and repellents are situated on the antennae, and prompted use of electroantennography (EAG) as a bioassay tool in identification of attractants.

Cattle odor was collected on Porapak resin and analyzed by gas chromatography directly linked to EAG. One minor component of the ox odor gave a strong EAG response from *G. morsitans morsitans*, and was identified as 1-octen-3-ol (19). This compound caused upwind flight of tsetse in a laboratory wind tunnel (20), and doubled the number of tsetse attracted by ox odor in the field (21). Rates of production of acetone and 1-octen-3-ol by cattle were measured, and a mixture of these two compounds with carbon dioxide, all at the natural rate of production, was nearly as attractive as ox odor for *G. morsitans morsitans* but only half as effective for *G. pallidipes* (21). Tenfold increase in the amounts of the synthetic attractants gave catches of tsetse double that with ox odor (21).

Carbon dioxide, acetone, and 1-octen-3-ol also increased the catches of tsetse in traps, as well as at electrified nets. Carbon dioxide increased catches by up to six times, the increase being dose-dependent up to 2 L/min and then flattening out above this dose. Acetone increased catches by six to eight times at release rates of 0.5 to 5 g/hr. 1-Octen-3-ol increased catches by 50% at moderate doses but could reduce catches at high doses. The effects of the three components were multiplicative, so that combination of all three gave increases in catch of up to 64 times the catch in an unbaited trap (22). The number of tsetse caught in a trap is a function of attraction from a distance and stimulation of the flies to enter the trap. These two effects can be separated by surrounding the trap with an incomplete ring of electric nets and comparing the number of flies attracted to the vicinity of the trap with the number actually caught in the trap (23). By using this method, it was found that all three of the above compounds attracted tsetse flies from a distance, and carbon dioxide also increased the trap-entering response (22).

More recent work on attractants has shown that cattle urine is a powerful attractant for savannah species of tsetse, particularly *G. pallidipes* (24,25), and that urine further increases the catches of tsetse in traps already baited with acetone and 1-octen-3-ol (26). Solvent extracts of the urine were also shown to be attractive and, after chromatographic (27) or chemical (28) fractionation of these extracts, the attractiveness was found to be associated with the phenolic fraction. This was shown to contain a mixture of up to eight simple phenols – phenol, 3- and 4-methylphenol, 3- and 4-ethylphenol, 3- and 4-propylphenol, and 2-methoxyphenol (27,28) – all of which, except 2-methoxyphenol, were active by EAG. In a laboratory bioassay, 3- and 4-methylphenol, 3-ethylphenol, 3-propylphenol, and 2-methoxyphenol stimulated upwind flight by *G. morsitans morsitans*, and 3-methylphenol also increased takeoff frequency (28). Field testing showed that none of the phenols was particularly attractive alone, but that combination of 4-methylphenol and 3-propylphenol could increase catches of *G. pallidipes* by up to six times. 2-Methoxyphenol, a very minor component of the urines analyzed, reduced trap catches of both *G. pallidipes* and *G. morsitans morsitans* (29).

Mixtures of 4-methylphenol and 3-propylphenol also increased by three to four times the numbers of *G. pallidipes* attracted to simple cloth screens covered by electrified nets and already baited with acetone and 1-octen-3-ol. This increase is slightly less than that with traps, indicating that the phenols also have an effect on increasing trap-entering responses (29).

The phenols seem to account for most of the attractiveness of the urine, but some minor components of the neutral fraction were found to cause intense EAG responses from *G. morsitans morsitans* and *G. pallidipes*. These were identified as indole, 3-methylindole, and two carotenoid metabolites, isomers of 3,3,5-trimethyl-4-hydroxy-4-(3-oxobutyl)cyclohexanone (30). No attractiveness has yet been demonstrated for these compounds in the field under a variety of conditions.

Acetone and 1-octen-3-ol have been shown to increase catches of *G. morsitans submorsitans* in biconical traps in Burkina Faso (27). Combining the two gave an even greater effect, increasing catches by 6.7 times in the wet season but only 2.8 times in the dry season, when traps were more visible and olfactory cues less important.

Most of these attractants have been tested against the riverine tsetse species *G. palpalis palpalis* in Ivory Coast, but none have proved consistently attractive, even carbon dioxide (Laveissière, personal communication). However, in Burkina Faso, it has been shown that the riverine species *G. tachinoides* is attracted by odors of host animals (31). Carbon dioxide (32) and a combination of 1-octen-3-ol and the phenols found in cattle urine have been identified as attractants for this species (Mérot, personal communication).

III. MONITORING

Of the known tsetse attractants, carbon dioxide at doses of several liters per minute is inconvenient and expensive to use on a large scale. However, baiting traps with acetone released at 100 mg/hr, 1-octen-3-ol at 0.5 mg/hr, 4-methylphenol at 1.5 mg/hr, and 3-propylphenol at 0.2 mg/hr can increase catches of *G. pallidipes* by 20 to 30 times, making the traps much more sensitive monitoring devices than traditional ox rounds for this species. Addition of odors also increases catches of *G. morsitans morsitans*, but this species responds very poorly to stationary baits, and the ox round is still the better method for this species.

Studies of the trap-orientated behavior of the savannah tsetse species, *G. pallidipes* and *G. morsitans morsitans*, in Zimbabwe (33) led to design of a trap that was twice as effective at capturing tsetse as the biconical trap and minimized the capture of nontsetse species (34). This "beta" trap was triangular in plan, white on the outside with an entrance and black target at the base to focus attracted flies. The inside was white at the base and black at the top with a cone of netting leading upward into a retaining cage. This trap was subsequently modified so that it was square and collapsible for easy transport (35). This "F2" trap was three times more effective than a beta trap for *G. pallidipes*.

Laboratory and field studies of the responses of tsetse to different colors defined by their spectral reflectivities (36,37) showed that blue and black were more attractive colors than white. Black also encouraged alighting responses, which decreased the numbers of flies entering a trap but increased the number contacting a simple screen. Thus, the color giving highest trap catches is a royal blue reflecting blue-green strongly but little ultraviolet or green-yellow-orange (37). Blue F2 traps are now in use, known as "F3" traps. A design based on the F3 trap but more easily made up from materials locally available in Kenya has been produced (38).

Such traps are now in use in many African countries for monitoring the presence of tsetse in surveys before and after control operations. Current dispensers for the synthetic attractants are sealed polythene sachets containing 4 ml of a 4:8:1 mixture of 1-octen-3-ol, 4-methylphenol, and 3-propylphenol, and a bottle with 2-mm diameter aperture for the acetone. Work is in progress to develop sealed dispensing systems for acetone or a substitute such as butanone. Alternatively, natural cow urine in an open bottle can be used in place of the phenols (39).

A particularly impressive use of the power of odor-baited traps was in testing the effectiveness against tsetse of various aerially sprayed insecticides on small plots. Marked flies were released at the center of a plot 1.8 km × 1.8 km and, when using traps baited with carbon dioxide and acetone in and around the plot, 9.3% of the *G. pallidipes* released could be captured during the period 22 to 110 hr later. Aerosol application of endosulfan between 1 and 4 hr after release reduced the cumulative recapture rate to 0.3%, indicating 97% mortality. Because of the high recapture rate, this provided a precise estimate of the effectiveness of the insecticide. The use of marked flies in this way made it possible to use very small plots for this assessment, greatly reducing the cost compared with that of conventional trials that use the natural population of unmarked flies on very large plots to minimize the effects of immigration of flies into the sprayed area (40).

IV. CONTROL

Although unbaited traps and simple insecticide-impregnated screens have been used for control of riverine species of tsetse, it is only since the development of powerful attractants that it has been economically feasible to apply these concepts to control of savannah species (4). It has been shown, theoretically, that the most efficient way of reducing a tsetse population is by trapping, sterilizing, and releasing both sexes back into the population. This is more effective than simply killing attracted flies, especially under conditions of high population growth or low trapping rates (3). Analysis of catches of *G. pallidipes* in traps baited with the odor of 13,000 kg of cattle indicated that these could trap 2.5% of the available population at a density of only one per 10 km^2. If these were sterilized and released, this should cause a population reduction of 95%/yr (41). Initial attempts to exploit this involved complex, automatically operated aerosol chemosterilant chambers attached to traps (42), and the use of decoys dosed with the tsetse contact pheromone and a chemosterilant was reported (43). More recently, the chemosterilant effect of bisazir vapor has been used in much simpler devices attached to standard F3 traps (44), and new, more-effective and safer chemosterilants are being studied (Langley, personal communication).

In the absence of suitable technology, the most significant control operations to date have been based on traps and insecticide-impregnated screens or "targets" baited with odors. Starting in 1981, the effects of various trapping devices were tested using artificially introduced populations of *G. pallidipes* and *G. morsitans morsitans* on an island of 4.5-km^2 area in Lake Kariba, Zimbabwe. In a first phase of 9 months, four traps fitted with automatic capture/sterilize/release devices (42) were baited with carbon dioxide at 2 L/min and acetone at 5 g/hr and run for the 3 hr before sunset. In a second phase, the sterilizing devices were then replaced by simple retaining cages and, after 8 months, in a third phase, the traps were run also for 3 hr after sunrise. After 8 months of the third phase, population levels of *G. pallidipes* were estimated to be decreasing by 3%/day and those of *G. morsitans morsitans* by 0.3%/day, and populations of the two species were estimated to have declined to 1% and 10% of their original levels, respectively.

In a fourth phase, the traps were replaced by 20 insecticide-impregnated screens or "targets" baited with acetone at 100 mg/hr and 1-octen-3-ol at 0.5 mg/hr. The targets consisted of 1 m^2 of black cloth and 0.5 m^2 of terylene netting on a frame pivotting about a vertical axis, with a PVC roof to protect the insecticide from sunlight and rain. The targets were sprayed twice with an effective dose of approximately 12 g of dieldrin at 3-month intervals during the dry season and then twice with 0.3 g of deltamethrin at 4-month intervals during the rainy season. Previous studies (46) had shown that such levels of dieldrin were active for over 140-days exposure to sunlight but were washed off by rainfall, whereas deltamethrin formulations were rainfast but were inactivated by prolonged exposure to sunlight. After installation of the targets, the remaining population of *G. pallidipes* disappeared after 11 weeks, and no more *G. morsitans morsitans* were found after 9 months. Estimated rates of population reduction were 9%/day and 1.9%/day, respectively for the two species.

The success of this experiment prompted a large-scale field trial with a natural population. Targets were used rather than traps because they are cheaper to build and maintain, and they are more effective for *G. morsitans morsitans*. Furthermore, whereas carbon dioxide is important for optimal operation of the traps because it increases trap-entering responses, it is less important with the targets that were baited with acetone and 1-octen-3-ol only (47).

The trial area was a 600-km^2 triangle of bush, bordered by the Zambesi river, the Zambesi escarpment, and a road. From March to May 1984, a barrier of 1200 targets was placed around the area at intervals of 100 m. Between May and July, a further 1227 traps were deployed in a second barrier on the two sides defined by the escarpment and the road at 150-m intervals and within the area along a network of tracks at 300-m intervals. From September to October, a further 471 targets were placed in situations identified as particularly good habitats for tsetse, giving an overall mean density of four targets per square kilometer. Tsetse populations in the area were assessed by comparison of catches at baited traps, electrified nets, and oxen in the treated area, with catches at the research station approximately 25-km outside the area. Before the start of the operation, population levels were similarly high inside and outside the area.

Tsetse population started to decline soon after installation of the targets began, and it was estimated that *G. morsitans morsitans* and *G. pallidipes* populations were being reduced at rates of 2% and 10%, respectively. By early 1986, tsetse had disappeared within the treated area, 5 to 15 km from the main invasion front. From zero to 5 km inside this front, populations were only 1% of those outside the area, and only upon going 5 to 10 km outside the area did populations reach the control level.

The targets were fully maintained until July 1985 with respraying of insecticide, replenishment of odors, and repair every 2 to 3 months. After this time, the targets were allowed to deteriorate to study the effects of reduced maintenance on effectiveness and the rate of reinfestation from surrounding areas. A test herd of cattle placed in the area in May stayed trypanosomiasis-free until September when monitoring traps also detected the first reinvading flies.

It was found that much simpler targets consisting of a 1 m^2 of black cloth flanked by 0.5 m^2 of black netting were twice as effective as the previous design. The roof used to protect the insecticide deposit on the target actually discouraged tsetse from visiting the target, and it was dispensed with when it was found that the deltamethrin could be made much more persistent in sunlight by incorporating a black dye and 2-hydroxy-3-methoxybenzophenone-5-sulfonic acid as a UV screener. The latter two additives also prevented fading of the black cloth, which markedly reduced attractiveness (47).

Since this operation, the effectiveness of the targets has been further increased, particularly for *G. pallidipes*, by addition of 4-methylphenol and 3-propylphenol to the odor bait. Targets baited with acetone and polythene sachets containing 1-octen-3-ol and the two phenols have been used in several further operations in Zimbabwe, including one in which tsetse were effectively eradicated in a 1300 km^2 block bounded by areas treated by aerial and ground spraying of insecticide.

In Western Province, Zambia, a trial area of 500 km^2 has been treated with targets baited with acetone and 1-octen-3-ol, at a density of four per square kilometer, against *G. morsitans centralis*. The tsetse population in the center of the area had collapsed 4 months after completion of installation of the targets. The treated area is being extended to 2000 km^2 and the phenols will be incorporated into the baits (Willemse, personal communication).

In Kenya, local tribesmen have been involved in a project to clear *G. pallidipes* from the Nguruman valley (Dransfield, personal communication). Locally constructed traps (38) are being used, baited with acetone and natural cattle urine, thus avoiding the use of foreign exchange for purchase of chemicals. In an initial trial, traps have been deployed over 100 km^2 at a density of one per square kilometer, and the tsetse population has been reduced by 99%. This area is being enlarged to counter the effects of flies immigrating from surrounding areas.

V. CONCLUSIONS

The demonstration that savannah species of tsetse fly are strongly attracted by the odor of host animals, and the identification of some of the components responsible for this attraction, have made possible development of a technology based on odor-baited traps and insecticide-impregnated targets for monitoring and control of these savannah species. This technology is still being improved. Although acetone, 1-octen-3-ol, 4-methylphenol, and 3-propylphenol greatly increase the attractiveness of traps and targets to several species of tsetse, there are indications that other attractants remain to be identified (21). Further improvements in the effectiveness and economy of the traps and targets can be envisaged, and simpler systems for catching, sterilizing, and releasing flies are under investigation. Nevertheless, the existing technology constitutes a valuable complement to control methods based on ground spraying of residual insecticide and aerial spraying of nonpersistent insecticides, and has many advantages.

Ground spraying is restricted to certain seasons of the year and makes heavy demands on labor and organization. This generally has to be repeated in successive years to prevent reinfestation. Aerial spraying has proved successful in terrain suited to low flying, but application is costly and requires highly trained staff operating to a rigidly defined schedule to cover pupal emergence. Eradication has been achieved in certain situations, but small areas left incompletely treated can provide population nuclei for reinfestation. On the other hand, the installation of traps and targets can be carried out by relatively untrained staff without the need to keep to rigorous timetables and there is an even demand for labor throughout most of the year. Mistakes can be rectified subsequently without jeopardizing the whole operation and, once installed, traps and targets can be used to maintain an area tsetse-free or as a barrier to reinvasion. The amounts of insecticide used on the targets are small and applied to discrete, sparsely distributed, man-made structures, so environmental contamination is minimal. Furthermore, the odors employed as baits and the designs of the traps and targets have proved to be very selective for tsetse (47).

The actual relative costs of the different control measures depend very much on local circumstances and availabilities of labor and equipment, but costs of control operations, to date, have been estimated to be between those of ground spraying and the more expensive aerial spraying. Problems still remain in the installation and maintenance of targets in poorly accessible terrain. It has often been difficult to ensure regular maintenance of the targets with replenishment of the odor baits and insecticide impregnation, clearing of vegetation, and protection from animals and bushfires. Improving the effectiveness of the individual targets, particularly by increasing the attractiveness of the baits used, would mean that fewer targets per unit area are required to be set out and maintained, and less accessible areas might be treated with targets around the perimeter, thus reducing the cost and increasing the practicability of control operations even further.

REFERENCES

1. Hagan, D. H. and Wilmshurst, E. C. The disease spread by tsetse. *World Crops* pp. 248–250 (1975).
2. Allsopp, R. Control of tsetse flies (Diptera: Glossinidae) using insecticides: A review and future prospects. *Bull. Entomol. Res. 74*:1–23 (1984).
3. Langley, P. A. and Weidhaas, D. Trapping as a means of controlling tsetse, *Glossina* spp. (Diptera: Glossinidae): The relative merits of killing and sterilization. *Bull. Entomol. Res. 76*: 89–95 (1986).
4. Vale, G. A., Bursell, E., and Hargrove, J. W. Catching-out the tsetse fly, *Parasitol. Today 1*:106–110 (1985).

5. Challier, A. and Laveissière, C. Un nouveau piège pour la capture des glossines: Description et essais sur le terrain. *Cah. ORSTOM. Sér. Entomol. Méd. Parasitol.* *11*:251–262 (1973).
6. Takken, W., Oladunmade, M. A., Dengwat, L., Feldmann, H. U., Onah, J. A., Tenabe, S. O., and Hamann, H. J. The eradication of *Glossina palpalis palpalis* (Robineau-Desvoidy) (Diptera: Glossinidae) using traps, insecticide-impregnated targets and the sterile insect technique in central Nigeria. *Bull. Entomol. Res.* *76*:275–286 (1986).
7. Politzar, H. and Cuisance, D. An integrated campaign against riverine tsetse, *Glossina palpalis gambiensis* and *Glossina tachinoides*, by trapping and the release of sterile males. *Insect Sci. Appl.* *5*:439–442 (1984).
8. Politzar, H. and Cuisance, D. A trap-barrier to block reinvasion of a river system by riverine tsetse species. *Rev. Elev. Méd. Vét. Pays Trop.* *36*:364–370 (1983).
9. Politzar, H. and Merot, P. Attraction of the tsetse fly *Glossina morsitans submorsitans* to acetone, 1-octen-3-ol, and the combination of these compounds in West Africa. *Rev. Elev. Méd. Vét. Pays Trop.* *37*:468–473 (1984).
10. Turner, D. A. Olfactory perception of live hosts and carbon dioxide by the tsetse fly *Glossina morsitans orientalis* Vanderplank. *Bull. Entomol. Res.* *61*:75–96 (1971).
11. Vale, G. A. The responses of tsetse flies (Diptera, Glossinidae) to mobile and stationary baits. *Bull. Entomol. Res.* *64*:545–588 (1974).
12. Vale, G. A. New field methods for studying the responses of tsetse flies (Diptera, Glossinidae) to hosts. *Bull. Entomol. Res.* *64*:199–208 (1974).
13. Vale, G. A. The interaction of men and traps as baits for tsetse flies (Diptera: Glossinidae). *Zimbabwe J. Agric. Res.* *20*:179–183 (1982).
14. Hargrove, J. W. and Vale, G. A. The effect of host odour concentration on catches of tsetse flies (Glossinidae) and other Diptera in the field. *Bull. Entomol. Res.* *68*:607–612 (1978).
15. Vale, G. A. An effect of host diet on the attraction of tsetse flies (Diptera: Glossinidae) to host odour. *Bull. Entomol. Res.* *71*:259–265 (1981).
16. Vale, G. A. Field studies of the responses of tsetse flies (Glossinidae) and other Diptera to carbon dioxide, acetone, and other chemicals. *Bull. Entomol. Res.* *70*:563–570 (1980).
17. Vale, G. A. Field responses of tsetse flies (Diptera: Glossinidae) to odours of men, lactic acid and carbon dioxide. *Bull. Entomol. Res.* *69*:459–467 (1979).

18. Vale, G. A. The role of the antennae in the availability of tsetse flies (Diptera: Glossinidae) to population sampling techniques. *Trans. Zimbabwe Sci. Assoc.* *61*:33–40 (1982).
19. Hall, D. R., Beevor, P. S., Cork, A., Nesbitt, B. F., and Vale, G. A. 1-Octen-3-ol: A potent olfactory stimulant and attractant for tsetse isolated from cattle odours. *Insect Sci. Appl.* *5*:335–339 (1984).
20. Bursell, E. Effects of host odour on the behaviour of tsetse. *Insect Sci. Appl.* *5*:345–349 (1984).
21. Vale, G. A. and Hall, D. R. The role of 1-octen-3-ol, acetone and carbon dioxide in the attraction of tsetse flies, *Glossina* spp. (Diptera: Glossinidae), to ox odour. *Bull. Entomol. Res.* *75*:209–217 (1985).
22. Vale, G. A. and Hall, D. R. The use of 1-octen-3-ol, acetone, and carbon dioxide to improve baits for tsetse flies, *Glossina* spp. (Diptera: Glossinidae). *Bull. Entomol. Res.* *75*:219–231 (1985).
23. Vale, G. A. and Hargrove, J. W. A method of studying the efficiency of traps for tsetse flies (Diptera: Glossinidae) and other insects. *Bull. Entomol. Res.* *69*:183–193 (1979).
24. Owaga, M. L. A. Preliminary observations on the efficacy of olfactory attractants derived from wild hosts of tsetse. *Insect Sci. Appl.* *5*:87–90 (1984).
25. Owaga, M. L. A. Observations on the efficacy of buffalo urine as a potent olfactory attractant for *Glossina pallidipes* Austen. *Insect Sci. Appl.* *6*:561–566 (1985).
26. Vale, G. A., Flint, S., and Hall, D. R. The field responses of tsetse flies, *Glossina* spp. (Diptera: Glossinidae), to odours of host residues. *Bull. Entomol. Res.* *76*:685–693 (1986).
27. Hassanali, A., McDowell, P. G., Owaga, M. L. A., and Saini, R. K. Identification of tsetse attractants from excretory products of a wild host animal, *Syncerus caffer*. *Insect Sci. Appl.* *7*:5–9 (1986).
28. Bursell, E., Gough, A. J. E., Beevor, P. S., Cork, A., Hall, D. R., and Vale, G. A. Identification of components of cattle urine attractive to tsetse flies, *Glossina* spp. (Diptera: Glossinidae). *Bull. Entomol. Res.* *78*:281–291 (1988).
29. Vale, G. A., Hall, D. R., and Gough, A. J. E. The olfactory responses of tsetse flies, *Glossina* spp. (Diptera: Glossinidae), to phenols and urine in the field. *Bull. Entomol. Res.* *78*:293–300 (1988).
30. Gough, A. J. E., Hall, D. R., Beevor, P. S., Cork, A., Bursell, E., and Vale, G. A. Attractants for tsetse from cattle urine. In *Proceedings of the 19th Meeting of the International Scientific Council for Trypanosomiasis Research and Control.* Lome, Togo, 1987.

31. Mérot, P., Galey, J. B., Politzar, H., Filledier, J., and Mitteault, A. Pouvoir attractif de l'odeur des hôtes nourriciers pour *Glossina tachinoides* en zone soudanoguinéene (Burkina Faso). *Rev. Elev. Med. Vet. Pays Trop.* *39*:345–350 (1986).
32. Galey, J. B., Mérot, P., Mitteault, A., Filledier, J., and Politzar, H. Efficacité du dioxyde de carbone comme attractif pour *Glossina tachinoides* en savane humide d'Afrique de l'Ouest. *Rev. Elev. Med. Vet. Pays Trop.* *39*:351–354 (1986).
33. Vale, G. A. The trap-orientated behaviour of tsetse flies (Glossinidae) and other Diptera. *Bull. Entomol. Res.* *72*:71–93 (1982).
34. Vale, G. A. The improvement of traps for tsetse flies (Glossinidae). *Bull. Entomol. Res.* *72*:95–106 (1982).
35. Flint, S. A comparison of various traps for *Glossina* spp. (Glossinidae) and other Diptera. *Bull. Entomol. Res.* *75*:529–534 (1985).
36. Green, C. H. and Flint, S. An analysis of colour effects in the performance of the F2 trap against *Glossina pallidipes* Austen and *G. morsitans morsitans* Westwood (Diptera: Glossinidae). *Bull. Entomol. Res.* *76*:409–418 (1986).
37. Green, C. H. Effects of colours and synthetic odours on the attraction of *Glossina pallidipes* and *G. morsitans morsitans* to traps and screens. *Physiol. Entomol.* *11*:411–421 (1986).
38. Brightwell, R., Dransfield, R. D., Kyorku, C., Golder, T. K., Tarimo, S. A., and Mungai, D. A new trap for *Glossina pallidipes*. *Trop. Pest Manage.* *33*:151–159 (1987).
39. Dransfield, R. D., Brightwell, R., Chaudhury, M. F., Golder, T. K., and Tarimo, S. A. R. The use of odour attractants for sampling *Glossina pallidipes* Austen (Diptera: Glossinidae) at Nguruman, Kenya. *Bull. Entomol. Res.* *76*:607–619 (1986).
40. Vale, G. A., Hursey, B. S., Hargrove, J. W., Torr, S. J., and Allsopp, R. The use of small plots to study populations of tsetse (Diptera: Glossinidae). Difficulties associated with population dispersal. *Insect Sci. Appl.* *5*:403–410 (1984).
41. Hargrove, J. W. and Vale, G. A. Aspects of the feasibility of employing odour-baited traps for controlling tsetse flies (Diptera: Glossinidae). *Bull. Entomol. Res.* *69*:283–290 (1979).
42. House, A. P. R. Chemosterilisation of *Glossina morsitans morsitans* Westwood and *G. pallidipes* Austen (Diptera: Glossinidae) in the field. *Bull. Entomol. Res.* *72*:65–70 (1982).
43. Langley, P. A., Coates, T. W., Carlson, D. A., Vale, G. A., and Marshall, J. Prospects for autosterilisation of tsetse flies, *Glossina* spp. (Diptera: Glossinidae), using sex pheromone and bisazir in the field. *Bull. Entomol. Res.* *72*:319–327 (1982).
44. Hall, M. J. R. and Langley, P. A. Development of a system for sterilising tsetse flies, *Glossina* spp., in the field. *Med. Vet. Entomol.* *1*:201–210 (1987).

45. Vale, G. A., Hargrove, J. W., Cockbill, G. F., and Phelps, R. J. Field trials of baits to control populations of *Glossina morsitans* Westwood and *G. pallidipes* Austen (Diptera: Glossinidae). *Bull. Entomol. Res.* *76*:179–193 (1986).
46. Torr, S. J. The susceptibility of *Glossina pallidipes* Austen (Diptera: Glossinidae) to insecticide deposits on targets. *Bull. Entomol. Res.* *75*:451–458 (1985).
47. Vale, G. A., Lovemore, D. F., Flint, S., and Cockbill, G. F. Odour-baited targets to control tsetse flies, *Glossina* spp. (Diptera: Glossinidae), in Zimbabwe. *Bull. Entomol. Res.* *78*: 31–49 (1988).

30

The Use of Pheromones and Other Attractants in House Fly Control

LLOYD E. BROWNE / Ecogen, Inc., Langhorne, Pennsylvania

I. INTRODUCTION

How useful are attractants in present day fly control technologies, and are they occupying a significant niche in the market place? The adult housefly *Musca domestica* is still a nuisance in almost all of the industrialized parts of the world, despite the extensive use of exclusion devices, such as screens and air-conditioning equipment; despite improved sanitation practices to eliminate fly-breeding areas; and despite an arsenal of insecticides. Although these factors greatly reduced the numbers of flies we encounter in our daily lives, there is still an occasional need for a flyswatter in the home. Furthermore, animal husbandmen are constantly seeking improved methods of fly control, as are managers of food-handling establishments and public places. In short, there appear to be ample markets for better fly control techniques. Over the past 20 years, researchers have revealed the role of several hydrocarbons isolated from adult flies that stimulate various aspects of mating behavior and, for the previous century or more, a considerable amount of research has been conducted on various concoctions that attract flies.

Fly attractants and fly traps are not new concepts. Snetsinger and Shetlar (1), in a very interesting history of animal traps, found references to sticky traps before the eighteenth century and mentioned cone traps baited with fermenting materials that were patented in the 1850s. Richardson, in 1916 and 1917 publications (2,3), reported on flies attracted to ammonia and fermentation products.

Brown and co-workers (4) reported on a number of chemical attractants for the adult housefly, including an enhanced attraction to multicomponent lures containing ethanol, skatole, acetal, and malt. Likewise, compounds that stimulate sexually oriented behavior have been described for some time. The chemical structure of (*Z*)-9-tricosene, the sex pheromone of the housefly, was reported in 1971 by Carlson et al. (5), and field evaluations of its attractiveness were reported in 1973 by Carlson and Beroza (6).

A patent disclosing the insecticidal composition containing (*Z*)-9-tricosene and methomyl was issued to Krinzer and McDaniel in 1978 (7). Data in this patent demonstrated that when the pheromone was incorporated into a sugar–plus–1% methomyl bait, 7 to 14 times as many dead flies were found in the bait-treated areas compared with similar areas treated with the sugar–methomyl bait without pheromone. A similar formulation under this patent has been marketed since 1973 under the trade name Golden Malrin by Starbar, a division of Zoecon Corp. The pheromone was listed on the label and was the first pheromone to be registered on an insecticide label with the United States Environmental Protection Agency (EPA). This abbreviated review of the literature shows that an impressive number of chemicals have the demonstrated ability to attract flies and that this information has been available for some time.

I will limit my discussion to those kinds of devices using fly attractants that are commercially available in the United States. I exclude a discussion of repellents, except for one brief observation that follows. Permanently installed automatic insecticide aerosol systems are used to control flies in various buildings. These devices, with the use of a timer, release short bursts of fogged insecticide into the interior air space of buildings, such as animal quarters, or into the air ducts of air-conditioned buildings. The insecticides are usually pyrethrum or synthetic pyrethroids, often mixed with a synergist such as piperonyl butoxide. I have been told that this synergist, a compound with little insecticidal value but quite repellent, was as effective alone, when used with timed aerosol releases, as it was in admixture with the insecticides in ridding dairy facilities of flies. I have not verified this effect, but I have noted the extreme fly-repellent behavior stimulated by both of these insecticides and the synergist. Repellency aside, I know of no attractant compound that has been used alone to manipulate fly populations. Attractants are combined with a population-control agent, usually a poison bait, as previously mentioned, or a trap. The July 1987 issue of *Sunset Magazine* (8) featured an article on fly control in which many of the noninsecticidal devices were shown. The devices shown were cone traps, jar traps, a disposable bag trap, an electric zapper, and sticky traps. With the single exception of the old-fashioned flypaper, all of these devices used some type of lure. Lures ranged from pictures of flies clustered together on the surface of sticky traps, to traps with instructions to provide your own raw meat bait. Other traps used UV lamps to attract flies to electrocution grids, and several used mixtures of ammonium carbonate and yeast. Many of these devices are limited in their distribution and are either available only through direct order from the manufacturer or through catalog sales. Cone traps have traditionally been merchandized in this manner. Jar traps have enjoyed a somewhat wider distribution through agriculture and hardware outlets. Sticky trap devices of various configurations and attraction claims have, from time to time, appeared in the mass-oriented household pesticide markets.

II. FLY-KILLING DEVICES USING ATTRACTANTS

A. Sticky Traps

Sticky traps or flypaper probably represent the first fly-catching devices. For most of their history, these traps only lured flies by providing them an ideal resting place. More recently, sticky traps have become available with attractants. J. T. Eaton and Co., Inc. and Farnam Companies, Inc. sell traps that include sex attractant baits formulated on large-grain sugar. Some attractant claims in products that I have tested give real meaning to the term *caveat emptor*, or "buyer beware." One product in particular I found to be exceptionally poor at trapping flies is no longer on the market. Several producers of sticky traps have incorporated visual cues into their trap designs. They are producing traps with rigid sticky surfaces, such as coated cardboard tubes, that are cleaner and easier to handle than flypaper. Many of these rigid type traps have pictures of clusters of flies on the sticky surface. I have noted that when these devices are first placed in use, the first flies caught are caught on or near these pictures. Sticky traps have the advantages of being generally cheap to manufacture, easy to use, and are safe. They have the disadvantage of not being highly effective in population control, and they are quite unsightly. Most products of this nature do not use attractive odors.

B. Jar Traps

"Jar trap" is a general term that refers to a trap consisting of a large jar, usually partially filled with liquid, to kill flies, and it is equipped with a special lid. This lid usually consists of a convex dome, darkened on the underside. Flies land on the upper surface of this dome and crawl to the underside and down a tube or inverted cone into the jar where they become trapped. This behavior goes against the flies' better judgment; thus attractants must be used. Some traps instruct the user to add decaying meat to the jar to provide attraction. These traps are also often called "stinky" traps. In the past few years, at least one manufacturer of jar traps, Farnam Companies Inc., has provided a small jar of attractive liquid with each of their Fly Terminator traps. This lure is a four-component mixture including (Z)-9-tricosene. These traps attract many more flies than they catch.

C. Cone Traps

Cone traps usually refer to traps made from screen shaped into a cone with a small hole at the apex and legs at the base to lift the device a few centimeters above the floor or other surface upon which it is placed. The apex hole is surrounded by an outer container that traps the flies as they fly up from the surface below to the under side of the cone and crawl through the hole. At least two companies, Beneficial BioSystems and Spalding Laboratories, sell these types of traps that are baited with an ammonium carbonate—yeast bait. A new product, a highly modified cone trap designed to be suspended from above, is now on the market. This trap incorporates several novel features. In addition to the side-entry ports, the trap is baited with a controlled-release packet of fermenting materials and a separate patch releasing (Z)-9-tricosene. The trap has a disposable plastic bag attached to the bottom to hold trapped flies in a manner similar to the design of the popular Japanese beetle traps. From the compactness of the design and the packaging, it appears that this trap is designed to appeal to home and garden market. The objectionable odor of the BioLure will probably prevent this product from developing a market in the household pesticide area. The trap is manufactured by Consep Membranes, Inc. and marketed by Safer, Inc.

D. Electrocution Traps

Electric "zapper" traps are popular insect-killing devices that represent one of the major growth areas of the various consumer insecticide markets. These devices are enjoying good sales for industrial uses especially in food-processing and handling facilities. Electrocution traps usually provide good customer satisfaction because of the sound effects emitted during the killing process and the resulting accumulations of dead insects. These elctrocutors are particularly useful for fly control in food production areas because the kill is contained in the device and is prevented from falling into the food. There are numerous manufacturers of electrocution traps, all of which use UV light as an attractant. However, one brand, Flowtron, sells a plastic strip containing (Z)-9-tricosene that is manufactured by Hercon Division of Health-Chem Corp. and is used for baiting electrocution traps.

E. Poison Baits

Poison baits, usually in a granular form, have been popular in fly control since the advent of organic phosphate insecticides during the 1950s. Insecticides that give rapid knockdown, causing large numbers of flies to die near the baited areas, give good consumer satisfaction whether or not they control fly problems. Selection pressure on fly populations by the extensive use of scatter baits containing methomyl, a restricted-use carbamate insecticide, has resulted in significant resistance to this insecticide in fly populations, especially in the eastern part of the United States. In my own work recently, I have measured knockdown times, (knockdown time being defined as the lapsed time from the onset of feeding until the beginning of uncontrolled spasms), in fly populations in California, Arkansas, Florida, Georgia, South Carolina, and Pennsylvania. I found knockdown times for a laboratory-reared pesticide-susceptible strain of flies to be 45 sec when presented a sugar bait laced with 2% technical methomyl. The knockdown time for flies at several dairies in California was approximately 90 sec. At the eastern United States sites, average knockdown times varied considerably but ranged from 2 to 8 min. What was more noticeable between the western flies and those in the east, was the feeding behavior. Susceptible flies sustained feeding, whereas those that exhibited longer knockdown times displayed signs of methomyl avoidance. If methomyl was in the bait, those resistant flies would break proboscis contact with the bait surface within 1 to 3 sec. Methomyl is not used in fly control except as scatter baits. Therefore, the development of avoidance behavior to methomyl is indicative of its widespread use.

A second concept in the use of poison baits is that of the Synerid Fly Control B product produced by Hilton-Davis. This product uses the lure pack, BioLure, manufactured by Consep Membranes, Inc., attached to a large reservoir of sugar—water containing erythrosin B, a blood stain that has been found to be insecticidal when the insect having ingested it is exposed to strong light.

III. CONCLUSIONS

In the United States, scatter baits containing 1% methomyl and the pheromone (*Z*)-9-tricosene represent, by a considerable increment, the greatest use of pheromones in fly control products. The unique methomyl bait avoidance mechanism that is now characteristic of many fly populations in the eastern United States, and which was not detected in fly populations in these same areas 6 years ago, is testimony to the scale of methomyl-based, scatter-bait use.

In my opinion, traps using attractive chemical lures, until recently, have not played a significant role in fly control. This is because they have not been very effective. They have been sold as gimmicks and not as serious products. None carry label instructions for the proper deployment and trap density to be used to achieve population control under various conditions. In fact, I know of no published data for which a trap-out strategy to control fly populations has been demonstrated. I believe this situation will soon change because several companies are actively engaged in the development of improved lures and killing devices.

I believe strongly that there is a need for a cheap and efficacious alternative to broadcast insecticides to reduce the numbers of adult houseflies. The use of methomyl scatter baits, a product already highly restricted in where it can be used, will decline in the future as methomyl resistance intensifies. This resistance, in conjunction with resistance to other insecticides, leaves a void in the arsenal of fly-control technologies around fly-breeding sites. The use of expensive electrocution traps in commercial areas indicates a market niche for a cheaper device that would offer similar efficacy in trapping flies and holding the carcasses. The household pesticide market consists, almost exclusively, of flyswatters and insecticide spray cans. By conservative estimate, these three markets offer a potential annual sales in excess of 100 million dollars. This market potential will not go unexploited forever. But to develop a good product, more technology than just the knowledge of attractive compounds is needed. Odor-release technology, fly-killing technology, product safety, cleanliness, aesthetics, and use-pattern efficacies, all pose difficult problems to be solved.

APPENDIX

A partial list of companies selling products containing fly attractants in the United States

Beneficial Biosystems
Box 8461
Emeryville, CA 94662

Consep Membranes, Inc.
P.O. Box 6059
Bend, OR 97708

Farnam Companies, Inc.
P.O. Box 12068
Omaha, NE 68112

Hercon Division
Health-Chem Corp.
200B Corporate Court
S. Plainfield, NJ 07080

Hilton-Davis
P.O. Box 37869
Cincinnati, OH 45222

J. T. Eaton & Co., Inc.
Twinsburg, OH 44087

Safer, Inc.
60 William Street
Wellesley, MA 02181

Spalding Laboratories
760 Printz Road
Arroyo Grande, CA 93420

This partial list of companies is provided for the benefit of the reader and does not constitute an endorsement of any product.

REFERENCES

1. Snetsinger, R. and Shetlar, D. J. Traps for Animal Pest Control. *Melsheimer Entomol. Ser. 32*:12--18 (1982).
2. Richardson, C. H. The response of the house fly to ammonia and other substances. *N. J. Agric. Exp. Sta. Bull.* 292 (1916).
3. Richardson, C. H. The response of the house fly to certain foods and their fermentation products. *J. Econ. Entomol. 10*:102--109 (1917).
4. Brown, A. W. A., West, A. S., and Lockley, A. S. Chemical attractants for the adult house fly. *J. Econ. Entomol. 54*:670–674 (1961).
5. Carlson, D. A., Mayer, M. S., Silhacek, D. L., James, J. D., Beroza, M., and Bierl, B. A. Sex attractant pheromone of the house fly: Isolation, identification, and synthesis. *Science 174*: 76–78 (1971).
6. Carlson, D. A. and Beroza, M. Field evaluations of (*Z*)-9-tricosene, a sex attractant pheromone of the house fly. *Environ. Entomol. 2*:555–559 (1973).
7. Kinzer, D. R. and McDaniel, J. D. Insecticidal composition containing *cis*-9-tricosene and methomyl. U.S. Patent No. 4,133,165 (1978).
8. *Sunset Magazine.* Winning the battle against those wily files. Lane Pub. Co., Menlo Park, CA 94025, July edition: 72–73 (1987).

Part VI
Development, Registration, and Use

31

Commercial Development: Mating Disruption of the European Grape Berry Moth

ULRICH NEUMANN / BASF Aktiengesellschaft, Limburgerhof, Federal Republic of Germany

I. INTRODUCTION

For 5 years, BASF has been carrying out trials on mating disruption in the grape berry moth, *Eupoecilia ambiguella*, using the sex pheromone (Z)-9-dodecenylacetate (Z-9-DDA), under conditions as close as possible to those encountered in practice. Submission of results on dosage, number of applications, and site differences obtained from trials (mainly carried out on a total trial area of more than 300 ha in 1985) for registration led to commercial recommendations for the control of the second generation of *E. ambiguella* by the mating-disruption technique. The method is now registered for use on the second generation in the Federal Republic of Germany and in Switzerland.

The first use on a commercial scale, on 850 ha, took place in 1986 and, in 1987, the method was employed in both countries on a total of more than 3000 ha.

II. MATERIAL AND METHODS

It was considered necessary to have relatively large experimental plots to obtain as much information as possible. Generally, the trial areas were about 4 ha or larger (one area was 120 ha). Some were isolated fields not adjacent to other vineyards, and others were situated in the middle of vinegrowing areas.

The number of pheromone sources (Hercon Dispenser) was eventually increased to 500 per hectare. The dispensers were affixed by hand to the training canes and wires, using a stapler, at the precise time when the moths began to fly. The rate of pheromone now used is 50 g active ingredient (AI) per hectare.

To determine the effects accurately, the dispensers must be placed throughout the affected area at the same time, within 1 to 2 days. On a commercial scale, it is recommended that in a particular vineyard region, a group of responsible and interested vine-growers undertake the operation. This entails checking the flight of the moths, using pheromone traps set up shortly before the flying period begins, and also checking the progress of the attack so that sprays with insecticide may be organized, if this should become necessary.

III. RESULTS

The following figures show the levels of attack obtained between 1985 and 1987 in the Federal Republic of Germany and in Switzerland after pheromone applications against the second-generation *E. ambiguella*.

It should be noted that the threshold level for damage on grapes in Germany and Switzerland is 4% to 6% for the second generation.

Figures 1 and 2 show that, in most of the trails, the population densities were medium to low, as can be deduced from the larval infestation figures for untreated control plots.

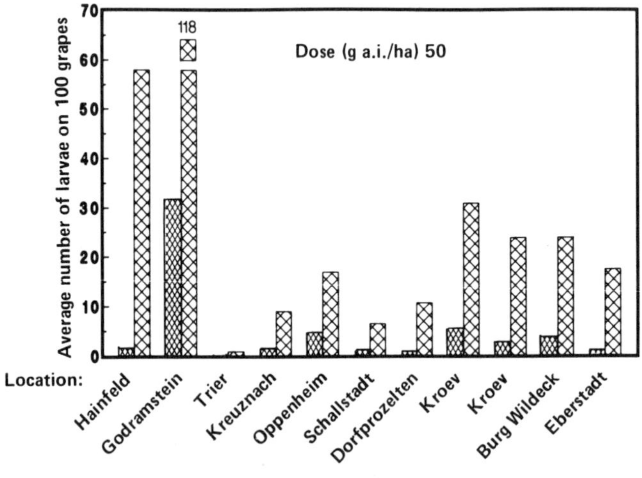

FIGURE 1 Control of *E. ambiguella* by the mating-disruption technique, FRG 1985.

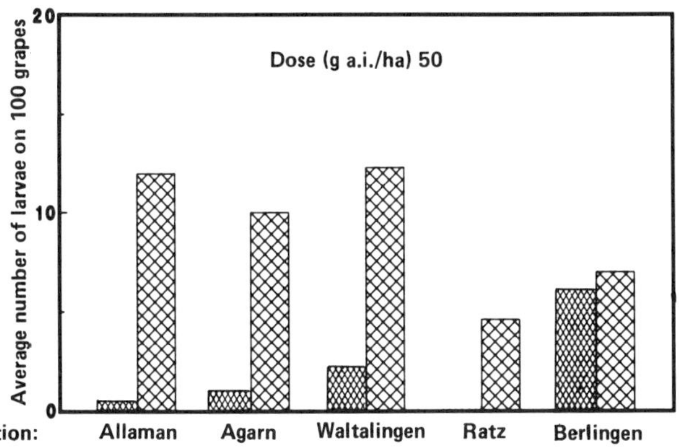

FIGURE 2 Control of *E. ambiguella* by the mating-disruption technique, CH 1985.

In such a situation, the mating disruption technique works extremely well, and the threshold level of damage is practically never exceeded. Very high population densities (Fig. 1, location "Godramstein"), such as may result from incorrectly carried out or inadequate insecticide treatment before the second-generation pheromone treatment, may cause the mating disruption procedure to fail.

Official advisory services are searching for new, more-specific control measures, so that, in spite of variable experimental results, official registration was obtained in both countries in 1986. It was made clear in the instructions that under certain (exceptional) conditions the mating disruption technique may be less effective and thus lead to some loss or damage, and corresponding instructions about how to avoid such damage are given.

Selected results from areas treated commercially in Germany in the first year after registration (1986) show a high level of success in all areas (850 ha in all were treated), with a relatively low-population density also in control areas (Figs. 3–5).

Results in 1987 were basically the same. The majority of the applications, on a total of more than 3000 ha, showed good results, comparable with those obtained when using insecticides (Figs. 6–8).

However, from single locations, there were also failures reported in 1987 (Fig. 9). Once again, these exceptions are related primarily to very high population densities, which might have resulted in the occurrence of too many random matings, thus drastically reducing the effects of mating disruption.

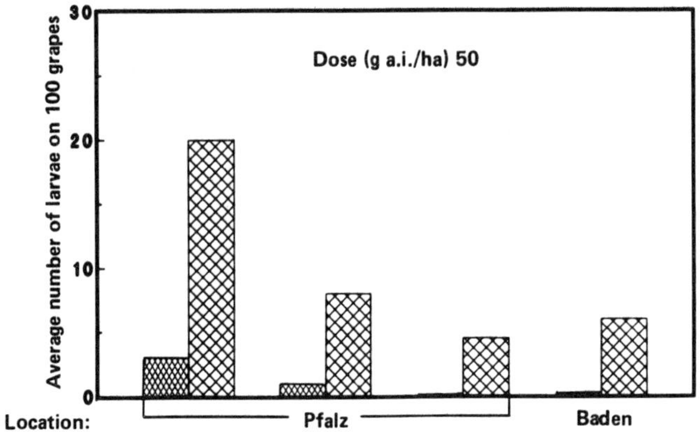

FIGURE 3 Control of *E. ambiguella* by the mating-disruption technique, FRG 1986.

In Figure 9 this can be demonstrated in the results from the "Siebeldingen" area belonging to a scientific institute. In this area, no control measures for the grape berry moth, *E. ambiguella* had been carried out for 10 years, which made conditions for the mating disruption technique extremely difficult. Insecticides can also fail in areas of high-population density, and this can clearly be seen in the "Kuernbach" area.

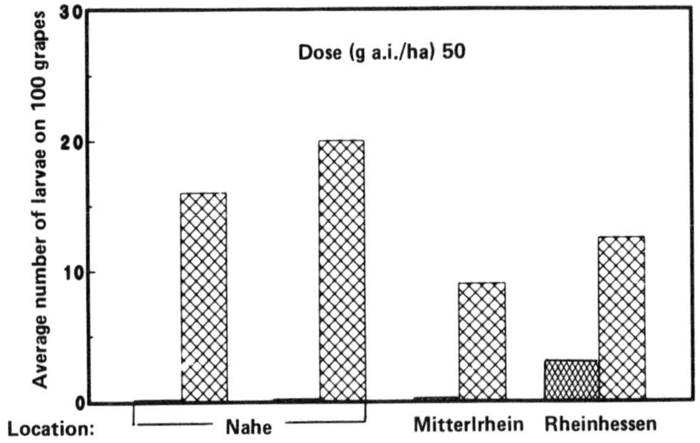

FIGURE 4 Control of *E. ambiguella* by the mating-disruption technique, FRG 1986.

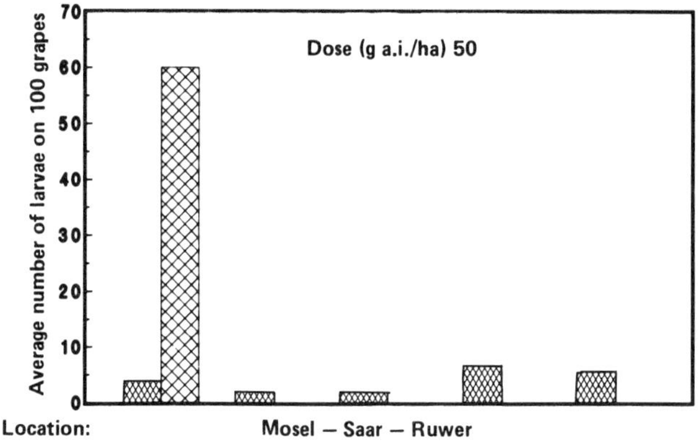

FIGURE 5 Control of *E. ambiguella* by the mating-disruption technique, FRG 1986.

It can also be seen from the "Flein" area (Fig. 10) that high-population density can induce poor results. In this location, the poor results were confined to the corners of the trial area surrounded by high trees. These segments had received inadequate insecticide coverage in the previous treatment for the first generation because application had been made by helicopter. Thus varying initial population levels were present when the pheromone was applied.

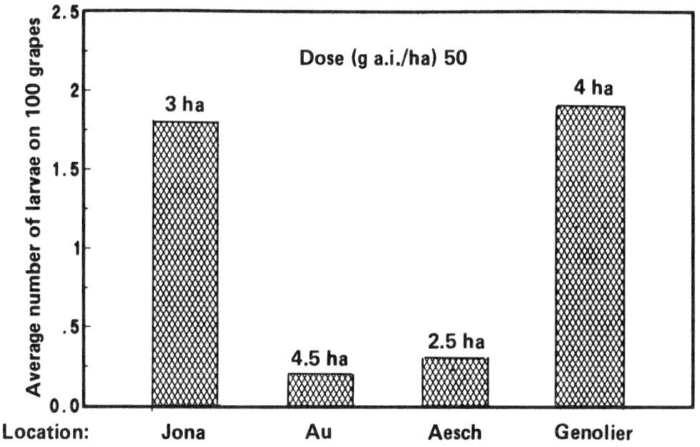

FIGURE 6 Control of *E. ambiguella* by the mating-disruption technique, CH 1987.

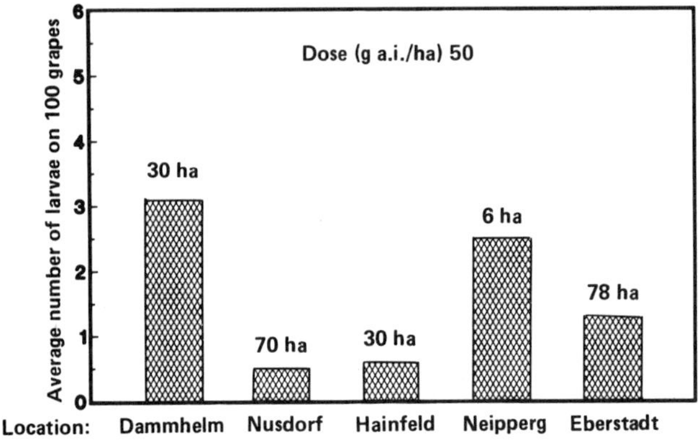

FIGURE 7 Control of *E. ambiguella* by the mating-disruption technique, FRG 1987.

IV. CONCLUSIONS

Two years of use on a commercial scale have shown that, under the conditions described, the mating-disruption technique can be regarded as a suitable approach for the control of the second-generation *E. ambiguella*. However, if failure is to be avoided, the population density as a limiting factor must be taken into consideration. If there is

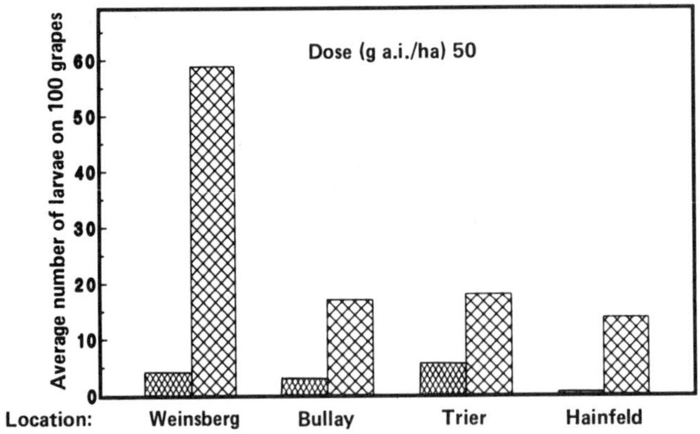

FIGURE 8 Control of *E. ambiguella* by the mating-disruption technique, FRG 1987.

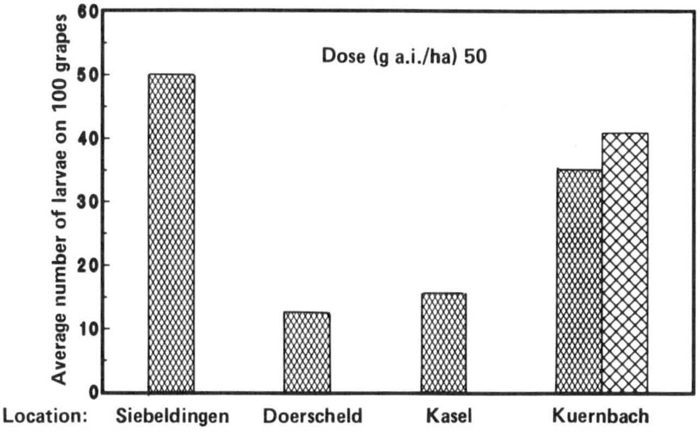

FIGURE 9 Control of *E. ambiguella* by the mating-disruption technique, FRG 1987.

any suspicion that the initial population density is high, then attempts must first be made to bring down the high level before pheromones are employed in a given area. Further experiments will show if this can be done by carrying out specific insecticide treatments.

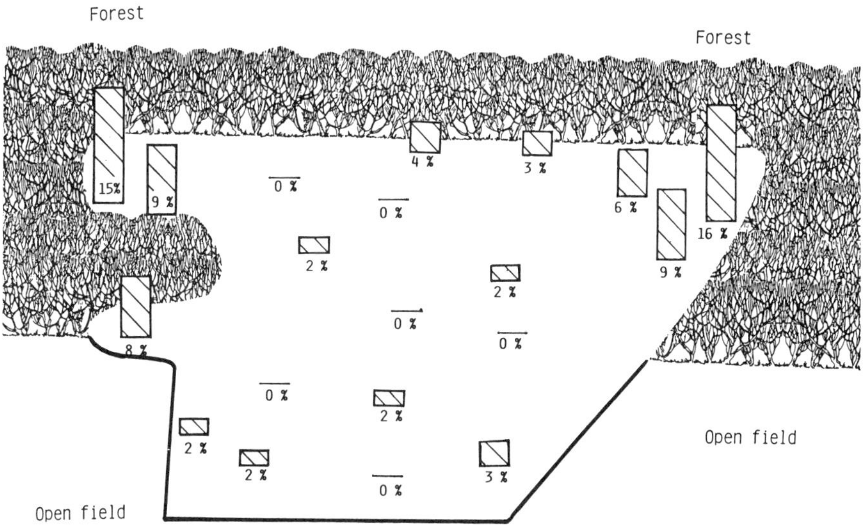

FIGURE 10 Control of *E. ambiguella* by the mating-disruption technique, Flein FRG 1987.

Furthermore, experiments will continue to find out if and how the mating-disruption technique can be used to control the first generation. Trials carried out so far have had limited success, to a large extent because, at this time, the vineyards are continuously open to the wind, as there is no leaf cover.

Under further study, is how other occasional pests, which were previously destroyed by the usual insecticide sprays, will develop if the mating-disruption technique against the grape berry moth is employed in an area over a period of years.

In practice, as well as in theory, the mating-disruption technique is a viable approach; however, under practical conditions, insurmountable hurdles may prevent the general introduction of the technique and may even make its employment impossible in certain crops.

32

Commercial Development: Mating Disruption of Tea Tortrix Moths

KINYA OGAWA / Shin-Etsu Chemical Company Ltd., Tokyo, Japan

I. INTRODUCTION

The tea tortrix and smaller tea tortrix are the most serious pests on tea plants in Japan. These pests have developed a strong resistance to methomyl and, today, in some areas, control is not effective, even at a dosage 10 times higher than that used previously.

Researchers at national institutes and universities in Japan identified the pheromones of both pests (Table 1) and have studied the activities of each component. These studies have shown that one common component, (Z)-11-tetradecenyl acetate, has the highest activity.

II. MATING DISRUPTION STUDIES

To try to control both pests simultaneously, we have studied the effectiveness of the mating disruption method at actual field sites, using different compositions.

Figure 1 shows that (Z)-11-tetradecenyl acetate alone has a higher activity than natural compositions. At the same release rate, the (Z)-11-tetradecenyl acetate formulation can maintain a lower mating ratio. When release rates are converted to the basis of (Z)-11-tetradecenyl acetate, the mating ratio has a linear relation to the release rate of this compound (Fig. 2).

TABLE 1 Composition of Pheromones (%)

Pheromone	Tea tortrix	Smaller tea tortrix
(*Z*)-11-Tetradecenyl acetate	88	31
(*Z*)-9-Tetradecenyl acetate	0	63
(*Z*)-9-Dodecenyl acetate	9	0
(*E*)-11-Tetradecenyl acetate	0	4
11-Dodecenyl acetate	3	0
10-Methyldodecenyl acetate	0	2

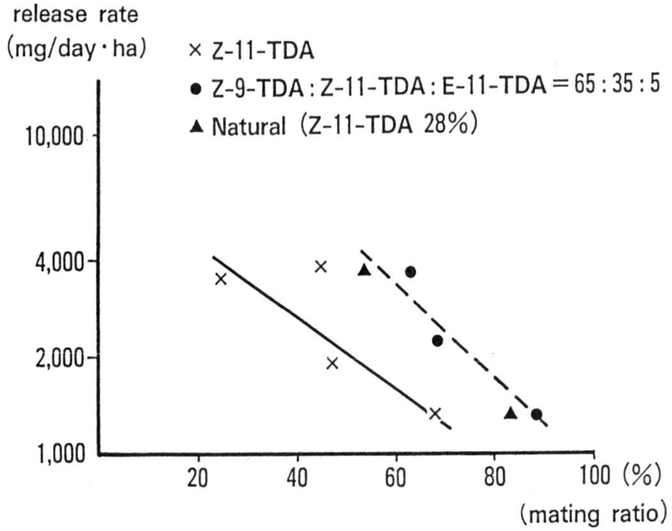

FIGURE 1 Efficacy of various pheromones at the same release rate.

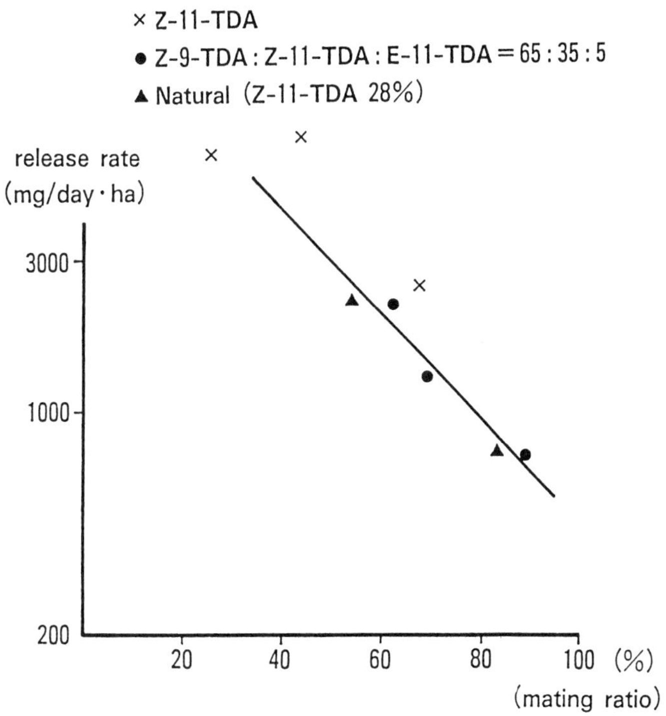

FIGURE 2 Efficacy of various pheromones converted to the basis of (Z)-11-tetradecenyl acetate.

III. DIFFICULTIES IN SELLING PHEROMONE

We sell (Z)-11-tetradecenyl acetate under the trade name of Hamakicon for the control of both pests. However, we face the following difficulties:

1. A group application is important.
2. The concept of pheromones is different from that of insecticides, and is hard for farmers to understand.
3. Farmers have little understanding of the disadvantages of insecticides and the advantages of pheromones.
4. Different dose rates and application methods are needed in different areas.
5. Price.
6. Registration.

A. Need for Areawide Application

First, for mating disruption by pheromone, a group application is indispensable. In Japan, an average field is less than 1 acre. But in such a small field, we cannot expect good results, partly because the distribution of pheromone in the field is uneven. At the center of the field, the pheromone concentration is about 20 ng/m^3, but at the border area, it is about 10 ng/m^3. Even above a road, where there is no pheromone treatment, the concentration is 5 to 8 ng/m^3. Therefore, to reduce border effects, a large-area application is very important. A group application is being conducted in 1000 acres at Shimada in Suizuoka, one of the most famous tea production areas.

B. Farmers' Attitude Toward Pheromones

Another problem is the farmers' attitude toward pheromones. The concept of pheromones is very different from that of insecticides and farmers sometimes find it difficult to understand the concept. For pheromones, an early application is most important because the population density of pests is one of the most important factors, but farmers sometimes apply it at the same schedule appropriate for insecticides.

C. Necessary Variations in Dose Rates and Application Methods

Pheromones are very specific, and it is difficult to use them in fields that suffer from many kinds of insects. Insecticides can be used at the same dosage level in different countries, but dosage levels of pheromones are determined by prevailing conditions.

Table 2 shows the results of the application of the same pheromone in different areas in China and in the United States. In China, a dosage less than half of that used in the Imperial Valley is effective and has a longer life.

TABLE 2 Dosages and Lifetimes of the Same Pheromone in China and in the United States

Test area	Dosage (AI)	Lifetime
China (Zhejiang prov.)	130 (g/acre)	60 (days)
United States (Imperial Valley)	320	45

D. Cost of Pheromones

The price of pheromones is also important. Prices are now somewhat high, but in the future the price of pheromones is expected to decrease as the market grows.

E. Pheromone Registration

To promote the application of pheromones, the registration system is important. Nowadays, each country has a different system, which requires us to conduct a number of different safety studies. Even for a small pheromone market, such as the light-brown apple moth, we must apply for a new registration. Therefore, we would like to propose a new worldwide registration system. One example would be the following. All lepidoptera pheromones should be classified into several groups by chemical structure, and safety studies should be conducted as groups. This would allow us to use pheromones without applying for a new registration for each new pheromone application as long as all the pheromone components belong to the specific groups. Finally, we must recognize that the assistance of university and government researchers and administrators is indispensable to overcome these difficulties and to develop pheromone applications.

33

Pheromones: A Marketing Opportunity?

KURT NABHOLZ / Sandoz Ltd., Basel, Switzerland

I. TECHNICAL VERSUS MARKETING VIEWS

Are all the technical possibilities that we have been describing in these proceedings also marketing opportunities?

When comparing the technique viewpoints versus marketing viewpoints on pheromones, we can state that pheromones are interesting products, being new, safe, price-worthy, technically feasible, and having environmental advantages. On the other side, a new product that is really a new concept in crop protection demands very high promotion costs; many man-hours have to be invested to convey the message of the new concept to the end user.

Safety, for a marketing company, is an internal positive factor and may be used to build up the image of the company. It definitely helps to reduce risks in production and logistics. But for the end user, safety is a factor of less importance because he is used to hard chemicals.

The price-worthiness is true for single pest situations. However, as soon as we must deal with more than one pest at the same time in a crop that normally can be controlled with a single hard chemical, pheromones become too expensive.

The technical feasibility is limited to people who are technically familiar with the new concept and with the ways and means to apply it. For the end user who has been using pesticides, this will mean a number of changes in timing and equipment and for the marketing company, it will demand supervision of the end user to make sure that the new concept is correctly used.

Environmental advantages are, unfortunately, only a qualitative aspect for the time being. We would like to have quantitative figures that show that the use of pheromones reduces the number of applications needed and the number of secondary pests.

In addition to this list, a marketing company will miss patent protection to cover the marketing investments. A difficulty in marketing is also that the pheromones are used preventively and are not appropriate to remedy a situation. If pheromones are not applied correctly, on time, and in relatively low infestations, they also may fail. Last, but not least, a problem area in pheromone marketing is logistics: the relatively long time needed for production and the short shelf-life of inventories.

II. MARKET POTENTIAL

The base for any marketing project is the evaluation of the market potential. It is relatively easy to define the infested area and to estimate the area treated with actual products. From these figures the theoretically available area for pheromone application can be calculated.

It becomes more difficult when a sales target has to be proposed. Because of the novelty of the concept, a number of emotional aspects must be considered. We all hope to find enthusiastic farmers who love this new idea. But we also will find farmers who are suspicious and will not believe the story of mating disruption. Others may react with fear of unnecessary expenditures owing to the preventive mode of use or with reluctance to invest in new equipment or to learn something new.

The time frame needed from initial trials and demonstration until the new concept becomes common practice has to be estimated. Innovations in this field will be numerous, and new competitors may appear on the scene and may need less time to introduce their product based on the same concept. Finally, experience in the marketing of pheromones is lacking in the plant protection industry. Everybody has replaced an older hard chemical with a newer one and has experience in the estimation of the time needed to introduce the new insecticide. With a totally new concept this will be quite different.

III. ATTITUDE OF PESTICIDE MARKETING COMPANIES

How does a pesticide marketing company look at the opportunity of selling pheromones? Many of them have decided to stand back and merely observe the ongoing activities on the market.

SANDOZ believes in biological pest control and has been selling, for a number of years, biological products such as B.T.-products, viruses, pheromone traps, and seeds with which biological agents can be included. Our entomologists are sure that pheromones have a future in crop protection. Therefore, we have decided to actively participate in pheromone marketing and to gather practical experience and inside information. However, SANDOZ has also decided not to overdo things, to follow only existing leads, and to use existing capacities in manufacturing and field force. Practical experience also leads us to participate only in official, authorized projects. We have seen that a sales organization alone cannot promote this new concept in a profitable way, without the help and the cooperation of government extension personnel or of cooperatives. In research, we also will depend on basic results obtained by universities and institutes. We see the role of a private company as developing and adapting basic research leads. However, we hope to motivate research people with input and ideas coming from the practical use of pheromones in the field.

IV. CONCLUSIONS

In practice, there are still too many limitations for private marketing and distribution of pheromones without the help and cooperation of official bodies. Experimental stations, universities, and extension services are expected to do the promotion of the pheromone concept and the education of the farming community in the use of this concept. The marketing company, then, still has enough to do to promote its own product.

The large-scale use of pheromones by the government will be the best demonstration of the feasibility of this concept and will be essential for the progress of pheromones in agriculture and forestry.

Finally, I have the strong feeling that there is plenty of room for new systems and new devices in the field of pheromones that will help to improve the impact of this new and beneficial technology.

34

Registration Requirements and Status for Pheromones in Europe and Other Countries

ALBERT K. MINKS / Research Institute for Plant Protection, Wageningen, The Netherlands

I. INTRODUCTION: HISTORY OF EUROPEAN INVOLVEMENT

A. The 1978 Seminar on Odor Communication in the Netherlands

In 1978 an international seminar called *Advanced Research Institute on Chemical Ecology: Odour Communication in Animals* took place in the Netherlands under the auspices of the NATO Science Committee. The participants came from the United States, Canada, Japan, Israel, and West European countries. One of the highlights was the report on experiences with the first commercial pheromone communication disruptant, gossyplure HF (pink bollworm pheromone in hollow fiber formulation), that had been granted registration by the U.S. Environmental Protection Agency (EPA) just half a year earlier (1). Also at this seminar, the late Dr. John Siddall described the enormous uncertainty in both the regulatory agencies and commercial organizations interested in pheromones, concerning what exactly should be required to achieve commercial registration (2). Apart from the risk of failure to achieve registration was another risk of finding at a late stage of development that more and more tests were asked for by regulatory agents before determination of "no unreasonable effects on the environment" could be made. Siddall called this the "regulatory treadmill effect."

The seminar produced a number of conclusions and recommendations that were included in the proceedings of the meeting published in 1979 (3). Two of the recommendations were specifically concerned with the registration of pheromones:

Regulatory agencies of all countries should publish special guidelines for registration of behavior-modifying chemicals for pest control. Such chemicals, which must provide insectistasis without insecticidal action, should clearly be distinguished from insecticides. Suitable terminology should be adopted internationally.

The measurement of the efficacy of a pheromone treatment should conform to internationally recognized methods and standards, as befits the particular situation, similar to those stated in the Study Team Report of the American Institute of Biological Sciences (AIBS) (4).

B. The NATO-CCMS Pilot Study on the Registration of Pheromones

These forenoted recommendations were used as basis for a pilot study in the framework of NATO-CCMS (Committee on the Challenges of Modern Society). The Netherlands accepted the task of acting as pilot leader and Dr. F. J. Ritter was appointed as chairman of the Study Group. The countries participating were the Netherlands, the United States, the United Kingdom, and France.

The aims of the study were to collect information on the status of registration procedures for insect pheromones in as many countries as possible; to analyze the contents of these procedures; and to formulate recommendations for improving these procedures.

The activities of the Pilot Study Group took place in the period 1980 to 1983 and were concluded with the publication of a report in April 1983 (5). From the collected information it became evident that only the United States has worked out special procedures for the registration of pheromones. It is safe to say that these guidelines were used as a basis for all further considerations of the Pilot Study Group. It was not surprising to note that one of the most important recommendations was that the guidelines of the U.S. Environmental Protection Agency (EPA) for biorational pesticides, including pheromones (see Fig. 1), summarized in its Subdivision M, are a good example of how registration procedures might be developed.

Other recommendations were the following:

A scheme of tier testing consisting of different phases or grades should be an important part of the procedures. The scheme could be used in four fields—residue chemistry, toxicology, nontarget organism hazard, and environmental fate.

If the use of pheromones as an alternative means of pest management is to be stimulated, not only should their registration be facilitated, but governments should also try to harmonize

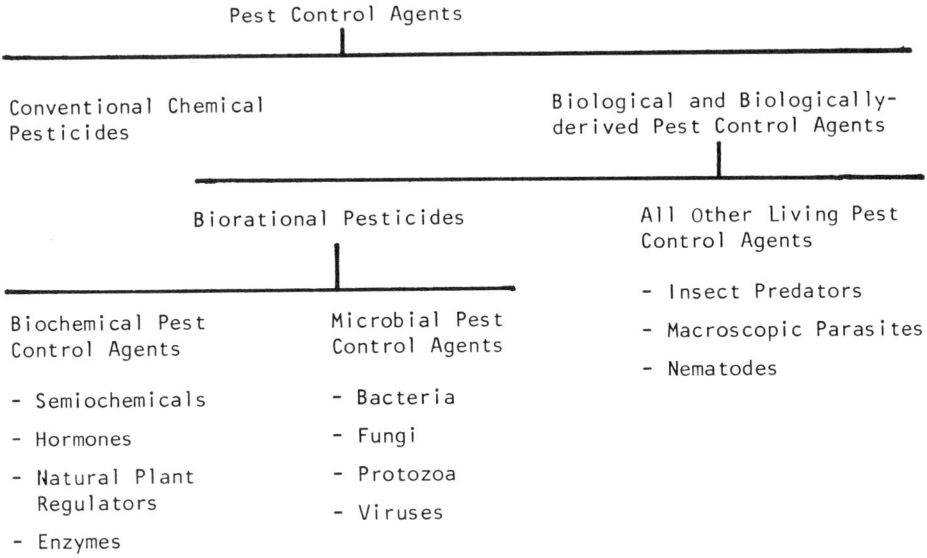

FIGURE 1 Relationships between conventional pesticides, biological control agents, and biorational pesticides (from Ref. 6).

their registration procedures to promote equivalent treatment of pheromones in different countries and, thus, increase opportunities for their application.

After its completion, the report of the Pilot Study Group was distributed widely to registration offices and research workers in the field of pheromones in more than 20 countries. It is noteworthy that a pilot study of the NATO/CCMS, by its nature, cannot formulate *demands* that should be adhered to by governments. What the pilot study could try to achieve is to put forward proposals for facilitating the use of pheromones by suggesting special registration procedures, to describe these procedures, and to bring them to the attention of the governments, e.g., registration offices. That is the (rather limited) status of the pilot study.

II. REGISTRATION PROCEDURES FOR PHEROMONES IN THE UNITED STATES

Because it is my opinion that the U.S. guidelines for pheromone registration should be used as a basis for the development of similar rules in other countries and that these other countries should be grateful for the pioneering work done by the United States in this respect, I would like to present some of the major points of the U.S. guidelines in this contribution dealing with the European situation. For more details I refer to other sources of information, mentioned in the reference list (6–10) and to the next two chapters in this book.

Pheromones used in agriculture to control or reduce the insect pest population must be registered by the U.S. EPA under the provisions of the Federal Insecticide, Fungicide, and Rodenticide Act (FIFRA). The requirements for registration depend on the way the pheromones are to be used.

Pheromones used for trapping (monitoring as well as mass trapping) are not subject to registration requirements of FIFRA, provided that the pheromones are the only active ingredients in the traps, but those intended for mating disruption must be registered. Pheromones are considered as *biochemical pesticides* and fall under the U.S. Pesticide Assessment Guidelines - Subdivision M-Biorational Pesticides (6). As is pointed out in Figure 1, biorational pesticides include microbials (viruses, bacteria, protozoa, and fungi) and biochemicals (insect and plant growth regulators, and semiochemicals such as pheromones). Of crucial importance are the following considerations when pheromones are classified as biochemical pesticides: Because of the nature of biochemical pesticides and their use, the data requirements for registration on occasion can be different and less extensive than those for conventional pesticides. The most important differences between naturally occurring biochemical and conventional pesticides are the target species specificity, low use concentrations, and the generally nontoxic mode of action.

These factors have led to the expectation that many biochemical pest control agents pose a lower potential hazard than conventional pesticides. Following this idea, the EPA developed a so-called *tier-testing scheme* for this category of control agents, with maximum hazard testing as well as mode of action and exposure criteria as important new elements.

A tier-testing scheme is a staged series of tests, progressing from simple and relatively inexpensive, to complex and more costly tests. It is designed to ensure that only the minimum data necessary to make a scientifically sound regulatory decision are required. The scheme is used in four sections: namely, residue chemistry, toxicology, nontarget organism hazard, and environmental fate. It is self-evident that along with the tier-testing program, data on product chemistry, which are not tiered, are required but are treated similarly to the requirements for conventional pesticides.

Five years after the official publication of the EPA guidelines on biorational pesticides, it can be concluded that these guidelines apparently work in a quite satisfactory manner, although there are pleas from industry for more flexibility and speed in the procedures (see Chap. 36). To my knowledge 11 behavior-modifying chemicals have thus far been granted permission for use on various crops in the United States (10).

III. PHEROMONE REGISTRATION IN EUROPE AND IN SOME OTHER COUNTRIES

What is the state of pheromone registration in Europe? In the Addendum a more detailed survey is given for 14 European countries and three countries outside of Europe. A number of general conclusions can be drawn from the survey:

In all but one country (Czechoslovakia), registration is not required for pheromone traps when used for detection and monitoring.

In all countries pheromone formulations used for mass trapping or mating disruption need registration when applied commercially.

Although in some countries control measures using pheromones are characterized as biological measures, all countries classify pheromones as chemical compounds that, for registration, basically should be treated in the same way as chemical insecticides. This is a principal difference with the United States' interpretation of pheromones as biochemical compounds.

In most of the countries, readiness has been expressed to demand "lighter" registration requirements for pheromones than for the classic insecticides, particularly for toxicology. However, there is still the same uncertainty, as has been mentioned by Siddall (2), about how stringent these demands should be. In most countries, this will be decided on a case-by-case basis. This seems disappointing to me because it indicates that not much has been done with the regulations designed by the EPA and with the experience they have gathered for these regulations in the past years, nor with the information collected and distributed by the NATO/CCMS Pilot Study Group.

Another important reason for the uncertainty in many countries is simply that, as yet, no application for pheromone registration has been made; hence, no experience could be collected.

To date, one pheromone has been registered for mass trapping in two European countries. For mating disruption purposes three pheromones have been granted definitive permission for commercial use in two countries, whereas this is provisional

TABLE 1 Pheromone Formulations Granted Registration for Commercial Use. Situation as of December 1987

Country	Type of use	Insect involved	Commercial name
Europe			
France	Mass trapping	*Ips typographus*	Pheroprax
Norway	Mass trapping	*I. typographus*	Ipslure
Federal Republic of Germany	Mating disruption	*Eupoecilia ambiguella*	RAK I Pheromon
USSR	Mating disruption	*Grapholita molesta*	
USSR	Mating disruption	*G. funebrana*	
Switzerland (provisional registration)	Mating disruption	*Cydia pomonella*	Bocep Carpo
	Mating disruption	*E. ambiguella*	Bocep Viti
Far East			
Australia	Mating disruption	*Grapholita molesta*	Isomate M
Japan	Mass trapping	*Spodoptera litura*	Pherodin SL
Japan	Mating disruption	*Adoxophyes* sp. and *Homona* sp.	Hamakicon
Japan	Mating disruption	*Carposina niponensis*	Shinkuikon

for two pheromones in one other country (Table 1). In Australia and in Japan some pheromones for mass trapping and mating disruption have also passed the registration offices (Table 1).

As yet, one of the important experiences of the applicants for pheromone registration is that the speed of the procedures is strongly related to the type of formulation. For nonsprayable delivery systems, such as the larger flakes, fewer tests are generally required than for sprayable formulations, such as microcapsules.

To my knowledge, no serious attention has been given to the consolidation of the rules for registration of pheromones and other biorational pesticides, except perhaps for the Scandinavian countries, where I could notice some preliminary arrangements. After an initiative taken in this area by the European Community in 1981, no further development could be recorded.

ACKNOWLEDGMENTS

I wish to thank all my colleagues in the area of pheromone research and the registration offices in the various countries for their assistance in collecting information for this contribution. My special thanks are due to Dr. A. R. Jutsum (UK) who so kindly gave me the opportunity to preexamine a relevant chapter of his recently published book.

ADDENDUM

Survey of the state of registration of insect pheromones as pest control agents in a number of European countries and in some other countries elsewhere.

The following data have been arranged and summarized from the responses of research workers in the area of pheromones and of the registration offices to an inquiry that I made concerning pheromone registration in a selection of European countries and in some other countries.

This summary follows (the countries are arranged in alphabetic order):

Austria: Pheromones are considered as plant protection agents under the Plant Protection Law, and as a separate category of compounds. They do not have to pass through all toxicological tests required for conventional pesticides. In 1988, approval is expected for commercial use of two pheromones, namely, for mating disruption of the codling moth, *Cydia pomonella*, (48 g AI/ha) and the grape berry moth, *Eupoecilia ambiguella*, 50 g AI/ha).

Belgium: Pheromones for mating disruption fall under the Royal Decree concerning the keeping, sale, and use of pest control agents. They are considered as chemicals. As yet, no applications for registration have been filed. Large-scale experimentation in the field has been conducted for the summerfruit tortrix moth, *Adoxophyes orana*.

Bulgaria: No specific official regulations for the registration of pheromones are available. No large-scale field experimentations with pheromones are being conducted.

Czechoslovakia: Pheromones are included in the Directions of the Federal Ministry of Agriculture and Food on Testing and Permission of Agents for Plant Protection, together with all other compounds of chemical or biological nature used for control of pests, weeds, and diseases. The Central Checking and Testing Institute of Agriculture is responsible. Synthetic pheromones used for mass trapping or mating disruption *in agriculture* are classified as chemicals of a specific nature, but pheromones used in forestry are listed under biological agents.

Pheromones used for monitoring also require registration, but only tests on efficacy and quality of the trapping system are required. Pheromones of the following insects have been registered: gypsy moth, *Lymantria dispar*, and larch bud moth, *Zeiraphera diniana*, in forestry; and codling moth, *Cydia pomonella*, in agriculture. For about 15 other moths special registration is expected soon.

Large-scale field tests on mating disruption of the codling moth, *C. pomonella*, are in progress. However, no application for registration is expected in the near future.

Denmark: The Ministry of Environment has made the basic decision that pheromones used for direct control of insects are to be considered as chemical control agents like all pesticides. Certain pesticides such as pheromones, that are regarded as less risky than conventional pesticides have to meet fewer requirements for registration on a case-by-case basis. As yet, no applications for registration have been filed.

Federal Republic of Germany: Pheromones are considered as plant protection agents within the Law for the Protection of Cultivated Plants. They are characterized as chemical compounds in terms of registration procedures. Registration is obligatory for all agents, depending on the intended use--for pheromones, if they are used for mass trapping and mating disruption. To date, one pheromone, namely, "RAK I Pheromone Einbindiger Traubenwickler," has been registered for commercial use as a mating disruption agent (50 g AI/ha) against the second generation of the grape berry moth, *Eupoecilia ambiguella*.

France: Pheromones are considered as pesticides and subject to the general registration requirements designed for all plant protection agents used directly for control in agriculture (for pheromones: for mass trapping and mating disruption). The "Service d'homologation" of INRA (National Institute for Agricultural Research) is responsible. Only after a positive decision by the official services can an authorization for commercialization be issued. As yet, this is for only one "specialty," namely, the control of the spruce bark beetle, *Ips typographus*, by mass trapping.

Greece: All substances that modify insect behavior are considered to be agricultural chemicals. For registration of these chemicals very stringent procedures for registration are required. As yet, no applications for pheromone registration have been submitted.

Italy: Legislation has not, as yet, distinguished pheromones as a separate group. They are in the same category as all chemical products used for the control of pests, diseases, and weeds. Registration is necessary when pheromones are used for mass trapping and mating disruption. There are indications that for biorational compounds, an opportunity will be granted to examine the possibility of reduced requirements before starting a registration procedure. As yet, no applications have been filed.

The Netherlands: Pheromones are included in the Law on Pesticides and are considered as chemical compounds. Registration is needed when they are used for mass trapping or mating disruption. The Bureau of Pesticides of the Committee for Phytopharmacy is responsible. As yet, no application for registration of sex pheromones has been submitted, but there is one in progress for a product containing aphid alarm pheromone. Although there are fewer requirements than for conventional pesticides, the procedure has already taken more than 2 years.

Norway: Pheromones are considered as chemicals under the Norwegian legislation, but they are subdivided as biological control agents. The Norwegian Pesticide Board, in general, agrees that the same stringent registration requirements as those for conventional pesticides are not necessary for pheromones. The Pesticide Board welcomes a dialogue with applicants for the registration of biological control agents before the procedure starts. In the period of 1979 to 1983, permission was granted for the very extended mass trapping campaign against *Ips typographus*.

Switzerland: Pheromones are considered as "Hilfstoffe" similar to all pesticides, including other biological control agents. In terms of registration, they basically are treated no differently than are classic pesticides. There is no special checklist of registration requirements for pheromones, but these can be negotiated between the applicants and the registration officials on a case-by-case basis. This country makes possible a provisional authorization of "Hilfstoffe," including plant protection compounds and plant-regulating substances; such authorization must be renewed each year. This provides an opportunity for the applicants to collect more data from large-scale operations in the field. Two provisional authorizations have been issued for mating disruption with sex pheromones: (a) for the codling moth, *Cydia pomonella*, (25 g AI/ha) and (b) for the grape berry moth, *Eupoecilia ambiguella* (50 g AI/ha).

United Kingdom: Recently, a new statutory approval system came into operation under the Control of Pesticide Regulations (Food and Environmental Protection Act 1986). Within these regulations, approval is needed for the commercial use of pheromones for mating disruption programs, and data must be submitted just as for chemical pesticides. As yet, no applications for direct control with pheromones have been made.

USSR: Pheromones are included in the legislation on pest control agents, and they are considered to be chemicals. They must go through the same registration procedures as traditional pesticides. For experimentation on 1 to 5 ha no special permission from the Ministry of Health is necessary. As yet, in two instances, permission has been granted for the commercial use of sex pheromones for mating disruption of the oriental fruit moth, *Grapholita molesta*, and the plum moth, *Grapholita funebrana*.

The status in some other countries outside Europe follows:

Israel: Pheromones are considered to be chemicals and are defined as pesticides. They fall under the Plant Protection Law and must be registered for commercial use and distribution. The use of pheromone traps for timing of chemical control is not subject to registration. A 1-year temporary registration has been granted for the use of the pink bollworm, *Pectinophora gossypiella*, pheromone as a mating disruptant. This is a standard procedure for any new product, and it is meant to generate more evidence for definitive registration; it is, more or less, comparable with an Experimental Use Permit (not existing in Israel). As yet, no objections have been raised about efficacy and safety. Registration of this pheromone can be expected soon.

Australia: Pheromones are considered as pest control agents within the Australian Pesticide Legislation. When used for mating disruption they are treated as a special case of chemical insecticides, which must go through only a limited number of safety tests, decided on a case-by-case basis. Requirements for residue data depend greatly on the way the pheromones are formulated: direct contact of the pheromone with the crop products (especially food) may require more tests. A convincing package of efficacy data is also required. Pheromones for monitoring purposes are not covered by legislation. The pheromone of the oriental fruit moth, *Grapholita molesta*, was granted registration, in 1985, for use as mating disruptant, and it is now applied commercially by 30% of the Australian peach growers.

Japan: Under Japanese legislation, a pheromone application is considered as a biological means of pest control. No permission is needed for experimental use, not even for large-scale tests, up to several hundreds of hectares. However, for commercial application, registration is required. The procedure is basically the same as for other chemical compounds such as insecticides, but fewer data are required, particularly on toxicological aspects. There are some pheromone formulations registered for direct control including (a) for mass trapping of the armywork moth, *Spodoptera litura*; (b) for mating disruption of the leafroller moths, *Adoxophyes* sp. and *Homona* sp. (combined) and (c) for mating disruption of *Carposino niponensis*.

REFERENCES

1. Brooks, T. W., Doane, C. C., and Staten, R. T. Experience with the first commercial pheromone communication disruptive for suppression of an agricultural insect pest. In *Chemical Ecology: Odour Communication in Animals*. (F. J. Ritter, ed.). Elsevier / North-Holland Biomedical Press, Amsterdam, 1979, pp. 375–388.
2. Siddall, J. B. Commercial production of insect pheromones – problems and prospects. In *Chemical Ecology: Odour Communication in Animals*. (F. J. Ritter, ed.). Elsevier /North-Holland Biomedical Press, Amsterdam, 1979, pp. 389–402.
3. Ritter, F. J. (ed.). *Chemical Ecology: Odour Communication in Animals*. Proceedings of the Advanced Research Institute on Chemical Ecology, Noordwijkerhout, the Netherlands, September 1978. Elsevier /North-Holland Biomedical Press, Amsterdam, 1979, 427 pp.
4. Roelofs, W. L. (ed.). *Establishing Efficacy of Sex Attractants and Disruptants for Insect Control*. Entomological Society of America, College Park, Md., 1979, 97 pp.
5. NATO Committee on the Challenges of Modern Society 1983. Pilot Study on the Registration of Pheromones No. 140 (plus annexes), 1983, 19 pp.

6. United States Environmental Protection Agency Pesticide Assessment Guidelines: PB83-153965, EPA 540/9-82-082, November 1982, Subdivision M: Biorational Pesticides.
7. United States Federal Register, Volume 48, No. 165, August 24, 1983, pp. 38572–74; Part II, Environmental Protection Agency; Products containing pheromone attractants: Exemption from FIFRA Requirements.
8. United States Federal Register, Volume 49, No. 207, October 24, 1984, pages 42856–905. Part II, Environmental Protection Agency, 40 CFR Part 158: Data requirements for pesticide registration: Final Rule.
9. Zweig, G., Cohen, S. Z., and Betz, F. S. EPA registration requirements for biochemical pesticides, with special emphasis on pheromones. In *Insect Suppression with Controlled Release Pheromone Systems*, Vol. 1. (A. F. Kydonieus and M. Beroza, eds.). CRC Press, Boca Raton, 1982, pp. 159–167.
10. Punja, N. The registration of pheromones. In *Insect Pheromones in Plant Protection* (A. R. Jutsum and R. F. S. Gordon, eds.). John Wiley & Sons, New York, 1989, pp. 295–302. 1988.

35

Regulation of Pheromones and Other Semiochemicals in the United States

EDWIN F. TINSWORTH / U.S. Environmental Protection Agency, Washington, D.C.

I. INTRODUCTION

This discussion focuses on the regulation of pheromones and other semiochemical pesticides in the United States. Pheromones are chemicals produced by an organism (e.g., insect species) that modify the behavior of other individuals of the same species. Pheromones are the best known examples of a much broader class of substances known as semiochemicals. *Semiochemicals* (1) are substances or mixtures of substances emitted by one species that modify the behavior of receptor organisms of like or different species. In addition to the pheromones, semiochemicals also include:

Kairomones (1): Chemicals emitted by one species that modify the behavior of a different species to the benefit of the receptor species

Allomones (1): Chemicals emitted by one species that modify the behavior of a different species to the benefit of the emitting species

Synomones (1): Chemicals emitted by one species that modify the behavior of a different species to the benefit of both the emitting and receptor species

In 1979, the U.S. Environmental Protection Agency (EPA), Office of Pesticide Programs (OPP), recognizing that semiochemicals and similar naturally occurring substances were inherently different from most broad-spectrum conventional toxic chemicals, published a policy statement encouraging the development and registration of this class of pesticide products as potentially safer alternatives to conventional pesticide products.

II. REGULATORY STATUS OF SEMIOCHEMICALS

The Federal Insecticide, Fungicide, and Rodenticide Act (FIFRA) (2) defines a *pesticide* as any substance or mixture of substances intended for preventing, destroying, repelling, or mitigating any pest. Section 3 of the FIFRA requires a pesticide to be registered with the Administrator of the EPA before it can be sold or distributed in the United States, except as otherwise provided for under other provisions of the act.

Title 40 of the Code of Federal Regulations (40 CFR) (3) identifies *attractants* as a class of pesticide active ingredients. The term *attractant* is defined as any substance or mixture of substances that through their property of attracting certain animals are intended to mitigate the population of a pest species. Sensory stimulants such as pheromones are identified as examples of substances included under the definition.

It can be seen from the foregoing definitions that the intent of use of a product determines its regulatory status under the FIFRA. If pheromones or other semiochemicals are intended to be used to mitigate the effects of a pest population, then they are, by definition, a pesticide and as such subject to regulation under the provisions of the FIFRA.

III. USES OF SEMIOCHEMICALS EXCLUDED FROM REGULATION

The EPA has excluded from regulation under the provisions of the FIFRA, substances or mixtures of substances intended to attract vertebrate or invertebrate animals exclusively for survey or detection purposes [40 CFR §162.3 (ff)(3)(ii)] (3). Such uses of semiochemicals are not intended to mitigate the effect of a pest population and, therefore, do not meet the definition of a pesticide under the FIFRA.

IV. EXEMPTION OF CERTAIN PESTICIDAL USES OF PHEROMONES

The EPA has exempted from the registration requirements of the FIFRA the use of pheromones and identical, or substantially similar compounds, labeled for use only in pheromone traps in which they are the sole active ingredient(s). The exemption is further qualified by a requirement that pest control be achieved by the process of attraction and removal of the target organisms from their natural environment, without resulting in increased levels of the pheromone(s) or identical, or substantially similar, compounds over a significant fraction of the treated area [40 CFR 162.5 (d)(2)] (3).

The agency has concluded that such uses of pheromones are of a character such that it is unnecessary that they be regulated under the provisions of the FIFRA.

Uses of pheromones or other semiochemicals in traps in conjunction with conventional toxic substances do not qualify for inclusion under the provisions of this exemption from the registration requirements of the FIFRA. The use of pheromones under other application technologies to reduce pest populations are now subject to regulation under the provisions of the FIFRA. The EPA maintains an open mind and will continue to review the available data with the intent of expanding the scope of this exemption as the facts warrant.

V. AGENCY LOGIC IN ESTABLISHING THE DATA REQUIREMENTS

The kinds of data required to be developed and submitted in support of applications for registration of semiochemical pesticide products are specified in 40 CFR Part 158.690(a) through (d) (4).

In formulating these regulations the EPA took into consideration that semiochemical pesticides are inherently different from conventional pesticides. Some of the characteristics that typically distinguish semiochemicals from conventional pesticides are their unique (nontoxic) mode of action, low use volume, target species specificity, and natural occurrence (1). The EPA expects that most semiochemical pesticides will pose low potential risk when compared with conventional pesticides.

Semiochemical pesticides must be naturally occurring substances. If the chemicals are synthesized by man, then they must be structurally identical to a naturally occurring substance. Minor differences in stereochemical isomer ratios between the naturally occurring compound(s) and the synthetic compound(s) normally will not rule out a chemical being classified as a semiochemical pesticide. However, if isomers demonstrate significantly different toxicological properties, then the EPA will handle the pesticide on a case-by-case basis. In making this assessment, the EPA will apply the following criteria: (a) consider the chemical and toxicological significance of the difference in chemical structure, (b) consider the mode of action of the synthetic analogue in the target species, compared with the mode of action of the naturally occurring compound, and (c) compare differences in toxicity (1).

Although semiochemical pesticide registrants will not be relieved of the burden of proof for the safety of their products, the agency will take into account the fundamentally different modes of action of their products.

Three elements form the basis of the EPA's approach to the assessment of chemicals. They are exposure potential, maximum hazard testing, and use of a tier-testing scheme. These factors provide a basis for the agency's criteria for reduced data requirements.

A. Exposure Potential

Certain factors frequently associated with semiochemical pesticide products or their use significantly limit the potential for human and nontarget organism exposure and, therefore, potential hazard. These factors provide the basis for reduced data requirements. They include pesticide formulations with low potential for exposure, low rate of application, nonaquatic sites of use, and high volatility (1).

1. Low-Exposure Pesticide Formulation

Semiochemicals frequently are formulated in passive dispensers such as hollow fibers, tapes, and traps that present limited opportunity for oral or dermal exposure of humans or direct exposure of other nontarget organisms.

2. Low Rate of Application

Semiochemicals are typically used in the field at very low rates (i.e., 20 g/acre or less). Such low rates of application present equally low opportunities for significant exposure to humans and other nontarget organisms.

3. Nonaquatic Use Sites

Semiochemicals are most commonly applied on terrestrial sites and, consequently, pose less risk to nontarget aquatic organisms. Therefore, they may logically qualify for reduced testing to estimate the hazard to nontarget organisms.

4. High Volatility

The high volatility of many semiochemicals reduces the potential for residues persisting on foods or feed crops or being available for ingestion by nontarget terrestrial organisms.

B. Maximum Hazard Testing

The EPA has applied the concept of maximum hazard testing (1) in their toxicology and nontarget organisms risk assessment strategy. By using this approach, the agency is able to develop a tiered approach to testing. Under this concept a maximum hazard challenge in terms of dose, concentration, or route of administration is used in a first tier of testing. From these sets of test results the agency can gain a high degree of confidence that no adverse effects will be likely to occur from use of the semiochemical pesticide, provided that the test results are negative.

C. Tier-Testing Scheme

The EPA has devised a tier-testing scheme (1) for the four major categories of data: residue chemistry, toxicology, nontarget organism hazard, and environmental fate and expression. The scheme has been established to ensure that only the minimum data set necessary to make a sound regulatory decision are required to be developed. There are three tiers of data requirements. The first tier of data is required to be developed when use conditions make the data relevant. Given the results of the first tier of studies, the second tier of studies may be required. However, if the first tier of studies demonstrate the low order of hazard of the semiochemical pesticide, then the second and third level tests are not required.

If, however, adverse effects are detected in the maximum challenge experiments of the first-tier tests then the second and third tiers of tests are intended to evaluate and quantify the actual hazard posed by the agent.

VI. DATA REQUIREMENTS

A. Product Analysis

Semiochemical pesticide products proposed for registration usually consist of synthetically produced chemicals or combinations of chemicals that are designed to simulate naturally occurring counterparts. The prospective registrant of a new semiochemical pesticide should submit to the EPA, at the very minimum, product analysis data as described in 40 CFR §158.690(a) (4). This information is absolutely essential for the agency to determine whether or not the substance in question is substantially similar to the naturally occurring counterpart. This information is also used by the EPA to determine what other data will be required to support registration of the product.

By using the best available technology, the applicant should provide as much of the following information in as much detail, as possible:

Manufacturing process, including flow charts where applicable, and demonstration of the chemical structure of each active ingredient; purification steps, and the like

Identification and quantification of the active ingredients and of the impurities present as part of the manufacturing process

A complete description of the analytical method used in establishing chemical structure and in quantifying the active ingredients

Title 40 CFR Sections 158.50 and 158.100 through 158.102 (4) describe how to determine product analysis data requirements and how to determine the appropriate substance(s) to be tested. Table 1 is adapted from 40 CFR 158.690(a) (4) and specifies the product analysis data requirements for semiochemical pesticides.

B. Residue Chemistry

The EPA must either establish a tolerance (amount of allowable residue) or issue an exemption from the requirements of a tolerance for any pesticide or metabolites thereof whenever the pesticide is registered for use on food or feed crops in such a way that residues are likely to occur on the treated food or feed commodity. If a residue is likely, data requirements are identified in 40 CFR §158.690(b). The registrant of a new semiochemical pesticide must identify what residues are likely to occur from the use of the product. If no residues are likely to occur, the registrant must provide a scientific rationale to convince the agency that this is so.

It is important to note that reduced pesticide residue data requirements are based on a low application rate (0.7 oz or 20 g of active ingredients (AI) per application per acre or less) and a finding of low hazard based upon Tier I toxicology data results.

Title 40 CFR 158.50 and 158.100 through 158.102 (4) describe how to determine pesticide residue data requirements and how to determine the appropriate substance to be tested. Table 2 is adapted from 40 CFR 158.690(b) (4) and specifies the pesticide residue data requirements for semiochemical pesticide products and the conditions under which they are required.

C. Toxicology

Toxicology data requirements are described in 40 CFR §158.690(c) (4). The EPA reviews semiochemical pesticides on a product-by-product basis, using a tiered-testing scheme. Depending on the uses proposed for a product and the risk that those uses may pose, the applicable requirements may range from a subset of the Tier I studies to Tier II or Tier III studies. Those requirements depend, in large part, on how well the registrant addresses the product analysis data requirements and the residue considerations.

Title 40 CFR 158.50 and 158.100 through 158.102 (4) describe how to determine toxicology data requirements for semiochemicals and how to determine the appropriate substance to be tested. Table 3 is adapted from 40 CFR 158.690(c) (4) and specifies the toxicology data requirements for semiochemical pesticide products and the conditions under which they are required.

Once again it is important to note that reduced data requirements are based upon a maximum challenge of the semiochemical pesticide under Tier I studies. If these data confirm that the substance presents minimum risk, then subsequent data in Tiers II and III are not to be required. The major concerns of the EPA for the use of pheromones and other semiochemical pesticides are their potential acute toxicity, possible irritation, or sensitization; their potential for mutagenic, teratogenic, or oncogenic activity; and their potential effects on the immune system. In the first series of tests, data on acute toxicity, irritation, hypersensitivity, mutagenicity, and immunotoxicity are generated. If the studies conducted in this first tier of tests generate all negative results, no further testing is required. However, if positive effects are indicated, subsequent tests, including subchronic and chronic exposure tests of animals and teratogenicity, oncogenicity, and additional mutagenicity and cellular immune response tests, are required to evaluate and quantify the nature of the hazard uncovered in the first-tier tests.

D. Nontarget Organism Hazard, and Environmental Fate, and Expression

The data requirements are identified in 40 CFR §158.690(d) (4). The applicant for a new semiochemical pesticide product should address those requirements in terms of the risks that the uses proposed for the product will pose. If the uses are likely to pose definite risks to nontarget organisms the applicable data will be required. It is up to the registrant to establish if use of the product will result in any risks to nontarget organisms or the environment.

Title 40 CFR 158.50 and 158.100 through 158.102 (4) describe how nontarget organism effect, and environment fate, and expression data requirements for semiochemical pesticides are determined and how to determine the appropriate substances to be tested. Table 4 is adapted from 40 CFR 158.690(d) and specifies nontarget organism, fate, and expression data requirements for semiochemical pesticide products, and the conditions under which the data are required.

TABLE 1 Product Analysis Data Requirements

Kinds of data required	General use: Terrestrial, Food crop	Terrestrial, Non-food	Aquatic, Food crop	Aquatic, Non-food	Greenhouse, Food crop	Greenhouse, Non-food
Product identity	[R]	[R]	[R]	[R]	[R]	[R]
Manufacturing process[a]	[R]	[R]	[R]	[R]	[R]	[R]
Discussion of formation of unintentional ingredients[b]	[R]	[R]	[R]	[R]	[R]	[R]
Analysis of samples[c]	[CR]	[CR]	[CR]	[CR]	[CR]	[CR]
Certification of limits	[R]	R	[R]	R	[R]	R
Analytical methods	R	R	R	R	R	R
Physical and chemical properties	[R]	[R]	[R]	[R]	[R]	[R]
Submittal of samples[d]	[CR]	[CR]	[CR]	[CR]	[CR]	[CR]

Abbreviations: R, required; CR, conditionally required; MP, manufacturing-use product; EP*, end-use product (asterisk identifies those data requirements that end-use applicants (i.e., "formulators") must satisfy, provided that their active ingredient(s) (are) purchased from a registered source); TGAI, technical grade of the active ingredient; brackets (i.e., [R], [CR]) indicate data requirements that apply when an experimental use permit is being sought.
[a]If an EUP is being sought, a schematic diagram or description of the manufacturing process or both will suffice if the product is not already under full-scale production.

Again negative results in the first tier of tests preclude the need for conducting Tier II and III tests.

It can be seen from Table 4 that testing to evaluate hazard to non-target organisms is divided into four general areas; hazard to terrestrial wildlife, aquatic animals, plants, and beneficial insects. Each area of testing is arranged in a hierarchial or tier system. It should be noted that the second tier of tests consists of a series of environmental fate studies rather than safety tests. The Tier II studies are designed to determine if there is a potential for expression of the effects that may have been identified through the Tier I studies. The

patterns			Test substance		
Forestry	Domestic outdoor	Indoor	Data to support MP	Data to support EP	Guidelines (see Ref. 1)
[R]	[R]	[R]	MP	EP*	151-10
[R]	[R]	[R]	MP and TGAI	EP* and TGIA	151-11
[R]	[R]	[R]	MP and TGAI	EP* and TGIA	151-12
[CR]	[CR]	[CR]	MP and TGAI	EP* and TGIA	151-13
R	R	R	MP	EP*	151-15
R	R	R	MP	EP*	151-16
[R]	[R]	[R]	MP and TGAI	EP* and TGIA	151-17
[CR]	[CR]	[CR]	MP and TGAI, PAI	EP*, TGAI, and PAI	151-18

[b]If the product is not already under full-scale production and an EUP is being sought, a discussion of unintentional ingredients shall be submitted to the extent this information is available.
[c]Required to support registration of each manufacturing-use product and end-use products produced by an integrated formulation system. Data on other end-use products will be required on a case-by-case basis. For pesticides in the production stage, a rudimentary product analytical method and data will suffice to support an EUP.
[d]Routinely required for products by an integrated formulation system. Required on a case-by-case basis for other products or materials.

Tier III studies are intended to further evaluate the nature of the risk and to quantify the extent of hazard posed by the intended use(s) of the substance.

VII. EXPERIMENTAL USE PERMITS

Registrants of semiochemical pesticide products are required to obtain an experimental use permit (EUP) to conduct field tests with their products. They should apply for the permit well in advance

TABLE 2 Pesticide Residue Data Requirements

	General use					
	Terrestrial		Aquatic		Greenhouse	
Kinds of data required	Food crop	Non-food	Food crop	Non-food	Food crop	Non-food
Chemical identity[a,b,n]	[CR]	[CR]	[CR]	[CR]	[CR]	[CR]
Direction for use[a,c,n]	[CR]	[CR]	[CR]	[CR]	[CR]	[CR]
Nature of the residue						
plants[a,n]	[CR]		[CR]		[CR]	
livestock[a,d,n]	[CR]		[CR]		[CR]	
Residue analytical method[a,e,n]	[CR]		[CR]		[CR]	
Magnitude of the residue crop field trials[a,n]	[CR]		[CR]		[CR]	
processed food/feed[a,f]	[CR]		[CR]		[CR]	
meat/milk/poultry/eggs[a,g]	[CR]		[CR]		[CR]	
potable water[a,h]			[CR]	[CR]		
fish[a,i]			[CR]	[CR]		
irrigated crops[a,j]			[CR]	[CR]		
food handling[a,k]						
Reduction of residue[a,l]	[CR]		[CR]		[CR]	
Proposed tolerance[a,m]	[CR]		[CR]		[CR]	
Reasonable grounds in support of the petition	[CR]		[CR]		[CR]	

Abbreviations: CR, conditionally required data; TGAI, technical grade of the active ingredient; PAIRA, pure active ingredient, radiolabeled; TEP, typical end-use product; MP, manufacturing-use product; brackets (i.e., [CR]) indicate data requirements that apply when an EUP is being sought; EP, end-use product.

[a]Residue chemistry data requirements shall apply to semiochemical pesticide products when any one or more of the following conditions apply:

patterns			Test substance		
Forestry	Domestic outdoor	Indoor	Data to support MP	Data to support EP	Guidelines (see Ref. 1)
[CR]	[CR]	[CR]	TGAI	TGAI	153-3
[CR]	[CR]	[CR]			153-3
	[CR]		PAIRA	PAIRA	153-3
	[CR]		PAIRA and plant metabolites	PAIRA and plant metabolites	153-3
	[CR]		TGAI and metabolites	TGAI and metabolites	153-3
	[CR]		TEP	TEP	153-3
			EP	EP	153-3
		[CR]	TGAI or plant metabolites	TGAI or plant metabolites	153-3
			EP	EP	153-3
			EP	EP	153-3
			EP	EP	153-3
		[CR]	EP	EP	153-3
			Residue of concern	Residue of concern	153-3
			Residue of concern	Residue of concern	153-3
					153-3

1. Tier II or III toxicology data are required, as specified for semiochemical agents in (3) of this section.
2. The application rate of the product exceeds 0.7 oz (20 g) active ingredient per acre per application.
3. The application rate of the product exceeds a level determined to be comparable with 0.7 oz active ingredient per application but the application rate is not expressible in terms of ounces per acre per application.

(continued)

[Footnotes to Table 1 (cont.)]

[b]The same chemical identity data as required in (a) of this section are required, with emphasis on impurities that could constitute a residue problem.
[c]Required information includes crops to be treated, rate of application, number and timing of applications, preharvest intervals, and relevant restrictions.
[d]Data on metabolism in livestock are required when residues occur on a livestock feed, or the pesticide is to be applied directly to livestock.
[e]A residue method suitable for enforcement of tolerances is needed whenever a numeric tolerance is proposed. Exemptions from the requirement of a tolerance will also usually require an analytical method.
[f]Data on the nature and level of residue in processed food/feed are required when detectable residues could be concentrated in processing and thus require establishment of a food additive tolerance.
[g]Livestock feeding studies are required whenever a pesticide occurs as a residue in a livestock feed. Uses involving direct application to livestock will require animal treatment residue studies.
[h]Data on residues in potable water are required whenever a pesticide is to be applied directly to water, unless it can be determined that the treated water would not be used (eventually) for drinking purpose, by man or animals.
[i]Data on residues in fish are required whenever a pesticide is to be applied directly to water.
[j]Data on residues in irrigated crops are required when a pesticide is to be applied directly to water that could be used for irrigation or to irrigation facilities such as irrigation ditches.
[k]Data on residues in food/feed in food handling establishments are required whenever a pesticide is to be used in food/feed-handling establishments.
[l]Reduction of residue data are required when the assumption of tolerance level residues results in an unsafe level of exposure. Data on the level of residue in food as consumed will be used to obtain a more precise estimate of potential dietary exposure.
[m]The proposed tolerance must reflect the maximum residue likely to occur in crops and meat, milk, poultry, or eggs.
[n]Residue data for outdoor domestic uses are required if home gardens are to be treated and the home garden-use pattern is different from the use pattern on which the tolerances were established.

of the initiation of the proposed field tests. The conditions and requirements for obtaining an EUP are described in 40 CFR Part 172 (5). Usually, the EUP for a semiochemical is designed to determine whether or not the product is effective in altering a specific pest's behavior, and to obtain the required residue data if food crops are involved. When food crops are involved, a temporary tolerance or an exemption from a tolerance must be obtained before an EUP may be issued, unless the applicant destroys the treated crop. An application for an EUP should clarify whether the crop is intended to be consumed by humans or by animals intended for human consumption.

It should be noted that some of the data requirements for a semiochemical, such as product analysis data, residue data, and toxicity data, are just as essential for an EUP as they are for a registration under the FIFRA. In the previously discussed tables, the data requirements for an EUP are indicated by placing these requirements in brackets.

VIII. DATA WAIVER REQUESTS

If an applicant believes that any of the data requirements cited do not apply to a particular product proposed for registration, he or she has the option of submitting requests to waive certain items of data. The rationale and policy for consideration of data waiver requests by the EPA are described at 40 CFR §158.45 (4). It should be noted that for the agency to waive any required item of data, the registrant must initiate the request and provide a scientific rationale to substantiate waiving each of the items of data in question. Requests to summarily waive all of the data requirements based on general assumptions about semiochemicals or testimonials to the generally recognized attributes of semiochemicals as a class of pesticides cannot be used by the EPA to waive data requirements.

The EPA recognizes the general merits of semiochemicals, that is, low relative mammalian toxicity, low use volume, pest specificity, relatively minor impact on the environment, and that pheromones are desirable alternatives to conventional (broad-spectrum) pesticides. In reviewing applications for registration of semiochemicals, the agency takes those factors into consideration and, thus, gives applications for registration of pheromones priority, consistent with other competing priorities in terms of time and resources.

Notwithstanding those considerations, the agency is obligated to review applications for registration of new biochemical pesticides (including semiochemicals) on a product-by-product basis against the applicable data requirements. In many situations the product in question will likely be registered if the registrant provides the minimal

TABLE 3 Toxicology Data Requirements

Kinds of data required	General use: Terrestrial Food crop	Terrestrial Non-food	Aquatic Food crop	Aquatic Non-food	Greenhouse Food crop	Greenhouse Non-food
Tier I						
acute oral toxicity[a]	[R]	[R]	[R]	[R]	[R]	[R]
acute dermal toxicity[a,b]	[R]	[R]	[R]	[R]	[R]	[R]
acute inhalation[n]	[R]	[R]	[R]	[R]	[R]	[R]
primary eye irritation[b]	[R]	[R]	[R]	[R]	[R]	[R]
primary dermal irritation[a,b]	[R]	[R]	[R]	[R]	[R]	[R]
hypersensitivity study[c]	CR	CR	CR	CR	CR	CR
hypersensitivity incidents[d]	CR	CR	CR	CR	CR	CR
studies to detect genotoxicity[e]	[R]	[CR]	[R]	[CR]	[R]	[CR]
immune response	[R]	R	[R]	R	[R]	R
90-day feeding (1 sp.)[f]	CR	CR	CR	CR	CR	CR
90-day dermal (1 sp.)[g]	CR	CR	CR	CR	CR	CR
90-day inhalation (1 sp.)[h]	CR	CR	CR	CR	CR	CR
teratogenicity (1 sp.)[i]	CR	CR	CR	CR	CR	CR
Tier II						
mammalian mutagenicity tests[j]	CR	CR	CR	CR	CR	CR
immune response[k]	CR	CR	CR	CR	CR	CR
Tier III						
chronic exposure[l]	CR		CR		CR	
oncogenicity[m]	CR		CR		CR	

Abbreviations: R, required, CR, conditionally required; MP, manufacturing-use product; EP*, end-use product (asterisk identifies those data requirements that end-use applicants (i.e., "formulators") must satisfy, provided that their active ingredient(s) is (are) purchased from a registered source; TGAI, technical grade of the active ingredient; brackets (i.e., [R], [CR]) indicate data requirements that apply when an experimental use permit is being sought.

patterns			Test substance		
Forestry	Domestic outdoor	Indoor	Data to support MP	Data to support EP	Guidelines (see Ref. 1)
[R]	[R]	[R]	MP and TGAI	EP* or EP dilution* and TGAI	152-10
[R]	[R]	[R]	MP and TGAI	EP* or EP dilution* and TGAI	152-11
[R]	[R]	[R]	MP and TGAI	EP* and TGAI	152-12
[R]	[R]	[R]	MP	EP	152-13
[R]	[R]	[R]	MP	EP	152-14
CR	CR	CR	MP	EP	152-15
CR	CR	CR			152-16
[CR]	[CR]	[CR]	TGAI	TGAI	152-17
R	R	R	TGAI	TGAI	152-18
CR	CR	CR	TGAI	TGAI	152-20
CR	CR	CR	TGAI	TGAI	152-21
CR	CR	CR	TGAI	TGAI	152-22
CR	CR	CR	TGAI	TGAI	152-23
CR	CR	CR	TGAI	TGAI	152-19
CR	CR	CR	TGAI	TGAI	152-24
		CR	TGAI	TGAI	152-26
		CR	TGAI	TGAI	152-29

[a]Not required if test material is a gas or is highly volatile.
[b]Not required if test material is corrosive to skin or has pH less than 2 or greater than 11.5; such a product will be classified toxicity category 1 on the basis of potential eye and dermal irritation effects.
[c]Required if repeated contact with human skin results under condition of use.
[d]Incidents must be reported, if they occur.

(continued)

[Footnotes to Table 3 (Cont.)]

[e]Required to support nonfood uses if use is likely to result in significant human exposure, or if the active ingredient or its metabolites is (are) structurally related to a known mutagen or belong(s) to any chemical class of compounds containing known mutagens.
[f]Required if the use requires a tolerance or an exemption from the requirement for a tolerance, or its use requires a food additive regulation or if the use of the product is otherwise likely to result in repeated human exposure by the oral route.
[g]Required if pesticidal use will involve purposeful application to the human skin or will result in comparable prolonged human exposure to the product (e.g., swimming pool algaecides, pesticides for impregnating clothing), and if either of the following criteria are met:

1. Data from a subchronic oral study are not required.
2. The active ingredient of the product is known or is expected to be metabolized differently by the dermal route of exposure than by the oral route, and a metabolite of the active ingredient is the toxic moiety.

[h]Required if pesticidal use may result in repeated inhalation exposure at a concentration that is likely to be toxic.
[i]Required if any of the following criteria are met:

1. Use of the product under widespread and recognized practice may reasonably be expected to result in significant exposure to female humans.
2. Its use requires a tolerance or an exemption from the requirements for a tolerance, or its use requires issuance of a food additive regulation.

[j]Required if results from any one of the Tier I mutagenicity tests are positive.
[k]Required if adverse effects are observed in the Tier I immune response studies.
[l]Required if the potential for adverse chronic effects are indicated based on the following:

1. The subchronic effect levels established in the Tier I subchronic oral toxicity studies, the Tier I subchronic dermal toxicity studies, or the Tier I subchronic inhalation toxicity studies.
2. The pesticide use pattern (e.g., rate, frequency and site of application).
3. The frequency and level of repeated human exposure that is expected.

[m]Required if the product meets either of the following criteria:

1. The active ingredient(s) or any of its (their) metabolites, degradation products, or impurities produce(s) in Tier I subchronic studies demonstrate a morphological effect (e.g., hyperplasia, metaplasia) in any organ that potentially could lead to neoplastic change.
2. If adverse cellular effects suggesting oncogenic potential are observed in Tier I or Tier II immune response studies or in Tier II mammalian mutagenicity assays.

[n]Required if the product consists of, or under conditions of use results in, an inhalable material (e.g., gas, volatile substance, or aerosol/particulate).

TABLE 4 Nontarget Organisms: Fate and Expression Data Requirements

Kinds of data required	General use					
	Terrestrial		Aquatic		Greenhouse	
	Food crop	Non-food	Food crop	Non-food	Food crop	Non-food
Tier I						
avian acute[a,b]	[R]	[R]	[R]	[R]	CR	CR
avian dietary[a,b,d]	[R]	[R]	[R]	[R]	CR	CR
freshwater fish LC_{50}[a,b,e]	[R]	[R]	[R]	[R]	CR	CR
freshwater invertebrate[a,b,g]	[R]	[R]	[R]	[R]	CR	CR
nontarget plant studies[c]		R		R		
nontarget insect testing[d,e]	CR	CR	CR	CR	CR	CR
Tier II						
volatility[h]	CR	CR	CR	CR		
dispenser-water leaching[i]	CR	CR	CR	CR		
adsorption-desorption[j]	CR	CR	CR	CR		
octanol/water partition[j]	CR	CR	CR	CR		
UV absorption[k]	CR	CR	CR	CR		
hydrolysis[j]	CR	CR	CR	CR		
aerobic soil metabolism[j]	CR	CR	CR	CR		
aerobic aquatic metabolism[j]	CR	CR	CR	CR		
soil photolysis[j]	CR	CR	CR	CR		
aquatic photolysis[j]	CR	CR	CR	CR		
Tier III						
terrestrial wildlife testing[l]	CR	CR	CR	CR		
aquatic animal testing[m]	CR	CR	CR	CR		
nontarget plant studies[n]						
nontarget insect testing[d]	CR	CR	CR	CR		

Abbreviations: R, required; CR, conditionally required; brackets (i.e., [R], [CR]) indicate data requirements that apply to products for which an EUP is being sought; MP, manufacturing-use product; TEP, typical end-use product; TGAI, technical grade of the active ingredient; EP, end-use product; PAI, "pure" active ingredient.

patterns			Test substance		
Forestry	Domestic outdoor	Indoor	Data to support MP	Data to support EP	Guidelines (see Ref. 1)
[R]	[R]	CR	TGAI	TGAI	154-6
[R]	[R]	CR	TGAI	TGAI	154-7
[R]	[R]	CR	TGAI	TGAI	154-8
[R]	[R]	CR	TGAI	TGAI	154-9
R			TGAI metabolites	TGAI metabolites	154-10
CR	CR		TGAI	TGAI	154-11
CR	CR		TEP	TEP	155-4
CR	CR		EP	EP	155-5
CR	CR		TGAI	TGAI	155-6
CR	CR		TGAI	TGAI	155-7
CR	CR		PAI	PAI	155-8
CR	CR		TGAI	TGAI	155-9
CR	CR		TGAI	TGAI	155-10
CR	CR		TGAI	TGAI	155-11
CR	CR		TGAI	TGAI	155-12
CR	CR		TGAI	TGAI	155-13
CR	CR		TGAI	TGAI	154-12
CR	CR		TGAI	TGAI	154-13
			TGAI	TGAI	154-14
CR	CR		TGAI	TGAI	154-15

[a]Tests for pesticides intended solely for indoor application will be required on a case-by-case basis, depending on use pattern, production volume, and other pertinent factors.
[b]Preferable test species are bobwhite quail or mallard for avian acute oral and avian dietary studies; rainbow trout for freshwater fish

(continued)

[Footnotes to Table 4 (Cont.)]

studies; and *Daphnia* for freshwater invertebrate studies on semiochemicals.

[c]Data are required for pesticides to be used in forests and natural grasslands. Data are also required when the supported products are to be used in other locations when any of the following conditions are met:

1. Phytotoxicity problems arise and open literature data are not available.
2. The product may pose hazards to endangered or threatened species.
3. A Rebuttal Presumption Against Registration Special Review has been initiated on the product.

[d]Required depending on pesticide mode of action and results of any available product performance data.

[e]Biochemicals introduced directly into an aquatic environment when used as directed shall be tested as specified in 40 CFR 158.145.

[f]Not required if pesticide is highly volatile (estimated volatility greater than 5×10^5 atm^3/mol).

[g]If the pesticide will be introduced directly into an aquatic environment when used as directed, then it must be tested as indicated in 40 CFR 158.145.

[h]Required when results of any one or more of the tier I tests indicate potential adverse effects on nontarget organisms and the biochemical agent is to be applied on land.

[i]Required when results of any one or more of Tier I tests indicate potential adverse effects on nontarget organisms and the semiochemical agent is to be applied on land in a passive dispenser.

[j]Required on a case-by-case basis when results of Tier I tests indicate environmental fate data are needed.

[k]Required when results of Tier I tests indicate potential adverse effects on beneficial insects and the intended route of exposure of the pesticide is through vapor-phase contact.

[l]Required if either of the following criteria are met:

1. Environmental fate characteristics indicate that the estimated concentration of the semiochemical pesticide in the terrestrial environment is equal to or greater than one-third the avian dietary LC_{50} or the avian single dose oral LD_{50} (converted to ppm).
2. The pesticide or any of its metabolites or degradation products are stable in the environment to the extent that potentially toxic amounts may persist in the avian feed.

[m]Required if environmental fate characteristics indicate that the estimated environmental concentration of the biochemical agent in the

aquatic environment is equal to or greater than 0.01 of any EC_{50} or LC_{50} determined in testing required by Tier 1 aquatic tests.
[n]Required if the product is expected to be transported from the site of application by air, soil, or water. The extent of movement will be determined by the Tier II environmental fate tests.
[o]Required when results of Tier I tests indicate potential adverse effects on nontarget insects and results of Tier II tests indicate exposure of nontarget insects.

information described in the foregoing. How long it takes to register a given new semiochemical and how much data the applicant must provide depend on the nature of the semiochemical, the proposed uses, the use pattern, and whether or not the applicant provides, at the very least, an adequate label describing the uses, product analysis data, and a justification for the EPA to waive any of the data requirements that do not apply.

IX. EXAMPLES OF DATA REQUIRED TO REGISTER SEMIOCHEMICAL PESTICIDES

The following are two examples of pheromone and other semiochemical pesticide products that have been registered by the U.S. Environmental Protection Agency. The data submitted and evaluated in each case are identified.

These examples should provide insights into the way the EPA implements the policies that have been discussed previously.

Example A

1. *Stirrup-M: Biochemical Pesticide (Pheromone)*

Generic names: (*Z*,*E*) 3,7,11-Trimethyl-2,6,10-dodecatriene-1-ol
3,7,11-Trimethyl-1,6,10-dodecatriene-3-ol
Common names: farnesol and nerolidol
Trade name: Stirrup-M
EPA/OPP chemical code (Shaughnessy) numbers:
(*Z*,*E*) 3,7,11-Trimethyl-2,6,10-dodecatriene-1-ol: 128911
3,7,11-Trimethyl-1,6,10-dodecatriene-3-ol: 128910
Chemical Abstracts Service (CAS) number: 801E
Year of initial registration: 1987

a. Biochemical pesticide characteristics of Stirrup-M: Stirrup-M – a selective mite pheromone to mitigate tetranychid mite infestation. It acts to permeate the surrounding area to which it is applied, giving off an olfactory stimulant that attracts mites to the crop surfaces that have been treated with both the pheromone and a conventional miticide (in a tank mix). The quantity and frequency of application of the conventional miticide may thus be reduced.

b. Use pattern and formulations:

1. Application site: To be tank-mixed with other conventional miticides for use on raw agricultural commodities on which the conventional pesticide is registered for use to control tetranychid mites

2. Pest controlled: Species of tetranychid mites
3. Type of formulation: Liquid (for tank-mixing with registered miticides)

2. Data Requirements

a. Product analysis: Product chemistry data as per 40 CFR Section 158.690(a) (4) (see Table 1) were submitted and reviewed.

b. Residue chemistry data requirements: Residue chemistry data are applicable only if Tier II or Tier III toxicology data are required (or if application rate exceeds 20 g/acre application) as specified by 40 CFR 158.165 (b) (4). This was not the case with this product. Therefore, no residue data were required to register Stirrup-M.

C. Toxicology data requirements: The company submitted a battery of studies as required by 40 CFR 158.690(c) (4). These studies included the following data which were reviewed and classified as follows:

1. Acute oral LD_{50}: rat (LD_{50} = >5050 mg/kg)
   Toxicity category: IV
   Classification: Core Minimum Data
2. Acute Dermal LD_{50}: rat (LD_{50} = >2020 mg/kg)
   Toxicity category: III
   Classification: Core Minimum Data
3. Primary dermal irritation: rabbit; using Stirrup-M (0.923% active ingredient); the test material caused slight irritation at 1 hr that cleared within 24 hr.
   Toxicity category: IV
   Classification: Core Guideline Data
4. Primary eye irritation: rabbit; using Stirrup-M (0.923% active ingredient). The test material produced mild ocular irritation in the washed and unwashed rabbit eyes.
   Toxicity category: III
   Classification: Core Guideline Data
5. Acute inhalation toxicity: rat (LC_{50} = 3.37 mg/L) from 2 to 4 hr
   Toxicity category: III
   Classification: Core Minimum Data
6. Ames mutagenicity assay: No mutagenic potential shown by this assay.
   Classification: Acceptable study
7. DNA repair test with $PolA^+$ and $PolA^-$ *Escherichia coli* bacteria, using Stirrup-M. No mutagenic potential.
   Classification: Acceptable study

d. Ecological effects data requirements: The applicable data consist of the following:

1. Avian single-dose oral toxicity test: bobwhite quail (Guideline No. 154-6) (1)
2. Avian single-dose oral toxicity test: mallard duck (Guideline No. 154-6) (1)
3. Fish acute bioassay: rainbow trout (Guideline No. 154-8).
4. Aquatic invertebrate acute bioassay: *Daphnia magna* (Guideline No. 154-9) (1)

The foregoing tests were conducted with technical material. The protocols, including formulas for calculating maximum dose levels, for conducting the studies were as found in Pesticide Assessment Guidelines, Subdivision M, Biorational Pesticides, A-540/9-82-028, October 1982 (1).

3. Summary of Regulatory Position and Rationale

The EPA concluded that all data requirements were met for Stirrup-M and granted an unconditional registration.

The following information indicated the potential safety of the product:

Both active components of Stirrup-M have been cleared under 21 CFR 172.515 for use as synthetic flavoring agents and adjuvants in food.

Nerolidol and farnesol, the active ingredients in Stirrup-M, are naturally occurring terpene alcohols that are widely distributed in nature in plant materials. Both components have been used as natural constituents of essential oils in perfumes. They are also found in orange juice.

The low application rate of 3.04 g/acre mixed with registered miticides suggests that the possibility of adverse effects to humans from exposure to Stirrup-M would be negligible or nonexistent.

Tier I biochemical pesticide acute toxicity tests demonstrate that Stirrup-M exhibits very low toxicity. In addition, the battery of Stirrup-M mutagenicity tests reviewed showed no mutagenicity potential.

Stirrup-M is a relatively low-toxicity pesticide that is estimated to have an exposure level well under the lowest aquatic toxicity value (1.8 ppm), by more than 1000-fold. Thus, no deleterious effects from its use is expected. This product will be used as part of an integrated pest management (IPM) system. This will result in the reduction of the use of broad-spectrum conventional pesticides of relatively higher toxicity.

The development and use of these innovative methods of pest control are deemed to be in the public interest because they may serve to alleviate the use of larger quantities and more frequent use of the more conventional pesticides.

Example B

1. Nomate Blockaide: Biochemical Pesticide
(Aggregation pheromone and plant volatiles combination)

Generic names:

(Z)-2-Isopropenyl-1-methylcyclobutaneethanol (Z)-3,3-Dimethyl-$\Delta^{1,\beta}$-cyclohexaneethanol (Z)-3,3-Dimethyl-$\Delta^{1,\alpha}$-cyclohexaneethanal (E)-3,3-Dimethyl-$\Delta^{1,\alpha}$-cyclohexaneethanal	grandlure
Cyclic decadiene Cyclic decene Cyclic pentadecatriene	plant volatiles

EPA/OPP chemical code (Shaughnessy Number(s):

Grandlure	112401-5
Plant Volatiles	126001-6

Chemical Abstracts Service (CAS) number: 605B
Year of initial registration: 1983

a. Biochemical pesticide characteristics of Nomate Blockaide: Nomate Blockaide contains grandlure and plant volatiles as a pesticide for use in the control of the cotton boll weevil on cotton. Nomate Blockaide is intended for use as an olfactory adjuvant to enhance the efficiency of conventional insecticide treatments.

The Nomate Blockaide application rate of 2 g/acre is designed to be released over a 21-day period. The effectiveness of conventional insecticide treatments traditionally used for boll weevil control is enhanced by causing the weevils to aggregate in designated areas. Boll weevils naturally migrate from overwintering places to edges of cotton fields. Once at the edge of the cotton fields, male boll weevils release an aggregation pheromone that attracts additional males and females. The product mimics this action.

b. Use pattern and formulation

1. Application site: cotton
2. Pest controlled: cotton boll weevil
3. Type of formulation: liquid (for mixing with cottonseed oil)

2. Data Requirements

a. Product analysis: Product chemistry data as per 40 CFR Section 158.690(a) (4) (see Table 1) were submitted and reviewed.

b. Residue chemistry data requirements: Residue chemistry data are applicable only if Tier II or Tier III toxicology data are required (or if application rate exceeds 20 g/acre application) as specified by 40 CFR 158.690(b) (4). This was not the case with this product. Therefore, no residue data were required to register Nomate Blockaide.

c. Toxicology data requirements: The company submitted a battery of studies as required by 40 CFR 158.690(c) (4) (see Table 3). The studies included:

For plant volatile chemicals

1. Acute oral LD_{50}: rat
   Toxicity category: IV
   Classification: Core Minimum Data
2. Acute dermal LD_{50}: rat
   Toxicity category: III
   Classification: Core Minimum Data
3. Primary dermal irritation: rabbit
   Toxicity category: IV
   Classification: Core Guidelines Data
4. Primary eye irritation: rabbit
   Toxicity category: IV
   Classification: Core Guidelines Data
5. Acute inhalation toxicity: rat
   Toxicity category: III
   Classification: Core Guidelines Data
6. Ames mutagenicity assay
   Classification: Acceptable study – no mutagenicity
7. Cellular immune response studies
   Classification: Acceptable study

For grandlure (boll weevil pheromone)

Registrant cited the acute oral LD_{50}, acute dermal LD_{50}; primary dermal, primary eye irritation, and acute inhalation toxicity LC_{50} studies previously submitted in support of grandlure registration. In addition the registrant submitted:

1. Ames mutagenicity assay
   Classification: Acceptable study – no mutagenicity
2. Cellular immune response study
   Classification: Acceptable study

d. Ecological effects data requirements: These data have been reviewed as per Tier I for ecological effects including

1. Fish LC_{50}: rainbow trout, using plant volatile
2. Acute LC_{50}: aquatic invertebrate (*Daphnia sp.*) using grandlure

3. Acute LC_{50}: aquatic invertebrate (*Daphnia sp.*) using plant volatile

3. *Summary of Regulatory Position and Rationale*

On the basis of the available data, the EPA concluded that the proposed use provides for a minimal potential risk to nontarget organisms. The subject semiochemical will be used as part of an IPM program. The EPA is committed to encourage the development of such biological pesticides. The development and use of this innovative method of pest control was deemed to be in the public interest because it could serve to alleviate the use of more toxic insecticides.

These two examples of data required to register semiochemical pesticides should provide some insight in how the EPA applies these regulations. For an additional discussion of the implementation of these regulations refer to Hodosh et al. (6). This document treats this subject from the outsider's perspective.

X. LIST OF SEMIOCHEMICALS REGISTERED BY EPA (7)

Table 5 presents a list of the pheromones and other semiochemicals registered by EPA through November 16, 1987. The list does not necessarily indicate that the products are available in the market place.

TABLE 5 List of Semiochemicals Registered by the EPA

Common name (trade name) and target insect	Chemical name	Product brand name	EPA reg. no.
Muscalure (*Muscamone*); house fly, *musca domestica* ([a]contains pesticide with muscalure)	(*Z*)-9-Tricosene	Lure 'em II Fly Attractant	270-129
		Attract Fly Stick	270-176
		Flystop Sticky Fly Trap	270-181
		Special Golden Marlin Sugar Bait[a]	2724-227
		Thuron Fly Bait[a]	2724-228
		Thuron Kennel and Yard Fly Killer[a]	2724-250
		Zoecon RF 199 Golden Fly Belt[a]	2724-296
		Fly Attractant FA 5000	34473-2
		Fly Attractant FA 5050	34473-3
		Golden Fly Belt[a]	46200-1
		Stimukil Fly Bait[a]	53871-7
Gossyplure; (Gossyplure H.F.) (Nomate); pink bollworm, *Pectinophora gossypiella*	(*Z,Z*)-7,11-Hexadecadien-1-ol acetate (*Z,E*)-7,11-Hexadecadien-1-ol acetate	Hercon Disrupt Pink Bollworm	8730-21
		Hercon Disrupt/Lure'N Kill PBW	8730-41
		Hercon Disrupt Plus Pink Bollworm	8730-44
		Hercon Disrupt II Pink Bollworm	8730-45
		Attract'N Kill Pink Bollworm	36638-1
		Pink Bollworm Pheromone Technical	50675-2
		Technical Pheromone Gossyplure	53901-2

Disparlure (*Disparlure*); gypsy moth; *Lymantria dispar* ([b]contains pesticide with disparlure)	*cis*-7,8-Epoxy-2-methyloctadecane	Bag-A-Bug Gypsy Moth Sex Lure	562-23 (8845-53)
		Gypsy Moth Monitor Trap[b]	5887-139
		Black Leaf Gypsy Moth Insecticide Strip[b]	5887-140
		Hercon Luretape w/5% Disparlure Gypsy Moth Trap Lure	8730-28
		Hercon Lure Tape Plus	8730-31
		Hercon Disrupt II Gypsy Moth Mating Disruptant	8730-46
		Luretape GM Disparlure Disruptant	8730-47
		Gypsy Moth Pheromone Sachet	36488-4
		Nomate Biogard Gypsy Moth Suppressant	36638-5
		N'Trap TM Gypsy Moth Pheromone Lure	36638-9
		Zoecon Gypsy Moth Pheromone Technical	55947-119
None (Shootgard); western pine shoot borer, *Eucosma sonomana*	(*Z*)-9-Dodecen-1-ol acetate (*E*)-9-Dodecen-1-ol acetate	Nomate Shootgard (Attract'N Kill Shootgard)	36638-2
None (Nomate Borer-Gard); peachtree borer, *Synanthedon exitiosa*	(*E,Z*)-3,13-Octadecadien-1-ol acetate (*Z,Z*)-3,13-Octadecadien-1-ol acetate	Nomate Borer-Gard (Attract'N Kill Borer Gard)	36638-3

(continued)

TABLE 5 (Cont.)

Common name (trade name) and target insect	Chemical name	Product brand name	EPA reg. no.
None (Luretape Bagworm); bagworm, *Thyridopteryx ephemeraeformis*	(*R*)-1-Methylbutyl decanoate	Hercon Luretape; Bagworm	8730-33
None (Nomate Chokegard); artichoke plume moth, *Platyptilia carduidactyla*	(*Z*)-11-Hexadecenal	Hercon Disrupt Artichoke Plume Moth	8730-34
		Nomate Chokegard	36638-10
		Z,11 HDAL Technical	50675-1
		Technical Pheromone Z-11	53901-1
None (Nomate Suppres); tomato pinworm, *Keiferia lycopersicella*	(*E*)-4-Tridecen-1-ol acetate	Nomate Suppres-2	36638-7
	Tetraiodofluorescin	Nomate Suppres	36638-8
None (Nomate Vantage); tobacco budworm, *Heliothis virescens*, and cotton bollworm, *Heliothis zea*	(*Z*)-11-Hexadecenal (*Z*)-9-Tetradecenal	Nomate Vantage	36638-6 (new)
None (Kontrol HV); tobacco budworm *Heliothis virescens*	(*Z*)-11-Hexadecenal (*Z*)-9-Tetradecenal (*Z*)-11-Hexadecen-1-ol (*Z*)-9-Hexadecenal	Kontrol HV	36638-6 (old)

	(*Z*)-7-Hexadecenal Hexadecanal Tetradecanal		
None (Isomate-M); oriental fruit moth, *Grapholita molesta*	(*Z*)-8-Dodecen-1-ol acetate (*E*)-8-Dodecen-1-ol acetate (*Z*)-8-Dodecen-1-ol	Isomate-M	53575-1
Grandlure (Grandlure); boll weevil, *Anthonomus grandis*	(*Z*)-2-Isopropenyl-1-methyl-cyclobutaneethanol (*Z*)-3,3-Dimethyl-$\Delta^{1,\beta}$-cyclohexaneethanol (*Z*)-3,3-Dimethyl-$\Delta^{1,\alpha}$-cyclohexaneethanal (*E*)-3,3-Dimethyl-$\Delta^{1,\alpha}$-cyclohexaneethanal	Hercon Luretape Boll Weevil Pheromone Dispenser	8730-15
Grandlure plus plant volatiles (Nomate Blockaide); boll weevil, *Anthonomus grandis*	(*Z*)-2-isopropenyl-1-methyl-cyclobutaneethanol (*Z*)-3,3-Dimethyl-$\Delta^{1,\beta}$-cyclohexaneethanol (Z)-3,3-Dimethyl-$\Delta^{1,\alpha}$-cyclohexaneethanal (*E*)-3,3-Dimethyl-$\Delta^{1,\alpha}$-cyclohexaneethanol plus cyclic decadiene cyclic decene cyclic pentadecatriene	Nomate Blockaide	36638-11

(continued)

TABLE 5 (Cont.)

Common name (trade name) and target insect	Chemical name	Product brand name	EPA reg. no.
None (Multilure); smaller European elm bark beetle, *Scolytus multistriatus*	(-)-4-Methyl-3-heptanol (-)-α-Cubebene (2-*endo*,4-*endo*)-5-Ethyl-2,4-dimethyl-6-8-dioxabicyclo[3.2.1]octane[c] [c][(-) α-Multistriatin]	Hercon Luretape with Multilure N-Trap Elm Bark Beetle Lures	8730-23 36638-4
None (Stirrup-M Tetranychid Mite Pheromone); two-spotted spider mite, carmine spider mite, European red mite, *Tetranychus* spp.	(*Z*,*E*)-3,7,11-Trimethyl-2,6,10-dodecatrien-1-ol[d] [d][(*Z*,*E*)-Farnesol] 3,7,11-Trimethyl-1,6,10-dodecatrien-3-ol[e] [e](Nerolidol)	Stirrup-M	53871-1
None (American Cockroach Lures, Lure'N Kill[a]); American cockroach, *Periplaneta americana* ([g]contains pesticide with Periplanone B)	[1*R*-(1*R**),2*R*,5*S*,6*E*,10*R**]-(±)-8-Methylene-5-(1-methylethyl)spiro[11-oxabicyclo[8.1.0]undec-6-ene-2,2-oxiran]-3-one[f] [f](Periplanone B)	Lure'N Kill Roach and Ant Killer Insecticidal Baits with Sex Lure[g] Hercon Topbait Cockroach Insecticide Efficacy Enhancer Lure'N Kill Roach and Ant Killer[g]	8730-35 8730-36 8730-38

Japonilure (Japonilure); Japanese beetle, *Popilia japonica*	[*R*-(*Z*)]-5-(1-Decenyl)dihydro-2-(3*H*)-furanone	Black Leaf Japanese Beetle Sex pheromone Lure	5887-141
		Perfect Garden Japanese Beetle Sex Attractant	36488-2
		Cassco Japanese Beetle Sex Lure Bait	46692-1
		Zoecon Japanese Beetle Pheromone Lure	55947-120
Japonilure plus (Integra Lure; combines Japonilure with Floral lure); Japanese beetle, *Popilia japonica*	[*R*-(*Z*)]-5-(1-Decenyl)dihydro-2-(3*H*)-furanone plus 2-Phenylethyl propionate Eugenol	Bag-A-Bug	8845-48
Japonilure plus (Japonilure combined with Floral Lure and oil of geranium); Japanese beetle; *Popilia japonica*	[*R*-(*Z*)]-5-(1-Decenyl)dihydro-2-(3*H*)-furanone plus 2-Phenylethyl propionate Eugenol plus Oil of geranium	Ellisco Double-Lure Japanese Beetle Bait	24067-3
		Zoecon Japanese Beetle Trap	55947-118
None (Floral Lure); Japanese beetle, *Popilia japonica*	2-Phenylethylpropionate Eugenol	Hercon Floratape Japanese Beetle Food Lure Dispense	8730-24
		Ellisco Granular Japanese Beetle Bait	24067-1
		Ellisco's Aromabar Japanese Beetle Lure	24067-2

(continued)

Table 5 (Cont.)

Common name (trade name) and target insect	Chemical name	Product brand name	EPA reg. no.
None (Floral Lure with oil of geranium); Japanese beetle, *Popilia japonica*	2-Phenylethyl propionate Eugenol plus Oil of Geranium	Japanese Beetle Flora-Lure with Geraniol	55947-21
		Japanese Beetle Floral-Lure II with Geraniol	55947-22

(Trade names in parentheses)
[a]Contains pesticide with muscalure
[b]Contains pesticide with disparlure
[c][(-)α-Multistriatin]
[d][(*Z*,*E*)-Farnesol]
[e](Nerolidol)
[f](Periplanone B)
[g]Contains pesticide with Periplanone B

REFERENCES

1. United States Environmental Protection Agency, Pesticide Assessment Guidelines, Subdivision M, Biorational Pesticides, EPA-54019-82-028, October 1982.
2. An Act of Congress, The Federal Insecticide, Fungicide and Rodenticide Act; Public Law 92-516 October 12, 1972 as amended by Public Law 94-140 November 28, 1975.
3. Code of Federal Regulations, Title 40 Protection of Environment, Part 162—Regulations For the Enforcement of the Federal Insecticide, Fungicide and Rodenticide Act, July 1, 1987.
4. Code of Federal Regulations, Title 40 Protection of Environment, Part 158—Data Requirements for Registration, July 1, 1987, as revised by Federal Register, May 4, 1988 (53 FR 15993).
5. Code of Federal Regulations, Title 40 Protection of Environment, Part 172—Experimental Use Permits, July 1, 1987.
6. Hodosh, R. J., Keough, E. M., and Luthra, Y. Toxicological evaluation and regulation requirements for biorational pesticides. In *CRC Handbook of Natural Pesticides: Methods*, Vol. I. CRC Press, Boca Raton, 1985, p. 632.
7. United States Environmental Protection Agency Product Label File.

36

Registration of Pheromones in Practice

CHARLES A. O'CONNOR III / McKenna, Conner & Cuneo, Washington, D.C.

I. INTRODUCTION

Pheromones are "substances emitted by members of one species that modify the behavior of others within the same species (1). Pheromones, when used as pesticides, are regulated by the U.S. Environmental Protection Agency (EPA) primarily under the Federal Insecticide, Fungicide, and Rodenticide Act (FIFRA) (2) and the Federal Food, Drug, and Cosmetic Act (FFDCA) (3). FIFRA requires pesticides to be registered by EPA and authorizes the agency to prescribe conditions for their use. FFDCA requires the agency to establish maximum acceptable levels of pesticide residues in foods. In establishing the conditions for a pesticide's use, both under FIFRA and FFDCA, EPA is required to consider both the benefits derived from the use of a pesticide and any risks that it may pose to public health or the environment.

Pheromones are used to disrupt reproduction of insects and to attract insects to a trap or other defined area at which conventional pesticides can be used in smaller quantities. EPA describes pheromones as *biorational* pesticides because they provide a safer method of pest control than conventional pesticides. Biorational pesticides are naturally occurring substances and are target-specific so other species are less likely to be affected by their use. Because application requires only a small amount of the active ingredient, biorational pesticides simply raise the level of a substance that already exists in nature. In contrast, conventional insecticides are innately toxic (lethal) to insects and less selective. Human exposure generally also is higher with conventional pesticides.

EPA acknowledged the benefits of biorational pesticides in a 1979 policy statement which proposed that the agency should "[f]acilitate the registration of environmentally acceptable biorational pesticides as alternatives to conventional pesticides by assuring that requirements for the registration of biorational agents are appropriate to their nature and are not unduly burdensome" (4). The two major categories of biorational pesticides are biochemical pest control agents and microbial pest control agents (5). Pheromones are biochemicals that belong to the biologically functional class of *semiochemicals* (chemicals emitted by plants or animals that modify the behavior of receptor organisms of like or different kinds (6). The Preamble to the agency's 1984 *Federal Register* publication of the data requirements for biochemicals also reflects a favorable assessment of biochemical pesticides. EPA intended to provide an incentive for faster development of biochemicals by imposing data requirements for biochemicals which are less extensive than the data requirements for conventional pesticide chemicals (7).

Despite the apparent safety of pheromones and EPA's professed desire to encourage their use, the registration process remains a significant disincentive to the development of new pheromones and pheromone products. The registration process is slow, the cost of generating data is high, and the market for pheromones is limited. Due to the expense and time required for registration, pheromone development is cost-efficient only for use in conjunction with conventional pesticides or on major crops. This paper presents a guide to the registration process with suggestions for proceeding as smoothly as possible. The discussion also focuses on the major problems encountered by companies involved in the development and registration of pheromones and pheromone products.

II. REGISTRATION PROCESS

The registration process involves several stages: product development, testing under an experimental use permit (EUP), and registration. The product development stage involves testing of the product to determine its pesticidal value, toxicity, and other qualities. Thereafter, an EUP allows for large-scale use of unregistered pesticides for testing purposes when a benefit in pest control is expected. Finally, the pheromone may be registered if adequate data is submitted to the agency that indicates that it will not cause "unreasonable adverse effects"on human health or the environment (8).

A. Product Development Before an Experimental Use Permit

1. *Purpose of Product Development*

Initial development of a pheromone requires identification of the chemical structure of the insect pheromone. Researchers then synthesize the chemical structure of the pheromone. Synthesized forms of chemicals generally fit within the scope of biorational pesticides as defined by EPA (9). The EPA's Pesticide Assessment Guidelines Subdivision M: Biorational Pesticides provides the following guidance on the degree to which a synthesized chemical must resemble a natural compound to be classified as a biorational pesticide (10):

> [T]he molecular structure(s) of the major component(s) of the synthetic chemical(s) must be the same as the molecular structure(s) of the naturally occurring analogue(s). Minor differences between the stereochemical isomer ratios (found in the naturally occurring compound compared to the synthetic compound) will normally not rule out a chemical being classified as biorational unless an isomer is found to have significantly different toxicological properties than another isomer.

The agency, on a case-by-case basis, evaluates whether a chemical is substantially similar to a natural pheromone to be classified as a biorational pesticide (11).

After identification and synthesis of the pheromone, the manufacturer should complete tests to determine the usefulness of the pheromone for pest control. If tests indicate that the pheromone will be marketable, the developer must then generate the data required to obtain an EUP.

2. *FIFRA Regulation of Product Development*

The EPA does not require an EUP for certain product testing. If a product is tested solely to determine its pesticidal value, toxicity, or other properties, and no benefit for pest control is expected, EPA does not consider the product to be a pesticide within the meaning of FIFRA and the product is not subject to regulation under the act (12). The EPA presumes that no benefit in pest control is expected when *fewer than 10 acres* are used to test a particular substance or mixture of substances (13). Unless a tolerance or exemption has been established, any food or feed crops affected by the tests must be either destroyed or consumed only by experimental animals (14). Thus, FIFRA jurisdiction over product development rests primarily on the intent of the test; if a benefit from the product is expected, an EUP must be obtained.

3. *TSCA Regulation of Product Development*

The Toxic Substances Control Act (TSCA) (15) regulates "chemical substances." TSCA section 8(b) requires EPA to maintain an inventory of chemical substances manufactured, imported, or processed in the United States for a commercial purpose (16). TSCA section 5 requires any company intending to manufacture or import a "new" chemical substance—that is, one not on the TSCA section 8(b) inventory—to give EPA 90 days prior "premanufacture notice" (17). Importantly, TSCA section 5(h)(3) exempts from this premanufacture notice requirement new chemicals intended solely for research and development (R&D). TSCA section 8 also requires manufacturers to perform recordkeeping and reporting (18). For example, TSCA section 8(e) requires manufacturers to report information concerning substantial risk (19).

TSCA may apply to the distribution and use of a pheromone pesticide before the manufacturer applies for an EUP or a registration. In TSCA section 3 a *chemical substance* is defined such as to exclude any pesticide (as defined in FIFRA) when manufactured, processed, or distributed in commerce for use as a pesticide (20). The EPA regulations implementing the EUP provisions of FIFRA section 5 provide that potential pesticides undergoing laboratory tests and limited field tests (up to 10 acres) in which researchers expect no benefit in pest control are not pesticides under FIFRA (21). Thus, a product undergoing R&D to determine its pesticidal value is not a pesticide under FIFRA and is subject to TSCA.

Two TSCA provisions potentially apply to the distribution of pheromone pesticides for limited testing purposes: premanufacture notices under TSCA section 5 and reporting and recordkeeping under TSCA section 8. The TSCA section 5(h)(3) exempts from premanufacture notice requirements small quantities of new chemical substances produced solely for R&D as long as the manufacturer complies with the risk assessment, notification, and recordkeeping requirements of EPA's implementing regulations (22). As a matter of policy, however, EPA relieves chemicals manufactured exclusively for use as a pesticide from the risk assessment, notification, and reporting requirements of the R&D rule. Thus, pheromone product development is exempt from these requirements as well as from premanufacture notice requirements *as long as* the exclusive intention of the R&D activities is to develop the pheromone as a pesticide, the quantities of pheromone manufactured are no greater than reasonably necessary for such R&D, and any product distributed or sold is labeled "for R&D use only" (23). Certain provisions of TSCA section 8, however, do apply to pheromone pesticides.

Under TSCA section 8(c), pheromone manufacturers and processors must maintain and, upon request from EPA, must submit records of allegations of significant adverse reactions on health or to the environment. Further, TSCA section 8(e) requires any person who manufactures, processes, or distributes a chemical substance and obtains information "which reasonably supports the conclusion" that such substance presents substantial risk of injury to health or the environment shall immediately report such information to EPA. The TSCA section 8(e) reporting obligation also applies to all nonexempt chemicals, including catalysts and intermediates, used in manufacturing the pheromones (24). According to EPA, TSCA jurisdiction over product development ends and FIFRA jurisdiction begins only when the manufacturer applies for an EUP or a registration (25).

B. Use of Pheromone Product Under an Experimental Use Permit

1. Data Requirements

The FIFRA section 5 contains the requirements for an EUP (26). The agency will issue an EUP "only if the administrator determines that the applicant needs such permit in order to accumulate information necessary to register a pesticide under Section 3 of this Act" (27). FIFRA section 5 states that EPA may require studies to detect whether the use of the pesticide under the permit may cause unreasonable adverse effects on the environment. The following data may be required by the EPA to support an application for an EUP: product analysis data, residue data, Tier I toxicology data, and nontarget organisms data (28). Moreover, depending on the results of Tier I testing, EPA may require further studies.

If the pheromone will be used on food or feed crops, the applicant must obtain a temporary tolerance or an exemption from tolerance (29). To do so, EPA requires the applicant for an EUP to submit not only Tier I acute toxicity data (oral, dermal, and inhalation; primary skin and eye) but also subchronic data (oral) and teratology data (one species) (30). These are the very same data required for full registration of the pheromone, except that EPA may require more data depending on the results of these Tier I tests and on whether the applicator has significant dermal or inhalation exposure (31). Thus, the manufacturer must face the testing costs at the EUP stage even before he makes the final decision to seek registration.

2. Time for Processing

Section 5 of FIFRA expressly requires the agency to complete its processing of an EUP within 120 days (32). The agency indicates, however, that processing of an EUP takes a minimum of six months—EPA's Hazard Evaluation Division alone takes four months to evaluate the application. Therefore, to ensure a timely response from EPA, the applicant wishing to test a pheromone during the spring and summer growing seasons should submit his EUP application early in the fall of the prior year. Apparently, EPA receives a number of EUP applications in the spring and has difficulty processing them before the start of the growing season. Indeed, EPA reviewers indicate that they should receive the EUP application a full year before registration is needed.

3. *Meeting With EPA*

The agency informally suggests that applicants for a registration contact the agency *before* undertaking significant testing. The applicant may find such a meeting useful either before initiating testing for data needed to obtain an EUP or in conjunction with the EUP application before initiating testing for additional data necessary for registration. At the meeting, the applicant may seek answers to questions concerning the data requirements or information on the required format for submitting data, or, alternatively, may request a waiver from particular data requirements. As indicated in EPA's PR Notice 85-4 and Title 40 of the Code of Federal Regulations (CFR) §158.40, the agency welcomes meetings with prospective registrants.

PR Notice 85-4 governs communication between registrants and the Office of Pesticide Programs. Meetings must be scheduled; EPA no longer permits drop-in visits. A prospective registrant should assemble a package of materials including the formulation of the product, labeling, mode of action, and an agenda containing specific questions for the agency, and should submit these materials to the product manager with the request for a meeting.

4. *Waivers*

EPA has authority to issue waivers from data requirements when "particular data requirements [would be] inappropriate, either because it would not be possible to generate the required data or because the data would not be useful in the Agency's evaluation of the risks or benefits of the product" (33). The agency issues waivers on a case-by-case basis (34).

The Pesticide Assessment Guidelines suggest that EPA usually waives some data requirements for pheromones. For example, EPA waives the full set of residue chemistry data when Tier I tests show the pheromone is relatively nontoxic, and its application rate does not exceed 0.7 oz active ingredient (AI) per acre (20 g AI/acre) (35).

The applicant should contact the product manager to discuss the availability of a waiver (36). The applicant must submit the waiver request to the product manager in writing. The request should include the following information:

Data requirement for which a waiver is requested
Rationale for waiver
Information to support the request
If possible, alternative means of obtaining the data (37)

The product manager generally will agree to meet with the applicant together with a representative of the Hazard Evaluation Division to discuss the waiver. Applicants for a pheromone registration should review the data requirements carefully and assess the advisability of applying for a waiver. If waivers are frequently granted for a group of pesticides, EPA will announce the practice in a PR Notice. Thus far, EPA has not issued a PR Notice identifying waivers available for pheromones.

C. Registration

Registration of the pheromone should proceed smoothly if the data has been prepared in consultation with the agency and the registration package is complete and adheres to the format required by the agency.

1. Application Packet

The requirements for submitting applications for registration are found at 40 CFR §162.6. The application packet for a new registration should include:

Application form
Complete labeling
Supporting data
Complete formula

Detailed guidance for preparing each of the above documents is available in the regulations (38).

Clear, precise labeling is particularly important to the product manager handling pheromone registration. The application rate, sites, and pests should be clearly stated. Most advertising claims should be put in promotional literature rather than on the label.

The PR Notice 86-5 contains the standard format for data submitted to EPA under FIFRA and FFDCA. Industry representatives, and even EPA personnel, complain that PR Notice 86-5 is difficult to understand. Applicants should review the notice carefully to be certain their application complies with it. The applicant should direct specific questions about PR Notice 86-5 to EPA's Information Services Branch, Program Management and Support Division, at (703) 557-2315.

2. *Length of Process*

According to EPA, a decision on the request for a new biochemical pesticide registration will take eight months. It takes four months for the Hazard Evaluation Division alone to complete its review. In comparison, EPA requires ten months to process a new conventional pesticide registration application. Thus, EPA in practice has done little to streamline registration of pheromones.

D. Exclusion from Registration Process

Pheromones for use only in pheromone traps are excluded from regulation under FIFRA owing to "the character" of the pesticide (39). *Pheromone trap* is defined by EPA (40) as

> [A] device containing a pheromone or identical or substantially similar compounds that: (A) Is used for the sole purpose of attracting and trapping or killing target arthropods; (B) Achieves pest control through removal of target organisms from their natural environment; and (C) Does not result in increased levels of pheromones or identical or substantially similar compounds over a significant fraction of a treated area.

The EPA excluded pheromone traps from regulation because exposure to the pheromone is limited and pheromones are relatively safe (41).

III. PROBLEMS WITH THE REGISTRATION PROCESS

An informal survey of companies actively involved in the registration of pheromones and companies that have considered and rejected pheromone development uncovered several common complaints.

A. Data Requirements

Industry contends that the data requirements for pheromone registration are excessive in light of the relatively safe nature of pheromones and the extremely low rate of application. Because pheromones are not persistent chemicals, the duration of exposure is very limited. Yet, Tier I testing requires several batteries of tests on the pheromone that will cost hundreds of thousands of dollars. Furthermore, the applicant must incur these costs to get an EUP--a stage in product development often too early to tell whether the pheromone is a viable product.

Upon recommendation of the FIFRA Scientific Advisory Panel, moreover, EPA expanded its data requirement to include teratogenicity and subchronic toxicity testing (42). Additionally, EPA may require, on a case-by-case basis, subchronic toxicity (inhalation ahd dermal) studies, depending on anticipated exposure routes.

Not including the cost of the test chemical, the genotoxicity and acute toxicity studies may cost between 25,000 and 50,000 dollars. The subchronic studies could run as high as 250,000 dollars (150,000 dollars for inhalation and 100,000 dollars for dermal), and the teratology study is likely to cost 25,000 to 30,000 dollars. The hypersensitivity studies, which are often waived, are the least expensive (4000–5000 dollars) because these studies can be coupled with the acute package. No data were available on the cost of cellular immune response studies. Smaller companies that produce fewer pesticides may find that their testing costs approach the high end of these estimates because they are less likely to receive volume discounts from testing laboratories.

These cost estimates do not include the cost of the test product itself. Because EPA requires toxicology tests on the active ingredient—the pheromone—such tests can be prohibitively expensive. The cost of producing active ingredient may run in the order of hundreds of dollars per gram. Industry representatives estimate that 50 to 100 g would be necessary to meet the requirements of the acute toxicity test package alone, and an entire kilogram may be necessary to conduct subchronic inhalation studies. Substantial additional quantities would be needed for the genotoxicity, teratology, hypersensitivity, and cellular immune response studies. Thus, if full Tier I testing is required, the cost of providing active ingredient may increase the total cost of meeting EPA's data requirements by several hundred thousand dollars. As a result, indusery has suggested several changes in EPA rules and procedures that would aid the development of pheromones.

1. Expand the Exemption for Traps

The EPA promulgated the exemption for pheromones used in traps because "the toxicity of pheromones in general and the level of potential exposure of man and the environment due to the exempted uses is very low" (43). Commenters on the proposed pheromone trap exclusion argued that the exemption should be extended to trap cropping. Trap cropping involves use of pheromone lures to attract an insect pest to a specific area at which conventional pesticides may then be used to kill the pest.

EPA's decision not to include trap cropping in the exclusion reflects its concern sbout uses of pheromones that raise the background levels of the pheromone and disrupt mating behavior (44). The agency is concerned that use of pheromones to disrupt mating behavior of insect pests also might disrupt the mating behavior of beneficial insects having similar pheromones. However, because pheromones are species-specific, this EPA concern may not be justified for many pheromone products.

2. *Permit Use of Data From Similar Pheromones*

The chemical structure of many pheromones is very similar, yet EPA requires a complete set of toxicology data for each pheromone. Many industry representatives argue that similar pheromones should be grouped together for purposes of generating data. But EPA insists on assessing pheromone registrations on a case-by-case basis. One EPA official stated that the tests are necessary because there is no information available on the environmental effects of artificially high levels of pheromones from their use as pesticides. The EPA contends that in the absence of data the environmental results cannot be predicted.

Yet, with the class of antimicrobial pesticides, EPA has indicated a willingness to permit the industry to test representative products rather than each distinct active ingredient. Perhaps the EPA can apply the same rationale to pheromones, which, like antimicrobials, cannot withstand heavy testing costs and remain economically viable products.

3. *Allow Testing on End-Use Product Rather than Active Ingredient*

Industry representatives argue that EPA should permit testing on the formulated product rather than on the active ingredient. Most pheromone products contain only a small percentage of active ingredient and require a low application ratio. For example, one company reported that its product requires application of only 1 or 2 g of active ingredient per acre. Thus, exposure to the active ingredient is very low. Industry argues that EPA's policy of testing the active ingredient rather than the end-use pheromone product unnecessarily prevents the development of new pheromones because of the amount of high cost material necessary to perform the tests.

The EPA's justification for this requirement is that any deleterious effects of the pheromone will be picked up only in tests performed at the highest dosage level of active ingredient. Nonetheless, the rigid adherence to the requirement of testing the active ingredient may be unnecessarily restrictive in light of the available data which indicates that pheromones are safe. Furthermore, because it takes an enormous concentration of pheromones to produce any deleterious effect, testing at the maximum tolerated dose may be an unrealistic testing requirement for this class of chemical.

B. Delay

The companies uniformly complain that EPA's registration process is too slow. Estimates of the true length of the EUP process ranged up to nine months, whereas the law imposes a four-month limitation. One company indicated that even on data from the prior growing season it was unable to obtain an EUP in time for the start of the succeeding growing season.

EPA frankly admits that only rarely does it meet the statutory deadline of 120 days for reaching a decision on an EUP. This is true even though EPA supposedly gives EUPs priority over other types of applications. Although EPA claims that it processes complete submissions as quickly as possible, delays persist and present a costly problem for companies involved in registration of pheromone pesticides. Perhaps EPA would agree to an administrative program that places pheromones in general, or at least those structurally similar to already registered pheromones, on a fast track.

C. Confusion Regarding Requirements for Experimental Use Permits and Registrations

EPA cited lack of knowledge and resultant failure to comply with the requirements for registration as the most common reason that registrations are denied. In fact, one official stated that no company has been denied a pheromone registration because toxicology testing indicated that the pheromone was unsafe. Rather, an estimated 15% of submissions are returned for failure to format the data correctly and assemble the full registration packet. Conversely, frequent submitters, because they comply with these requirements, have virtually no problems with acceptance of their registration applications.

The best solution is to review carefully the registration kit, the data requirements, the Pesticide Assessment Guidelines, and the PR Notices that contain the requirements. Then, specific questions should be directed to the product manager. Registration information is available from the following sources:

1. 40 CFR §§150-189
   Write or call:
   Superintendent of Documents
   Government Printing Office
   Washington, D.C. 20402
   (202) 783-3238
2. Registration Kit
   Call Bernadine Usen at
   (703) 557-7781
3. Subpart M, Regulatory Guidelines for
   Microbials and Biochemical Pesticides

Write or call:
National Technical Information Service
Attention: Order Desk
5285 Port Royal Road
Springfield, VA 22161
(703) 487-4350

4. PR Notices: To receive notices, call:
Ferial Bishop, (703) 557-7700 or
Summer Gardner, (703) 557-1293

IV. THE FUTURE OF PHEROMONE REGISTRATION

EPA has registered at least 16 different pheromones as active ingredients, as well as many end-use pheromone products incorporating them. Pheromones have been registered for houseflies, pink bollworm on cotton, gypsy moths in residential and forest areas, cotton boll weevil, elm bark beetles in elm trees, and cockroaches. The EPA stated that new registrations are frequently submitted for the pink bollworm pheromone, gossyplure. For example, one company has registered gossyplure for application on cotton in the form of tiny flakes, for release with a conventional pesticide, and in a concentrated formulation. Many pheromone products rely on pheromones to attract insect pests to a location at which conventional pesticides are applied.

Because of the time and cost required for pheromone registration, pheromones will probably be developed for use on only major crops or in conjunction with conventional pesticides. Thus, pheromones will be most valuable for integrated pest management. Pheromone use could decrease reliance on conventional pesticides. The pheromones would attract the insect pests to a smaller area where they would be killed with conventional pesticides. Some companies felt that the potential usefulness of pheromones provided adequate incentive to continue efforts to develop pheromone products and to take on the hurdles posed by the registration process. Others thought that the registration process was too unwieldy and questioned the practicality of developing pheromone products until EPA streamlines the process and revises the data requirements to reflect the relative safety of pheromones for general use. In particular, small companies have been disadvantaged by the high cost and the time required for pheromone registration.

ACKNOWLEDGMENT

I acknowledge with thanks the assistance of Karen E. Harrison, a former associate with McKenna, Conner, and Cuneo, for her help in the preparation of this article.

REFERENCES

1. Pesticide Assessment Guidelines Subdivision M: Biorational Pesticides 2 (U.S. Department of Commerce, National Technical Information Service, 1982).
2. Federal Insecticide, Fungicide, and Rodenticide Act (FIFRA), 7 USC §§136–136y (1982 and Suppl. IV, 1986).
3. Federal Food, Drug and Cosmetic Act, 21 USC §§301–392 (1982 and Suppl. 111, 1985).
4. Regulation of "biorational" Pesticides; Policy Statement and Notice of Availability of Background Document, 44 Fed. Reg. 28,093–28,095 (1979).
5. Pesticide Assessment Guidelines, *supra*, at 1.
6. Ibid. at 2.
7. 49 Fed. Reg. 42,876 (1984).
8. FIFRA §3, 7 USC §136a.
9. Pesticide Assessment Guidelines, *supra*, at 1.
10. Ibid at 2.
11. Ibid.
12. Title Code of Federal Regulations (CFR) §172.3(a).
13. 40 CFR §172.3(a)(1).
14. Ibid.
15. Toxic Substances Control Act, 15 USC §§2601–2629 (1982 and Suppl. IV, 1986).
16. TSCA §8, 15 USC §2607.
17. TSCA §5, 15 USC §2604.
18. TSCA §8, 15 USC §2607.
19. Ibid.
20. TSCA §3, 15 USC §2602.
21. 40 CFR §172.3(a).
22. 40 CFR §720.36 and §720.78(b).
23. 51 Fed. Reg. 15,098 (1986).
24. Ibid.
25. Ibid; see also 42 Fed. Reg. 64,585 (1977).
26. FIFRA §5, 7 USC §136c.
27. Ibid.
28. Pesticide Assessment Guidelines, *supra*, at 301–304.
29. FIFRA §5, 7 USC §136c; see also 40 CFR §180.7.
30. 40 CFR §158.165.
31. Ibid.
32. FIFRA §5, 7 USC §136c.
33. 40 CFR §158.45(a)(1).
34. 40 CFR §158.45(a)(2).
35. Pesticide Assessment Guidelines, *supra*, at 31.
36. 40 CFR §158.45(b)(1).
37. 40 CFR §158.45(b)(2).

38. 40 CFR §162.6.
39. 40 CFR §162.5(d)(2).
40. Ibid.
41. 48 Fed. Reg. 38,572 (1983).
42. 49 Fed. Reg. 42,876, 42,879 (1984).
43. 48 Fed. Reg. 38,572 (1983).
44. Ibid.

37

Use of Pheromones and Attractants by Government Agencies in the United States

CHARLES P. SCHWALBE and VICTOR C. MASTRO / U.S. Department of Agriculture, Otis Air National Guard Base, Massachusetts

1. INTRODUCTION

Sex pheromones and chemical attractants are used widely throughout the United States for surveying the whereabouts of insect pests and, to an increasing extent in control programs, employing methods such as mass trapping, attracticide, and mating disruption. Most of these activities are associated with governmental agencies at either the federal, state, or local level. This paper will address the application of pheromone technology to survey programs and show how such activities are conducted and, ultimately, shaped within this governmental framework.

It is often thought that involvement of various agencies in the U.S. Federal Government is fundamental to the conduct of programs that use pheromones. Indeed, considering the diversity of pest problems throughout the country and the localized nature of control approaches, a great amount of local detail in an overall national picture of pest management activities is evident. As we witness increased interest in regionwide management programs (e.g., *Heliothis zea* and *Spodoptera frugiperda* migration), it becomes essential that mechanisms exist for coordinating the accumulation of pest population information. State and federal bodies are helpful in coordinating program activities to ensure that results from area to area are complimentary. Nonetheless, in keeping with the principles of state sovereignty, all management and survey programs are planned and executed in very close cooperation with state agencies. Finally, each management unit (farm, cooperative, county, state, region) may have needs, philosophies, and authorities that are quite distinctive and unique. Clearly, the successful incorporation of pheromone technology into this array of pest management activities will be dependent on our sensitivity to these local needs.

II. TYPES OF PROGRAMS INVOLVING PHEROMONES

Table 1 lists various programs that have pheromones as part of their management strategies. Pests of quite local dimensions are handled almost entirely by the private sector, that is, by individual growers or groups of growers. The mating disruption system for tomato pinworm, *Keiferia lycopersicella*, is a good example of pheromone technology that could be used commercially. In such cases, the technology is advanced to the point at which there is no longer a need for governmental involvement. Obviously, to be attractive to growers, such commercialized technology must keenly focus on their specific management needs (direct control, timing for pesticide applications). The preponderance of pheromone usage is in the area for which the principal responsibility and operational involvement is at the state level.

TABLE 1 Operational Involvement by Various Governmental Levels in Selected Chemical Attractant-Based Survey or Control Programs

Pest	Objective	Federal	State	County	Private
Summer fruit tortrix	Detection	X	X		
Khapra beetle	Detection	X	X		
Gypsy moth	Monitoring/ detection	X	X	X	
Boll weevil	Monitoring/ detection	X	X	X	X
Codling moth	Pesticide timing		X	X	X
Oriental fruit moth	Monitoring		X	X	X
Tomato pinworm	Control				X

Many states have active survey programs for which statewide monitoring networks provide an overview of pest situations throughout the state. Here, county level and private cooperation is present. Generally, these activities are conducted independently of adjoining states. As regional approaches to dealing with such pests evolve, a multistate consortium or, perhaps, federal role in coordinating these activities may emerge. Finally, there are certain pests that are of interest to almost every state within the cropping area and, in such examples, the federal government plays an important role in organizing and coordinating appropriate survey activities. Gypsy moth and boll weevil are good examples of such programs.

There is another class of survey activities that has been receiving increasing attention in recent years: The use of pheromones for detecting the introduction of exotic pests. This is an example of the classic federal role in survey activities, because there is little or no immediate economic value to the private sector in conducting such surveys. Accordingly, their involvement is generally minor from an operational standpoint. On the other hand, federal and state bodies have a traditional responsibility for diligence in exclusion and quarantine activities.

III. SURVEY ACTIVITIES IN THE UNITED STATES

In 1984 and 1986, questionnaires were distributed to designated individuals in all 50 states to determine the extensiveness and diversity of pheromone-based survey activities in the United States. Responses were received from all states and, although it is likely that certain survey activities were not reported on the questionnaires, we believe the results give an approximation of countrywide use of pheromones for survey. The 1986 questionnaire sought information concerning the governmental level (federal, state, county, other) that was responsible for trapping programs that year. Those results are reported in Figure 1. In that year, it was reported that approximately 225,000 traps were used. This almost certainly is an underestimate, particularly because it is difficult to obtain such data from private sources. If we recognize the limitations of the data, the survey revealed that state agencies were responsible for 51.5% of the traps placed. Federal and county leadership led to the deployment of 22.5% and 27.7% of the traps, respectively, and private or local users accounted for 3.3% of the total traps used. This trend illustrates that, currently, leadership at the state level is necessary to implement trapping programs for specific pest management purposes.

Table 2 lists the number of species in six orders for which traps were placed; the most diverse trapping was done for Lepidoptera, and

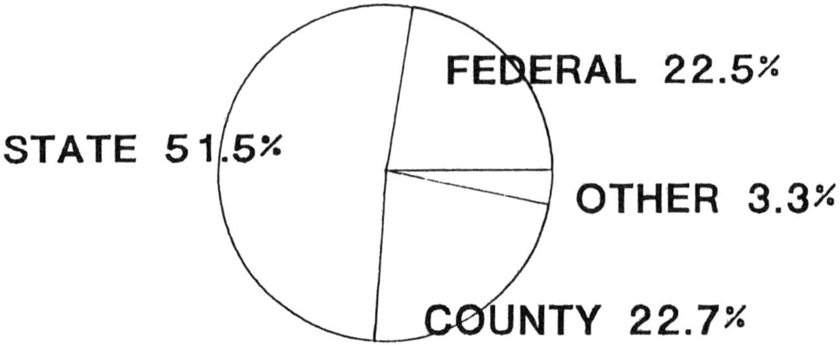

FIGURE 1 Level of responsibility for 1986 trapping programs.

60 species were surveyed in 1986, compared with 47 in 1984. This increased diversity in Lepidoptera trapping largely contributed to the national increase of species trapped from 68 in 1984 to 84 in 1986. One can speculate that this changing usage pattern represents an increasing awareness of, and confidence in, the type of population monitoring information derived from pheromone-baited traps. It is also noted that boll weevil, gypsy moth, and tephritids were the targets of most traps deployed in those years.

Because of their sensitivity in monitoring low-density insect populations, pheromones are ideally suited for detection survey. In fact, most of the traps used in the large boll weevil and gypsy moth survey programs are for detection purposes. Similarly, introductions of pests not known to occur in the United States (exotic species) can be detected promptly by means of a network of traps. It was noted from the 1984 questionnaire results that the only exotic species for which traps were extensively used were exotic tephritids, mainly in California, Texas, and Florida. Experience has shown that these trapping programs are effective in detection of introductions of medfly, oriental fruit fly, and others sufficiently early that successful eradication programs can be conducted.

IV. DETECTION OF EXOTIC PESTS

International trade and travel provide countless opportunities for exotic species to be introduced into the country. In 1984, for example, over 28,000 interceptions of pest insects were made during the course

TABLE 2 Number of Species for Which Traps Were Placed in the United States in 1984 and 1986

	Number of species		Total traps placed (%)	
Order	1984	1986	1984	1986
Coleoptera	7	10	43	13
Diptera	9	9	20	36
Hemiptera	0	1		<1
Homoptera	4	2	<1	<1
Hymenoptera	1	2	<1	<1
Lepidoptera	47	60	37	51
Total	68	84	100	100

of agricultural quarantine inspections at ports of entry. While recognizing that inspectional and quarantine activities are clearly effective, but of finite value in completely preventing new pest introductions, it is advantageous to organize an ongoing program of exotic pest detection aimed at key pests so that appropriate containment and control actions can be undertaken when populations are established.

The mounting of meaningful and effective detection surveys for exotic species is a costly undertaking. The greatest expense in such programs is the time required to place and monitor traps; by comparison, the cost of traps and lures is minor. Therefore, superimposing exotic pest surveys on field activities already in place is an appealing approach. As an example, orchards that are being monitored for the seasonal occurrence of codling moth, *Cydia pomonella*, could also be trapped for the exotic light brown apple moth, *Epiphyas postvittana*. Combining these two activities would not appreciably add to personnel costs and, in this way, an exotic pest survey program could be incorporated into ongoing routine field work. This, in fact, is the way most exotic detection trapping is currently performed.

Another approach to merging exotic and domestic pest surveys is to consider baiting traps for more than one species. This has been done successfully with some fruit tree Lepidoptera (1) and stored grain beetles (2). Other attempts with *Autographa gamma* and *Mamestra brassicae* (3), *Heliothis zea*, *H. virescens*, and *Spodoptera frugiperda* (4) and ambrosia beetles (5) demonstrated that inhibitory effects of certain pheromone components rendered the technique of multibaiting of traps unworkable with some insects. Baiting of traps for more than one species may be particularly efficient in connection with the current gypsy moth survey, which employs about 250,000 traps annually. Because the gypsy moth is cosmopolitan, traps are placed in a wide variety of situations where the pest may occur. This large national detection trapping network offers a unique logistic framework into which exotic survey could be incorporated. Needless to say, baiting single traps with two (or more) pheromones is fraught with potential complications, particularly the potential inhibitory effects of pheromone components on either species. Similarly, behavior of insects that have common components in different proportions may be negatively affected by distortion of the final ratio of components released from the trap. Such blend distortions may render the final "mixture" unattractive to one or both target species. Trap design and placement in the crop and pest phenology must also be carefully scrutinized in considering usefulness of multibaited traps. Nonetheless, the potential usefulness and efficiency of this approach justifies investigations to assess feasibility; field tests have been performed with 29 species to evaluate compatibility of attractants. Table 3 describes the pheromone dispensers used in the experiments.

Tables 4 and 5 are representative of the types of experiments that have been conducted to determine compatibility of attractants. When available, commercially prepared dispensers were used in the tests. Data summarized in Table 4 reveal that gypsy moth trap catch is not negatively affected when traps also include lures for *Cydia pomonella*, *C. funebrana*, *Lobesia botrana*, *Pectinophora gossypiella*, and *Heliothis zea*. However, inclusion of baits for *Cryptophlebia leucotreta*, *Heliothis punctiger*, *H. armigera*, and *H. virescens* reduced the number of gypsy moths captured. There are several details of the results that cannot be explained with available data. *Cydia funebrana* and *Cryptophlebia leucotreta* dispensers contain *Z*8- and *E*8-12:Ac at 95:5 and 1:1 ratios, respectively. *Cryptophlebia leucotreta* lure dispensers were loaded with 5.68 mg, whereas *C. funebrana* baits carried only 0.2 mg. Accordingly, it is likely that the inhibitory effect of *C. leucotreta* lures was due to their higher loading rate, rather than to blend differences. Similarly, *H. zea* and *H. armigera* dispensers were loaded with slightly different ratios of *Z*11- and *Z*9-16Al, but *H. armigera* baits had five times more material than *H. zea*, again possibly accounting for reduced catch with *H. armigera*. *Heliothis punctiger* and *H. virescens* pheromones, also inhibitory to the gypsy moth, were provided at very high loading rates (24 and 48 mg/dispenser, respectively).

Because *Lobesia botrana* is an exotic species of interest to the United States and, from results in Table 4, its pheromone is not inhibitory to attraction of gypsy moths, field tests were performed to

TABLE 3 Composition, Loading Rates, and Dispenser Designs Used in Field Tests Evaluating the Compatibility of Lures for Two Species

Species	Components	Ratio	Loading rate	Dispenser
Adoxophyes orana	Z9-Z11-14:Ac	9:1	2.0 mg	Polycap
Argyrotaenia velutinana	Z11-14:Ac/*E*11-14:Ac/12:Ac	31:3:66	28.1 mg	Fibers
Clysia ambiguella	Z9-12:Ac		6.0 mg	Septa
Cryptophlebia leucotreta	Z8/*E*8-12:Ac	1:1	5.68 mg	Fibers
Cydia funebrana	Z8/*E*8-12:Ac	95:5	0.2 mg	Septa
C. pomonella	*E*,*E*8,10-12:OH		4.8 mg	Fibers
Grapholita molesta	Z8-12:Ac/*E*8-12:Ac/Z8-12:OH	88:7:5	4.3 mg	Fibers
Heliothis armigera	Z11/Z9-16:Al	9:1	24.0 mg	Fibers
H. punctiger	Z11-16:Al/Z11-16:Ac/Z9-14:Al	50:50:1	24.0 mg	Fibers
H. virescens	Z11-16:Al/Z9-14:Al	16:1	48.0 mg	Fibers
H. zea	Z11/Z9-16:Al	24:1	4.8 mg	Fibers
Lobesia botrana	*E*,Z7,9-12:Ac		0.1 mg	Septa
Lymantria dispar	(+)*cis*7,8-epo-2-Me-18:Hy		0.5 mg	Laminate
Paralobesia viteana	Z9-12:Ac		1.7 mg	Fibers
Pectinophora gossypiella	Z,*E*/Z,Z7,11-16:Ac	1:1	5.0 mg	Fibers
P. scutigera	Z,Z7,11-16:Ac		5.0 mg	Fibers
Spodoptera littoralis	Z,*E*9,11/Z,*E*9,12-14:Ac	99.5:0.5	2.0 mg	Septa
S. litura	Z,*E*9,11/Z,*E*9,12-14:Ac	88:12	2.0 mg	Septa

TABLE 4 Mean Numbers of Gypsy Moth, *Lymantria dispar*, Males Captured in Traps Baited with Its Pheromone Alone and in Paired Combination with Various Other Attractants[a]

Attractant combinations		$\bar{x}$ number *L. dispar* males captured/trap/reading[b]
1	2	
L. dispar	*Cydia pomonella*	24.94
L. dispar	*C. funebrana*	24.60 a
L. dispar	*Control*	20.37 a b c
L. dispar	*Lobesia botrana*	18.43 a b c
L. dispar	*Pectinophora gossypiella*	16.34 b c
L. dispar	*Heliothis zea*	14.60 c
L. dispar	*Cryptophlebia leucotreta*	9.31 d
L. dispar	*H. punctiger*	5.77 e
L. dispar	*H. armigera*	3.09 f g
L. dispar	*H. virescens*	1.54 g

[a]Composition of test dispensers is given in Table 3. Test was performed at Otis Methods Development Center, Massachusetts, 7/20--7/27/83.
[b]Numbers followed by the same letter are not significantly different.

determine the effects of pheromones of other species on *L. botrana* trap catch. Results summarized in Table 5 indicate that *Cydia pomonella*, *Lymantria dispar*, and *Paralobesia viteana* lures do not interfere with *L. botrana* trap catch; the pheromones of *Clysia ambiguella*, *Argyrotaenia velutinana*, *Adoxophyes orana*, *Cryptophlebia leucotreta*, *Cydia funebrana*, and *Grapholita molesta* all rendered traps baited for *L. botrana* less attractive.

Combining the results of Tables 4 and 5 reveals that traps can be cobaited for *Lymantria dispar* and *Lobesia botrana* without negatively affecting attraction of either species.

Lobesia botrana infests grapes, olives, privet, black currants, and persimmons. Traps for gypsy moth detection surveys are deployed throughout forested, residential, and agricultural areas where host trees are found. Trap sites could be specified in areas where hosts of both species overlap or abut, and a detection survey for *L.*

TABLE 5 Mean Numbers of European Grapevine Moth, *Lobesia botrana*, Males Captured in Traps Baited with Its Pheromone Alone and in Paired Combination with Various Other Attractants[a]

Attractant combinations 1	2	x number *L. botrana* males captured/trap/reading[b]
L. botrana	*Cydia pomonella*	25.00 a
L. botrana	*Lymantria dispar*	21.67 a
L. botrana	*Control*	20.00 a
L. botrana	*Paralobesia viteana*	11.33 a b
L. botrana	*Clysia (Eupoecilia) ambiguella*	6.00 b c
L. botrana	*Argyrotaenia velutinana*	2.33 c d e
L. botrana	*Adoxophyes orana*	1.00 d e
L. botrana	*Cryptophlebia leucotreta*	0.67 e
L. botrana	*Cydia funebrana*	0.67 e
L. botrana	*Grapholita molesta*	0.00 e

[a]Composition of test dispensers is given in Table 3. Test was performed at Tain L'Hermitage, France, 7/26–8/13/84.
[b]Numbers followed by the same letter are not significantly different.

botrana could thus be merged into the ongoing gypsy moth survey program.

Our screening of pheromones of exotic insects has thus far identified ten pairs of target pests for which attractants of both species can be combined within a single trap without compromising the attraction of either species (Table 6). In some cases, both species are exotic and, although this precludes incorporation of detection survey into an established domestic pest trapping program, greater "return" is gained from multibaited traps deployed specifically for exotic detection.

It should be emphasized that most field studies have been done with pheromone dispensers manufactured by commercial concerns. Because even trace contamination by nonpheromonal components may render baits unattractive or inhibitory to one or both target species, it is essential to conduct preseason testing of lures that will be used in

TABLE 6 Species for Which Pheromones Are Compatible When Used Together in the Same Trap[a]

Attractant combinations	
1	2
Pectinophora scutigera[b]	*Cryptophlebia leucotreta*[b]
P. scutigera[b]	*Heliothis punctiger*[b]
P. scutigera[b]	*Spodoptera litura*[b]
Cydia pomonella	*C. funebrana*[b]
C. pomonella	*Lobesia botrana*[b]
Lymantria dispar	*C. funebrana*[b]
L. dispar	*L. botrana*[b]
L. dispar	*P. gossypiella*
L. dispar	*C. pomonella*
S. litura[b]	*S. littoralis*[b]

[a]Composition of dispensers used in the evaluations is summarized in Table 3.
[b]Species that are exotic to the United States.

the program. Similarly, new research results may lead to changes in bait formulations used by manufacturers. Incorporating additional components may enhance the effectiveness of a lure for its target species, but those components may be inhibitory to the other species for which the trap will be cobaited. Careful bioassay of the baits to be combined should always precede operational deployment. Therefore, even though the approach has been shown effective on a limited basis, wide-scale application of multibaited traps for detection of exotics is being introduced slowly and carefully.

In 1983, before this work was initiated, very few traps were in use in the United States for detecting exotic insect introductions. This program has heightened the interest in surveying for exotic pests to the point that, in 1986, several thousand traps were placed specifically for that purpose. Table 7 illustrates the expansion of this effort; state involvement and trapping extensiveness and diversity have increased steadily since 1984. We are now developing handbooks to guide field activities to ensure detection of infestations (above certain threshold

TABLE 7 State Involvement in Detection Trapping for Exotic Insect Pests (Excluding Tephritidae)

	1984	1985	1986
No. states	16	32	37
No. species	6	6	13
No. traps	853	3020	3892

levels). Our program currently focuses on 14 exotic species (exclusive of Tephritidae).

The technical and organizational infrastructure now evolving within the responsible federal, state, and local agencies can be applied to exotic detection efforts in a general sense. Obviously, if detection surveys for a particular species are negative, and there are no important pathways for future introductions, it is probably unnecessary to continue trapping for that species, other than in high-risk areas. However, provided the technical and organizational framework of the program is suitable, resources can be refocused on other species that may have become established in the past. Thus, the program should be considered as constantly evolving. As evidence mounts that a species does not occur in the survey area, emphasis on it should decline while efforts to detect a different species are phased in. In this way, the use of pheromones for early detection of new pest introductions becomes integrated as a basic component of pest survey.

V. CONCLUSIONS

Initiatives into new or expanded applications of pheromone technology for pest management are worthwhile only if the agricultural community recognizes real or potential advantages. Federal involvement is often necessary or desirable, in the formative stages to provide funding, expertise, and coordination. However, unless state agencies and the growers themselves understand, value, and support such programs, it is unlikely the technology will be fully accepted and used to its fullest advantage. It is the responsibility of scientific institutions to see that technology is developed, adapted, and demonstrated with usage patterns in mind that match the practical needs of the ultimate user.

REFERENCES

1. Sato, R. Simultaneous trapping of several fruit-tree pest insects by one trap baited with multiple sex lures. *Jpn. J. Appl. Entomol. Zool.* *25*:176–181 (1981).
2. Lindgren, B. S., Borden, J. H., Pierce, A. M., Pierce, H. D., Jr., Oehlschlager, A. C., and Wong, J. W. A potential method for simultaneous semiochemical-based monitoring of *Cryptolestes ferrugineus* and *Tribolium castaneum* (Coleoptera:Cucujidae and Tenebrionidae). *J. Stored Prod. Res.* *21*:83–87 (1985).
3. Subchev, M. A. Interaction between synthetic sex pheromones of the cabbage moth and the gamma moth in traps. *Ekologiya (Sophia)* *11*:80–83 (1983).
4. Carpenter, J. E., Pair, S. D., and Sparks, A. N. Trapping of different noctuid moth species by one trap baited with two lures. *J. Georgia Entomol. Soc.* *19*:120–124 (1984).
5. Borden, J. H., Chong, L., Slessor, K. N., Oehlschlager, A. C., Pierce, H. D., Jr., and Lindgren, B. S. Allelochemic activity of aggregation pheromones between three sympatric species of ambrosia beetles (Coleoptera:Scolytidae). *Can. Entomol.* *113*:557–564 (1981).

38

Commercial Availability of Insect Pheromones and Other Attractants

MAY N. INSCOE, BARBARA A. LEONHARDT, and RICHARD L. RIDGWAY / Agricultural Research Service, U.S. Department of Agriculture, Beltsville, Maryland

I. INTRODUCTION

For over four decades, reliance on conventional insecticides has dominated insect control in the United States and most other developed countries. Environmental and health concerns associated with pesticide use emerged in the late 1950s and became highly visible in the 1960s. Although these concerns had significant effects on the kinds of pesticides used, reliance on broad-spectrum pesticides for control of many pests has continued. But now, perhaps, scientific and technological advances and public interest concerns have reached the point at which significant changes in insect pest management are in the offing. There is considerable evidence that behavior-modifying chemicals can be a significant part of these changes. On the basis of the practical applications of pheromones and other attractants documented in the first 37 chapters of this volume, we can anticipate continued increase in the use of these materials as components in insect management systems, provided that cooperative efforts among all interested parties continue and are supported by research, development, and information transfer efforts. As a part of the needed informational effort, in this chapter we will survey the commercial availability of pheromones and other attractants. We are hopeful that this review will contribute to an improved understanding of patterns of discovery and use, and that it will provide a convenient reference that developers and practitioners can use to determine the commercial availability of pheromones or other attractants for specific insects.

The numbers of known insect behavior-modifying chemicals and their applications have increased greatly since the first chemical structure of a pheromone was announced in 1959. By 1970, for example, 20 insect sex attractant and aggregation pheromones had been identified (1); in 1975, this number had increased to 80 (2), and in 1980, to 210 (3). By the end of 1985, sex and aggregation pheromones had been identified for an estimated 342 insect species (4). Other compounds, for example, the analogue of the pea moth pheromone (see Chap. 2), have been found to have pheromonelike activity, without having been shown to be present in the specific insect involved. Many of these compounds, known as *parapheromones*, are effective attractants, and some have been, or eventually may be, demonstrated to be pheromones. In some reports, the distinction between pheromones and parapheromones has become blurred. For example, traps baited with pheromones are often included in the term *pheromone traps*. This has given rise to occasional discrepancies in numbers of identified pheromones in the published literature. In 1980, when pheromones had been reported for 210 insects, parapheromones had been found for and additional 356 insects (3); thus, at the beginning of this decade, pheromones and similar attractants were known for over 550 insects. The number may well reach 1000 by the early 1990s.

Another group of behavior-modifying chemicals, which bear no apparent relationship to pheromones in their action, are also important as species-specific attractants. These compounds, many of which were found by empirical screening, are termed *synthetic attractants*. They generally are used in much larger quantities than are needed with attractant pheromones. Although the mechanisms of action of these materials are not known, some, such as methyl eugenol for the oriental fruit fly, are found in plants and may be regarded as kairomones (5). These compounds are important for trapping several insects for which no effective attractant pheromone is known. By 1980, 89 insect species had been reported to respond to synthetic attractants (3). Although this is not as active an area of research as the pheromone field, investigations are continuing, and patent applications for new synthetic attractants for two important tephritid species, the Mediterranean fruit fly, *Ceratitis capitata*, and the Malaysian fruit fly, *Dacus latifrons*, have recently been filed (T. P. McGovern, personal communication).

Other insect behavior-modifying chemicals of possible practical significance include alarm pheromones, oviposition attractant or deterrent pheromones, insect kairomones, and other plant volatiles (or plant kairomones). Uses for these materials are increasing. For example, plant volatiles are frequently used as pheromone synergists for coleopteran species; the host plant terpene, myrcene, synergizes the response of the western pine beetle to its pheromone, *exo*-brevicomin (6), and more Japanese beetles are caught by the combination of the pheromone, japonilure, and a food attractant (or "floral lure") than by either attractant alone (7). Also, alarm pheromones can be used to flush aphids out of protected feeding sites, thus increasing contact with pesticides or biological control agents (8,9).

The many applications discussed in this volume illustrate the variety of ways in which behavior-modifying chemicals are being applied in the management of a number of insect species. However, widespread practical application of a behavior-modifying chemical can be achieved only when the compound is readily available commercially. Many of the known behavior-modifying compounds are not now of commercial interest. Some are for insects of little economic importance and, consequently, there is little incentive for their development. For others, the extensive knowledge of insect behavior and response required for their effective use has not yet been generated. Other requirements that must be met before a behavior-modifying chemical is widely used include the development of suitable traps with effective dispensing systems and the availability of materials at a reasonable cost; deficiencies in these areas will obviously limit development and practical use. A few references present further discussions on the subject of commercialization of pheromones (10--13; see Chaps. 31 through 33).

Nevertheless, the numbers of commercially available pheromones are increasing. In 1982, there were 86 species of insects for which pheromones, parapheromones, and attractants were commercially available in the United States, all from six suppliers (14); lures for monitoring were offered for all of these species, but formulations for actual insect control by mass trapping or mating disruption were available for only 12 species. As interest and awareness increased, pheromones and parapheromones for more insects were offered, and a list prepared early in 1986 with the assistance of colleagues in the United States and overseas (15) contained 138 insect species. Updated information in a readily accessible form is needed to encourage expanded use of pheromones and other attractants and to provide input into the planning of research and development for further expansion.

II. INFORMATION SOURCES

In the course of our work with pheromones, we have developed a list of enterprises that are involved in some way with insect pheromones and other attractants. This information has come from a variety of sources, including visits, personal contacts, discussions with other workers in this field, literature references, advertisements, and chemical directories. Most of these organizations are commercial enterprises; however, a few, such as the Institut National de la Recherche Agronomique in France, better known as INRA, are public-sector organizations that have made materials available when commercial organizations would not do so for lack of sufficient profit incentive. An updated version of this list is presented in the Appendix to this volume. To obtain information on the commercial availability of products and services, we contacted those enterprises of which we had knowledge, referred to here as "suppliers." The information received from 41 suppliers forms the basis of this chapter.

III. PRODUCTS AND SERVICES

The 41 suppliers providing information offer a variety of products and services. About half of the companies most directly involved with pheromone products specialize in a single type of application; 12 companies offer synthetic chemicals or custom syntheses, 3 of the companies provide formulation services, and 1 company specializes in traps and specialized trap design. The other 25 companies offer two or more types of products; 18 offer traps and dispensers sold either separately or as monitoring or detection systems, 9 provide pheromone systems for mass trapping, 7 offer formulations for mating disruption, and 5 offer attracticides (formulations containing both an attractant and a pesticide) or *bioirritants* (materials that stimulate an insect to greater movement and thus cause more contact with a concurrently applied pesticide). Major distributors offer all of these products, in addition to consulting services. One company offers complete insect management services, including provision of traps and dispensers, trap placement, interpretation of trap catches, and insect management consultation.

IV. APPLICATIONS

The information provided by the 41 companies on the various applications for which behavior-modifying chemicals are supplied and the numbers of arthropod species involved are summarized in Table 1. More than one application was frequently reported for a given insect and, for some insects, the applications were not specified. As might be expected, the chief application is trapping. Various applications of trapping, including detection, survey, study of population dynamics, and others, are included under "Monitoring" in this table. Lepidoptera predominate in the species involved.

Far fewer behavior-modifying chemicals are offered for control than for monitoring. Mass-trapping systems are offered for 19 insects (12 Lepidoptera, 4 Coleoptera, 2 Diptera, and 1 Hymenoptera). Formulations for mating disruption are offered for 18 lepidopteran insects. "Attracticide" formulations are offered to seven species (3 Lepidoptera, 3 Diptera, 1 Blattodea). Stimulant or bioirritant formulations, for use with pesticides, are listed for three species. Finally, antiaggregants, which tend to prevent aggregation, are available for two coleopteran species, and pheromone dispensers for bait trees, in which a pheromone is used to attract the insects to a specific tree where they can be controlled by pesticide applications or by destroying the tree, are listed for four coleopteran species.

The commodities or activities affected by the various arthropod species for which products are available are presented in Table 2. The information provided should not be regarded as complete, because only commodities specifically mentioned by a supplier are included, and many of the suppliers did not specify commodities. As in Table 1, many insects are represented in more than one column of the table.

V. ARTHROPODS AND CHEMICALS

The 257 arthropod species mentioned by the suppliers are presented in Tables 3 through 5. Table 3 is an alphabetical list of the scientific names and authors, with cross-references to order and family and to the common names used by the suppliers or approved by the Entomological Society of America (16). All scientific names given by suppliers are listed; in those cases for which a name other than the one used by the supplier is regarded as more authoritative, reference is made to the preferred name.

To facilitate locating an insect when the scientific name is not known, an alphabetical list of common names used by suppliers in their responses is given in Table 4, with cross-references to scientific name and to order and family. Some of the common names used by the suppliers are general terms, rather than specific names; for example, "fruit tree tortix" is used for both *Archips podanus* and *Argyrotaenia pulchellana* and "tignola del cacao" is used for both *Cadra cautella* and *Ephestia elutella.*

Table 5 presents details of the information summarized in Tables 1 and 2 and provides chemical information on the materials involved. The various arthropods are arranged here alphabetically by order and family. Of the 257 insects and mites listed, by far the largest group is the Lepidoptera, with 202. There are 33 Coleoptera, 8 Diptera, 6 Homoptera, 4 Acari, 3 Blattodea, and 1 Hymenoptera. The preponderance of lepidopterans may be partially accounted for by the fact that chemical structures of lepidopteran pheromones are often less complex than those found in other insect orders, and they often consist of mixtures of alcohols, esters, and aldehydes of similar structure. The large number of economically important lepidopteran pests is, undoubtedly, another factor.

Chemical information on pheromone components of the various species and the ratios in which they occur, when available, was taken from published scientific literature. In addition to the comprehensive lists of insect pheromones and other behavior-modifying chemicals used in Table 5 (3,17,18), additional compilations can be found in appropriate chapters (37–41) of the work edited by Morgan and Mandava (42).

It is quite probable that components of commercial formulations for a specific insect will differ from the published attractant composition. There are many possible reasons for this, aside from problems of quality control. In some pheromone lures, a component may be added to increase specificity by inhibiting attraction of another insect. A more stable, but somewhat less attractive, parapheromone may be used instead of the pheromone of a specific insect. Economics may also come into play; reduced amounts of an expensive component may be used to obtain a less-than-optimum formulation that is still effective, but costs considerably less, or a mixture of isomers may be used in place of a pure active isomer to avoid separation costs, if the added isomer does not adversely affect biological activity. Indeed, Wall has said that "it is evident that most monitoring systems use traps containing attractants that differ from the natural pheromone to varying degrees" (see Chap. 2).

The commodities and applications in Table 5 are those listed by suppliers. Although the suppliers of formulations for specific insects are not designated, reference to the list of suppliers in the Appendix to this book and a few direct inquiries should lead to a source of supply for a given material.

VI. CONCLUSIONS

The numbers of insects for which pheromones and other attractant products are being offered and the numbers of commercial suppliers have increased greatly in the last decade. Trap use for detection and other types of monitoring has expanded much more rapidly than have applications for population suppression or control, such as mass trapping, mating disruption, and attracticides. As our knowledge of insect behavior and of mechanisms of processes such as chemoreception and communication disruption grows and as communication and cooperation among researchers, suppliers, and users improves, effective applications of behavior-modifying chemicals in insect management programs should continue to increase. However, because of the species-specificity of these products, markets for the new materials are likely to be relatively small; therefore, considerable continued involvement of the public sector in research and development will be required to offset the limited profit incentives for the private investor.

ACKNOWLEDGMENTS AND DISCLAIMER

We gratefully acknowledge the assistance of the following scientists of the Systematic Entomology Laboratory and the Smithsonian Institution for verifying insect names in the specified orders: Donald M. Anderson (Scolytidae); Don R. Davis (Gracillariidae, Lyonetiidae, Psychidae); James R. Dogger (Elateridae); Raymond J. Gagné (Muscidae): Robert D. Gordon (Bostrichidae, Scarabaeidae, Tenebrionidae); Ronald W. Hodges (Acrolepiidae, Argyresthiidae, Carposinidae, Coleophoridae, Cossidae, Gelechiidae, Plutellidae, Sesiidae, Sphingidae, Thaumetopoeidae, Yponomeutidae, Zygaenidae); John M. Kingsolver (Dermestidae); Douglass R. Miller (Diaspididae, Pseudococcidae): David A. Nickle (Blattellidae, Blattidae); Allen L. Norrbom (Tephritidae): Robert W. Poole (Arctiidae, Lymantriidae, Noctuidae, Notodontidae); Theodore J. Spillman (Cucujidae); Richard E. White (Chrysomelidae); and Donale R. Whitehead (Curculionidae).

Although attempts have been made to use the correct scientific names for all arthropods listed, it has not been possible to verify the names of some Acari (Ixodidae and Tetranychidae) and a number of Lepidoptera (Cochylidae, Geometridae, Lasiocampidae, Pterophoridae, Pyralidae, Saturniidae, and Tortricidae).

Mention of commercial names and products is for information only and does not constitute a recommendation by the U.S. Department of Agriculture.

TABLE 1 Uses Listed by Suppliers in 1988 for Commercially Available Pheromones and Other Behavior-Modifying Chemicals

Order	Numbers of arthropod species							
	Monitoring	Mating disruption	Mass trapping	Attracticide	Bioirritant	Anti-aggregant	Bait tree	Use not specified
Acari					2			2
Blattodea	3			1				
Coleoptera	27		4			2	4	5
Diptera	7		2	3				1
Homoptera	4							2
Hymenoptera			1[a]					
Lepidoptera	189	18	12	3	1	—	—	9
Totals	230	18	19	7	3	2	4	19

[a]Swarm trapping.

TABLE 2 Commodities and Activities Listed by Suppliers That Are Affected by Arthropods for Which Pheromones and Other Behavioral Chemicals Were Commercially Available in 1988

Order	Numbers of arthropod species								
	Field crops	Vegetables	Orchard	Vineyard	Forest	Horticultural crops	Bee keeping	Stored products	Animals
Acari	3		3						1
Blattodea								3	
Coleoptera	6	3			19			10	
Diptera		1	5						3
Homoptera			6		1				
Hymenoptera							1		
Lepidoptera	80	59	65	12	35	4	2	6	—
Totals	89	63	79	12	55	4	3	19	4

TABLE 3 Alphabetical List of Arthropods for Which Pheromones and Other Behavioral Chemicals Were Commercially Available[a] in 1988.

Arthropods	Order and family	Common names used
Achroia grisella (Fabricius)	Lepidoptera: Pyralidae	lesser wax moth*
Acrolepiopsis assectella (Zeller)	Lepidoptera: Acrolepiidae	leek moth,* Lauchmotte, teigne du poireau
Actebia fennica (Tauscher)	Lepidoptera: Noctuidae	black army cutworm*
Adoxophyes fasciata, see *Adoxophyes* sp. (in Japan)		
A. orana (Fischer von Roslerstamm)	Lepidoptera: Tortricidae	summerfruit tortrix, capua, Fruchtschalenwickler
A. reticulana, see *Adoxophyes orana*		
Adoxophyes sp. (in Japan)	Lepidoptera: Tortricidae	smaller tea tortrix
Agrotis comes, see *Noctua comes*		
A. exclamationis (L.)	Lepidoptera: Noctuidae	agrotide, exclamator cutworm, heart and dart moth
Agrotis fucosa Butler	Lepidoptera: Noctuidae	common cutworm
A. ipsilon (Hufnagel)	Lepidoptera: Noctuidae	black cutworm,* dark sword grass moth, greasy cutworm, nottua ipsilon
A. orthogonia Morrison	Lepidoptera: Noctuidae	pale western cutworm*
A. segetum (Denis & Schiffermüller)	Lepidoptera: Noctuidae	turnip moth, nottua dei seminati, noctuelle des moissons
A. ypsilon, see *Agrotis ipsilon*		
Amathes c-nigrum, see *Xestia dolosa*		
Amblyomma americanum (L.)	Acari: Ixodidae	lone star tick*
Amyelois transitella (Walker)	Lepidoptera: Pyralidae	navel orangeworm,* navel orange worm

Anagasta kuehniella (Zeller)	Lepidoptera: Pyralidae	Mediterranean flour moth,* Mehlmotte
A. kuhniella, see *A. kuehniella*		
Anarsia lineatella Zeller	Lepidoptera: Gelechiidae	peach twig borer,* peach twigborer, petite mineuse du pêcher, tignola del pesco
Antheraea polyphemus (Cramer)	Lepidoptera: Saturniidae	polyphemus moth*
Anthonomus grandis Boheman	Coleoptera: Curculionidae	boll weevil*
Anthrenus flavipes LeConte	Coleoptera: Curculionidae	furniture carpet beetle*
Anticarsia gemmatalis Hübner	Lepidoptera: Noctuidae	velvetbean caterpillar,* velvet bean caterpillar
Aonidiella aurantii (Maskell)	Homoptera: Diaspididae	California red scale*
A. citrina (Coquillett)	Homoptera: Diaspididae	yellow scale*
Apis mellifera L.	Hymenoptera: Apidae	honeybee,* honey bee
Archippus breviplicanus, see *Archips breviplicanus*		
Archips argyrospila, see *A. argyrospilus*		
A. argyrospilus (Walker)	Lepidoptera: Tortricidae	fruittree leafroller*
A. breviplicanus Walsingham	Lepidoptera: Tortricidae	Asiatic leafroller
A. crataeganus	Lepidoptera: Tortricidae	tordeuse des bourgeons
A. podana, see *A. podanus*		
A. podanus (Scopoli)	Lepidoptera: Tortricidae	fruit tree tortrix, cacoecia dei frutti, cacoecia delle gemme, braunlicher Obstbaumwickler
A. rosana, see *A. rosanus*		
A. rosanus (L.)	Lepidoptera: Tortricidae	filbert leafroller, European leafroller, cacoecia delle gemme, rose twist moth

[a]Names marked with an asterisk are those approved by the Entomological Society of America (16).

(continued)

TABLE 3 (Cont.)

Arthropods	Order and family	Common names used
A. semiferana, see *A. semiferanus*		
A. semiferanus (Walker)	Lepidoptera: Tortricidae	oak leafroller*
A. sorbianus, see *Choristoneura sorbiana*		
A. xylosteana, see *A. xylosteanus*		
A. xylosteanus (L.)	Lepidoptera: Tortricidae	tordeuse des bourgeons, Apfelblatt-wickler
Argyresthia ephippella, see *A. pruniella*		
A. pruniella (Clerck)	Lepidoptera: Argyresthiidae	Kirschblütenmotte
Argyroploce leucotreta, see *Pseudogalleria leucotreta*		
Argyrotaenia citrana (Fernald)	Lepidoptera: Tortricidae	orange tortrix*
A. pulchellana (Hübner)	Lepidoptera: Tortricidae	eulia, fruit tree tortrix, fruit tree tortrix moth, tordeuse de la pelure
A. velutinana (Walker)	Lepidoptera: Tortricidae	redbanded leafroller,* red-banded leafroller
Attagenus flavipes, see *Anthrenus flavipes*		
A. megatoma, see *A. unicolor*		
A. unicolor (Brahm)	Coleoptera: Dermestidae	black carpet beetle*
Autographa californica (Speyer)	Lepidoptera: Noctuidae	alfalfa looper*
A. gamma (L.)	Lepidoptera: Noctuidae	silver Y moth, nottuelle Gamma
Blatta orientalis L.	Blattodea: Blattidae	oriental cockroach*
Blattella germanica (L.)	Blattodea: Blattellidae	German cockroach*
Brachmia macroscopa, see *Helcystogramma macroscopum*		

Bryotropha similis (Stainton)	Lepidoptera: Gelechiidae	
Busseola fusca (Fuller)	Lepidoptera: Noctuidae	maize stalk-borer
Cacoecia musculana, see *Syndemis musculana*		
C. pronubana, see *Cacoecimorpha pronubana*		
Cacoecimorpha pronubana (Hübner)	Lepidoptera: Tortricidae	carnation tortrix moth, bega del garofano, tordeuse européenne de l'oeillet
Cadra cautella (Walker)	Lepidoptera: Pyralidae	almond moth,* fig moth, tignola dei fichi secchi, Dattelmotte
C. figulilella (Gregson)	Lepidoptera: Pyralidae	raisin moth,* tignola del cacao, tignola della frutta secca
Carpocapsa pomonella, see *Cydia pomonella*		
Carposina niponensis Walsingham	Lepidoptera: Carposinidae	peach fruit moth
Ceratitis capitata (Wiedemann)	Diptera: Tephritidae	Mediterranean fruit fly*
Chilo partellus (Swinhoe)	Lepidoptera: Pyralidae	spotted stalk borer
C. sacchariphagus	Lepidoptera: Pyralidae	
C. suppressalis (Walker)	Lepidoptera: Pyralidae	Asiatic rice borer,* rice stem borer, striped rice borer
C. zacconius (Blesz.)	Lepidoptera: Pyralidae	African rice borer moth, pyrale du riz
Choristoneura fumiferana (Clemens)	Lepidoptera: Tortricidae	spruce budworm,* eastern spruce budworm
C. hebenstreitella, see *C. sorbiana*		
C. lafauryana Rag.	Lepidoptera: Tortricidae	soy-bean tortrix moth, tortrice della soia e della fragola
C. murinana (Hübner)	Lepidoptera: Tortricidae	European fir budworm moth, tordeuse du sapin

(continued)

TABLE 3 (Cont.)

Arthropods	Order and family	Common names used
C. occidentalis (Freeman)	Lepidoptera: Tortricidae	western spruce budworm*
C. rosaceana (Harris)	Lepidoptera: Tortricidae	obliquebanded leafroller,* oblique banded leaf roller
C. sorbiana (Hübner)	Lepidoptera: Tortricidae	tordeuse rouge (de la pelure), Vogelbeerwickler
Chrysodeixis chalcites (Esper)	Lepidoptera: Noctuidae	green garden looper,* tomato looper, nottua del pomodoro
Chrysoteuchia topiaria Zeller	Lepidoptera: Pyralidae	cranberry girdler,* sod webworm
Clepsis spectrana (Treitschke)	Lepidoptera: Tortricidae	cabbage leafroller, cyclamen tortrix, tortrice della vite, geflammter Reberwickler
Clysia ambiguella, see *Eupoecilia ambigue la*		
Cnephasia longana (Haworth)	Lepidoptera: Tortricidae	omnivorous leaftier,* Getreidewickler
C. pumicana Zeller	Lepidoptera: Tortricidae	cereal tortrix moth, tordeuse des céréales, Ahrenwickler
Cochylis hospes Walsingham	Lepidoptera: Cochylidae	banded sunflower moth*
Coleophora laricella (Hübner)	Lepidoptera: Coleophoridae	larch casebearer*
Cossus cossus (L.)	Lepidoptera: Cossidae	European goat moth, goat moth, rodilegno rosso
Crymodes devastator (Brace)	Lepidoptera: Noctuidae	glassy cutworm*
Cryptophlebia leucotreta, see *Pseudogalleria leucotreta*		
Cydia caryana (Fitch)	Lepidoptera: Tortricidae	hickory shuckworm*
C. funebrana (Treitschke)	Lepidoptera: Tortricidae	plum fruit moth, verme delle susine, Pflaumenwickler, carpocapse

C. janthinana(Duponchel)	Lepidoptera: Tortricidae	hawthorne leafroller, tordeuse de l'aubépine
C. molesta, see *Grapholita molesta*		
C. nigricana (Fabricius)	Lepidoptera: Tortricidae	pea tortrix moth,* tortrice dei piselli, Erbsenwickler
C. pomonella (L.)	Lepidoptera: Tortricidae	codling moth,* carpocapsa delle mele, verme delle mele, Apfelwickler
Cylas formicarius elegantulus (Summers)	Coleoptera: Curculionidae	sweetpotato weevil,* sweet potato weevil
Dacus cucurbitae Coquillett	Diptera: Tephritidae	melon fly,* melon fruit fly
D. dorsalis Hendel	Diptera: Tephritidae	oriental fruit fly*
D. oleae (Gmelin)	Diptera: Tephritidae	olive fruit fly,* olive fly, mosca olearia, mouche de l'olivier
D. tryoni (Froggatt)	Diptera: Tephritidae	Queensland fruit fly
Dendroctonus brevicomis LeConte	Coleoptera: Scolytidae	western pine beetle*
D. frontalis Zimmermann	Coleoptera: Scolytidae	southern pine beetle*
D. ponderosae Hopkins	Coleoptera: Scolytidae	mountain pine beetle*
D. pseudotsugae Hopkins	Coleoptera: Scolytidae	Douglas-fir beetle*
D. rufipennis (Kirby)	Coleoptera: Scolytidae	spruce beetle*
D. terebrans (Olivier)	Coleoptera: Scolytidae	black turpentine beetle*
Dendrolimus spectabilis Butler	Lepidoptera: Lasiocampidae	pine moth
Diabrotica undecimpunctata howardi Barber	Coleoptera: Chrysomelidae	southern corn rootworm*
D. undecimpunctata, see *D. undecimpunctata howardi*		
D. virgifera virgifera LeConte	Coleoptera: Chrysomelidae	western corn rootworm*
D. virgifera, see *D. virgifera virgifera*		
Diaphania nitidalis (Stoll)	Lepidoptera: Pyralidae	pickleworm*

(continued)

TABLE 3 (Cont.)

Arthropods	Order and family	Common names used
Diatraea grandiosella (Dyar)	Lepidoptera: Pyralidae	southwestern corn borer*
D. saccharalis (Fabricius)	Lepidoptera: Pyralidae	sugarcane borer*
Dicentria semirufescens, see *Oligocentria semirufescens*		
Dioryctria abietella (Schiffermüller)	Lepidoptera: Pyralidae	cone pyralid
Diparopsis castanea Hampson	Lepidoptera: Noctuidae	red bollworm*
D. watersi (Rothschild)	Lepidoptera: Noctuidae	
Discestra trifolii (Hufnagel)	Lepidoptera: Noctuidae	clover cutworm,* nottua del trifoglio
Dryocoetes confusus Swaine	Coleoptera: Scolytidae	western balsam bark beetle*
Earias insulana (Boisduval)	Lepidoptera: Noctuidae	spiny bollworm
E. vittella (Fabricius)	Lepidoptera: Noctuidae	
Elasmopalpus lignosellus (Zeller)	Lepidoptera: Pyralidae	lesser cornstalk borer*
Enarmonia formosana (Scopoli)	Lepidoptera: Tortricidae	cherrybark tortrix moth, cherry bark tortrix, Rinderwickler, tordeuse de l'écorce des arbres fruitiers
Endopiza viteana Clemens	Lepidoptera: Tortricidae	grape berry moth*
Endothenia quadrimaculana	Lepidoptera: Tortricidae	stachis tortrix moth, tortrice degli stachis
Ephestia cautella, see *Cadra cautella*		
E. elutella (Hübner)	Lepidoptera: Pyralidae	tobacco moth,* cocoa moth, warehouse moth, tignola del cacao, Speichermotte
E. figulilella, see *Cadra figulilella*		
E. kuehniella, see *Anagasta kuehniella*		
E. kuhniella, see *A. kuehniella*		

Epichoristodes acerbella (Walker)	Lepidoptera: Tortricidae	South African carnation tortrix, bega sud-africana del garofano
Epinotia tedella (Clerck)	Lepidoptera: Tortricidae	spruce leafroller
Epiphyas postvittana (Walker)	Lepidoptera: Tortricidae	light brown apple moth*
Eucosma sonomana Kearfott	Lepidoptera: Tortricidae	western pine shoot borer, western pineshoot borer
Eucosoma sonomana, see *E. sonomana*		
Eupoecilia ambiguella Hübner	Lepidoptera: Tortricidae	European grape berry moth, European grapevine moth, grape moth, einbindiger Traubenwickler
Euxoa auxillaris (Grote)	Lepidoptera: Noctuidae	army cutworm*
E. messoria (Harris)	Lepidoptera: Noctuidae	darksided cutworm*
E. ochrogaster (Guenée)	Lepidoptera: Noctuidae	redbacked cutworm*
Euzophera bigella	Lepidoptera: Pyralidae	verme della mela cotogna
Evergestis forficalis (L.)	Lepidoptera: Pyralidae	garden pebble moth, Kohlzunsler
Fannia canicularis (L.)	Diptera: Muscidae	little house fly,* lesser house fly
Galleria mellonella (L.)	Lepidoptera: Pyralidae	greater wax moth,* greater waxmoth, great wax moth
Gnathotrichus sulcatus (LeConte)	Coleoptera: Scolytidae	ambrosia beetle
Gortyna xanthenes (Germar)	Lepidoptera: Noctuidae	artichoke moth
Grapholita funebrana, see *Cydia funebrana*		
G. janthinana, see *C. janthinana*		
G. molesta (Busck)	Lepidoptera: Tortricidae	oriental fruit moth,* Pfirsischwickler, tordeuse orientale du pecher
G. packardi Zeller	Lepidoptera: Tortricidae	cherry fruitworm*
G. prunivora (Walsh)	Lepidoptera: Tortricidae	lesser appleworm,* lesser apple worm
Grapholitha funebrana, see *Cydia funebrana*		
Grapholitha janthinana, see *C. janthinana*		

(continued)

TABLE 3 (Cont.)

Arthropods	Order and family	Common names used
Grapholitha molesta, see *Grapholita molesta*		
Grapholitha packarki, see *Grapholita packardi*		
Grapholitha pomonella, see *Cydia pomonella*		
Grapholitha prunivora, see *Grapholita prunivora*		
Gypsonoma aceriana (Duponchel)	Lepidoptera: Tortricidae	poplar shoot moth, poplar shoot-borer, gemmaiola del pioppo
Harrisina brillians Barnes & McDunnough	Lepidoptera: Zygaenidae	western grapeleaf skeletonizer*
Hedya nubiferana (Haworth)	Lepidoptera: Tortricidae	green budworm,* tordeuse verte (de la pelure), tortrice delle gemme, grauer Knosperwickler
Helcystogramma macroscopum (Meyrick)	Lepidoptera: Gelechiidae	sweet potato leaf folder, sweet potato leafroller
Helicoverpa armigera, see *Heliothis armigera*		
Heliocoverpa virescens, see *Heliothis virescens*		
Heliocoverpa zea, see *Heliothis zea*		
Heliothis armiger, see *H. armigera*		
H. armigera (Hübner)	Lepidoptera: Noctuidae	Old World bollworm, African bollworm, American bollworm, American cotton bollworm
H. punctiger, see *H. punctigera*		
H. punctigera (Wallengren)	Lepidoptera: Noctuidae	Australian bollworm

H. virescens (Fabricius)	Lepidoptera: Noctuidae	tobacco budworm*
H. zea (Boddie)	Lepidoptera: Noctuidae	corn earworm,* tomato fruitworm,* bollworm, cotton bollworm
Holomelina aurantiaca (Hübner)	Lepidoptera: Arctiidae	tiger moth
Homoeosoma electellum (Hulst)	Lepidoptera: Pyralidae	sunflower moth*
Homona coffearia (Neitner)	Lepidoptera: Tortricidae	tea tortrix
H. magnanima Diakonoff	Lepidoptera: Tortricidae	tea tortrix
Ips cembrae (Heer)	Coleoptera: Scolytidae	larch bark beetle
I. confusus (Le Conte)	Coleoptera: Scolytidae	pine bark beetle
I. pini (Say)	Coleoptera: Scolytidae	pine engraver,* pine engraver beetle
I. typographus (L.)	Coleoptera: Scolytidae	eight-toothed engraver beetle, bark beetle, Buchdrucker, pine bark beetle
Keiferia lycopersicella (Walsingham)	Lepidoptera: Gelechiidae	tomato pinworm*
Lacanobia oleracea (L.)	Lepidoptera: Noctuidae	noctuelle potagère
L. pisi (L.)	Lepidoptera: Noctuidae	noctuelle du pois
L. suasa (Denis & Schiffermüller)	Lepidoptera: Noctuidae	
L. oleracea, see *Lacanobia oleracea*		
Laspeyresia caryana, see *Cydia caryana*		
L. funebrana, see *C. funebrana*		
L. medicaginis (Kuznetsov)	Lepidoptera: Tortricidae	
L. molesta, see *Grapholita molesta*		
L. nigricana, see *Cydia nigricana*		
L. pomonella, see *C. pomonella*		
Leucoptera scitella (Zeller)	Lepidoptera: Lyonetiidae	mineuse cerclée
Limonius californicus (Mannerheim)	Coleoptera: Elateridae	sugarbeet wireworm*
Lithocolletis blancardella, see *Phyllonorycter blancardella*		
L. corylifoliella, see *P. corylifoliella*		

(continued)

TABLE 3 (Cont.)

Arthropods	Order and family	Common names used
Lobesia botrana (Denis & Schiffermüller)	Lepidoptera: Tortricidae	European grapevine moth, grape vine moth, eudemis de la vigne, tignoletta dell'uva, bekreuzter Traubenwickler
Loxagrotis albicosta (Smith)	Lepidoptera: Noctuidae	western bean cutworm*
Loxostege sticticalis (L.)	Lepidoptera: Pyralidae	beet webworm*
Lymantria dispar (L.)	Lepidoptera: Lymantriidae	gypsy moth,* limantria dispar, bombice dispari, Schwammspinner
Lymantria monacha (L.)	Lepidoptera: Lymantriidae	nun moth, monaca, la nonne, Nonne
Macdunnoughia confusa (Stephens)	Lepidoptera: Noctuidae	chrysanthemum moth, nottua dei crisantemi
Malacosoma disstria Hübner	Lepidoptera: Lasiocampidae	forest tent caterpillar,* lackey moth
Mamestra brassicae (L.)	Lepidoptera: Noctuidae	cabbage armyworm, nottua dei cavoli, noctuelle du chou
M. configurata	Lepidoptera: Noctuidae	bertha armyworm,* beet armyworm
M. oleracea, see *Lacanobia oleracea*		
M. pisi, see *Lacanobia pisi*		
M. suasa, see *Lacanobia suasa*		
Manduca sexta (L.)	Lepidoptera: Sphingidae	tobacco hornworm,* sphinx moth
Melissopus latiferreanus (Walsingham)	Lepidoptera: Torticidae	filbertworm*
Mocis latipes (Guenée)	Lepidoptera: Noctuidae	
Monima gothica, see *Orthosia gothica*		
M. incerta, see *O. incerta*		
Musca autumnalis De Geer	Diptera: Muscidae	face fly*
M. domestica L.	Diptera: Muscidae	house fly,* common house fly

Mythimna unipuncta, see *Pseudaletia unipuncta*		
Naranga aenescens (Moore)	Lepidoptera: Noctuidae	rice green caterpillar
Noctua comes Hübner	Lepidoptera: Noctuidae	Bandeule
N. fimbriata (Schreber)	Lepidoptera: Noctuidae	gelbe Bandeule
Odites leucostola (Meyrick)	Lepidoptera: Gelechiidae	
Oligocentria semirufescens (Walker)	Lepidoptera: Notodontidae	
Operophtera brumata (L.)	Lepidoptera: Geometridae	winter moth,* Kleiner Frostspanner
Orgyia antiqua (L.)	Lepidoptera: Lymantriidae	rusty tussock moth,* bombice antico, bombix antico, orgia antico
O. pseudotsugata (McDunnough)	Lepidoptera: Lymantriidae	Douglas-fir tussock moth,* Douglas fir tussock moth
Orthosia gothica (L.)	Lepidoptera: Noctuidae	braungraue Obstbaumeule
O. hibisci (Guenée)	Lepidoptera: Noctuidae	speckled green fruitworm
O. incerta (Hufnagel)	Lepidoptera: Noctuidae	Obstbaumeule
Oryzaephilus mercator (Fauvel)	Coleoptera: Cucujidae	merchant grain beetle*
O. surinamensis (L.)	Coleoptera: Cucujidae	sawtoothed grain beetle*
Ostrinia furnacalis (Guenée)	Lepidoptera: Pyralidae	Asian corn borer
O. nubilalis (Hübner)	Lepidoptera: Pyralidae	European corn borer,* Iowa strain, New York strain, pyrale du mais
Pammene rhediella (Clerck)	Lepidoptera: Tortricidae	fruitlet mining tortrix, Bodenswickler
Pandemis cerasana (Hübner)	Lepidoptera: Tortricidae	
P. heparana (Denis & Schiffermüller)	Lepidoptera: Tortricidae	fruit tree tortrix moth, pandemis, ricamatrice dei fruttiferi
P. limitata (Robinson)	Lepidoptera: Tortricidae	threelined leafroller*
P. pyrusana Kearfott	Lepidoptera: Tortricidae	
Panolis flammea (Denis & Schiffermüller)	Lepidoptera: Noctuidae	pine beauty moth, Forleule
Panonychus ulmi (Koch)	Acari: Tetranychidae	European red mite*

(continued)

TABLE 3 (Cont.)

Arthropods	Order and family	Common names used
Paralobesia viteana, see *Endopiza viteana*		
Parathrene simulans (Grote)	Lepidoptera: Sesiidae	oak borer
P. tabaniformis (Rottemburg)	Lepidoptera: Sesiidae	dusky clearwing moth, poplar waspworm, tarlo vespa del pioppo
Pectinophora gossypiella (Saunders)	Lepidoptera: Gelechiidae	pink bollworm*
Pennisetia hylaeiformis (Laspeyres)	Lepidoptera: Sesiidae	raspberry clearwing moth
Peridroma saucia (Hübner)	Lepidoptera: Noctuidae	variegated cutworm*, nottua dei garofani
Periplaneta americana (L.)	Blattodea: Blattidae	American cockroach*
Pexicopia malvella (Hübner)	Lepidoptera: Gelechiidae	mallow seed moth
Phthorimaea operculella (Zeller)	Lepidoptera: Gelechiidae	potato tuberworm,* potato tuber moth, potato tuberworm moth, tignola della patata
Phyllonorycter blancardella (Fabricius)	Lepidoptera: Gracillariidae	spotted tentiform leafroller, spotted tentiform leafminer, litocollette
P. corylifoliella (Hübner)	Lepidoptera: Gracillariidae	litocollete
Pityogenes chalcographus (L.)	Coleoptera: Scolytidae	Kupferstecher
Planococcus citri (Risso)	Homoptera: Pseudococcidae	citrus mealybug*
Platyedra subcinerea (Haworth)	Lepidoptera: Gelechiidae	cotton stem moth*
Platynota flavedana Clemens	Lepidoptera: Tortricidae	variegated leafroller
P. idaeusalis (Walker)	Lepidoptera: Tortricidae	tufted apple bud moth, tufted apple budmoth
P. sultana Walsingham	Lepidoptera: Tortricidae	omnivorous leafroller, omnivorous leaf roller
Platyptilia carduidactyla (Riley)	Lepidoptera: Pterophoridae	artichoke plume moth*

Plodia interpunctella (Hübner)	Lepidoptera: Pyralidae	Indianmeal moth,* Indian meal moth, tignola fasciata, Kupferrote Dorrobstmotte
Plusia chalcites, see *Chrysodeixis chalcites*		
P. gamma, see *Autographa gamma*		
Plutella maculipennis, see *P. xylostella*		
P. xylostella (L.)	Lepidoptera: Plutellidae	diamondback moth,* diamond back moth, diamond-backed cutworm, Kohlschabe
Podosesia syringae (Harris)	Lepidoptera: Sesiidae	lilac borer,* ash borer*
Popillia japonica Newman	Coleoptera: Scarabaeidae	Japanese beetle*
Porthetria dispar, see *Lymantria dispar*		
Prays citri (Milliere)	Lepidoptera: Plutellidae	citrus flower moth, citrus moth, teigne des agrumes, teigne des citrus
Prays oleae (Bernard)	Lepidoptera: Plutellidae	olive moth, tignola dell'olivo, teigne de l'olivier
Prionoxystus robiniae (Peck)	Lepidoptera: Cossidae	carpenterworm,* carpenter worm
Prodenia eridania, see *Spodoptera eridania*		
Prostephanus truncatus (Horn)	Coleoptera: Bostrichidae	larger grain borer*
Protobathra leucostola, see *Odites leucostola*		
Pseudaletia unipuncta (Haworth)	Lepidoptera: Noctuidae	armyworm*
Pseudaulacaspis pentagona (Targioni-Tozzetti)	Homoptera: Diaspididae	white peach scale*

(continued)

TABLE 3 (Cont.)

Arthropods	Order and family	Common names used
Pseudococcus comstocki (Kuwana)	Homoptera: Pseudococcidae	Comstock mealybug*
Pseudogalleria leucotreta (Meyrick)	Lepidoptera: Tortricidae	false codling moth
Pseudoplusia includens (Walker)	Lepidoptera: Noctuidae	soybean looper,* soyabean looper
Ptycholoma lecheana (L.)	Lepidoptera: Tortricidae	orchard tortrix moth
P. lecheana circumclusana (Christoph)	Lepidoptera: Tortricidae	
P. lecheanum, see *P. lecheana*		
Quadraspidiotus perniciosus (Comstock)	Homoptera: Diaspididae	San Jose scale*
Raphia frater Grote	Lepidoptera: Noctuidae	
Rhopobota unipunctana (Haworth)	Lepidoptera: Tortricidae	blackheaded fireworm*
Rhyacionia buoliana (Denis & Schiffermüller)	Lepidoptera: Tortricidae	European pine shoot moth,* tortrice delle gemme dei pini, Kieferntrieb-wickler
R. frustrana (Comstock)	Lepidoptera: Tortricidae	Nantucket pine tip moth*
R. neomexicana (Dyar)	Lepidoptera: Tortricidae	southwestern pine tip moth,* southern pine tip moth
R. rigidana (Fernald)	Lepidoptera: Tortricidae	pitch pine tip moth
Sanninoidea exitiosa, see *Synanthedon exitiosa*		
Schizura semirufescens, see *Oligocentria semirufescens*		
Scolytus multistriatus (Marsham)	Coleoptera: Scolytidae	smaller European elm bark beetle,* elm bark beetle, European elm bark beetle*
S. pygmaeus (Fabricius)	Coleoptera: Scolytidae	

S. scolytus (Fabricius)	Coleoptera: Scolytidae	larger European elm bark beetle, large elm bark beetle
Scotia ipsilon, see *Agrotis ipsilon*		
S. segetum, see *A. segetum*		
Scotogramma trifolii, see *Discestra trifolii*		
Scrobipalpa ocellatella (Boyd)	Lepidoptera: Gelechiidae	sugar beet moth, teigne de la betterave
Scobipalpopsis solanivora Povolný	Lepidoptera: Gelechiidae	
Sesamia cretica Lederer	Lepidoptera: Noctuidae	dura stem borer
S. inferens (Walker)	Lepidoptera: Noctuidae	purple stem borer
S. nonagrioides (LeFebvre)	Lepidoptera: Noctuidae	sesamie
Sitotroga cerealella (Olivier)	Lepidoptera: Gelechiidae	Angoumois grain moth,* Getreidemotte, alucite des céréales
Sparganothis pilleriana (Denis & Schiffermüller)	Lepidoptera: Tortricidae	woodbine leafroller moth, tortrice delle vite, pyrale de la vigne
S. sulfureana (Clemens)	Lepidoptera: Tortricidae	
Spilonota ocellana (Fabricius)	Lepidoptera: Tortricidae	eye-spotted budmoth,* eyespotted budmoth, tortrice rossa delle gemme
Spodoptera eridania (Cramer)	Lepidoptera: Noctuidae	southern armyworm*
S. exempta (Walker)	Lepidoptera: Noctuidae	nutgrass armyworm*
S. exigua (Hübner)	Lepidoptera: Noctuidae	beet armyworm,* African armyworm, lesser armyworm, nottua della bietola
S. frugiperda (J. E. Smith)	Lepidoptera: Noctuidae	fall armyworm*
S. littoralis (Boisduval)	Lepidoptera: Noctuidae	Egyptian cotton leafworm, cotton leafworm, noctuelle, nottua dei seminati
S. litura (Fabricius)	Lepidoptera: Noctuidae	tobacco cutworm
S. sumia Guenée	Lepidoptera: Noctuidae	

(continued)

TABLE 3 (Cont.)

Arthropods	Order and family	Common names used
Synanthedon bibionipennis (Boisduval)	Lepidoptera: Sesiidae	strawberry crown moth*
S. exitiosa (Say)	Lepidoptera: Sesiidae	peachtree borer,* greater peachtree borer, peach tree borer
S. hector (Butler)	Lepidoptera: Sesiidae	cherrytree borer
S. myopaeformis, see *S. myopiformis*		
S. myopiformis (Borkhausen)	Lepidoptera: Sesiidae	apple-clearwing moth, apple clearwing moth, apple root borer, sesia del melo, sésie du pommier
S. pictipes (Grote & Robinson)	Lepidoptera: Sesiidae	lesser peachtree borer,* lesser peach tree borer
S. rhododendri (Beutenmueller)	Lepidoptera: Sesiidae	rhododendron borer*
S. tipuliformis (Clerck)	Lepidoptera: Sesiidae	currant borer,* Johannisbeerglasfluegler, sésie des framboisiers
S. vespiformis (L.)	Lepidoptera: Sesiidae	
Syndemis musculana (Hübner)	Lepidoptera: Tortricidae	
Tetranychus cinnabarinus (Boisduval)	Acari: Tetranychidae	carmine spider mite*
T. urticae Koch	Acari: Tetranychidae	twospotted spider mite*
Thaumetopoea pityocampa (Denis & Schiffermüller)	Lepidoptera: Thaumetopoeidae	pine processionary moth, processionnaire du pin, processionaria del pino
Thyridopteryx ephemeraeformis (Haworth)	Lepidoptera: Psychidae	bagworm*

Tortrix viridana L.	Lepidoptera: Tortricidae	green oak tortrix moth, tordeuse du chêne, tordeuse verte du chêne, Eichenwickler
Tribolium castaneum (Herbst)	Coleoptera: Tenebrionidae	red flour beetle*
T. confusum Jacquelin du Val	Coleoptera: Tenebrionidae	confused flour beetle*
Trichoplusia ni (Hübner)	Lepidoptera: Noctuidae	cabbage looper,* ni moth, nottua delle crucifere
Trogoderma granarium Everts	Coleoptera: Dermestidae	khapra beetle*
T. inclusum LeConte	Coleoptera: Dermestidae	
T. variabile Ballion	Coleoptera: Dermestidae	warehouse beetle*
Trypodendron lineatum (Olivier)	Coleoptera: Scolytidae	striped ambrosia beetle,* ambrosia beetle, gestreifter Nutzholzborkenkäfer
Vitacea polistiformnis (Harris)	Lepidoptera: Sesiidae	grape root borer*
Xestia c-nigrum (L.)	Lepidoptera: Noctuidae	c-nigrum moth, nottua c-nigrum
X. dolosa Franclemont	Lepidoptera: Noctuidae	large spotted cutworm
Yponomeuta malinellus Zeller	Lepidoptera: Yponomeutidae	hyponomeute du pommier
Zeiraphera diniana (Guenée)	Lepidoptera: Trotricidae	larch bud moth

TABLE 4 Arthropod Common Names Used by Suppliers[a]

Common name	Scientific name	Order and family
African armyworm	*Spodoptera exigua*	Lepidoptera: Noctuidae
African bollworm	*Heliothis armigera*	Lepidoptera: Noctuidae
African rice borer moth	*Chilo zacconius*	Lepidoptera: Pyralidae
agrotide	*Agrotis exclamationis*	Lepidoptera: Noctuidae
Ahrenwickler	*Cnephasia pumicana*	Lepidoptera: Tortricidae
alfalfa looper*	*Autographa californica*	Lepidoptera: Noctuidae
almond moth*	*Cadra cautella*	Lepidoptera: Pyralidae
alucite des cereales	*Sitotroga cerealella*	Lepidoptera: Gelechiidae
ambrosia beetle	*Gnathotrichus sulcatus*	Coleoptera: Tenebrionidae
ambrosia beetle	*Trypodendron lineatum*	Coleoptera: Scolytidae
American bollworm	*Heliothis armigera*	Lepidoptera: Noctuidae
American cockroach*	*Periplaneta americana*	Blattodea: Blattidae
American cotton bollworm	*Heliothis armigera*	Lepidoptera: Noctuidae
Angoumois grain moth*	*Sitotroga cerealella*	Lepidoptera: Gelechiidae
Apfelbaumglasfluegler	*Synanthedon myopiformis*	Lepidoptera: Sesiidae
Apfelblattfaltenmotte	*Phyllonorycter blancardella*	Lepidoptera: Gracillariidae
Apfelblattwickler	*Archips xylosteanus*	Lepidoptera: Tortricidae
Apfelwickler	*Cydia pomonella*	Lepidoptera: Tortricidae
apple clearwing moth	*Synanthedon myopiformis*	Lepidoptera: Sesiidae
apple root borer	*Synanthedon myopiformis*	Lepidoptera: Sesiidae
army cutworm*	*Euxoa auxiliaris*	Lepidoptera: Noctuidae
armyworm*	*Pseudaletia unipuncta*	Lepidoptera: Noctuidae
artichoke moth	*Gortyna xanthenes*	Lepidoptera: Noctuidae
artichoke plume moth*	*Platyptilia carduidactyla*	Lepidoptera: Pterophoridae
ash borer*	*Podosesia syringae*	Lepidoptera: Sesiidae
Asian corn borer	*Ostrinia furnacalis*	Lepidoptera: Pyralidae
Asiatic leafroller	*Archips breviplicanus*	Lepidoptera: Tortricidae

Asiatic rice borer*	*Chilo suppressalis*	Lepidoptera: Pyralidae
Austalian bollworm	*Heliothis punctigera*	Lepidoptera: Noctuidae
bagworm*	*Thyridopteryx ephemeraeformis*	Lepidoptera: Psychidae
banded sunflower moth*	*Cochylis hospes*	Lepidoptera: Cochylidae
Bandeule	*Noctua comes*	Lepidoptera: Noctuidae
bark beetle	*Ips typographus*	Coleoptera: Scolytidae
beet armyworm	*Mamestra configurata*	Lepidoptera: Noctuidae
beet armyworm*	*Spodoptera exigua*	Lepidoptera: Noctuidae
beet webworm*	*Loxostege sticticalis*	Lepidoptera: Pyralidae
bega del garofano	*Cacoecimorpha pronubana*	Lepidoptera: Tortricidae
bega sud-africana del garofano	*Epichoristodes acerbella*	Lepidoptera: Tortricidae
bekreuzter Traubenwickler	*Lobesia botrana*	Lepidoptera: Tortricidae
bertha armyworm*	*Mamestra configurata*	Lepidoptera: Noctuidae
black army cutworm*	*Actebia fennica*	Lepidoptera: Noctuidae
black carpet beetle*	*Attagenus unicolor*	Coleoptera: Dermestidae
black cutworm*	*Agrotis ipsilon*	Lepidoptera: Noctuidae
black turnpentine beetle*	*Dendroctonus terebrans*	Coleoptera: Scolytidae
blackheaded fireworm*	*Rhopobota unipunctana*	Lepidoptera: Tortricidae
Bodenseewickler	*Pammene rhediella*	Lepidoptera: Tortricidae
boll weevil*	*Anthonomus grandis*	Coleoptera: Curculionidae
bollworm	*Heliothis zea*	Lepidoptera: Noctuidae
bombice antico	*Orgyia antiqua*	Lepidoptera: Lymantriidae
bombice dispari	*Lymantria dispar*	Lepidoptera: Lymantriidae
bombix antico	*Orgyia antiqua*	Lepidoptera: Lymantriidae
bombyx disparate	*Lymantria dispar*	Lepidoptera: Lymantriidae
braungraue Obstbaumeule	*Orthosia gothica*	Lepidoptera: Noctuidae
braunlicher Obstbaumwickler	*Archips podanus*	Lepidoptera: Tortricidae
Buchdrucker	*Ips typographus*	Coleoptera: Scolytidae

[a]Names marked with an asterisk are those approved by the Entomological Society of America (16).

(continued)

TABLE 4 (Cont.)

Common name	Scientific name	Order and family
c-nigrum moth	*Xestia c-nigrum*	Lepidoptera: Noctuidae
cabbage armyworm	*Mamestra brassicae*	Lepidoptera: Noctuidae
cabbage leafroller	*Clepsis spectrana*	Lepidoptera: Tortricidae
cabbage looper*	*Trichoplusia ni*	Lepidoptera: Noctuidae
cacoecia dei frutti	*Archips podanus*	Lepidoptera: Tortricidae
cacoecia delle gemme	*Archips podanus*	Lepidoptera: Tortricidae
cacoecia delle gemme	*Archips rosanus*	Lepidoptera: Tortricidae
California red scale*	*Aonidiella aurantii*	Homoptera: Diaspididae
capua	*Adoxophyes orana*	Lepidoptera: Tortricidae
carmine spider mite*	*Tetranychus cinnabarinus*	Acari: Tetranychidae
carnation tortrix moth	*Cacoecimorpha pronubana*	Lepidoptera: Tortricidae
carpenterworm*	*Prionoxystus robiniae*	Lepidoptera: Cossidae
carpocapsa delle mele	*Cydia pomonella*	Lepidoptera: Tortricidae
carpocapse des prunes	*Cydia funebrana*	Lepidoptera: Tortricidae
cereal tortrix moth	*Cnephasia pumicana*	Lepidoptera: Tortricidae
chematobie hiemale	*Operophtera brumata*	Lepidoptera: Geometridae
cherry bark tortrix	*Enarmonia formosana*	Lepidoptera: Tortricidae
cherry fruitworm*	*Grapholita packardi*	Lepidoptera: Tortricidae
cherrybark tortrix moth	*Enarmonia formosana*	Lepidoptera: Tortricidae
cherrytree borer	*Synanthedon hector*	Lepidoptera: Sesiidae
chrysanthemum moth	*Macdunnoughia confusa*	Lepidoptera: Noctuidae
citrus flower moth	*Prays citri*	Lepidoptera: Plutellidae
citrus mealybug*	*Planococcus citri*	Homoptera: Pseudococcidae
clover cutworm*	*Discestra trifolii*	Lepidoptera: Noctuidae
cochylis de la vigne	*Eupoecilia ambiguella*	Lepidoptera: Tortricidae
cocoa moth	*Ephestia elutella*	Lepidoptera: Pyralidae

codling moth*	*Cydia pomonella*	Lepidoptera: Tortricidae
common cutworm	*Agrotis fucosa*	Lepitoptera: Noctuidae
common house fly	*Musca domestica*	Diptera: Muscidae
Comstock mealybug*	*Pseudococcus comstocki*	Homoptera: Pseudococcidae
cone pyralid	*Dioryctria abietella*	Lepidoptera: Pyralidae
confused flour beetle*	*Tribolium confusum*	Coleoptera: Tenebrionidae
corn earworm*	*Heliothis zea*	Lepidoptera: Noctuidae
cotton bollworm	*Heliothis zea*	Lepidoptera: Noctuidae
cotton leafworm	*Spodoptera littoralis*	Lepidoptera: Noctuidae
cotton stem moth*	*Platyedra subcinerea*	Lepidoptera: Gelechiidae
cranberry girdler*	*Chrysoteuchia topiaria*	Lepidoptera: Pyralidae
currant borer*	*Synanthedon tipuliformis*	Lepidoptera: Sesiidae
cyclamen tortrix	*Clepsis spectrana*	Lepidoptera: Tortricidae
dark sword grass moth	*Agrotis ipsilon*	Lepidoptera: Noctuidae
darksided cutworm*	*Euxoa messoria*	Lepidoptera: Noctuidae
Dattelmotte	*Cadra cautella*	Lepidoptera: Pyralidae
diamond-backed cutworm	*Plutella xylostella*	Lepidoptera: Plutellidae
diamondback moth*	*Plutella xylostella*	Lepidoptera: Plutellidae
Douglas-fir beetle*	*Dendroctonus pseudotsugae*	Coleoptera: Scolytidae
Douglas-fir tussock moth*	*Orgyia pseudotsugata*	Lepidoptera: Lymantriidae
dura stem borer	*Sesamia cretica*	Lepidoptera: Noctuidae
dusky clearwing moth	*Paranthrene tabaniformis*	Lepidoptera: Sesiidae
eastern spruce budworm	*Choristoneura fumiferana*	Lepidoptera: Tortricidae
Egyptian cotton leafworm	*Spodoptera littoralis*	Lepidoptera: Noctuidae
Eichenwickler	*Tortrix viridana*	Lepidoptera: Tortricidae
eight-toothed engraver beetle	*Ips typographus*	Coleoptera: Scolytidae
einbindiger Traubenwickler	*Eupoecilia ambiguella*	Lepidoptera: Tortricidae
elm bark beetle	*Scolytus multistriatus*	Coleoptera: Scolytidae

(continued)

TABLE 4 (Cont.)

Common name	Scientific name	Order and family
Erbsenwickler	*Cydia nigricana*	Lepidoptera: Tortricidae
eudemis de la vigne	*Lobesia botrana*	Lepidoptera: Tortricidae
eulia	*Argyrotaenia pulchellana*	Lepidoptera: Tortricidae
European corn borer* ("Z," "E" strains)	*Ostrinia nubilalis*	Lepidoptera: Pyralidae
European corn borer* (N.Y., Iowa strains)	*Ostrinia nubilalis*	Lepidoptera: Pyralidae
European elm bark beetle	*Scolytus multistriatus*	Coleoptera: Scolytidae
European fir budworm moth	*Choristoneura murinana*	Lepidoptera: Tortricidae
European goat moth	*Cossus cossus*	Lepidoptera: Cossidae
European grape berry moth	*Eupoecilia ambiguella*	Lepidoptera: Tortricidae
European grapevine moth	*Lobesia botrana*	Lepidoptera: Tortricidae
European grapevine moth	*Eupoecilia ambiguella*	Lepidoptera: Tortricidae
European leafroller	*Archips rosanus*	Lepidoptera: Tortricidae
European pine shoot moth*	*Rhyacionia buoliana*	Lepidoptera: Tortricidae
European red mite*	*Panonychus ulmi*	Acari: Tetranychidae
exclamator cutworm	*Agrotis exclamationis*	Lepidoptera: Noctuidae
eye-spotted budmoth*	*Spilonota ocellana*	Lepidoptera: Tortricidae
face fly*	*Musca autumnalis*	Diptera: Muscidae
fall armyworm*	*Spodoptera frugiperda*	Lepidoptera: Noctuidae
false codling moth	*Pseudogalleria leucotreta*	Lepidoptera: Tortricidae
fig moth	*Cadra cautella*	Lepidoptera: Pyralidae
filbert leafroller	*Archips rosanus*	Lepidoptera: Tortricidae
filbertworm*	*Melissopus latiferreanus*	Lepidoptera: Tortricidae
forest tent caterpillar*	*Malacosoma disstria*	Lepidoptera: Lasiocampidae
Forleule	*Panolis flammea*	Lepidoptera: Noctuidae

Fruchtschalenwickler	*Adoxophyes orana*	Lepidoptera: Tortricidae
fruit tree tortrix	*Archips podanus*	Lepidoptera: Tortricidae
fruit tree tortrix	*Argyrotaenia pulchellana*	Lepidoptera: Tortricidae
fruit tree tortrix moth	*Pandemis heparana*	Lepidoptera: Tortricidae
fruit tree tortrix moth	*Argyrotaenia pulchellana*	Lepidoptera: Tortricidae
fruitlet mining tortrix	*Pammene rhediella*	Lepidoptera: Tortricidae
fruittree leafroller*	*Archips argyrospilus*	Lepidoptera: Tortricidae
furniture carpet beetle*	*Anthrenus flavipes*	Coleoptera: Dermestidae
garden pebble moth	*Evergestis forficalis*	Lepidoptera: Pyralidae
gestreifter Nutzholzborkenkäfer	*Trypodendron lineatum*	Coleoptera: Scolytidae
geflammter Reberwickler	*Clepsis spectrana*	Lepidoptera: Tortricidae
gelbe Bandeule	*Noctua fimbriata*	Lepidoptera: Noctuidae
gemmaiola del pioppo	*Gypsonoma aceriana*	Lepidoptera: Tortricidae
German cockroach*	*Blatella germanica*	Blattodea: Blattellidae
Getreidemotte	*Sitotroga cerealella*	Lepidoptera: Gelechiidae
Getreidewickler	*Cnephasia longana*	Lepidoptera: Tortricidae
gialla del pomodoro e del mais	*Heliothis armigera*	Lepidoptera: Noctuidae
glassy cutworm*	*Crymodes devastator*	Lepidoptera: Noctuidae
goat moth	*Cossus cossus*	Lepidoptera: Cossidae
grape berry moth	*Eupoecelia ambiguella*	Lepidoptera: Tortricidae
grape berry moth*	*Endopiza viteana*	Lepidoptera: Tortricidae
grape moth	*Eupoecilia ambiguella*	Lepidoptera: Tortricidae
grape root borer*	*Vitacea polistiformis*	Lepidoptera: Sessiidae
grape vine moth	*Lobesia botrana*	Lepidoptera: Tortricidae
grauer Knosperwickler	*Hedya nubiferana*	Lepidoptera: Tortricidae
greasy cutworm	*Agrotis ipsilon*	Lepidoptera: Noctuidae
great wax moth	*Galleria mellonella*	Lepidoptera: Pyralidae
greater peachtree borer	*Synanthedon exitiosa*	Lepidoptera: Sesiidae
greater wax moth*	*Galleria mellonella*	Lepidoptera: Pyralidae

(continued)

TABLE 4 (Cont.)

Common name	Scientific name	Order and family
green budworm*	*Hedya nubiferana*	Lepidoptera: Tortricidae
green garden looper*	*Chrysodeixis chalcites*	Lepidoptera: Noctuidae
green oak tortrix moth	*Tortrix viridana*	Lepidoptera: Tortricidae
gypsy moth*	*Lymantria dispar*	Lepidoptera: Lymantriidae
hawthorne leafroller	*Cydia janthinana*	Lepidoptera: Tortricidae
heart and dart moth	*Agrotis exclamationis*	Lepidoptera: Noctuidae
Heckenwickler	*Archips rosanus*	Lepidoptera: Tortricidae
hickory shuckworm*	*Cydia caryana*	Lepidoptera: Tortricidae
honeybee*	*Apis mellifera*	Hymenoptera: Apidae
house fly*	*Musca domestica*	Diptera: Muscidae
hyponomeute du pommier	*Yponomeuta malinellus*	Lepidoptera: Yponomeutidae
Indianmeal moth*	*Plodia interpunctella*	Lepidoptera: Pyralidae
Japanese beetle*	*Popillia japonica*	Coleoptera: Scarabaeidae
Johannisbeerglasfluegler	*Synanthedon tipuliformis*	Lepidoptera: Sesiidae
khapra beetle*	*Trogoderma granarium*	Coleoptera: Dermestidae
Kieferntriebwickler	*Rhyacionia buoliana*	Lepidoptera: Tortricidae
Kirschblütenmotte	*Argyresthis pruniella*	Lepidoptera: Argyresthiidae
kleiner Frostspanner	*Operophtera brumata*	Lepidoptera: Geometridae
Kohlschabe	*Plutella xylostella*	Lepidoptera: Plutellidae
Kohlzunsler	*Evergestis forficalis*	Lepidoptera: Pyralidae
kupferrote Dorrobstmotte	*Plodia interpunctella*	Lepidoptera: Pyralidae
Kupferstecher	*Pityogenes chalcographus*	Coleoptera: Scolytidae
la nonne	*Lymantria monacha*	Lepidoptera: Lymantriidae
lackey moth	*Malacosoma disstria*	Lepidoptera: Lasiocampidae
larch bark beetle	*Ips cembrae*	Coleoptera: Scolytidae
larch bud moth	*Zeiraphera diniana*	Lepidoptera: Tortricidae

larch casebearer*	*Coleophora laricella*	Lepidoptera: Coccinellidae
large elm bark beetle	*Scolytus scolytus*	Coleoptera: Scolytidae
large spotted cutworm	*Xestia dolosa*	Lepidoptera: Noctuidae
larger European elm bark beetle	*Scolytus scolytus*	Coleoptera: Scolytidae
larger grain borer*	*Prostephanus truncatus*	Coleoptera: Bostrichidae
Lauchmotte	*Acrolepiopsis assectella*	Lepidoptera: Acrolepiidae
leek moth*	*Acrolepiopsis assectella*	Lepidoptera: Acrolepiidae
lesser appleworm*	*Grapholita prunivora*	Lepidoptera: Tortricidae
lesser armyworm	*Spodoptera exigua*	Lepidoptera: Noctuidae
lesser cornstalk borer*	*Elasmopalpus lignosellus*	Lepidoptera: Pyralidae
lesser house fly	*Fannia canicularis*	Diptera: Muscidae
lesser peachtree borer*	*Synanthedon pictipes*	Lepidoptera: Sesiidae
lesser wax moth*	*Achroia grisella*	Lepidoptera: Pyralidae
light brown apple moth*	*Epiphyas postvittana*	Lepidoptera: Tortricidae
lilac borer*	*Podosesia syringae*	Lepidoptera: Sesiidae
limantria dispar	*Lymantria dispar*	Lepidoptera: Lymantriidae
litocollete	*Phyllonorycter corylifoliella*	Lepidoptera: Gracillariidae
litocollete	*Phyllonorycter blancardella*	Lepidoptera: Gracillariidae
little house fly*	*Fannia canicularis*	Diptera: Muscidae
lone star tick*	*Amblyomma americanum*	Acari: Ixodidae
Maiszunsler	*Ostrinia nubilalis*	Lepidoptera: Pyralidae
maize stalk-borer	*Busseola fusca*	Lepidoptera: Noctuidae
mallow seed moth	*Pexicopia malvella*	Lepidoptera: Gelechiidae
Mediterranean flour moth*	*Anagasta kuehniella*	Lepidoptera: Pyralidae
Mediterranean fruit fly*	*Ceratitis capitata*	Diptera: Tephritidae
Mehlmotte	*Anagasta kuehniella*	Lepidoptera: Pyralidae
melon fly*	*Dacus cucurbitae*	Diptera: Tephritidae
melon fruit fly	*Dacus cucurbitae*	Diptera: Tephritidae
merchant grain beetle*	*Oryzaephilus mercator*	Coleoptera: Cucujidae

(continued)

TABLE 4 (Cont.)

Common name	Scientific name	Order and family
mineuse cerclée	*Leucoptera scitella*	Lepidoptera: Lyonetiidae
mineuse marbrée des feuilles	*Phyllonorycter blancardella*	Lepidoptera: Gracillariidae
monaca	*Lymantria monacha*	Lepidoptera: Lymantriidae
mosca olearia	*Dacus oleae*	Diptera: Tephritidae
mouche de l'olivier	*Dacus oleae*	Diptera: Tephritidae
mountain pine beetle*	*Dendroctonus ponderosae*	Coleoptera: Scolytidae
Nantucket pine tip moth*	*Rhyacionia frustrana*	Lepidoptera: Tortricidae
navel orangeworm*	*Amyelois transitella*	Lepidoptera: Pyralidae
ni moth	*Trichoplusia ni*	Lepidoptera: Noctuidae
noctuelle	*Spodoptera littoralis*	Lepidoptera: Noctuidae
noctuelle des moissons	*Agrotis segetum*	Lepidoptera: Noctuidae
noctuelle du chou	*Mamestra brassicae*	Lepidoptera: Noctuidae
noctuelle du pois	*Mamestra pisi*	Lepidoptera: Noctuidae
noctuelle potagére	*Lacanobia oleracea*	Lepidoptera: Noctuidae
Nonne	*Lymantria monacha*	Lepidoptera: Lymantriidae
nottua c-nigrum	*Xestia c-nigrum*	Lepidoptera: Noctuidae
nottua dei cavoli	*Mamestra brassicae*	Lepidoptera: Noctuidae
nottua dei crisantemi	*Macdunnoughia confusa*	Lepidoptera: Noctuidae
nottua dei garofani	*Peridroma saucia*	Lepidoptera: Noctuidae
nottua dei seminati	*Agrotis segetum*	Lepidoptera: Noctuidae
nottua del cotone	*Spodoptera littoralis*	Lepidoptera: Noctuidae
nottua del pomodoro	*Chrysodeixis chalcites*	Lepidoptera: Noctuidae
nottua del trifoglio	*Discestra trifolii*	Lepidoptera: Noctuidae
nottua della bietola	*Spodoptera exigua*	Lepidoptera: Noctuidae
nottua delle crucifere	*Trichoplusia ni*	Lepidoptera: Noctuidae
nottua ipsilon	*Agrotis ipsilon*	Lepidoptera: Noctuidae

nottuelle Gamma	*Autographa gamma*	Lepidoptera: Noctuidae
nun moth	*Lymantria monacha*	Lepidoptera: Lymantriidae
nutgrass armyworm*	*Spodoptera exempta*	Lepidoptera: Noctuidae
oak borer	*Paranthrene simulans*	Lepidoptera: Sesiidae
oak leafroller*	*Archips semiferanus*	Lepidoptera: Tortricidae
obliquebanded leafroller*	*Choristoneura rosaceana*	Lepidoptera: Tortricidae
Obstbaumeule	*Orthosia incerta*	Lepidoptera: Noctuidae
Old World bollworm	*Heliothis armigera*	Lepidoptera: Noctuidae
olive fly	*Dacus oleae*	Diptera: Tephritidae
olive fruit fly*	*Dacus oleae*	Diptera: Tephritidae
olive moth	*Prays oleae*	Lepidoptera: Plutellidae
omnivorous leafroller	*Platynota stultana*	Lepidoptera: Tortricidae
omnivorous leaftier*	*Cnephasia longana*	Lepidoptera: Tortricidae
orange tortrix*	*Argyrotaenia citrana*	Lepidoptera: Tortricidae
orchard tortrix moth	*Ptycholoma lecheana*	Lepidoptera: Tortricidae
orgia antico	*Orgyia antiqua*	Lepidoptera: Lymantriidae
oriental cockroach*	*Blatta orientalis*	Blattodea: Blattidae
oriental fruit fly*	*Dacus dorsalis*	Diptera: Tephritidae
oriental fruit moth*	*Grapholita molesta*	Lepidoptera: Tortricidae
pale western cutworm*	*Agrotis orthogonia*	Lepidoptera: Noctuidae
pandemis	*Pandemis heparana*	Lepidoptera: Tortricidae
pea tortrix moth*	*Cydia nigricana*	Lepidoptera: Tortricidae
peach fruit moth	*Carposina niponensis*	Lepidoptera: Carposinidae
peach twig borer*	*Anarsia lineatella*	Lepidoptera: Gelechiidae
peachtree borer*	*Synanthedon exitiosa*	Lepidoptera: Sesiidae
petite mineuse du pêcher	*Anarsia lineatella*	Lepidoptera: Gelechiidae
Pfirsischwickler	*Grapholita molesta*	Lepidoptera: Tortricidae
Pflaumenwickler	*Cydia funebrana*	Lepidoptera: Tortricidae
pickleworm*	*Diaphania nitidalis*	Lepidoptera: Pyralidae

(continued)

TABLE 4 (Cont.)

Common name	Scientific name	Order and family
pine bark beetle	*Ips confusus*	Coleoptera: Scolytidae
pine bark beetle	*Ips typographus*	Coleoptera: Scolytidae
pine beauty moth	*Panolis flammea*	Lepidoptera: Noctuidae
pine engraver beetle	*Ips pini*	Coleoptera: Scolytidae
pine engraver*	*Ips pini*	Coleoptera: Scolytidae
pine moth	*Dendrolimus spectabilis*	Lepidoptera: Lasiocampidae
pine processionary moth	*Thaumetopoea pityocampa*	Lepidoptera: Thaumetopoeidae
pink bollworm*	*Pectinophora gossypiella*	Lepidoptera: Gelechiidae
piralide del mais	*Ostrinia nubilalis*	Lepidoptera: Pyralidae
pitch pine tip moth	*Rhyacionia rigidana*	Lepidoptera: Tortricidae
plum fruit moth	*Cydia funebrana*	Lepidoptera: Tortricidae
polyphemus moth*	*Antheraea polyphemus*	Lepidoptera: Saturniidae
poplar shoot moth	*Gypsonoma aceriana*	Lepidoptera: Tortricidae
poplar shoot-borer	*Gypsonoma aceriana*	Lepidoptera: Tortricidae
poplar wasp-worm	*Paranthrene tabaniformis*	Lepidoptera: Sesiidae
potato tuber moth	*Phthorimaea operculella*	Lepidoptera: Gelechiidae
potato tuberworm moth	*Phthorimaea operculella*	Lepidoptera: Gelechiidae
potato tuberworm*	*Phthorimaea operculella*	Lepidoptera: Gelechiidae
processionaria del pino	*Thaumetopoea pityocampa*	Lepidoptera: Thaumetopoeidae
processionnaire du pin	*Thaumetopoea pityocampa*	Lepidoptera: Thaumetopoeidae
purple stem borer	*Sesamia inferens*	Lepidoptera: Noctuidae
pyrale de la vigne	*Sparganothis pilleriana*	Lepidoptera: Tortricidae
pyrale du maïs	*Ostrinia nubilalis*	Lepidoptera: Pyralidae
pyrale du riz	*Chilo zacconius*	Lepidoptera: Pyralidae
Queensland fruit fly	*Dacus tryoni*	Diptera: Tephritidae
raisin moth*	*Cadra figulilella*	Lepidoptera: Pyralidae

raspberry clearwing moth	*Pennisetia hylaeiformis*	Lepidoptera: Sesiidae
red bollworm*	*Diparopsis castanea*	Lepidoptera: Noctuidae
red flour beetle*	*Tribolium castaneum*	Coleoptera: Tenebrionidae
redbacked cutworm*	*Euxoa ochrogaster*	Lepidoptera: Noctuidae
redbanded leafroller*	*Argyrotaenia velutinana*	Lepidoptera: Tortricidae
rhododendron borer*	*Synanthedon rhododendri*	Lepidoptera: Sesiidae
ricamatrice dei fruttiferi	*Pandemis heparana*	Lepidoptera: Tortricidae
rice green caterpillar	*Naranga aenescens*	Lepidoptera: Noctuidae
rice stem borer	*Chilo suppressalis*	Lepidoptera: Pyralidae
Rinderwickler	*Enarmonia formosana*	Lepidoptera: Tortricidae
rodilegno rosso	*Cossus cossus*	Lepidoptera: Cossidae
rose twist moth	*Archips rosanus*	Lepidoptera: Tortricidae
roter Knospenwickler	*Spilonota ocellana*	Lepidoptera: Tortricidae
rusty tussock moth*	*Orgyia antiqua*	Lepidoptera: Lymantriidae
San Jose scale*	*Quadraspidiotus perniciosus*	Homoptera: Diaspididae
sawtoothed grain beetle*	*Oryzaephilus surinamensis*	Coleoptera: Cucujidae
Schlehenspinner	*Orgyia antiqua*	Lepidoptera: Lymantriidae
Schwammspinner	*Lymantria dispar*	Lepidoptera: Lymantriidae
sesia del melo	*Synanthedon myopiformis*	Lepidoptera: Sesiidae
sésie des framboisiers	*Synanthedon tipuliformis*	Lepidoptera: Sesiidae
sésie du pommier	*Synanthedon myopiformis*	Lepidoptera: Sesiidae
silver Y moth	*Autographa gamma*	Lepidoptera: Noctuidae
six-toothed spruce bark beetle	*Pityogenes chalcographus*	Coleoptera: Scolytidae
smaller European elm bark beetle*	*Scolytus multistriatus*	Coleoptera: Scolytidae
smaller tea tortrix	*Adoxophyes* sp. (in Japan)	Lepidoptera: Tortricidae
sod webworm	*Chrysoteuchia topiaria*	Lepidoptera: Pyralidae
South African carnation tortrix	*Epichoristodes acerbella*	Lepidoptera: Tortricidae
southern armyworm*	*Spodoptera eridania*	Lepidoptera: Noctuidae
southern corn rootworm*	*Diabrotica undecimpunctata howardi*	Coleoptera: Chrysomelidae

(continued)

TABLE 4 (Cont.)

Common name	Scientific name	Order and family
southern pine beetle*	*Dendroctonus frontalis*	Coleoptera: Scolytidae
southern pine tip moth	*Rhyacionia neomexicana*	Lepidoptera: Tortricidae
southwestern corn borer*	*Diatraea grandiosella*	Lepidoptera: Pyralidae
southwestern pine tip moth*	*Rhyacionia neomexicana*	Lepidoptera: Tortricidae
soy-bean tortrix moth	*Choristoneura lafauryana*	Lepidoptera: Tortricidae
soyabean looper	*Pseudoplusia includens*	Lepidoptera: Noctuidae
soybean looper*	*Pseudoplusia includens*	Lepidoptera: Noctuidae
speckled green fruitworm	*Orthosia hibisci*	Lepidoptera: Noctuidae
Speichermotte	*Ephestia elutella*	Lepidoptera: Pyralidae
sphinx moth	*Manduca sexta*	Lepidoptera: Sphingidae
spiny bollworm	*Earias insulana*	Lepidoptera: Noctuidae
spotted stalk borer	*Chilo partellus*	Lepidoptera: Pyralidae
spotted tentiform leafminer	*Phyllonorycter blancardella*	Lepidoptera: Gracillariidae
spotted tentiform leafroller	*Phyllonorycter blancardella*	Lepidoptera: Gracillariidae
spruce bark beetle	*Ips typographus*	Coleoptera: Scolytidae
spruce beetle*	*Dendroctonus rufipennis*	Coleoptera: Scolytidae
spruce budworm*	*Choristoneura fumiferana*	Lepidoptera: Tortricidae
spruce leafroller	*Epinotia tedella*	Lepidoptera: Tortricidae
stachis tortrix moth	*Endothenia quadrimaculana*	Lepidoptera: Tortricidae
strawberry crown moth*	*Synanthedon bibionipennis*	Lepidoptera: Sesiidae
striped ambrosia beetle*	*Trypodendron lineatum*	Coleoptera: Scolytidae
striped rice borer	*Chilo suppressalis*	Lepidoptera: Pyralidae
sugar beet moth	*Scrobipalpa ocellatella*	Lepidoptera: Gelechiidae
sugarbeet wireworm*	*Limonius californicus*	Coleoptera: Elateridae
sugarcane borer*	*Diatraea saccharalis*	Lepidoptera: Pyralidae
summerfruit tortrix	*Adoxophyes orana*	Lepidoptera: Tortricidae
sunflower moth*	*Homoeosoma electellum*	Lepidoptera: Pyralidae

sweet potato leaf folder	*Helcystogramma macroscopa*	Lepidoptera: Gelechiidae
sweet potato leafroller	*Helcystogramma macroscopa*	Lepidoptera: Gelechiidae
sweetpotato weevil*	*Cylas formicarius elegantulus*	Coleoptera: Curculionidae
tarlo vespa del pioppo	*Paranthrene tabaniformis*	Lepidoptera: Sesiidae
tea tortrix	*Homona coffearia*	Lepidoptera: Tortricidae
tea tortrix moth	*Homona magnanima*	Lepidoptera: Tortricidae
teigne de l'olivier	*Prays oleae*	Lepidoptera: Plutellidae
teigne de la betterave	*Scrobipalpa ocellatella*	Lepidoptera: Gelechiidae
teigne des agrumes	*Prays citri*	Lepidoptera: Plutellidae
teigne des citrus/agrumes	*Prays citri*	Lepidoptera: Plutellidae
teigne des crucifères	*Plutella xylostella*	Lepidoptera: Plutellidae
teigne du poireau	*Acrolepiopsis assectella*	Lepidoptera: Acrolepiidae
threelined leafroller*	*Pandemis limitata*	Lepidoptera: Tortricidae
tiger moth	*Holomelina aurantiaca*	Lepidoptera: Arctiidae
tignola degli agrumi	*Prays citri*	Lepidoptera: Plutellidae
tignola dei fichi secchi	*Cadra cautella*	Lepidoptera: Pyralidae
tignola del cacao	*Cadra figulilella*	Lepidoptera: Pyralidae
tignola del cacao	*Ephestia elutella*	Lepidoptera: Pyralidae
tignola del pesco	*Anarsia lineatella*	Lepidoptera: Gelechiidae
tignola dell'olivo	*Prays oleae*	Lepidoptera: Plutellidae
tignola dell'uva	*Eupoecilia ambiguella*	Lepidoptera: Tortricidae
tignola della frutta secca	*Cadra figulilella*	Lepidoptera: Pyralidae
tignola della patata	*Phthorimaea operculella*	Lepidoptera: Gelechiidae
tignola fasciata	*Plodia interpunctella*	Lepidoptera: Pyralidae
tignoletta dell'uva	*Lobesia botrana*	Lepidoptera: Tortricidae
tobacco budworm*	*Heliothis virescens*	Lepidoptera: Noctuidae
tobacco cutworm*	*Spodoptera litura*	Lepidoptera: Noctuidae
tobacco hornworm*	*Manduca sexta*	Lepidoptera: Sphingidae
tobacco moth*	*Ephestia elutella*	Lepidoptera: Pyralidae
tomato fruitworm*	*Heliothis zea*	Lepidoptera: Noctuidae

(continued)

TABLE 4 (Cont.)

Common name	Scientific name	Order and family
tomato looper	*Chrysodeixis chalcites*	Lepidoptera: Noctuidae
tomato pinworm*	*Keiferia lycopersicella*	Lepidoptera: Gelechiidae
tordeuse de l'aubepine	*Cydia janthinana*	Lepidoptera: Tortricidae
tordeuse de l'écorce des arbres fruitiers	*Enarmonia formosana*	Lepidoptera: Tortricidae
tordeuse de la luzerne	*Laspeyresia medicaginis*	Lepidoptera: Tortricidae
tordeuse de la pelure	*Argyrotaenia pulchellana*	Lepidoptera: Tortricidae
tordeuse de la pelure	*Pandemis heparana*	Lepidoptera: Tortricidae
tordeuse des bourgeons	*Archips crataeganus*	Lepidoptera: Tortricidae
tordeuse des bourgeons	*Archips xylosteanus*	Lepidoptera: Tortricidae
tordeuse des bourgeons	*Archips rosanus*	Lepidoptera: Tortricidae
tordeuse des céréales	*Cnephasia pumicana*	Lepidoptera: Tortricidae
tordeuse du chêne	*Tortrix viridana*	Lepidoptera: Tortricidae
tordeuse du mélèze	*Zeiraphera diniana*	Lepidoptera: Tortricidae
tordeuse du pois	*Cydia nigricana*	Lepidoptera: Tortricidae
tordeuse du sapin	*Choristoneura murinana*	Lepidoptera: Tortricidae
tordeuse européenne d l'oellet	*Cacoecimorpha pronubana*	Lepidoptera: Tortricidae
tordeuse orientale du pêcher	*Grapholite molesta*	Lepidoptera: Tortricidae
tordeuse rouge (de la pelure)	*Choristoneura sorbiana*	Lepidoptera: Tortricidae
tordeuse rouge des bourgeons	*Spilonota ocellana*	Lepidoptera: Tortricidae
tordeuse sud-africaine de l'oeillet	*Epichoristodes acerbella*	Lepidoptera: Tortricidae
tordeuse verte (de la pelure)	*Hedya nubiferana*	Lepidoptera: Tortricidae
tordeuse verte des bourgeons	*Hedya nubiferana*	Lepidoptera: Tortricidae
tordeuse verte du chêne	*Tortrix viridana*	Lepidoptera: Tortricidae
tortrice degli stachis	*Endothenia quadrimaculana*	Lepidoptera: Tortricidae
tortrice dei piselli	*Cydia nigricana*	Lepidoptera: Tortricidae
tortrice della soia e della fragola	*Choristoneura lafauryana*	Lepidoptera: Tortricidae

tortrice della vite	*Clepsis spectrana*	Lepidoptera: Tortricidae
tortrice delle gemme	*Hedya nubiferana*	Lepidoptera: Tortricidae
tortrice delle gemme dei pini	*Rhyacionia buoliana*	Lepidoptera: Tortricidae
tortrice delle vite	*Sparganothis pilleriana*	Lepidoptera: Tortricidae
tortrice rossa delle gemme	*Spilonota ocellana*	Lepidoptera: Tortricidae
tufted apple budmoth	*Platynota idaeusalis*	Lepidoptera: Tortricidae
turnip moth	*Agrotis segetum*	Lepidoptera: Noctuidae
twospotted spider mite*	*Tetranychus urticae*	Acari: Tetranychidae
variegated cutworm*	*Peridroma saucia*	Lepidoptera: Noctuidae
variegated leafroller	*Platynota flavedana*	Lepidoptera: Tortricidae
velvetbean caterpillar*	*Anticarsia gemmatalis*	Lepidoptera: Noctuidae
ver rose des capsules du cotonnier	*Pectinophora gossypiella*	Lepidoptera: Gelechiidae
verme della mela cotogna	*Euzophera bigella*	Lepidoptera: Pyralidae
verme delle mele	*Cydia pomonella*	Lepidoptera: Tortricidae
verme della susine	*Cydia funebrana*	Lepidoptera: Tortricidae
Vogelbeerwickler	*Choristoneura sorbiana*	Lepidoptera: Tortricidae
warehouse beetle*	*Trogoderma variable*	Coleoptera: Dermestidae
warehouse moth	*Ephestia elutella*	Lepidoptera: Pyralidae
western balsam bark beetle*	*Dryocoetes confusus*	Coleoptera: Scolytidae
western bean cutworm*	*Loxagrotis albicosta*	Lepidoptera: Noctuidae
western corn rootworm*	*Diabrotica virgifera virgifera*	Coleoptera: Chrysomelidae
western grapeleaf skeletonizer*	*Harrisina brillians*	Lepidoptera: Zygaenidae
western pine beetle*	*Dendroctonus brevicomis*	Coleoptera: Scolytidae
western pineshoot borer	*Eucosma sonomana*	Lepidoptera: Tortricidae
western spruce budworm*	*Choristoneura occidentalis*	Lepidoptera: Tortricidae
white peach scale*	*Pseudaulacaspis pentagona*	Homoptera: Diaspididae
winter moth*	*Operophtera brumata*	Lepidoptera: Geometridae
woodbine leafroller moth	*Sparganothis pilleriana*	Lepidoptera: Tortricidae
y-Eule	*Agrotis ipsilon*	Lepidoptera: Noctuidae
yellow scale*	*Aonidiella citrina*	Homoptera: Diaspididae

TABLE 5 Arthropods for Which Pheromones and Other Behavioral Chemicals Were Commercially Available in 1988.

Arthropods	Compounds[a]	Ratio[a]	Ref.[b]	Commodity[c]	Application[c]
Acari: Ixodidae					
Amblyomma americanum	Phenol 2,6-Dichlorophenol 4-Methylphenol (*p*-cresol)		(3)	Animal	Synthetic chemical
Acari: Tetranychidae					
Panonychus ulmi				Field crops, orchard	Synthetic chemical
Tetranychus cinnabarinus				Field crops, orchard	Bioirritant
T. urticae	(*Z,E*)-3,7,11-Trimethyl-2,6,10-dodecatrien-1-ol((*Z,E*)-farnesol) 3,7,11-Trimethyl-1,6-10-dodecatrien-3-ol (nerolidol)		(3)	Field crops, orchard	Bioirritant
Blattodea: Blattellidae					
Blattella germanica	3,11-Dimethyl-2-nonacosanone (*S*)-3,11-Dimethyl-2-nonacosanone 29-Hydroxy-3,11-dimethyl-2-nonacosanone Propyl cyclohexaneacetate[d]		(3)	Stored products	Monitoring
Blattodea: Blattidae					
Blatta orientalis	8-Methylene-5-(1-methylethyl)-spiro[11-oxabicyclo[8.1.0]undec-		(19)	Stored products	Monitoring

	6-ene-2,2'-oxiran]-3-one(periplanone B)			
Periplaneta americana	8-Methylene-5(1-methylethyl)-spiro[11-oxa-bicyclo[8.1.0]undec-6-ene-2,2'-oxiran]-3-one (periplanone B) (*E*,*E*)-(-)1-Methyl-5-methylene-8-(1-methylethyl)-1,6-cyclodecadiene (Germacrene D)	(3)	Stored products	Monitoring, attracticide
Coleoptera: Bostrichidae				
Prostephanus truncatus	1-Methyl (*E*)-2-methyl-2-pentenoate	(See Chap. 28)	Stored products	Monitoring
Coleoptera: Chrysomelidae				
Diabrotica undecimpunctata howardi	(*R*)-10-Methyl-2-tridecanone	(20)	Field crops, vegetables	Synthetic chemical
D. virgifera virgifera	(2*R*,8*R*)-8-Methyl-2-decanol propanoate	(21)	Field crops, vegetables	Synthetic chemical

[a]As reported in literature.
[b]Reference giving chemical structure and ratio of components. Reviews or summaries are used when available.
[c]As given by commercial sources.
[d]Synthetic attractant.

(continued)

TABLE 5 (Cont.)

Arthropods	Compounds[a]	Ratio[a]	Ref.[b]	Commodity[c]	Application[c]
Coleoptera: Cucujidae					
Oryzaephilus mercator	(*Z*,*Z*)-3,6-Dodecadien-11-olide (*Z*)-3-Dodecen-11-olide		(See Chap. 28)	Stored products	Monitoring
O. surinamensis	(*Z*,*Z*)-3,6-Dodecadien-11-olide (*Z*,*Z*)-3,6-Dodecadienolide (*Z*,*Z*)-5,8-Tetradecadien-13-olide		(See Chap. 28)	Stored products	Monitoring
Coleoptera: Curculionidae					
Anthonomus grandis	(*cis*)-1-Methyl-2-(1-methylethenyl)cyclobutaneethanol (grandlure I; grandisol) (*Z*)-2-(3,3-Dimethylcyclohexylidene)ethanol (grandlure II) (*Z*)-(3,3-Dimethylcyclohexylidene)acetaldehyde (grandlure III) (*E*)-(3,3-Dimethylcyclohexylidene)acetaldehyde (grandlure IV) 4,11,11-Trimethyl-8-methylenebicyclo[7.2.0]-undec-4-ene(β-caryophyllene)		(3)	Field crops	Monitoring
Cylas formicarius elegantulus	(*Z*)-3-Docen-1-ol (*E*)-2-butenoate		(22)	Field crops, vegetables	Synthetic chemical
Coleoptera: Dermestidae					
Anthrenus flavipes	(*Z*)-3-Decenoic acid		(See Chap. 28)	Stored products	Monitoring

Attagenus unicolor	(*E*,*Z*)-3,5-Tetradecadienoic acid		(See Chap. 28)	Stored products	Monitoring
Trogoderma granarium	(*Z*)-14-Methyl-8-hexadecenal (*E*)-14-Methyl-8-hexadecenal	92:8	(See Chap. 28)	Stored products	Monitoring
T. inclusum	(*Z*)-14-Methyl-8-hexadecen-1-ol (*Z*)-14-Methyl-8-hexadecenal		(See Chap. 28)	Stored products	Monitoring
T. variabile	(*Z*)-14-Methyl-8-hexadecenal		(See Chap. 28)	Stored products	Monitoring
Coleoptera: Elateridae					
Limonius californicus	Valeric acid		(23)	Field crops	Synthetic chemical
Coleoptera: Scarabaeidae					
Popillia japonica	[*R*-(*Z*)]-5-(1-Decenyl)dihydro-2-(3*H*)-furanone Phenethyl propanoate[d] (benzene-ethanol propanoate) 2-Methoxy-4-(2-propenyl)phenol[d] (eugenol)		(3)	Field crops	Monitoring

[a]As reported in literature.
[b]Reference giving chemical structure and ratio of components. Reviews or summaries are used when available.
[c]As given by commercial sources.
[d]Synthetic attractant.

(continued)

TABLE 5 (Cont.)

Arthropods	Compounds[a]	Ratio[a]	Ref.[b]	Commodity[c]	Application[c]
Coleoptera: Scolytidae					
Dendroctonus brevicomis	*exo*-7-Ethyl-5-methyl-6,8-dioxabicyclo[3.2.1]octane (*exo*-brevicomin)		(3)	Forest	Monitoring
	1,5-Dimethyl-6,8-dioxabicyclo-[3.2.1]octane (frontalin)				
	7-Methyl-3-methylene-1,6-octadiene (myrcene)				
D. frontalis	4,6,6-Trimethylbicyclo[3.1.1]-hept-3-en-2-one (verbenone)		(3)	Forest	Monitoring
	1,5-Dimethyl-6,8-dioxabicyclo-[3.2.1]octane (frontalin)				
D. ponderosae	7-Methyl-3-methylene-1,6-octadiene (myrcene)		(See Chap. 19)	Forest	Monitoring, bait tree
	cis-4,6,6-Trimethylbicyclo[3.1.1]-hept-3-en-2-ol(*cis*-2-verbenol)				
	exo-7-Ethyl-5-methyl-6,8-dioxabicyclo[3.2.1]octane (*exo*-brevicomin)				
D. pseudotsugae	1,5-Dimethyl-6,8-dioxabicyclo-[3.2.1]octane (frontalin)		(3)	Forest	Monitoring, bait tree
	3-Methyl-3-cyclohexen-1-one		(3)	Forest	Antiaggregant

D. rufipennis	3-Methyl-2-cyclohexen-1-ol (seudenol)		(3)	Forest	Monitoring
	1,5-Dimethyl-6,8-dioxabicyclo-[3.2.1]octane (frontalin)				
	2-Methoxybiphenyl (*o*-phenyl anisole)				
	3-Methyl-3-cyclohexen-1-one		(3)	Forest	Antiaggregant
D. terebrans	1,5-Dimethyl-6,8-dioxabicyclo-[3.2.1]octane (frontalin)		(24)	Forest	Monitoring, mass trapping
	endo-7-Ethyl-5-methyl-6,8-dioxabicyclo[3.2.1]octane (*endo*-brevicomin)				
	Turpentine				
Dryocoetes confusus	*exo*-7-Ethyl-5-methyl-6,8-dioxabicyclo[3.2.1]octane (*exo*-brevicomin)		(25)	Forest	Monitoring
Gnathotrichus sulcatus	6-Methyl-5-hepten-2-ol (sulcatol)	50:50	(See Chap. 19)	Forest	Mass trapping
	Ethanol				
Ips cembrae	2-Methyl-6-methylene-7-octen-4-ol (ipsenol)		(3)	Forest	Monitoring
	2-Methyl-6-methylene-2,7-octadien-4-ol (ipsdienol)				

[a] As reported in literature.
[b] Reference giving chemical structure and ratio of components. Reviews or summaries are used when available.
[c] As given by commercial sources.

(continued)

TABLE 5 (Cont.)

Arthropods	Compounds[a]	Ratio[a]	Ref.[b]	Commodity[c]	Application[c]
Coleoptera: Scolytidae (Cont.)					
I. Confusus	2-Methyl-6-methylene-7-octen-4-ol (ipsenol) 2-Methyl-6-methylene-2,7-octadien-4-ol (ipsdienol) *cis*-4,6,6-Trimethylbicyclo-[3.1.1]hept-3-en-2-ol (*cis*-2-verbenol)		(3)	Forest	Synthetic chemical
I. pini	2-Methyl-6-methylene-7-octen-4-ol (ipsenol) (*S*)-2-Methyl-6-methylene-7-octen-4-ol		(3)	Forest	Monitoring
I. typographus	2-Methyl-6-methylene-2,7-octadien-4-ol (ipsdienol) *cis*-4,6,6-Trimethylbicyclo[3.1.1]-hept-3-en-2-ol (*cis*-2-verbenol) (1*S*,2*S*,5*S*)-4,6,6-Trimethylbicyclo [3.1.1]hept-3-en-2-ol		(3)	Forest	Monitoring, mass trapping
Pityogenes chalcographus	1-Hexanol 2-Ethyl-1,6-dioxaspiro[4,4]nonane (chalcogran)		(3)	Forest	Monitoring
Scolytus multistriatus	(-)-4-Methyl-3-heptanol		(3)	Forest	Monitoring

	(2-*endo*, 4-*endo*)-5-Ethyl-2,4-dimethyl-6,8-dioxabicyclo[3.2.1]-octane (α-multistriatin) (1*S*, 2*R*, 4*S*, 5*R*)-5-Ethyl-2,4-dimethyl-6,8-dioxabicyclo[3.2.1]-octane [(-)-α-multistriatin]			
S. pygmaeus	(-)-4-Methyl-3-heptanol (2-*endo*, 4-*endo*)-5-Ethyl-2,4-dimethyl-6,8-dioxabicyclo[3.2.1]-octane (α-multistriatin)	(3)	Forest	Monitoring
S. scolytus	(-)-4-Methyl-3-heptanol (2-*endo*, 4-*endo*)-5-Ethyl-2,4-dimethyl-6,8-dioxabicyclo[3.2.1]-octane (α-multistriatin)	(3)	Forest	Monitoring
Trypodendron lineatum	3,3,7-Trimethyl-2,9-dioxatricyclo-[3.3.1.0^{4,7}]nonane (lineatin) Ethanol	(See Chap. 19)	Forest	Monitoring, mass trapping
Coleoptera: Tenebrionidae				
Tribolium castaneum	4,8-Dimethyldecanal	(See Chap. 28)	Stored products	Monitoring
T. confusum	4,8-Dimethyldecanal	(See Chap. 28)	Stored products	Monitoring

[a]As reported in literature.
[b]Reference giving chemical structure and ratio of components. Reviews or summaries are used when available.
[c]As given by commercial sources.

(continued)

TABLE 5 (Cont.)

Arthropods	Compounds[a]	Ratio[a]	Ref.[b]	Commodity[c]	Application[c]
Diptera: Muscidae					
Fannia canicularis	(*Z*)-9-Pentacosene		(3)	Animals	Monitoring
Musca autumnalis	(*Z*)-13-Heptacosene (*Z*)-13-Nonacosene (*Z*)-14-Nonacosene		(3)	Animals	Monitoring
M. domestica	(*Z*)-9-Tricosene		(3)	Animals	Monitoring, attracticide
Diptera: Tephritidae					
Ceratitis capitata	*tert*-Butyl 4(or 5)-chloro-2-methylcyclohexanecarboxylate[d]		(3)	Orchard	Monitoring, attracticide, mass trapping
Dacus cucurbitae	4-[4-(Acetyloxy)phenyl]2-butanone[d] (4-*p*-hydroxyphenyl)-2-butanone acetate)		(3)	Orchard, vegetables	Monitoring
D. dorsalis	1,2-Dimethoxy-4-(2-propenyl)-phenol[d] (methyl eugenol)		(3)	Orchard	Monitoring
D. oleae	(*Z*)-6-Nonen-1-ol (*Z*)-9-Nonadecene (*Z*)-9-Tricosene		(3)	Orchard	Monitoring, attracticide, mass trapping

D. tryoni	4-[4-(Acetyloxy)phenyl]-2-butanone[d] (4-*p*-hydroxyphenyl-2-butanone acetate)	(3)	Orchard	Synthetic chemical
Homoptera: Diaspididae				
Aonidiella aurantii	(3*S*,6*R*)-3-Methyl-6-(1-methylethenyl)-9-decen-1-ol acetate [6*R*-(*Z*)]-3-Methyl-6-(1-methylethenyl)-3,9-decadien-1-ol acetate	(3)	Orchard	Monitoring
A. citrina	(*E*)-3,9-Dimethyl-6-(1-methylethenyl)-5,8-decadien-1-ol acetate	(3)	Orchard	Synthetic chemical
Pseudaulacaspis pentagona	[*R*-(*Z*)]-3,9-Dimethyl-6-(1-methylethenyl)-3,9-decadien-1-ol propanoate	(3)	Orchard	Synthetic chemical
Quadraspidiotus perniciosus	3-Methylene-7-methyl-7-octen-1-ol propanoate (*Z*)-3,7-Dimethyl-2,7-octadien-1-ol propanoate	(3)	Orchard, forest	Monitoring

[a]As reported in literature.
[b]Reference giving chemical structure and ratio of components. Reviews or summaries are used when available.
[c]As given by commercial sources.
[d]Synthetic attractant.

(continued)

TABLE 5 (Cont.)

Arthropods	Compounds[a]	Ratio[a]	Ref.[b]	Commodity[c]	Application[c]
Homoptera: Pseudococcidae					
Planococcus citri	(1*R*-*cis*)-2,2-Dimethyl-3-(1-methylethenyl)cyclobutanemethanol acetate		(3)	Orchard	Monitoring
Pseudococcus comstocki	2,6-Dimethyl-1,5-heptadien-3-ol acetate		(3)	Orchard	Monitoring
Hymenoptera: Apidae					
Apis mellifera	(*E*)- and (*Z*)-3,7-Dimethyl-2,6-octadienal (citral) (*E*)-3,7-Dimethyl-2,6-octadien-1-ol (geraniol) (*E*)- and (*Z*)-3,7-Dimethyl-2,6-octadienoic acids (nerolic and geranic acids)	1:1:1	(26)	Beekeeping	Swarm trapping
Lepidoptera: Acrolepiidae					
Acrolepiopsis assectella	(*Z*)-11-Hexadecenal		(18)	Field crops, vegetables	Monitoring
Lepidoptera: Arctiidae					
Holomelina aurantiaca	2-Methylheptadecane		(18)		Synthetic chemical
Lepidoptera: Argyresthiidae					
Agyresthia pruniella	(*Z*)-11-Hexadecenal		(18)	Orchard	Monitoring

Lepidoptera: Carposinidae					
Carposina niponensis	(*Z*)-7-Eicosen-11-one (*Z*)-7-Nonadecen-11-one	20:1	(18)	Orchard	Monitoring, mating disruption
Lepidoptera: Cochylidae					
Cochylis hospes	(*E*)-11-Tetradecen-1-ol acetate (*Z*)-11-Tetradecen-1-ol acetate	5:1	(27)	Field crops	Monitoring
Lepidoptera: Coleophoridae					
Coleophora laricella	(*Z*)-5-Decen-1-ol		(18)	Forest	Monitoring
Lepidoptera: Cossidae					
Cossus cossus	(*Z*)-3-Decen-1-ol acetate (*Z*)-5-Dodecen-1-ol acetate (*Z*)-5-Tetradecen-1-ol acetate	1:2:1	(18)	Orchard	Monitoring, mass trapping
Prionoxystus robiniae	(*Z,E*)-3,5-Tetradecadien-1-ol acetate		(18)	Forest	Monitoring
Lepidoptera: Gelechiidae					
Anarsia lineatella	(*E*)-5-Decen-1-ol acetate (*E*)-5-Decen-1-ol	85:15	(18)	Orchard	Monitoring
Bryotropha similis	(*Z*)-9-Tetradecen-1-ol acetate		(18)		Monitoring

[a]As reported in literature.
[b]Reference giving chemical structure and ratio of components. Reviews or summaries are used when available.
[c]As given by commercial sources.

(continued)

TABLE 5 (Cont.)

Arthropods	Compounds[a]	Ratio[a]	Ref.[b]	Commodity[c]	Application[c]
Lepidoptera: Gelechiidae					
Helcystogramma macroscopum	(*E*)-11-Hexadecen-1-ol acetate		(18)	Field crops, vegetables	Monitoring
Keiferia lycopersicella	(*E*)-4-Tridecen-1-ol acetate		(18)	Vegetables	Monitoring, attracticide
Odites leucostola	(*Z*)-7-Dodecen-1-ol		(3)		Synthetic chemical
Pectinophora gossypiella	(*Z*,*Z*)-7,11-Hexadecadien-1-ol acetate (*Z*,*E*)-7,11-Hexadecadien-1-ol acetate	1:1	(18)	Field crops	Attracticide, bioirritant, mating disruption
Pexicopia malvella	(*Z*,*E*)-7,11-Hexadecadien-1-ol acetate (*Z*,*Z*)-7,11-Hexadecadien-1-ol acetate	3:1	(18)	Field crops	Monitoring
Phthorimaea operculella	(*E*,*Z*)-4,7-Tridecadien-1-ol acetate (*E*,*Z*,*Z*)-4,7,10-Tridecatrien-1-ol acetate	1:1	(18)	Field crops, vegetables	Monitoring
Platyedra subcinerea	(*Z*,*Z*)-7,11-Hexadecadien-1-ol acetate		(18)	Field crops	Monitoring
Scrobipalpa ocellatella	(*E*)-3-Dodecen-1-ol acetate		(18)	Field crops	Monitoring

Scrobipalpopsis solanivora	(*E*)-3-Dodecen-1-ol acetate (*Z*)-3-Dodecen-1-ol acetate	98:2	(18)	Vegetables	Monitoring
Sitotroga cerealella	(*Z*,*E*)-7,11-Hexadecadien-1-ol acetate		(See Chap. 28)	Stored products	Monitoring
Lepidoptera: Geometridae					
Operophtera brumata	(*Z*,*Z*,*Z*)-3,6,9-Nonadecatriene		(18)	Forest, orchard	Monitoring
Lepidoptera: Gracillariidae					
Phyllonorycter blancardella	(*E*)-10-Dodecen-1-ol acetate		(18)	Orchard	Monitoring
P. corylifoliella	(*E*, *Z*)-4,7-Tridecadien-1-ol acetate		(18)	Orchard	Monitoring
Lepidoptera: Lasiocampidae					
Dendrolimus spectabilis	(*Z*,*E*)-5,7-Dodecadien-1-ol (*E*,*E*)-5,7-Dodecadien-1-ol	5:1	(18)	Forest	Monitoring
Malacosoma disstria	(*Z*,*E*)-5,7-Dodecenal (*Z*,*E*)-5,7-Dodecadien-1-ol	1:10	(18)	Forest	Monitoring
Lepidoptera: Lyonetiidae					
Leucoptera scitella	5,9-Dimethylheptadecane		(28)	Orchard	Monitoring

[a]As reported in literature.
[b]Reference giving chemical structure and ratio of components. Reviews or summaries are used when available.
[c]As given by commercial sources.

(continued)

TABLE 5 (Cont.)

Arthropods	Compounds[a]	Ratio[a]	Ref.[b]	Commodity[c]	Application[c]
Lepidoptera: Lymantriidae					
Lymantria dispar	(7*R*,8S)-*cis*-7,8-Epoxy-2-methyloctadecane((+)-disparlure)		(18)	Forest	Monitoring, mating disruption
L. monacha	(7*S*,8*R*)-*cis*-7,8-Epoxy-2-methyloctadecane (7*R*,8*S*)-*cis*-7,8-Epoxy-2-methyloctadecane	9:1	(18)	Forest	Monitoring
Orgyia antiqua	(*Z*)-6-Heneicosen-11-one		(18)	Forest, orchard	Monitoring
O. pseudotsugata	(*Z*)-6-Heneicosen-11-one		(18)	Forest	Monitoring
Lepidoptera: Noctuidae					
Actebia fennica	(*Z*)-11-Tetradecen-1-ol acetate (*Z*)-7-Dodecen-1-ol acetate	1:1	(18)	Forest	Monitoring
Agrotis exclamationis	(*Z*)-5-Tetradecen-1-ol acetate (*Z*)-9-Tetradecen-1-ol acetate	100:15	(18)	Field crops, vegetables	Monitoring
A. fucosa	(*Z*)-5-Decen-1-ol acetate (*Z*)-7-Decen-1-ol acetate	68:32	(18)	Field crops, vegetables	Monitoring
A. ipsilon	(*Z*)-7-Dodecen-1-ol acetate (*Z*)-9-Tetradecen-1-ol acetate (*Z*)-11-Hexadecen-1-ol acetate	15:20:10	(18)	Field crops, vegetables	Monitoring
A. orthogonia	(*Z*)-5-Dodecen-1-ol acetate (*Z*)-7-Dodecen-1-ol acetate	1:2	(18)	Field crops, vegetables	Monitoring

A. segetum	(*Z*)-5-Decen-1-ol acetate (*Z*)-7-Dodecen-1-ol acetate (*Z*)-9-Tetradecen-1-ol acetate	1:1:1	(18)	Field crops, vegetables	Monitoring
Anticarsia gemmatalis	(*Z,Z,Z*)-3,6,9-Eicosatriene (*Z,Z,Z*)-3,6,9-Heneicosatriene	5:3	(18)	Field crops, vegetables	Monitoring
Autographa californica	(*Z*)-7-Dodecen-1-ol acetate (*Z*)-7-Dodecen-1-ol	10:1	(18)	Field crops	Monitoring
A. gamma	(*Z*)-7-Dodecen-1-ol acetate (*Z*)-7-Dodecen-1-ol	20:1	(18)	Field crops, vegetables	Monitoring
Busseola fusca	(*Z*)-11-Tetradecen-1-ol acetate (*E*)-11-Tetradecen-1-ol acetate (*Z*)-9-Tetradecen-1-ol acetate	10:2:2	(18)	Field crops, vegetables	Monitoring
Chrysodeixis chalcites	(*Z*)-7-Dodecen-1-ol acetate (*Z*)-9-Tetradecen-1-ol acetate (*Z*)-9-Dodecen-1-ol acetate 1-Dodecanol acetate 1-Tetradecanol acetate 1-Hexadecanol acetate	67:14:5: 1.5:1:16	(18)	Field crops	Monitoring
Crymodes devastator	(*Z*)-11-Hexadecenal (*Z*)-11-Hexadecen-1-ol acetate (*Z*)-11-Hexadecen-1-ol	100:50:3	(18)	Field crops, vegetables	Monitoring

[a]As reported in literature.
[b]Reference giving chemical structure and ratio of components. Reviews or summaries are used when available.
[c]As given by commercial sources.

(continued)

TABLE 5 (Cont.)

Arthropods	Compounds[a]	Ratio[a]	Ref.[b]	Commodity[c]	Application[c]
Lepidoptera: Noctuidae (Cont.)					
Diparopsis castanea	(*E*)-9,11-Dodecadien-1-ol acetate (*Z*)-9,11-Dodecadien-1-ol acetate 11-Dodecen-1-ol acetate (*E*)-9-Dodecen-1-ol acetate 1-Dodecanol acetate	8:2:2.5:0.5:5	(18)	Field crops	Monitoring
D. watersi				Field crops	Monitoring
Dicestra trifolli	(*Z*)-11-Hexadecen-1-ol acetate (*Z*)-11-Hexadecen-1-ol	9:1	(18)	Field crops, vegetables	Monitoring
Earias insulana	(*E*,*E*)-10,12-Hexadecadienal		(18)	Field crops	Monitoring
E. vittella	(*E*,*E*)-10,12-Hexadecadienal (*Z*)-11-Hexadecenal (*Z*)-11-Octadecenal	10:2:2	(18)	Field crops	Monitoring
Euxoa auxillaris	(*Z*)-5-Tetradecen-1-ol acetate (*Z*)-7-Tetradecen-1-ol acetate (*Z*)-9-Tetradecen-1-ol acetate	100:1:10	(18)	Field crops, vegetables	Synthetic chemical
E. messoria	(*Z*)-7-Hexadecen-1-ol acetate (*Z*)-11-Hexadecen-1-ol acetate	1:20	(18)	Field crops, vegetables	Monitoring
E. ochrogaster	(*Z*)-5-Decen-1-ol acetate (*Z*)-5-Dodecen-1-ol acetate	1:500:5:2	(18)	Field crops, vegetables	Monitoring

	(*Z*)-7-Dodecen-1-ol acetate (*Z*)-9-Dodecen-1-ol acetate				
Gortyna xanthenes	(*Z*)-11-Hexadecenal (*Z*)-9-Hexadecenal Hexadecanal (*Z*)-11-Hexadecen-1-ol	500:15:15:12	(18)	Field crops, vegetables	Monitoring
Heliothis armigera	(*Z*)-11-Hexadecenal (*Z*)-9-Hexadecenal	25:1	(18)	Field crops, vegetables	Monitoring
H. punctigera	(*Z*)-11-Hexadecenal (*Z*)-11-Hexadecen-1-ol acetate (*Z*)-11-Hexadecen-1-ol	60:25:15	(18)	Field crops, vegetables	Monitoring
H. virescens	(*Z*)-11-Hexadecenal (*Z*)-9-Tetradecenal Hexadecanal (*Z*)-11-Hexadecen-1-ol	65:2:33:1	(18)	Field crops	Monitoring
H. zea	(*Z*)-11-Hexadecenal (*Z*)-11-Hexadecen-1-ol (*Z*)-7-Hexadecenal Hexadecanal (*Z*)-9-Hexadecenal	67:28:1:3:2	(18)	Field crops, vegetables	Monitoring, mating disruption
Laconobia oleracea	(*Z*)-11-Hexadecen-1-ol acetate		(3)	Field crops, vegetables	Monitoring

[a]As reported in literature.
[b]Reference giving chemical structure and ratio of components. Review or summaries are used when available.
[c]As given by commercial sources.

(continued)

TABLE 5 (Cont.)

Arthropods	Compounds[a]	Ratio[a]	Ref.[b]	Commodity[c]	Application[c]
Lepidoptera: Noctuidae (Cont.)					
L. pisi	(*Z*)-9-Tetradecen-1-ol acetate (*Z*)-11-Tetradecen-1-ol acetate	1:1	(18)	Field crops, vegetables	Monitoring
L. suasa	(*Z*)-11-Hexadecen-1-ol acetate (*Z*)-11-Hexadecenal	10:1	(18)	Field crops, vegetables	Monitoring
Loxagrotis albicosta	(*Z*)-5-Dodecen-1-ol acetate (*Z*)-7-Dodecen-1-ol acetate 11-Dodecen-1-ol acetate 1-Dodecanol acetate	5:1:5:5	(18)	Field crops, vegetables	Monitoring
Macdunnoughia confusa	(*Z*)-7-Dodecen-1-ol acetate		(18)	Horticultural	Monitoring
Mamestra brassicae	(*Z*)-11-Hexadecen-1-ol acetate (*E*)-11-Hexadecen-1-ol acetate 1-Hexadecanol acetate 1-Tetradecanol acetate	90:1:7:2	(18)	Field crops, vegetables	Monitoring
M. configurata	(*Z*)-11-Hexadecen-1-ol acetate (*Z*)-9-Tetradecen-1-ol acetate	19:1	(18)	Field crops, vegetables	Monitoring
Mocis latipes	(*Z,Z,Z*)-3,6,9-Heneicosatriene (*Z,Z*)-6,9-Heneicosadiene		(18)	Forest	Monitoring
Naranga aenescens	(*Z*)-11-Hexadecen-1-ol acetate (*Z*)-9-Hexadecen-1-ol acetate (*Z*)-9-Tetradecen-1-ol acetate	4:1:1	(18)	Field crops, vegetables	Monitoring, mass trapping

Noctua comes	(*Z*)-5-Tetradecen-1-ol acetate		(18)	Vineyard	Monitoring
N. fimbriata	(*Z*)-9-Tetradecen-1-ol acetate (*Z*)-11-Tetradecen-1-ol acetate	1:9	(18)	Vineyard	Monitoring
Orthosia gothica	(*Z*)-9-Tetradecen-1-ol acetate (*Z*)-11-Tetradecen-1-ol acetate	20:1	(18)	Orchard	Monitoring
O. hibisci	(*Z*)-9-Tetradecenal (*Z*)-11-Tetradecenal	100:1	(18)	Orchard	Monitoring
O. incerta	(*Z*)-9-Tetradecenal (*Z*)-11-Hexadecenal	9:1	(18)	Orchard	Monitoring
Panolis flammea	(*Z*)-9-Tetradecen-1-ol acetate (*Z*)-11-Tetradecen-1-ol acetate	100:5	(18)	Forest	Monitoring
Peridroma saucia	(*Z*)-11-Hexadecen-1-ol acetate (*Z*)-9-Tetradecen-1-ol acetate	7:3	(18)	Field crops, vegetables	Monitoring
Pseudaletia unipuncta	(*Z*)-11-Hexadecen-1-ol acetate (*Z*)-11-Hexadecen-1-ol (*Z*)-11-Hexadecenal (*Z*)-9-Tetradecen-1-ol acetate	1000:2:0.5:0.1	(18)	Field crops, vegetables	Monitoring
Pseudoplusia includens	(*Z*)-7-Dodecen-1-ol acetate		(18)	Field crops	Monitoring
Raphia frater	(*Z*)-7-Dodecen-1-ol		(18)		Synthetic chemical

[a]As reported in literature.
[b]Reference giving chemical structure and ratio of components. Reviews or summaries are used when available.
[c]As given by commercial sources.

(continued)

TABLE 5 (Cont.)

Arthropods	Compounds[a]	Ratio[a]	Ref.[b]	Commodity[c]	Application[c]
Lepidoptera: Noctuidae (Cont.)					
Sesamia cretica	(*Z*)-9-Tetradecen-1-ol acetate (*Z*)-9-Tetradecen-1-ol	1:3	(18)	Field crops, vegetables	Monitoring
S. inferens	(*Z*)-11-Hexadecen-1-ol acetate		(18)	Field crops, vegetables	Monitoring
S. nonagrioides	(*Z*)-11-Hexadecen-1-ol acetate (*Z*)-11-Hexadecen-1-ol	95:5	(18)	Field crops	Monitoring
Spodoptera eridania	(*Z*)-9-Tetradecen-1-ol acetate (*Z,E*)-9,12-Tetradecadien-1-ol acetate (*Z,Z*)-9,12-Tetradecadien-1-ol acetate (*Z,E*)-9,11-Tetradecadien-1-ol acetate (*Z*)-11-Hexadecen-1-ol acetate	61:17:15:5:3	(18)	Field crops, vegetables	Monitoring
S. exempta	(*Z,E*)-9,12-Tetradecadien-1-ol acetate (*Z*)-9-Tetradecen-1-ol acetate	1:20	(18)	Vegetables	Monitoring
S. exigua	(*Z,E*)-9,12-Tetradecadien-1-ol acetate (*Z*)-9-Tetradecen-1-ol	10:1	(18)	Field crops, vegetables	monitoring
S. frugiperda	(*Z*)-7-Dodecen-1-ol acetate	0.5:0.3:82:18	(18)	Field crops	Monitoring

	(*Z*)-9-Dodecen-1-ol acetate (*Z*)-9-Tetradecen-1-ol acetate (*Z*)-11-Hexadecen-1-ol acetate				
S. littoralis	(*Z,E*)-9,11-Tetradecadien-1-ol acetate (*Z,E*)-9,12-Tetradecadien-1-ol acetate	1000:3	(18)	Field crops	Monitoring, mass trapping
S. litura	(*Z,E*)-9,11-Tetradecadien-1-ol acetate (*Z,E*)-9,12-Tetradecadien-1-ol acetate	50:7	(18)	Field crops	Monitoring, mass trapping
S. sunia	(*Z*)-9-Tetradecen-1-ol acetate (*Z,E*)-9,12-Tetradecadien-1-ol acetate (*Z*)-9-Tetradecen-1-ol (*Z*)-11-Hexadecen-1-ol acetate	100:5:31:20	(29)	Field crops	Synthetic chemical
Trichoplusia ni	(*Z*)-7-Dodecen-1-ol acetate (*Z*)-7-Tetradecen-1-ol acetate	200:1	(18)	Field crops, vegetables	Monitoring
Xestia c-nigrum	(*Z*)-7-Tetradecen-1-ol acetate (*Z*)-5-Tetradecen-1-ol acetate	100:5	(18)	Field crops, vegetables	Monitoring
X. dolosa	(*Z*)-7-Tetradecen-1-ol acetate		(18)	Field crops, vegetables	Synthetic chemical

[a] As reported in literature.
[b] Reference giving chemical structure and ratio of components. Reviews or summaries are used when available.
[c] As given by commercial sources.

(continued)

TABLE 5 (Cont.)

Arthropods	Compounds[a]	Ratio[a]	Ref.[b]	Commodity[c]	Application[c]
Lepidoptera: Notodontidae					
Oligocentria semirufescens	(*Z*)-7-Dodecen-1-ol		(18)		Synthetic chemical
Lepidoptera: Plutellidae					
Plutella xylostella	(*Z*)-11-Hexadecenal (*Z*)-11-Hexadecen-1-ol acetate (*Z*)-11-Hexadecen-1-ol	10:10:0.1	(18)	Field crops, vegetables	Monitoring
Prays citri	(*Z*)-7-Tetradecenal		(18)	Orchard	Monitoring, mass trapping
P. oleae	(*Z*)-7-Tetradecenal		(18)	Orchard	Monitoring
Lepidoptera: Psychidae					
Thyridopteryx ephemeraeformis	(*R*)-1-Methylbutyl decanoate		(18)	Horticultural	Monitoring
Lepidoptera: Pterophoridae					
Platyptilia carduidactyla	(*Z*)-11-Hexadecenal		(18)	Vegetables	Attracticide
Lepidoptera: Pyralidae					
Achroia grisella	Undecanal (*Z*)-11-Octadecenal		(3)	Beekeeping	Monitoring
Amyelois transitella	(*Z,Z*)-11,13-Hexadecadienal		(18)	Orchard	Monitoring

Anagasta kuehniella	(*Z,E*)-9,12-Tetradecadien-1-ol acetate (*Z,E*)-9,12-Tetradecadien-1-ol		(See Chap. 28)	Stored products	Monitoring
Cadra cautella	(*Z,E*)-9,12-Tetradecadien-1-ol acetate (*Z,E*)-9,12-Tetradecadien-1-ol (*Z*)-9-Tetradecen-1-ol acetate		(See Chap. 28)	Stored products	Monitoring, mass trapping
C. figulilella	(*Z,E*)-9,12-Tetradecadien-1-ol acetate		(See Chap. 28)	Stored products	Monitoring, mass trapping
Chilo partellus	(*Z*)-11-Hexadecenal (*Z*)-11-Hexadecen-1-ol	20:3	(18)	Field crops, vegetables	Monitoring
C. sacchariphagus	(*Z*)-13-Octadecen-1-ol acetate (*Z*)-13-Octadecen-1-ol	49:4.5	(18)	Field crops, vegetables	Monitoring
C. suppressalis	(*Z*)-11-Hexadecenal (*Z*)-9-Hexadecenal (*Z*)-13-Octadecenal	50:5:6	(18)	Field crops, vegetables	Monitoring
C. zacconius	(*Z*)-11-Hexadecen-1-ol (*Z*)-13-Octadecen-1-ol 1-Hexadecanol	7:1:2	(18)	Field crops, vegetables	Monitoring

[a] As reported in literature.
[b] Reference giving chemical structure and ratio of components. Reviews or summaries are used when available.
[c] As given by commercial sources.

(continued)

TABLE 5 (Cont.)

Arthropods	Compounds[a]	Ratio[a]	Ref.[b]	Commodity[c]	Application[c]
Lepidoptera: Pyralidae (Cont.)					
Chrysoteuchia topiaria	(*Z*)-11-Hexadecenal		(18)	Field crops	Monitoring
Diaphania nitidalis	(*E*)-11-Hexadecenal (*Z*)-11-Hexadecenal (*E*)-11-Hexadecen-1-ol (*Z*)-11-Hexadecen-1-ol (*E*,*Z*)-10,12-Hexadecadienal		(34)	Vegetables	Synthetic chemical
Diatraea grandiosella	(*Z*)-9-Hexadecenal (*Z*)-11-Hexadecenal (*Z*)-13-Octadecenal	21:71:8	(31)	Field crops, vegetables	Monitoring
D. saccharalis	(*Z*,*E*)-9,11-Hexadecadienal		(18)	Field crops	Monitoring
Dioryctria abietella	(*Z*,*E*)-9,11-Tetradecadien-1-ol acetate (*Z*,*E*)-9,11-Tetradecadien-1-ol	5:1	(18)		Monitoring
Elasmopalpus lignosellus	(*Z*)-7-Tetradecen-1-ol acetate (*Z*)-9-Tetradecen-1-ol acetate (*Z*)-11-Hexadecen-1-ol acetate 1-Tetradecanol acetate	17:9:17:4	(18)	Field crops, vegetables	Monitoring
Ephestia elutella	(*Z*,*E*)-9,12-Tetradecadien-1-ol acetate (*Z*,*E*)-9,12-Tetradecadien-1-ol	10:1	(18)	Field crops, stored products	Monitoring, mass trapping

Euzophera bigella	(*Z*)-9-Tetradecen-1-ol (*Z*)-9-Tetradecen-1-ol acetate	3:1	(18)	Orchard	Monitoring
Evergestis forficalis	(*E*)-11-Tetradecen-1-ol acetate		(18)	Field crops, vegetables	Monitoring
Galleria mellonella	Nonanal Undecanal		(18)	Beekeeping	Monitoring
Homoeosoma electellum	(*Z*,*E*)-9,12-Tetradecadien-1-ol (*Z*)-9-Tetradecen-1-ol 1-Tetradecanol		(18)	Field crops	Monitoring
Loxostege sticticalis	(*E*)-11-Tetradecen-1-ol acetate		(18)	Field crops, vegetables	Monitoring
Ostrinia furnacalis	(*Z*)-1(2)-11-Tetradecen-1-ol acetate acetate (*E*)-1(2)-11 Tetradecen-1-ol acetate		(18)	Field crops, vegetables	Monitoring
O. nubilalis	(*Z*)-11-Tetradecen-1-ol acetate (*E*)-11-Tetradecen-1-ol acetate	*Z* strain = 97:3 *E* strain = 3:97	(18)	Field crops, vegetables	Monitoring

[a]As reported in literature.
[b]Reference giving chemical structure and ratio of components. Reviews or summaries are used when available.
[c]As given by commercial source.

(continued)

TABLE 5 (Cont.)

Arthropods	Compounds[a]	Ratio[a]	Ref.[b]	Commodity[c]	Application[c]
Lepidoptera: Pyralidae (Cont.)					
Plodia interpunctella	(*Z,E*)-9,12-Tetradecadien-1-ol acetate (*Z,E*)-9,12-Tetradecadien-1-ol	8:3	(See Chap. 28)	Stored products	Mass trapping, monitoring, mating disruption
Lepidoptera: Saturniidae					
Antheraea polyphemus	(*E,Z*)-6,11-Hexadecadien-1-ol acetate (*E,Z*)-6,11-Hexadecadienal	9:1	(18)		Monitoring
Lepidoptera: Sesiidae					
Paranthrene simulans	(*Z,Z*)-3,13-Octadecadien-1-ol acetate		(18)	Forest	Monitoring
P. tabaniformis	(*E,Z*)-3,13-Octadecadien-1-ol		(18)	Forest	Monitoring
Pennisetia hylaeiformis	(*E,Z*)-3,13-Octadecadien-1-ol acetate (*E,Z*)-3,13-Octadecadien-1-ol	1:1	(32)	Orchard	Monitoring
Podosesia syringae	(*Z,Z*),3-13-Octadecadien-1-ol acetate		(18)	Forest	Monitoring
Synanthedon bibionipennis	(*E,Z*)-3,13-Octadecadien-1-ol acetate (*E,Z*)-3,13-Octadecadien-1-ol	2:1	(18)	Orchard	Monitoring

S. exitiosa	(*Z*,*Z*)-3,13-Octadecadien-1-ol acetate (*E*,*Z*)-3,13-Octadecadien-1-ol acetate	96:4	(18)	Orchard	Monitoring, mating disruption
S. hector	(*Z*,*Z*)-3,13-Octadecadien-1-ol acetate (*E*,*Z*)-3,13-Octadecadien-1-ol acetate	1:1	(18)	Orchard	Monitoring
S. myopiformis	(*Z*,*Z*)-3,13-Octadecadien-1-ol acetate (*Z*,*E*)-3,13-Octadecadien-1-ol acetate (*E*,*Z*)-3,13-Octadecadien-1-ol acetate (*E*,*E*)-3,13-Octadecadien-1-ol acetate	95:2:2:9	(18)	Orchard	Monitoring, mating disruption
S. pictipes	(*E*,*Z*)-3,13-Octadecadien-1-ol acetate		(18)	Orchard	Monitoring, mating disruption
S. rhododendri	(*Z*,*Z*)-3,13-Octadecadien-1-ol acetate (*E*,*Z*)-3,13-Octadecadien-1-ol acetate	49:1	(18)	Orchard	Monitoring

[a]As reported in literature.
[b]Reference giving chemical structure and ratio of components. Reviews or summaries are used when available.
[c]As given by commercial source.

(continued)

TABLE 5 (Cont.)

Arthropods	Compounds[a]	Ratio[a]	Ref.[b]	Commodity[c]	Application[c]
Lepidoptera: Sesiidae (Cont.)					
S. tipuliformis	(*E*, *Z*)-2,13-Octadecadien-1-ol acetate (*Z*)-13-Octadecen-1-ol acetate	93:7	(18)	Orchard, vineyard	Monitoring
S. vespiformis	(*Z*, *Z*)-3,13-Octadecadien-1-ol acetate (*E*, *Z*)-3,13-Octadecadien-1-ol acetate	1:9	(18)	Orchard	Monitoring
Vitacea polistiformis	(*E*, *Z*)-2,13-Octadecadien-1-ol acetate		(18)	Vineyard	Monitoring
Lepidoptera: Sphingidae					
Manduca sexta	(*E*, *Z*)-10,12-Hexadecadienal		(18)	Field crops	Monitoring
Lepidoptera: Thaumetopoeidae					
Thaumetopoea pityocampa	(*Z*)-13-Hexadecen-11-yn-1-ol acetate		(18)	Forest	Monitoring, mass trapping
Lepidoptera: Tortricidae					
Adoxophyes orana	(*Z*)-9-Tetradecen-1-ol acetate (*Z*)-11-Tetradecen-1-ol acetate	9:1	(18)	Orchard	Monitoring, mating disruption

Adoxophyes sp.	(*Z*)-9-Tetradecen-1-ol acetate (*Z*)-11-Tetradecen-1-ol acetate (*E*)-11-Tetradecen-1-ol acetate 10-Methyl-1-dodecanol acetate	63:31:4:2	(18)	Field crops	Monitoring, mating disruption
Archips argyrospilus	(*Z*)-11-Tetradecen-1-ol acetate (*E*)-11-Tetradecen-1-ol acetate (*Z*)-9-Tetradecen-1-ol acetate 1-Dodecanol acetate	15:10:1:50	(18)	Orchard	Monitoring
A. breviplicanus	(*E*)-11-Tetradecen-1-ol acetate (*Z*)-11-Tetradecen-1-ol acetate 1-Tetradecanol acetate	20:9:3	(18)	Forest, orchard	Monitoring
A. crataeganus	(*Z*)-9-Tetradecen-1-ol acetate (*Z*)-11-Tetradecen-1-ol acetate	25:75	(33)	Orchard	Monitoring
A. podanus	(*Z*)-11-Tetradecen-1-ol acetate (*E*)-11-Tetradecen-1-ol acetate	1:1	(18)	Orchard	Monitoring
A. rosanus	(*Z*)-11-Tetradecen-1-ol acetate (*E*)-11-Tetradecen-1-ol acetate (*Z*)-11-Tetradecen-1-ol	100:0.1:10	(18)	Orchard	Monitoring
A. semiferanus	(*Z*)-11-Tetradecen-1-ol acetate (*E*)-11-Tetradecen-1-ol acetate 1-Tetradecanol acetate	33:67:50	(18)	Forest, orchard	Monitoring

[a] As reported in literature.
[b] Reference giving chemical structure and ratio of components. Reviews or summaries are used when available.
[c] As given by commercial source.

(continued)

TABLE 5 (Cont.)

Arthropods	Compounds[a]	Ratio[a]	Ref.[b]	Commodity[c]	Application[c]
Lepidoptera: Tortricidae					
A. xylosteanus	(*Z*)-11-Tetradecen-1-ol acetate (*E*)-11-Tetradecen-1-ol acetate	8:2	(18)	Orchard	Monitoring
Argyrotaenia citrana	(*Z*)-11-Tetradecen-1-ol acetate (*Z*)-11-Tetradecenal	15:1	(18)	Orchard	Monitoring
A. pulchellana	(*Z*)-11-Tetradecen-1-ol acetate (*E*)-11-Tetradecen-1-ol acetate 1-Tetradecanol acetate (*Z*)-11-Tetradecen-1-ol	76:19:5:2	(18)	Orchard, vineyard	Monitoring
A. velutinana	(*Z*)-11-Tetradecen-1-ol acetate (*E*)-11-Tetradecen-1-ol acetate 1-Tetradecanol acetate 11-Dodecen-1-ol acetate (*Z*)-9-Dodecen-1-ol acetate (*E*)-9-Dodecen-1-ol acetate 1-Dodecanol acetate	100:9:5:3:1:2:6	(18)	Orchard	Monitoring
Cacoecimorpha pronubana	(*Z*)-11-Tetradecen-1-ol acetate (*Z*)-11-Tetradecen-1-ol (*Z*)-9-Tetradecen-1-ol 1-Tetradecanol acetate	18:1:1:20	(18)	Horiticultural	Monitoring
Choristoneura fumiferana	(*E*)-11-Tetradecenal (*Z*)-11-Tetradecenal	95:5	(18)	Forest	Monitoring, mating disruption

C. lafauryana	(*Z*)-11-Tetradecen-1-ol (*E*)-11-Tetradecen-1-ol	9:1	(34)	Field crops, vegetables	Monitoring
C. murinana	(*Z*)-9-Dodecen-1-ol acetate (*Z*)-11-Tetradecen-1-ol acetate	10:1	(18)	Forest	Monitoring
C. occidentalis	(*E*)-11-Tetradecenal (*Z*)-11-Tetradecenal (*E*)-11-Tetradecen-1-ol acetate	200:18:4	(18)	Forest	Monitoring
C. rosaceana	(*Z*)-11-Tetradecen-1-ol acetate (*E*)-11-Tetradecen-1-ol acetate	98:2	(18)	Orchard	Monitoring
C. sorbiana	(*Z*)-11-Tetradecenal (*Z*)-11-Tetradecen-1-ol	1:1	(18)	Forest, orchard	Monitoring
Clepsis spectrana	(*Z*)-9-Tetradecen-1-ol acetate (*Z*)-11-Tetradecen-1-ol acetate	1:3	(18)	Orchard, vineyard	Monitoring
Cnephasia longana	(*Z*)-9-Dodecen-1-ol acetate		(18)	Field crops, vegetables	Monitoring
C. pumicana	(*Z*)-9-Dodecen-1-ol acetate (*E*)-9-Dodecen-1-ol acetate 1-Dodecanol acetate	4:6:4	(18)	Field crops, vegetables	Monitoring
Cydia caryana	(*E,E*)-8,10-Dodecadien-1-ol acetate (*Z*)-9-Dodecen-1-ol acetate		(35)	Forest, orchard	Monitoring

[a]As reported in literature.
[b]Reference giving chemical structure and ratio of components. Reviews or summaries are used when available.
[c]As given by commercial source.

(continued)

TABLE 5 (Cont.)

Arthropods	Compounds[a]	Ratio[a]	Ref.[b]	Commodity[c]	Application[c]
Lepidoptera: Tortricidae (Cont.)					
C. funebrana	(*Z*)-8-Dodecen-1-ol acetate (*E*)-8-Dodecen-1-ol acetate	100:1	(18)	Orchard	Monitoring
C. janthinana	(*Z*)-8-Dodecen-1-ol acetate (*E*)-8-Dodecen-1-ol acetate		(3)	Orchard	Monitoring
C. nigricana	(*E, E*)-8,10-Dodecadien-1-ol acetate		(18)	Field crops, vegetables	Monitoring
C. pomonella	(*E, E*)-8,10-Dodecadien-1-ol (*E*)-9-Dodecen-1-ol 1-Dodecanol 1-Tetradecanol	60:3:19:8	(18)	Orchard	Mass trapping, mating disruption
Enarmonia formosana	(*Z*)-9-Dodecen-1-ol acetate (*E*)-9-Dodecen-1-ol acetate	1:1	(18)	Orchard	Monitoring
Endopiza viteana	(*Z*)-9-Dodecen-1-ol acetate (*E*)-9-Dodecen-1-ol acetate	96:4	(18)	Vineyard	Monitoring
Endothenia quadrimaculana	(*E*)-10-Tetradecen-1-ol acetate (*Z*)-10-Tetradecen-1-ol acetate	9:1	(18)	Field crops, vegetables	Monitoring
Ephichoristodes acerbella	(*Z*)-11-Tetradecen-1-ol acetate (*Z*)-9-Tetradecen-1-ol acetate (*Z*)-11-Tetradecen-1-ol (*Z*)-9-Tetradecen-1-ol	8:2:1:1	(18)	Horticultural	Monitoring

Epinotia tedella	(*E*)-9-Dodecen-1-ol acetate (*Z*)-9-Dodecen-1-ol acetate	9:1	(18)	Forest	Monitoring
Epiphyas postvittana	(*E*)-11-Tetradecen-1-ol acetate (*E*,*E*)-9,11-Tetradecadien-1-ol acetate	13:0.7	(18)	Orchard	Monitoring
Eucosma sonomana	(*Z*)-9-Dodecen-1-ol acetate (*E*)-9-Dodecen-1-ol acetate	2:1	(18)	Forest	Monitoring, mating disruption
Eupoecilia ambiguella	(*Z*)-9-Dodecen-1-ol acetate 1-Dodecanol acetate	1:1	(18)	Vineyard	Monitoring, mating disruption
Grapholita molesta	(*Z*)-8-Dodecen-1-ol acetate (*E*)-8-Dodecen-1-ol acetate	100:7	(18)	Orchard	Monitoring, mating disruption
G. packardi	(*E*)-8-Dodecen-1-ol acetate		(18)	Orchard	Monitoring
G. prunivora	(*Z*)-8-Dodecen-1-ol acetate (*E*)-8-Dodecen-1-ol acetate	98:2	(18)	Orchard	Monitoring
Gypsonoma aceriana	(*E*)-10-Dodecen-1-ol acetate (*E*)-10-Dodecen-1-ol	3:1	(18)	Forest	Monitoring

[a]As reported in literature.
[b]Reference giving chemical structure and ratio of components. Reviews or summaries are used when available.
[c]As given by commercial source.

(continued)

TABLE 5 (Cont.)

Arthropods	Compounds[a]	Ratio[a]	Ref.[b]	Commodity[c]	Application[c]
Lepidoptera: Tortricidae (Cont.)					
Hedya nubiferana	(*E,E*)-8,10-Dodecadien-1-ol acetate (*Z*)-8-Dodecen-1-ol acetate (*E*)-8-Dodecen-1-ol acetate 1-Dodecanol acetate	55:32:5:9	(18)	Orchard	Monitoring
Homona coffearia	(*E*)-9-Dodecen-1-ol acetate 1-Dodecanol acetate 1-Dodecanol	1:1:3	(18)	Field crops	Synthetic chemical
H. magnanima	(*Z*)-11-Tetradecen-1-ol acetate (*Z*)-9-Dodecen-1-ol acetate 11-Dodecen-1-ol acetate	30:3:1	(18)	Field crops	Monitoring, mating disruption
Laspeyresia medicaginis	(*E,E*)-8,10-Dodecadien-1-ol acetate		(17)	Field crops, vegetables	Monitoring
Lobesia botrana	(*E,Z*)-7,9-Dodecadien-1-ol acetate (*E,E*)-7,9-Dodecadien-1-ol acetate		(18)	Vineyard	Monitoring, mating disruption
Melissopus latiferreanus	(*E,Z*)-8,10-Dodecadien-1-ol acetate (*E,E*)-8,10-Dodecadien-1-ol acetate	13:3	(18)	Orchard	Monitoring

Pammene rhediella	(*Z,E*)-8,10-Dodecen-1-ol		(18)	Orchard	Monitoring
Pandemis cerasana	(*Z*)-11-Tetradecen-1-ol acetate (*E*)-11-Tetradecen-1-ol acetate (*E*)-11-Tetradecen-1-ol (*Z*)-11-Tetradecen-1-ol 1-Tetradecanol acetate	21:64:10: 3:2	(18)	Orchard	Monitoring
P. heparana	(*Z*)-11-Tetradecen-1-ol acetate (*Z*)-9-Tetradecen-1-ol acetate (*Z*)-11-Tetradecen-1-ol 1-Dodecanol acetate 1-Tetradecanol acetate	90:5:5:1:1	(18)	Orchard	Monitoring
P. limitata	(*Z*)-11-Tetradecen-1-ol acetate (*Z*)-9-Tetradecen-1-ol acetate	91:9	(18)	Forest, orchard	Monitoring
P. pyrusana	(*Z*)-11-Tetradecen-1-ol acetate (*Z*)-9-Tetradecen-1-ol acetate	94:6	(18)	Forest, orchard	Monitoring
Platynota flavedana	(*E*)-11-Tetradecen-1-ol (*Z*)-11-Tetradecen-1-ol	9:1	(18)	Orchard	Monitoring
P. idaeusalis	(*E*)-11-Tetradecen-1-ol (*E*)-11-Tetradecen-1-ol acetate	2:1	(18)	Orchard	Monitoring
P. stultana	(*E*)-11-Tetradecen-1-ol acetate (*Z*)-11-Tetradecen-1-ol acetate	88:12	(18)	Orchard	Monitoring

[a]As reported in literature.
[b]Reference giving chemical structure and ratio of components. Reviews or summaries are used when available.
[c]As given by commercial source.

(continued)

TABLE 5 (Cont.)

Arthropods	Compounds[a]	Ratio[a]	Ref.[b]	Commodity[c]	Application[c]
Lepidoptera: Tortricidae (Cont.)					
Pseudogalleria leucotreta	(*E*)-7-Dodecen-1-ol acetate (*Z*)-8-Dodecen-1-ol acetate (*E*)-8-Dodecen-1-ol acetate		(3)	Orchard	Monitoring
Ptycholoma lecheana	(*Z*)-11-Tetradecen-1-ol		(18)	Orchard	Monitoring
P. lecheana circumclusana	(*Z*)-11-Tetradecen-1-ol (*Z*)-11-Tetradecen-1-ol acetate	9:1	(18)	Orchard	Monitoring
Rhopobota unipunctana	(*Z*)-11-Tetradecen-1-ol acetate (*Z*)-11-Tetradecen-1-ol	3:1	(36)	Field crops, vegetables	Monitoring
Rhyacionia buoliana	(*E*)-9-Dodecen-1-ol acetate (*E*)-9-Dodecen-1-ol 1-Dodecanol acetate 1-Dodecanol	50:5:5:1	(18)	Forest	Mass trapping, mating disruption
R. frustrana	(*E*)-9-Dodecen-1-ol acetate (*E*)-9,11-Dodecadien-1-ol acetate	96:4	(18)	Forest	Monitoring
R. neomexicana	(*E*)-9-Dodecen-1-ol acetate		(18)	Forest	Monitoring
R. rigidana	(*E,E*)-8,10-Dodecadien-1-ol acetate		(18)	Forest	Monitoring
Sparganothis pilleriana	(*E*)-9-Dodecen-1-ol acetate (*E*)-11-Tetradecen-1-ol acetate (*Z*)-11-Tetradecen-1-ol acetate	9:7:6	(18)	Vineyard	Monitoring

S. sulfureana	(*E*)-11-Tetradecen-1-ol acetate		(18)	Vineyard	Monitoring
Spilonota ocellana	(*Z*)-8-Dodecen-1-ol acetate (*E*)-8-Dodecen-1-ol acetate	9:1	(18)	Orchard	Monitoring
Syndemis musculana	(*E*)-11-Tetradecen-1-ol acetate (*Z*)-11-Tetradecen-1-ol acetate	9:1	(18)	Forest, orchard	Monitoring
Tortrix viridana	(*Z*)-11-Tetradecen-1-ol acetate		(18)	Forest, orchard	Monitoring
Zeiraphera diniana	(*E*)-11-Tetradecen-1-ol acetate 1-Tetradecanol acetate 1-Hexadecanol acetate	1:2:2	(18)	Forest	Monitoring,
Lepidoptera: Yponomeutidae					
Yponomeuta malinellus				Orchard	Monitoring
Lepidoptera: Zygaenidae					
Harrisina brillians	*d*-(1-Methylpropyl) (*Z*)-7-tetradecenoate		(18)	Vineyard	Monitoring

[a]As reported in literature.
[b]Reference giving chemical structure and ratio of components. Reviews or summaries are used when available.
[c]As given by commercial source.

REFERENCES

1. Beroza, M. Insect sex attractants. *Am. Sci.* *59*:320–325 (1971).
2. Inscoe, M. N., and Beroza, M. Insect-behavior chemicals active in field trials. In *Pest Management with Insect Sex Attractants and Other Behavior-Controlling Chemicals* (Beroza, M., ed.). *ACS Symp. Ser.* *23*:145–181 (1976).
3. Inscoe, M. N. Insect attractants, attractant pheromones, and related compounds. In *Insect Suppression with Controlled Release Pheromone Systems*, Vol. 2 (Kydonieus, A. F., and Beroza, M., eds.). CRC Press, Boca Raton, 1982, pp. 201–295.
4. Ridgway, R. L., Leonhardt, B. A., and Inscoe, M. N. Cooperative development and expanding uses of delivery systems for insect attractants. In *Proceedings and Abstracts of the 13th International Symposium on Controlled Release of Bioactive Materials*, Norfolk, Va., Aug. 3–6, 1986 (Chaudry, I. A., and Thies, C., eds.). Controlled Release Society, Lincolnshire, Ill., 1986, pp. 100-101.
5. Metcalf, R. L., Mitchell, W. C., and Metcalf, E. R. Olfactory receptors in the melon fly *Dacus cucurbitae* and the oriental fruit fly *Dacus dorsalis*. *Proc. Natl. Acad. Sci. USA* *80*:3142–3147 (1983).
6. Bedard, W. D., Tilden, P. E., Wood, D. L., Silverstein, R. M., Brownlee, R. G., and Rodin, J. O. Western pine beetle: Field response to its sex pheromone and a synergistic host terpene, myrcene. *Science* *164*:1284–1285 (1969).
7. Ladd, T. L., Klein, M. G., and Tumlinson, J. H. Phenethyl propionate + eugenol + geraniol (3:7:3) and japonilure: A highly effective joint lure for Japanese beetles. *J. Econ. Entomol.* *74*:665–667 (1981).
8. Griffiths, D. C., and Pickett, J. A. Novel chemicals and their formulation for aphid control. In *Proceedings 14th International Symposium on Controlled Release of Bioactive Materials*, Toronto, Ontario, Canada, Aug. 2–5, 1987 (Lee, P. I., and Leonhardt, B. A., Eds.). Controlled Release Society, Lincolnshire, Ill., 1987, pp. 243–244.
9. Pickett, J. A. The future of semiochemicals in pest control. *Aspects Appl. Biol.* *17*:397–406 (1988).
10. Siddall, J., and Olsen, C. M. Pheromones in agriculture–from chemical synthesis to commercial use. In *Pest Management with Insect Sex Attractants* (Beroza, M., ed.). *ACS Symp. Ser.* *23*: 17–41 (1976).
11. Kydonieus, A. F., and Beroza, M. Marketing and economic considerations in the use of pheromones for suppression of insect populations. In *Insect Suppression with Controlled Release*

Pheromone Systems, Vol. 2 (Kydonieus, A. F., and Beroza, M., eds.). CRC Press, Boca Raton, 1982, pp. 187–199.

12. Jones, O. T. Commercial development of pheromone-based monitoring systems for insect pests of stored products. In *Stored Products Pest Control*, University of Reading, Mar. 25–27, 1987 (Lawson, T. J., ed.). British Crop Protection Council, Thornton Heath, 1987, pp. 169–174.
13. Kirsch, P. Pheromones: Their potential role in control of agricultural insect pests. *Am. J. Alternative Agric.* *3*:83–97 (1988).
14. Inscoe, M. (compiler). *Insect Lures Available Commercially in the United States, from Company Literature Published in 1981–1982.* Unpublished table. (1983).
15. Leonhardt, B. A. (compiler). *Commercial Dispensers for Insect Pheromones.* Unpublished table. (1986).
16. Werner, F. G. (Chairman, Committee on Common Names of Insects). *Common Names of Insects and Related Organisms.* Entomological Society of America, College Park, Md., 1982, 132 pp.
17. Minks, A. K. *Attractants and Pheromones of Noxious Insects (Selected References)*. West Palaearctic Regional Section, IOBC, 1984, 176 pp.
18. Arn, H., Toth, M., and Priesner, E. *List of Sex Pheromones of Lepidoptera and Related Attractants.* West Palaearctic Regional Section, IOBC, 1986, 123 pp.
19. Seelinger, G. Interspecific attractivity of female sex pheromone components of *Periplaneta americana*. *J. Chem. Ecol.* *11*:137–148 (1985).
20. Guss, P. L., Tumlinson, J. H., Sonnet, P. E., and McLaughlin, J. R. Identification of a female-produced sex pheromone from the southern corn rootworm, *Diabrotica undecimpunctata howardi* Barber. *J. Chem. Ecol.* *9*:1363–1375 (1983).
21. Guss, P. L., Tumlinson, J. H., Sonnet, P. E., and Proveaux, A. T. Identification of a female-produced sex pheromone of the western corn rootworm. *J. Chem. Ecol.* *8*:545–556 (1982).
22. Heath, R. R., Coffelt, J. A., Sonnet, R. E., Proshold, F. E. D., and Tumlinson, J. H. Identification of sex pheromone produced by female sweetpotato weevil, *Cylas formicarius elegantulus* (Summers). *Entomol. Exp. Appl.* *12*:1489–1503 (1986).
23. Jacobson, M., Lilly, C. E., and Harding, C. Sex attractant of sugar-beet wireworm: Identification and biological activity. *Science* *159*:208 (1968).
24. Payne, T. L., Billings, R. F., Delorme, J. D., Andryszak, N. A., Bartels, J., Francke, W., and Vité, J. P. Kairomonal-pheromonal system in the black turpentine beetle, *Dendroctonus terebrans* (Ol.). *J. Appl. Entomol.* *103*:15–22 (1987).

25. Stock, A. J., and Borden, J. H. Secondary attraction in the western balsam bark beetle, *Dryocoetes confusus* (Coleoptera: Scolytidae). *Can. Entomol. 11*:539–550 (1983).
26. Free, J. B. *Pheromones of Social Bees*. Chapman & Hall, London, 1987, p. 129.
27. Underhill, E. W., Arthur, A. P., and Mason, P. G. Sex pheromone of the banded sunflower moth, *Cochylis hospes* (Lepidoptera: Cochylidae): Identification and field trapping. *Environ. Entomol. 15*:1063–1066 (1986).
28. Francke, W., Francke, S., Toth, M., Szocs, G., Guerin, P., and Arn, J. Identification of 5,9-dimethylheptadecane as a sex pheromone of the moth *Leucoptera scitella*. *Naturwissenschaften 74*: 143–144 (1987).
29. Bestmann, H. J., Attygalle, A. B., Schwarz, J., Vostrowsky, O., and Knauf, W. Identification of sex pheromone components of *Spodoptera sunia* Guenée (Lepidoptera: Noctuidae). *J. Chem. Ecol. 14*:683–690 (1988).
30. Klun, J. A., Leonhardt, B. A., Schwarz, M., Day, A., and Raina, A. K. Female sex pheromone of the pickleworm, *Diaphania nitidalis* (Lepidoptera: Pyralidae). *J. Chem. Ecol. 12*: 239–249 (1986).
31. Hedin, P. A., Davis, F. M., Dickens, J. C., Burks, M. L., Bird, T. G., and Knutson, A. E. Identification of the sex attractant pheromone of the southwestern corn borer *Diatraea grandiosella* Dyar. *J. Chem. Ecol. 12*:2051–2063 (1986).
32. Priesner, E., Witzgall, P., and Voerman, S. Field attraction response of raspberry clearwing moths, *Pennisetia hylaeiformis* Lasp. (Lepidoptera: Sesiidae), candidate chemicals. *J. Appl. Entomol. 102*:195–210 (1986).
33. Ghizdavu, I., Hodosan, F. P., and Oprean, I. Specific sex attractants for *Adoxophyes orana* F.v.R. and *Archips crataegana* Hb. *Rev. Roum. Biol. 32*:23–28 (1987).
34. Castellari, P. L. Feromone sessuale di *Choristoneura lafauryana* Rag. (Lep.: Tortricidae): Prove in campo sull'attrattivita di varie miscele di componenti. *Boll. Bologna Univ. Is. Entomol. 39*:243–260 (1985).
35. Smith, M. T., McDonough, L. M., Voerman, S., and Davis, H. G. Hickory shuckworm *Cydia caryana*: Electroantennogram and flight tunnel studies of potential sex pheromone components. *Entomol. Exp. Appl. 44*:23–30 (1987).
36. Slessor, K. N., Raine, J., King, G. G. S., Clements, S. J., and Allan, S. A. Sex pheromone of blackheaded fireworm, *Rhopobota naevana* (Lepidoptera: Tortricidae), a pest of cranberry. *J. Chem. Ecol. 13*:1163–1170 (1987).

37. Tamaki, Y. Pheromones of the Lepidoptera. In *Handbook of Natural Pesticides*, Vol. 4 *Pheromones*, Part A, (Morgan, E. D., and Mandava, N. B., eds.). CRC Press, Boca Raton, 1988, pp. 35–94.
38. Bestmann, H. J., and Vostrowsky, O. Pheromones of the Coleoptera. In *Handbook of Natural Pesticides*, Vol. 4 *Pheromones*, Part A (Morgan, E. D., and Mandava, N. B., eds.). CRC Press, Boca Raton, 1988, pp. 185–183.
39. Fletcher, B. S., and Bellas, T. E. Pheromones of Diptera. In *Handbook of Natural Pesticides*, Vol. 4 *Pheromones*, Part B, (Morgan, E. D., and Mandava, N. B., eds.). CRC Press, Boca Raton, 1988, pp. 1–57.
40. Fletcher, B. S., and Bellas, T. E. Pheromones of Hemiptera, Blattodea, Orthoptera, Mecoptera, Other Insects, and Acari. In *Handbook of Natural Pesticides*, Vol. 4 *Pheromones*, Part B, (Morgan, E. D., and Mandava, N. B., eds.). CRC Press, Boca Raton, 1988, pp. 207–271.
41. Wheeler, J. W., and Duffield, R. M. Pheromones of Hymenoptera and Isoptera. In *Handbook of Natural Pesticides*, Vol. 4 *Pheromones*, Part B (Morgan, E. D., and Mandava, N. B., eds.). CRC Press, Boca Raton, 1988, pp. 59–206.
42. Morgan, E. D., and Mandava, N. B. (eds.). *Handbook of Natural Pesticides*, Vol. 4 *Pheromones*, Parts A and B. CRC Press, Boca Raton, 1988, 203 pp. and 291 pp.

Part VII
Prospects

39

Pheromones: Prophecies, Economics, and the Ground Swell

HEINRICH ARN / Swiss Federal Agricultural Research Station, Wädenswil, Switzerland

I. THE FATE OF PROPHECIES

Projections into the future are nearly always a fascinating blend of right and wrong. We can easily extrapolate from the path of progress in the past and today's technical possibilities to future developments. My father had a book, published in 1928 (1), which described space flight in such detail that when it actually took place it was like a case of déjà vu to me. It is usually difficult to predict just what a particular technical solution will look like: some apparently trivial problems turn out to be big obstacles, whereas others vanish like snow in the sun. The least predictable element, however, is the human mind. Apart from food, sex, and power, humans are driven by ephemeral motivations and sorrows and, sometimes, are outright irrational. Therefore, it is impossible to foresee when mankind is ready for something new, be this a technical, political, or other change. Some fantasies will be put into practice at once, whereas others never see the light of day.

Scientists have long dreamed of using pheromones for insect control. After several years of research and mass trapping experiments with female-baited traps in the vineyards in the 1930s, the German entomologist Bruno Götz wrote:

---

*Reflections on the panel discussion, "Future Prospects and Needs."

> We have proof that the grape moth can be controlled by means of sex attractants. An introduction of this control method will require analysis and synthesis of the sex attractants. From previous experience, these compounds are probably all accessible to chemical analysis, and even their synthesis might be a possibility. The difficulty lies primarily in obtaining the starting material. Since the small amounts emitted by one or a few females produce such a dramatic effect, the application of large quantities should lead to irresistible attraction of all males that under normal conditions nothing could stop.....
>
> The use of attractants to create artificial sources of stimulation, unknown to nature in the same strength, could fundamentally change pest control, which today is based on toxic chemicals. The problem of replacing arsenic, still impending for many pests, would be solved in an elegant way. An additional advantage would be the replacement of human labor. Anyone living in a grape-growing area knows of the effort and time invested in spraying over and over against vineyard pests. Especially today, when both trained and unskilled labor in rural areas are scarce, automatic procedures would be most welcome (2).

From these lines, one reads the fascination created by an insect following an airborne trail to its mate, and the challenge of changing the world for the better by mastering the secrets of life. If Butenandt had turned to *Eupoecilia ambiguella* and *Lobesia botrana* instead of *Bombyx mori* when he started his pheromone research in the 1950s, pest management in European vineyards would perhaps be different today. That he thought about using pheromones for pest control is well documented in a series of patents. However, second-generation pesticides arrived on the scene and were successful; and, until their glamour had vanished, there was no more room for pheromones.

Today, attractants are widely used for insect monitoring and detection, although examples in which trap catch can be used to assess pest population or estimate damage are few. Actual pest control with pheromones is registered and in practical use in a handful of examples; many others are still in the experimental stage or under development. This book contains an impressive list of applications; it should encourage industries to engage in further development and marketing of semiochemicals and scientists, at universities and other research institutions, to provide support.

Some of the people attending the Boston symposium, including myself, entered pheromone research in the 1960s as an answer to the plea by Rachel Carson (3) for biorational pesticides, as they are called now. For most of us, progress has been slower than expected. I remember an electrifying talk by Harry Shorey (4) in which he announced that the problem of controlling matings of the cabbage looper was essentially solved, and that the rest would be a matter of technology. Cabbage looper control with pheromones is no longer an issue, but in the other pests investigated, pheromone technology is still far from perfect. Why has progress been so slow?

II. ECONOMICS, EFFICACY, AND RED TAPE

In the early days of pheromone research, the general assumption was that the chemical companies that were so strong in the pesticide field would also handle the development and marketing of pheromones. This was too optimistic a view, at least for quite some time. The reasons for a lack of commercial interest have been discussed by Silverstein (see Chap. 1): the high price of some pheromones, the cost of registration procedures, as well as uncertainties about patent protection, and the success of novel technologies. Siddall's (5) gloomy study, possibly intended as an apology to other scientists, may have been a major deterrent to industry executives. Further grief was caused by the increasing complexity of the field, both chemical and biological. One of the main obstacles, though, was that industry simply had little interest in developing chemicals that would compete with its own established products. The proof for this is that small companies or some that had no successful insecticides were the first to turn to this specialty market. George Rothschild, a scheduled panel member who had to send his regrets, wrote in his letter: "I still believe that the main constraint on widespread pheromone usage is the continued effectiveness of pesticides in terms of relative cheapness and ease of use."

One of the major worries was whether pheromones would really work. As, for example, the sex attractants of moths, most pheromones we know do not act on the developmental stage doing the damage. There can be no doubt today that we generally can suppress matings with pheromones and reduce the damage caused by the next generation. However, everyone familiar with the situation knows that pheromones sometimes fail, even though speakers at the symposium were reluctant to talk about it. Often, the formulation or unusual temperatures could take the blame, but there are persistent reports of occasional uncontrolled infestations, sometimes very localized, that do not find an easy explanation. Arrival of gravid females from the outside is one, although in some instances the expected damage gradient was missing (6). Either the insects know mechanisms of finding each other with which we are not yet familiar, or the pheromone "cloud" that we create is less uniform than we would like to believe.

Mani (7) demonstrated that pheromone traps placed near the tree canopy of disruption-treated orchards often catch codling moths in large numbers. Placing a dispenser in each tree will leave the borders less covered, depending upon the way the wind turns, and the tops of the trees will be the first to be swept by winds. The same author (Mani, personal communication) observed that on a hilltop, control of the codling moth with pheromone failed, whereas it succeeded in a nearby depression. These examples show that we cannot be professionals in this field before we can visualize the dispersal of the disruptant chemical. I can recall only one paper during the symposium in which the amount of airborne pheromone was monitored (see Chap. 32). The invention of the "sniffer," an instrument to continuously determine pheromone concentration in the air, will be a golden key to pheromone applications.

A concerted effort by industry and research scientists will be needed if pheromones are to find a permanent place in the toolbox of integrated control. We will have to continue studying the pheromone blends of local insect populations and the behavior of insects in the field. We cannot expect to succeed all the time, even against an insect whose chemistry and behavior are well known, but we must be able to tell the grower where he can work with pheromones and in which situations he should use something else.

A statement by Wendell Roelofs during an earlier talk set the tone for a large part of the panel discussion. He criticized the Environmental Protection Agency (EPA) for being slow and bureaucratic in registering pheromones. This "outcry" triggered a series of responses from the floor. It is indeed a sad alternative to either use the pheromone in a field test and destroy the crop or treat with a pesticide and sell the crop. John Borden suggested that: "sooner or later ... a grower who happens to be a chemist will become impatient and will synthesize pheromones for himself and start using them, registered or not."

III. THE GROUND SWELL

Everyone present at the symposium seemed to agree that the time is ripe to put insect control with pheromones into practice. Nobody thinks it will be a panacea, rather a tool for special situations in integrated pest management systems. Not only the scientific community agrees on this, but also everyone in industry I talked to. Growers appear to be prepared to accept the greater risk and labor associated with pheromones than with proven pesticides and, above all, the consumer wants the soft technology, and is ready to pay for it.

The discussions also made it clear that industries may be prepared to develop and sell pheromone products, but not to spend large amounts of money on toxicological and environmental studies. Small companies cannot afford them, and many pheromone applications are too small to be profitable for larger companies. As a consequence, we will have to get registration supported by the public. Compared with the money already spent on pheromone research over the past decades, this will not cost very much. And without it, the applied research done so far would be practically in vain. To publish the toxicological information and make the data available to all would even be more efficient than having every company provide the same information.

How much toxicology is really needed? In a disruption application, the airborne pheromone has a chance of being adsorbed on the waxy surface of leaves and fruits. The amount will be extremely small, probably little more than the residues of the fumes of a tractor pulling through the orchard. The public has a right to know if these chemicals pose no threat, and we hope that we can come up with satisfactory answers. Registration agencies should focus on the type of application, and not ask for data that have no relevance. Arnon Shani and Chuck Doane pressed for a change in the classification of pheromones to avoid the term "pesticide." This might help where registration is already bureaucratic; the best thing, however, would seem to handle all chemicals case by case.

A criticism often heard, formulated in the discussion by Jerry Klun, is that industry is out to make the fast buck and not interested in long-term investments. This may be true, to some extent, but there is also an opposite viewpoint to consider. Some companies make sincere efforts to enhance their credibility. An industry representative complained to me that once they moved into the field of pheromones, everything started to cost them money: researchers wanted to be paid for field testing and growers for damage, even though both had been enthusiastically at work before and just asked for *technical* help. It is difficult for industry to break away from the image of the big spender and polluter, but maybe there will be a time again when we can consider its activities a service.

Nevertheless, companies want to make money, and it would be wrong to rely entirely on their initiative. Wendell Roelofs said: "Companies come and go in pheromone research. We've found that over and over again. The economics keep changing year after year, but if the pheromone works, the entomologist has to continue working to get it on the market. I think that persistence is needed for every little insect around the world where there is a niche for pheromones."

Predictions on the success of pheromones have always been based on ecomonics. We can indeed think of two mechanisms by which semiochemicals could win the competition with insecticides: One is by public support—and today this means support of registration procedures—or through a ban of pesticides. George Rothschild considers the latter a possible scenario, at least for Australia. And he says "The main thing ... is to be prepared for that crisis and to have done the necessary R&D work beforehand." However, we know that human behavior cannot be explained by economics alone. Money is on the way to becoming the only renewable resource. There are many other reasons for using pheromones; one is that they are elegant.

ACKNOWLEDGMENT

I would like to thank the members of the panel, who contributed to the spirited discussion their varied perspectives and insights on the need and prospects for the practical application of pheromones. The panel members were: Donald V. Allemann, CIBA-GEIGY Corp., Greensboro, NC 27419; Philipp Kirsch, Biocontrol Ltd. (Australia), Davis, CA 95616; Reidar Lie, Borregaard Industries, Ltd., N-1701 Sarpsborg, Norway (present address, Norsk Hydro A. S., N-0203 Oslo 2, Norway); B. William Lingren, Trece, Inc., Salinas, CA 93915; and Wendell L. Roelofs, Department of Entomology, New York State Agric. Exp. Stn., Geneva, NY 14456.

REFERENCES

1. Gail, O. W. *Mit Raketenkraft ins Weltenall.* Thienemann, Stuttgart, 1928.
2. Götz, B. Sexualduftstoffe als Lockmittel in der Schädlingsbekämpfung. *Umschau 44*:794–796 (1940).
3. Carson, R. *Silent Spring.* Houghton Mifflin, Boston, 1962, 368 pp.
4. Shorey, H. H. and Gaston, L. K. Use of sex pheromones for behavioral control of the cabbage looper. *Abstr. Natl. Meet., Entomol. Soc. Am.*, Miami Beach, Fla., 1970, p. 55.
5. Siddall, J. B. Commercial production of insect pheromones: Problems and prospects. In *Chemical Ecology: Odour Communication in Animals* (F. J. Ritter, ed.), Elsevier, Amsterdam, 1979, pp. 389–402.
6. Winkelmann-Vogt, H. Untersuchungen zum Pheromoneinsatz bei der Bekämpfung des Einbindigen Traubenwicklers (*Eupoecilia ambiguella* Hbn.) im Weinbau. Dissertation Universität Kaiserslautern, 1986, p. 107.
7. Mani, E., Schwaller, F., and Riggenbach, W. Trap catches as indicators of disruption efficiency and uniformity of pheromone dispersal. In *Mating Disruption: Behaviour of Moths and Molecules* (H. Arn, ed.), Bulletin IOBC-WPRS, International Organisation of Biological Control, Paris, 1987, p. 17.

Appendix:

List of Commercial Suppliers

MAY N. INSCOE and RICHARD L. RIDGWAY / Agricultural Research Service, U.S. Department of Agriculture, Beltsville, Maryland

A Partial List of Suppliers of Insect Pheromones, Synthetic Attractants, and Traps[a]

Supplier	Product
Agrimont S.p.A. (formerly Farmoplant) Piazza della Repubblica, 14/16 20124 Milano, Italy Tel. 02-63331	Traps and dispensers for monitoring and mass trapping
AgriSense 4230 W. Swift St., Suite 106 Fresno, CA 93722 Tel (209) 276-7037	Formulations, dispensers
Agrisense-BCS, Limited Treforest Industrial Estate, Treforest Mid Glamorgan CF37, UK Tel. (044) 384-1155	Dispensers and traps for monitoring and mass trapping, attracticides, mating disruptants

(continued)

[a]Mention of a supplier does not constitute an endorsement or recommendation by the U.S. Department of Agriculture. Corrections and additions to this list will be welcomed.

A Partial List of Suppliers of Insect Pheromones, Synthetic Attractants, and Traps[a] (Cont.)

Supplier	Product
Alfrebro Inc. P.O. Box 15765 980 Redna Terr. Cincinnati, OH 45215 Tel. (513) 771-2030	Pheromone synthesis
Atomergic Chemetals Corp. 91 Carolyn Blvd. Farmingdale, NY 11735-1527 Tel. (516) 694-9000	Formulations
BASF Aktiengesellschaft Crop Protection Division 6703 Limburgerhof, West Germany Tel. (06 21) 60-0	Mating disruptant formulations, traps
Bedoukian Research Inc. Finance Drive Danbury, CT 06810 Tel. (203) 792-8153	Pheromone synthesis
Bend Research, Inc. 64550 Research Rd. Bend, OR 97701 Tel. (503) 3382-4400	Formulation
Bharat Pulverising Mills Pvt. Ltd. P.O. Box 11481 Shriniketan 14, Queens Road Churchgate Bombay 400 020, India Tel. 29-2877	Formulation
Bio-Control Services 2949 Chemin Ste-Foy Ste-Foy, Quebec G1X 1P3 Canada Tel. (306) 773-2131	Traps
BioControl Ltd. 148 Palmerin St. P.O. Box 515 Warwick 4730, Queensland, Australia Tel. (076) 61 4488	Mating disruptants

(continued)

A Partial List of Suppliers of Insect Pheromones, Synthetic Attractants, and Traps[a] (Cont.)

Supplier	Product
BioControl Ltd. (America) 719 2nd Street, Suite 12 Davis, CA 95616 Tel. (916) 757-2307	Mating disruptants
Borregaard Industries Ltd. Postboks 256 N-17-1 Sarpsborg, Norway Tel. (0)20-838811	Dispensers and traps for mass trapping
Brody Enterprises 9 Arlington Place Fair Lawn, NJ 077410 Tel. (201) 794-9618 or 1-(800) 458-8727	Distributor: traps, and dispensers
Chemia S.p.A. Via Statale 327 44040 Dosso (Ferrara), Italy Tel. 39 53-284-5085	Dispensers and traps
Chemolimpex Hungarian Trading Company for Chemicals P.O. Box 121 Budapest 1805, Hungary Tel. 36-118-3970	Dispensers, traps
Chemtech B.V. Vondelstraat 58 1054 GG Amsterdam, The Netherlands Tel. (0) 20-838811	Pheromones, traps
Consep Membranes Inc. P.O. Box 6059 Bend, OR 97708 Tel. (503) 388-3688	Dispensers and traps for monitoring and mass trapping, mating disruptants

(continued)

A Partial List of Suppliers of Insect Pheromones, Synthetic Attractants, and Traps[a] (Cont.)

Supplier	Product
Denka International bv. P.O. Box 337, 3770 AH Hanzeweg 1, 3771 NG Barneveld, The Netherlands Tel. 03420-12174	Synthesis, dispensers, and traps for monitoring and mass trapping, mating disruptants, bioirritants
Dewill Inc. 61 S. Herbert Rd. Riverside, IL 60546 Tel. (312) 442-6009	Traps, adhesives
Djinnii Industries, Inc. 302 Vermont Ave. Dayton, OH 45404 Tel. (513) 223-3607	Formulation
Elan Chemical Co. 268 Doremus Ave. Newark, NJ 07015 Tel. (201) 344-8014	Synthetic attractants
Eurand America 845 Center Drive Vandalia, OH 45377 Tel. (513) 898-9669	Formulation
Fermone Chemicals Inc. 1700 N. 7th Ave., Suite 100 Phoenix, AZ 85007 Tel. (602) 254-7946	Bioirritants, traps, house fly bait
Fersol Inc. e Com. Ltd. Rua Leopoldo Couto Magalhaes Jr., 1304/06 Sao Paulo SP Brazil Tel. 011 8133111	Distributor: traps, and dispensers
Frank Enterprises, Inc. 700 Rose Ave. Columbus, OH 43219 Tel. (614) 253-5519	Synthesis

(continued)

A Partial List of Suppliers of Insect Pheromones, Synthetic Attractants, and Traps[a] (Cont.)

Supplier	Product
Gowan Co. P.O. Box 5696 Yuma, AZ 85364 Tel. (602) 783-8844	Distributor: formulations
Great Lakes IPM 10220 Church Rd., NE Vestaburg, MI 48891 Tel. (517) 268-5693	Distributor: pheromone and synthetic attractand dispensers and traps
Hara Products Ltd. P.O. Box 134 1981 Chaplin St. West Swift Current, Saskatchewan S9H 3Y5 Canada Tel. (403) 465-7937	Traps
Helena Chemical Co. 5100 Poplar Ave., Suite 3200 Memphis, TN 38137 Tel. (901) 761-0050	Distributor: dispensers, traps
Hercon Laboratories Corp. Environmental Division P.O. Box 786 York, PA 17405 Tel. (717) 764-1191	Dispensers and formulations
Hoechst AG Verkauf Landwirtschaft/Berat. Postfach 800320 D-6230 Frankfurt am Main 80, West Germany Tel. (069) 305 0	Dispensers and traps for monitoring
ICI Plant Protection Division Jealott's Hill Research Station Bracknell Berkshire RG 12-6EY, UK Tel. (0344) 424701	Mating disruptants, formulation

(continued)

A Partial List of Suppliers of Insect Pheromones, Synthetic Attractants, and Traps[a] (Cont.)

Supplier	Product
I.N.R.A. Laboratoire des Mediateurs Chimiques Domaine de Brouessy Magny-Les-Hameaux 78470 Saint-Remy-Les-Chevreuse, France Tel. 30 44 35 54	Dispensers and traps for monitoring
Insects Limited, Inc. P.O. Box 40641 10505 N. College Ave. Indianapolis, IN 46280 Tel. (317) 846-5444	Dispensers and traps for monitoring and mass trapping
Institute for Pesticide Research (Instituut voor Onderzoek van Bestrijdingsmiddelen) Marijkeweg 22 6709 PG 4, Wageningen, The Netherlands Tel. (08370) 11821	Dispensers and traps for monitoring, synthetic pheromones
International Pheromones Systems Ltd. 15b, Broadway Bebington, Wirral Merseyside L63 5RQ, UK Tel. 051-608 2366	Dispensers and traps for monitoring, mating disruptants, attracticides
Istituto Guido Donegani S.p.A. Via G. Fauser 4 28100 Novara, Italy Tel. (0321) 4471	Synthetic pheromones
Kenco Chemical and Manufacturing Co. P.O. Box 6246 10 W. Adams St., Jacksonville, FL 32236 Tel. (904) 781-9622	Synthetic attractant traps
LIM Technology Laboratory 409 E. Main St. Richmond, VA 23219 Tel. (804) 780-3855	Formulation

(continued)

A Partial List of Suppliers of Insect Pheromones, Synthetic Attractants, and Traps[a] (Cont.)

Supplier	Product
Monterey Chemical Co. P.O. Box 5317 5150 N. 6th, Suite 156 Fresno, CA 93755 Tel. (209) 255-4470	Distributor: dispensers
New Brunswick Research & Productivity Council Fredericton, New Brunswick E3B 5H1 Canada	Formulations
Nitto Electric Industrial Co. Ltd. 1-2 Shimohozumi 1-Chome Ibaraki 567, Japan Tel. 0726 22-2981	Formulations
Oecos Ltd. Monitoring and Control Equipment 130 High Street Kimpton, Herts. SG4 8QP, UK Tel. (0438) 832481	Traps, adhesives
Orsynex, Inc. 1401 Broad Street Clifton, NJ 07015 Tel. (201) 713-6300	Pheromone and attractant synthesis
Pest Management Supply Co., Inc. P.O. Box 938 Amherst, MA 01004 Tel. (413) 253-3747	Distributor: pheromone traps and dispensers
Phero Tech Inc. 1140 Clark Drive Vancouver, British Columbia V5L 3K3 Canada Tel. (604) 255-7381	Dispensers and traps for monitoring and mass trapping, synthetic pheromones, tree baits, pest management services

(continued)

A Partial List of Suppliers of Insect Pheromones, Synthetic Attractants, and Traps[a] (Cont.)

Supplier	Product
Point Enterprise S.A. Agrochemical Division P.O. Box 48 1260 Nyon, Switzerland Tel. 022 619564	Distributor: pheromones, traps, dispensers
Productos OSA Sacifia Av. de Mayo 1161, 1 Piso 1085 Buenos Aires, Argentina Tel. 33 17340 OSA AR	Distributor: dispensers and traps
Raylo Chemicals Limited 8045 Argyll Road Edmonton, Alberta T6C 4A9 Canada Tel. (403) 468-6060	Dispensers, traps
Reanal Factory of Laboratory Chemicals P.O. Box 54 H-1141 Budapest 70, Hungary Tel. 635 849	Dispensers and traps for monitoring
Redell Industries, Inc. P.O. Box 4299 Pacoima, CA 91331 Tel. (818) 767-2038	Attractant synthesis
Reuter Laboratories 8450 Natural Way Manassas Park, VA 22111 Tel. (703) 361-2500 or 1-(800) 368-2244	Dispensers and traps for monitoring and mass trapping
Sandoz, Ltd. Agro Division CH-4002 Basle, Switzerland Tel. 061 24 11 11	Distributor: dispensers and traps for monitoring, mating disruptant, attracticide, synthetic pheromones

(continued)

A Partial List of Suppliers of Insect Pheromones, Synthetic Attractants, and Traps[a] (Cont.)

Supplier	Product
Scentry, Inc. 26405 West Highway 85 P.O. Box 426 Buckeye, AZ 85326 Tel. (602) 233-1772 or (602) 386-6737	Dispensers and traps for monitoring mass trapping, mating disruptants
Shell Agrar GmbH & Co. KG (formerly Celamerck GmbH & Co.) P.O. Box 200 D-6507 Ingelheim am Rhein, West Germany Tel. 06132/789-0	Dispensers and traps for mass trapping, dispensers for bait trees
Shin-Etsu Chemical Co., Ltd. Fine Chemicals Department 6-1. Ohtemachi 2-Chome Chiyoda-ku, Tokyo 100, Japan Tel. 03-246-5280	Dispensers and traps for monitoring, mating disruptant, synthetic pheromones
Siegfried Agro S.A. AGRO Division CH-4800 Zofingen, Switzerland Tel. 062 502293	Synthetic pheromone
Silwood Centre for Pest Management Imperial College Silwood Park Ascot, Berks SL5 7PY, UK Tel. Ascot (0990) 23911	Pheromone synthesis, formulation
Trece, Inc. 635 South Sanborn Rd. Suite 17 P.O. Box 5267 Salinas, CA 93901 Tel. (408) 758-0205	Dispensers and traps for monitoring

(continued)

A Partial List of Suppliers of Insect Pheromones, Synthetic Attractants, and Traps[a] (Cont.)

Supplier	Product
Uncommon Scents of America P.O. Box 670 1245 South 6th St. Coshocton, OH 43812 Tel. (614) 622-0755	Pheromone synthesis
University College Department of Chemistry Cardiff, Wales CF1 1XL, UK	Pheromone synthesis
Vioryl S.A. Viltaniotis St., 145 64 Kifisia, Greece Tel. 8074603, 8074452	Pheromone synthesis, dispensers
Wolfson Unit of Chemical Entomology Department of Chemistry The University Southampton SO9 5NH, UK Tel. 0703-559122, Ext. 3306	Pheromone synthesis

Index*

*Insect and chemical names mentioned in the text are indexed here. Additional insects for which pheromones are commercially available are listed in the tables of Chap. 38.